# Print Reading for Industry

Ninth Edition

## Write-In Text

**Walter C. Brown**

**Ryan K. Brown**
Department of Technology
Illinois State University
Normal, Illinois

Publisher
**The Goodheart-Willcox Company, Inc.**
Tinley Park, IL
www.g-w.com

Manufactured in the United States of America

Library of Congress Catalog Card Number 2010001429

ISBN: 978-1-60525-308-4

3 4 5 6 7 8 9 – 11 – 15 14 13 12 11

**The Goodheart-Willcox Company, Inc., Brand Disclaimer:** Brand names, company names, and illustrations for products and services included in this text are provided for educational purposes only and do not represent or imply endorsement or recommendation by the author or the publisher.

**The Goodheart-Willcox Company, Inc. Safety Notice:** The reader is expressly advised to carefully read, understand, and apply all safety precautions and warnings described in this book or that might also be indicated in undertaking the activities and exercises described herein to minimize risk of personal injury or injury to others. Common sense and good judgment should also be exercised and applied to help avoid all potential hazards. The reader should always refer to the appropriate manufacturer's technical information, directions, and recommendations; then proceed with care to follow specific equipment operating instructions. The reader should understand these notices and cautions are not exhaustive.

The publisher makes no warranty or representation whatsoever, either expressed or implied, including but not limited to equipment, procedures, and applications described or referred to herein, their quality, performance, merchantability, or fitness for a particular purpose. The publisher assumes no responsibility for any changes, errors, or omissions in this book. The publisher specifically disclaims any liability whatsoever, including any direct, indirect, incidental, consequential, special, or exemplary damages resulting, in whole or in part, from the reader's use or reliance upon the information, instructions, procedures, warnings, cautions, applications or other matter contained in this book. The publisher assumes no responsibility for the activities of the reader.

**Library of Congress Cataloging-in-Publication Data**

Brown, Walter Charles

Print reading for industry / by Walter C. Brown, Ryan K. Brown. -- 9th ed.

p. cm.

Includes bibliographical references and index.

ISBN 978-1-60525-308-4

1. Blueprints. I. Brown, Ryan K. II. Title.

T379.B764 2011

604.2'5--dc22 2010001429

# Introduction

The term print reading, as used in this textbook, means interpreting and visualizing drawings and prints. The textbook is based on actual prints from various industries. Most industry practices conform to national drafting standards. However, some companies vary from standards to meet their specific needs. In order to provide real-life print reading experiences, these variations are retained on the prints in this textbook.

*Print Reading for Industry* is a training course designed to assist beginning and intermediate students and workers to read and understand industrial prints. This textbook addresses the skills necessary in reading and understanding "the language of industry."

## Print Reading Procedures

The following steps should be used for reading a print. By performing these steps when reading prints, your skills will continue to develop.

- Step 1: Read the title.
- Step 2: Check the drawing number.
- Step 3: Read the title block and notes.
- Step 4: Read all callouts.
- Step 5: Read the revisions and changes.
- Step 6: Analyze the part or assembly.

## About the Authors

During his career, Dr. Walter C. Brown was a leading authority in the fields of drafting and print reading. He served as a consultant to industry on design and drafting standards and procedures. He authored several books in the fields of drafting, print reading, and mathematics and was a professor in the Division of Technology at Arizona State University, Tempe, Arizona.

Dr. Ryan K. Brown is an Associate Professor in the Department of Technology at Illinois State University, Normal, Illinois. He has been involved with drafting education for over thirty years. His experiences include secondary education at the junior and senior high level; faculty member at his alma mater, Eastern Kentucky University; drafter for Sargent & Greenleaf, Inc.; geometric dimensioning and tolerancing consultant; and summer experiences in a wide array of drafting and graphics-based positions, such as steel detailing and civil engineering applications. His current roles in drafting education also include the authorship of activities and problems for secondary-level drafting competitions and architectural drafting applications that include 2D drafting, 3D modeling, and rendering and animation. Feel free to contact him at: rkbrown@ilstu.edu

## Acknowledgment

Dr. Brown would like to acknowledge and thank Dr. Louis Reifschneider, a colleague in the Department of Technology at Illinois State University, for his review and writing of the unit on plastic parts. Dr. Reifschneider has experience in the plastics industry as an engineer and designer. He graciously brought his expertise to this unit.

## Prints Used in the Text

Throughout the textbook are illustrations and activities based on actual industry prints. A few companies have donated prints anonymously. The publisher and authors want to thank these companies and the companies listed below for their assistance.

Aerojet-General
AIR Corporation
AISIN Manufacturing Illinois
Barko Hydraulics
Barton Manufacturing
BC Design and Associates
Bell Aerosystems
Boston Gear
Brown and Sharpe
Clark Equipment Company
Cleveland Gear
Deere and Company
Detroit Diesel Allison
GenCorp Aerojet
Gleason Works
Grayhill, Inc.
Hydro-Gear
Intermatic, Inc.
Iron-A-Way
J.I. Case
Johnson and Towers, Inc.
Kennemetal
Marathon LeTourneau Company
MASCO Corporation
Master Spring and Wire Form Company
Monarch Sidney
Motorola Inc.
North American Aviation
OMC-Lincoln
Perkin Elmer
Pratt and Whitney Canada
RegO Cryo-Flow Products
Rockwell Manufacturing Company
Skil Corporation
Sono-Mag Corporation
Sperry Phoenix Company
Sterling Precision Corporation
Sunnen Products Company
Talley Industries
Unidynamics
United Technologies Otis
Wis·Con Total Power Corporation

# Brief Table of Contents

## Section Four
### Industrial Drawing Types



## Section Five
### Specialized Parts and Prints

# Expanded Table of Contents

UNIT 1

# Prints: The Language of Industry

*After completing this unit, you will be able to:*

**Identify** the importance of prints.

**Discuss** historical processes and technologies related to prints.

**Identify** and **define** terms related to prints.

**Explain** how prints are produced.

**Identify** two important elements of print reading.

**Identify** ways in which to care for prints.

**Identify** and **discuss** options for using prints in an electronic (digital) format.

**Discuss** the role of various organizations in the standardization of drawings.

You have probably heard the saying, "a picture is worth a thousand words." This is certainly true when referring to a drawing of a product. It would be next to impossible for an engineer or designer to describe in words the shape, size, and relationship of the various parts of a machine in sufficient detail for skilled workers to produce the object. Drawings are the universal language used by engineers, designers, technicians, and skilled workers to quickly and accurately communicate the necessary information to fabricate, assemble, and service industrial products. See Figure 1-1.

Figure 1-1.
Drawings are the universal language used by engineers, designers, technicians, and skilled workers to quickly and accurately communicate the necessary information to fabricate, assemble, and service industrial products.

## Importance of Print Reading

Within the context of this book, the word ***print*** will be simply defined as a copy of a drawing, but can generically be used the same as the word *drawing*. In some situations, drawings may not be printed, but nevertheless are a primary means of communication within the manufacturing enterprise. Many industrial products, such as automobiles, aircraft, and computers, consist of thousands of component parts. These parts may be manufactured in separated factories. The "moment of truth" in the manufacture of these products comes during final assembly or when a spare part is shipped for field installation. These parts must always fit. To meet manufacturing requirements, all industries need workers who can read and understand prints.

A drawing describes what an object should look like when it is completed. Prints provide workers with the details of size, shape, tolerances (allowable variation), materials used, finish, and other special treatments. In many cases, the print is also an important part of the contractual agreement within the industrial setting. The supplier of parts must meet the specifications dictated by the print. Often, purchasing agents have to ensure the print is contractually sufficient to ensure vendors supply quality parts without cutting corners. Quality-control inspectors have to verify that all parts, both those made by suppliers and those made in-house, match the print and continually review the print's role in controlling the precision and quality of the parts needed.

## History of Prints

The study of print reading is closely related to the study of ***drafting,*** the general term for creating drawings of objects in technical fields. Of course, those who have studied drafting not only know how to create prints, but also read prints. Courses, curricular programs, and textbooks for drafting have used a multitude of terms throughout the years. ***Engineering graphics*** is a common term used within engineering programs, while ***technical drawing*** is a common term in drafting programs that train technicians and technologists. In addition, terms such as ***mechanical drafting*** or ***instrumental drafting*** were used in the past to describe the process of creating industrial drawings. While the focus of this textbook is on reading the drawings, information about creating drawings will also be examined to provide insight into the standards the drafter should be following.

One of the earliest copying processes began in the middle of the 19th century. It involved exposing treated, photosensitive paper under an original drawing on translucent paper. This early method and the copy paper it used required submersion in a liquid to process the copy. The resulting print had a blue background and white lines. Thus, this type of print was called a ***blueprint,*** Figure 1-2. Even though the process that created "blue" prints has not been used for decades, the term blueprint has become a part of vocabulary to "mean any plan of action or detailed procedure to accomplish a task."

During the 1940s, a dry process evolved that used a paper coated with a type of organic compound referred to as a ***diazo*** compound. Exposure to light evaporated the background area while leaving the coating where lines blocked the light. Ammonia vapor then converted the remaining coating to a permanent blue. This type of print was sometimes called a *whiteprint* or a *blue-line print,* but often the term *blueprint* continued to be used. While the diazo process can still be found in use in a few settings today, most companies have replaced that technology with engineering photocopiers or plotters used to make paper prints.

Often, the copying process required an original that would be durable, yet would allow light to pass through. So, original drawings were produced on ***vellum*** or ***plastic film.*** Copies of the drawings, or prints, were made and distributed to those who needed them.

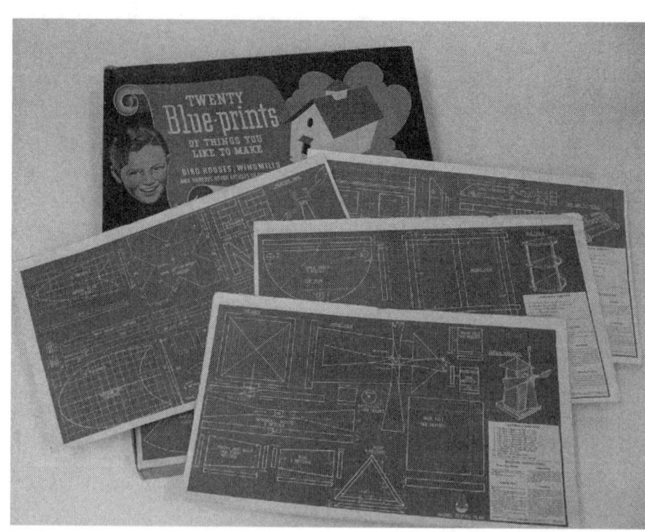


Figure 1-2.
Years ago, copies of drawings (prints) were made that featured a blue background and white lines. This is the origin of the term blueprint, which is still in our vocabulary today.

Throughout most of the twentieth century, companies produced paper copies of drawings of parts and assemblies, either drawn with traditional drafting equipment or plotted from computer-aided drafting (CAD) systems. These original drawings—created by the drafter or plotted by the CAD operator—were then stored in a file-drawer system supervised by the engineering or quality control department. Original drawings were seldom used in the plant or field. Instead, prints were made for distribution.

While some companies may still use these traditional systems, for many companies original drawings no longer exist in paper format. Prints can be created directly from the computer as needed. It is also possible for those who need prints to view the drawings in electronic (digital) format using viewing software. This type of software utility allows the drawing to be viewed by anyone in the company who does not have access to CAD software.

## How Prints Are Made

Today, most drawings are produced using a CAD system. See Figure 1-3. If needed, a hardcopy of the drawing is plotted with a laser or inkjet ***printer*** or a pen- or cartridge-based ***plotter.*** See Figure 1-4. Originally, the term *plotter* indicated a vector-based

**Figure 1-3.**
Most original drawings are now produced using a computer-aided drafting (CAD) system.

**Figure 1-4.**
Hard copies are generated from a CAD system using an output device, such as this cartridge-based plotter. (DesignJet Division, Hewlett-Packard)

output device that physically drew lines with pens, but today most plotters use an ink-cartridge system. The images are actually transferred to the paper as a raster image. The paper is roll-fed into the machine. A ***raster image*** is composed of tiny pixels or dots. Under high magnification you can see the dots, but the quality is more than adequate for industrial prints.

In some companies, approved originals may still be produced and filed. Prints can be made by authorized personnel using a photocopier. In other companies, approved originals may only exist in a protected directory of the computer network. Prints can be sent to output devices by authorized personnel.

## Print Reading

***Print reading*** is the process of analyzing a print to obtain information. This task involves two principal elements—visualization and interpretation. ***Visualization*** is the ability to envision or "see" the shape of the object from the various views shown on a print. See Figure 1-5. Scientifically, every view created by the drafter or designer is based on a projection of the object onto a two-dimensional plane, such as a sheet of paper or a computer screen. Learning the principles of projection will help the reader gain the ability to visualize objects from the views shown on a print. The ***interpretation*** of lines, symbols, dimensions, notes, and other information on a print is also an important factor in print reading. These factors are presented in this text. Actual industrial prints are provided in this text so you can learn print reading on "real-life" drawings.

## Care of Prints

Prints are valuable records of information. When working with prints, you should observe the following rules.

- Never write on a print unless you have been authorized to make changes.
- Keep prints clean, especially free of oil and dirt. Soiled prints are difficult to read and contribute to errors.
- If working with prints that are stored in filing cabinets, very carefully fold and unfold prints to avoid tearing.
- Do not lay sharp tools, machine parts, or similar objects on top of prints.

If the drawing border and title block conform to industry standards, the prints can be folded with the title block showing. A second print number is usually visible when stored in a drawer folder. See Figure 1-6. In many companies, prints are easily accessed from a Print Control Area or Engineering Department. In most organizations, tight control is maintained over all prints. This helps ensure the most-recent revision is always used. With CAD, "read-only" drawing files may be available through a company computer network. This allows the design and manufacturing teams to make their own prints as needed or to view the design on the computer screen.

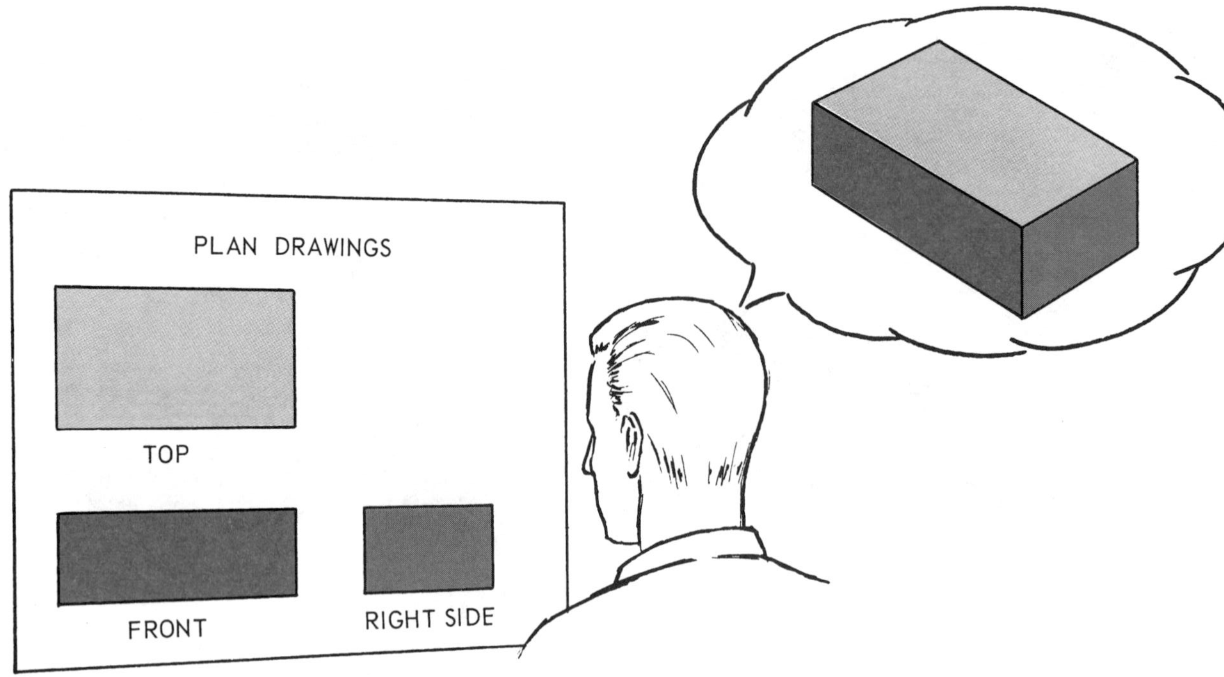


Figure 1-5.
Visualization is the ability to "see" the shape of an object from the various views shown on a print.

## Electronic Formats for Drawings

In recent years, digital file formats have become available that enhance the ability to electronically share prints without jeopardizing or supplying to others the actual CAD data. Two of these formats are the ***portable document format (PDF)*** format and the ***design web format (DWF)*** format. The DWF format was developed by Autodesk® to allow AutoCAD® drawings to be published in a format that can be web-based, but also a format that can be viewed in a free viewer program. This allows individuals that do not use or need AutoCAD to view facsimiles of AutoCAD drawings. The PDF format was developed by Adobe® to allow all computer users to easily share printable documents, regardless of installed software or even computer platform (Windows®, Macintosh®, Linux™, etc.). In both cases, the digital formats can be viewed with free viewer software. One exciting aspect of these technologies is the ability to view, pan, zoom, and dynamically rotate a 3D model of a part or assembly. This means drawings and "prints" can now be interactive on the computer screen, which has changed print reading for current and future generations, Figure 1-7.

## Importance of Standards for Engineering Drawings

Almost every aspect of an engineering drawing can conform to an established standard. By definition, a ***standard*** is a voluntary guideline. Companies can elect to have their own supplemental standards. While voluntary, standards are often incorporated into regulations and business contracts between manufacturers and their subcontractors and suppliers.

In the United States, the standardization of engineering drawings for prints has long been established through the ***American Society of Mechanical Engineers (ASME)***, Figure 1-8. This is an independent, not-for-profit organization that, for decades, has defined standards for engineering drawings. Prior to 1994, however, the standards were identified by American National Standards Institute (ANSI) document numbers, even though published by ASME. For example, the document entitled *Multi and Sectional View Drawings* was identified as ANSI Y14.3-1975. Currently, that document is identified as ASME Y14.3-2003 and is entitled *Multiview and Sectional View Drawings*. On the cover of the ASME booklet, it states it is "An American National Standard." Many drawings and documents created since 1994 may also use a joint acronym, such as ANSI/ASME B1.90-1973 (R2001) Buttress Inch Screw Threads.

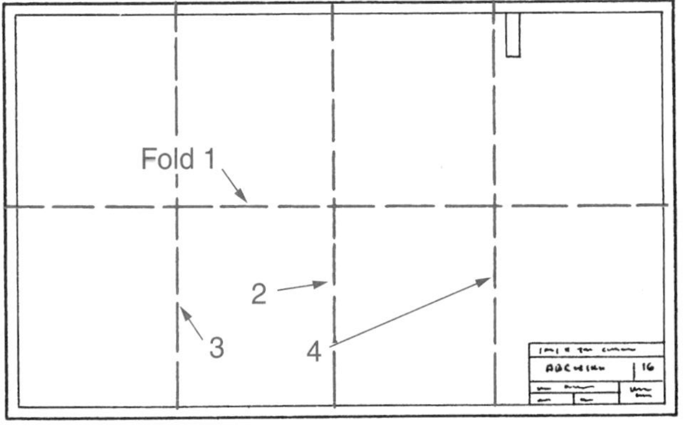

Figure 1-6.
If paper prints are filed in a file folder, larger prints should be folded in such a way as to keep the title block showing.

ANSI is the umbrella organization that serves as an overseeing body and is still relevant to manufacturing enterprises, but publications are now purchased through ASME, either in paper format or PDF format. Throughout this text, you will become familiar with some of the ASME standard publications that impact various aspects of print reading.

For companies with an international presence and product line, the ***International Organization for Standardization (ISO)*** may be the more appropriate source for standardization practices. For example, ISO 128-1:2003 is entitled *Technical drawings–General principles of presentation–Part 1: Introduction and index* and ISO 5456-2:1996 is entitled *Technical drawings—Projection methods—Part 2: Orthographic representations.* These standards are a major investment for engineering departments, but necessary for success in the 21st century.

The general purpose and scope of this text is focused on industrial prints that follow ASME standard practices, both current and recent. It is important to reiterate that not all companies follow all ASME guidelines, sometimes intentionally and sometimes unintentionally. As feasible, this text will identify situations on the "real" exercise prints in the text with respect to their deviation from current ASME standard practice.

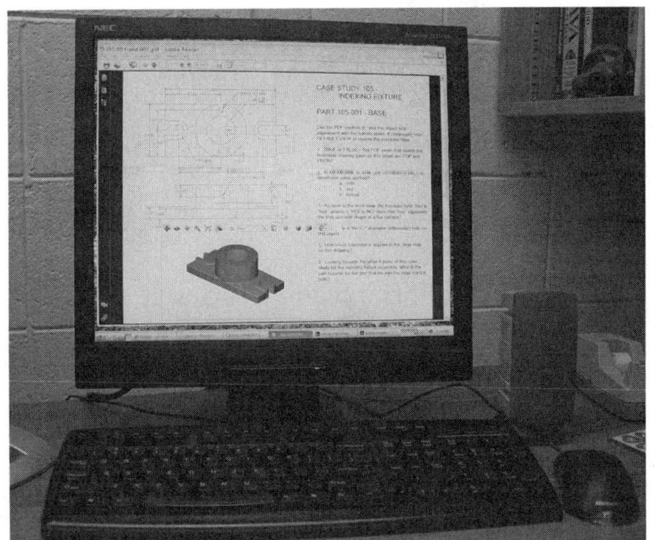

Figure 1-7.
Electronic documents can be viewed on the computer screen with zoom, pan, and turn-and-tilt (3D orbit) capabilities.

Figure 1-8.
The American Society of Mechanical Engineers oversees and publishes standards for engineering drawings.

## Review Questions

*Circle the answer of choice, fill in the blank, or write a short answer.*

1. A print can simply be defined as a(n) ______________________________ of a drawing.

2. *True or False?* Prints are often an important part of a contractual agreement.

3. The study of print reading is closely related to the study of _____.
   A. measurements
   B. drafting
   C. machining
   D. art

4. Briefly explain why *prints* are sometimes called *blueprints.* ______________________________

   ______________________________

   ______________________________

5. List two sheet materials that have been primary media for original drawings: ______________________________

   ______________________________

   ______________________________

6. What name is often applied to a large, roll-feed output device for a CAD system? ______________________________

   ______________________________

7. *True or False?* Scientifically, every view created by the drafter or designer is based on a projection.

8. The two principle elements of print reading are ____________________ and ____________________.

9. What does PDF stand for?
   A. Printable Document Format
   B. Plotted Drawing File
   C. Portable Document File
   D. Print Detail Freehand

10. When referring to engineering drawing standards, which of the following acronyms is *fictitious?*
    A. ANSI
    B. ADSN
    C. ISO
    D. ASME

# UNIT 2
# Line Conventions and Lettering

*After completing this unit, you will be able to:*

**Identify** the standard alphabet of lines.

**Describe** the types of lines by appearance and purpose.

**Identify** the style of lettering recommended for standard industrial drawings.

By definition, a ***convention*** is a generally accepted way of doing something. Before computers, prints were created by drafters with pencils and pens and lettering was primarily a freehand craft. Lines were made with lead pencils and the ability to create a line an exact width was a skill that took time to develop. ASME Y14.2, entitled *Line Conventions and Lettering*, sets forth the recommended appearance of lines, although in our current age of computers the guidelines no longer specify dash length and line width as precisely as in years past. Since drawings can easily be printed at a variety of scales and sizes, there is more flexibility today.

## Alphabet of Lines

There are several types of lines commonly used in engineering drawings, depending on the field of study. The current ASME standard illustrates these and identifies each by name. On some occasions, more than one option for the appearance of the line is given. Each line has a particular meaning to an engineer, designer, or drafter. A skilled technician must recognize and understand the meanings of these lines in order to correctly interpret an industrial print used in the manufacturing environment.

The list of line types, defined in many references as the ***alphabet of lines,*** is used throughout industry. See Figure 2-1. Each line has a definite form or dash pattern and a standard line weight, or width. In older standards, three line weights were recommended.

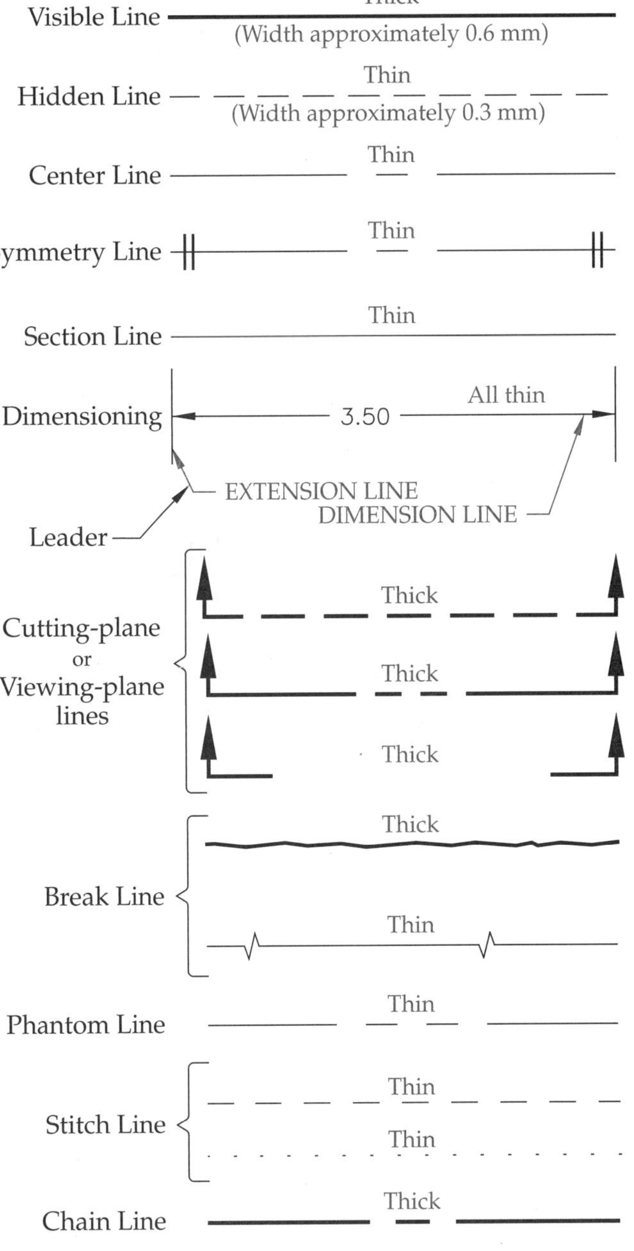


**Figure 2-1.**
The alphabet of lines is a standardized list of lines from which the drafter can choose.

These were thin, medium, and thick. In the last thirty years, the standards have only recommended two line weights—thick and thin. The standard thick line is recommended to be 0.6 mm in width. The standard thin line is recommended to be 0.3 mm in width. Of course, these lines may vary depending on the size of the drawing and whether or not the drawing is photographically reduced or enlarged. In all cases, however, the thin-to-thick ratio should remain 1:2. In most cases, CAD programs can be configured to print drawings at any size or scale while maintaining any desired line thickness. It is important to note that thinner lines should not be lighter. All linework should be black.

Be aware, many drawings throughout the years have been produced without attention to standard line weight. Many of the early CAD systems could not easily print variations in line weight. This often created a drawing that was harder to read and interpret. Within your company, you have the opportunity to be the champion for conformance to industry standards, thus promoting drawings that "speak with the proper language."

Lines in a drawing convey information essential to understanding the print. Therefore, to understand the print, you must know and understand the alphabet of lines. Refer to Figures 2-1 and 2-2 as you read the remainder of this unit.

## Primary View Lines

Of primary importance to the print reader are the three lines in the alphabet that are used in multiview drawings, discussed in Unit 5. Multiview drawings are characterized by a set of views. Within

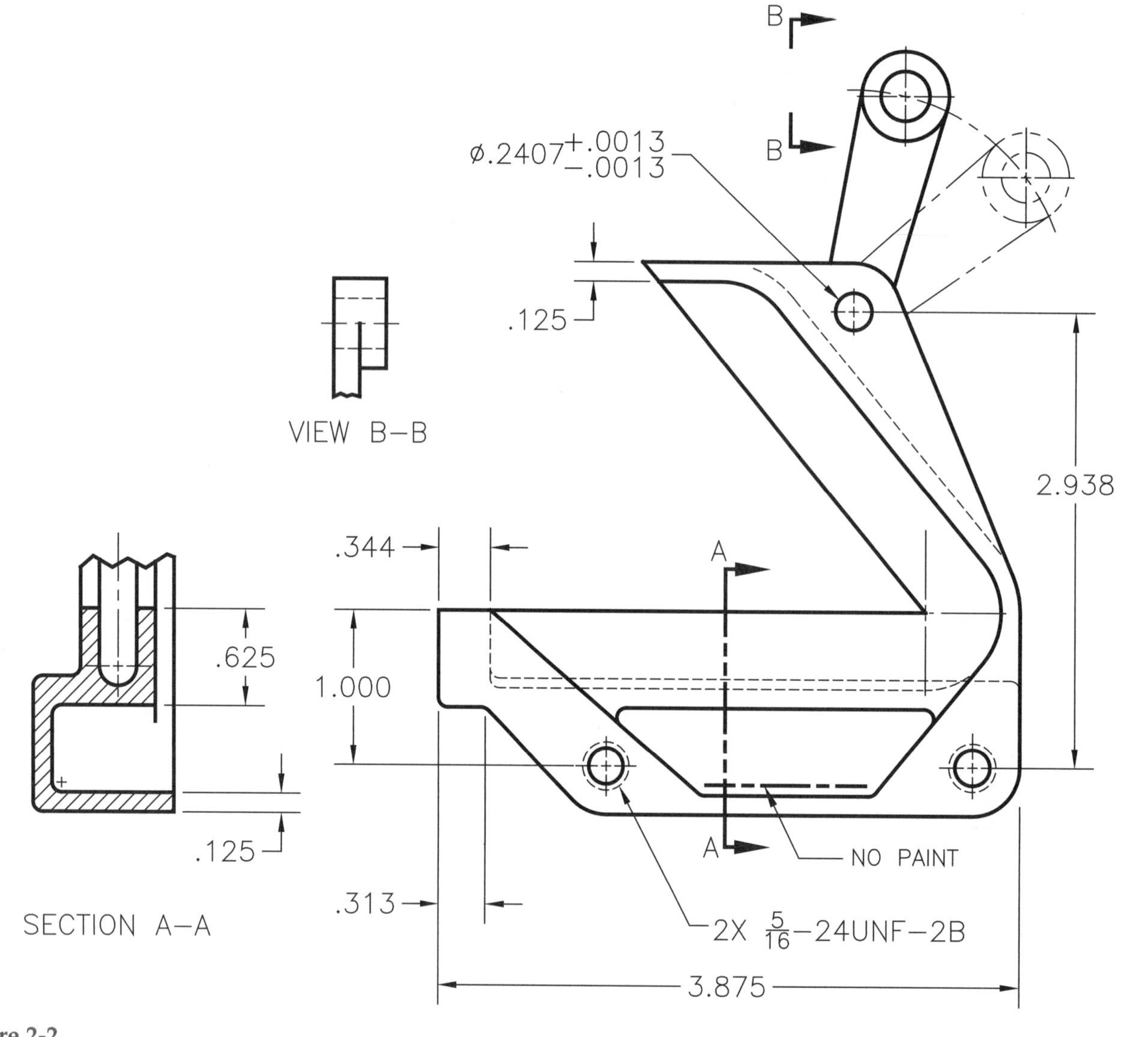


**Figure 2-2.**
This drawing features visible lines, hidden lines, center lines, phantom lines, section lines, a cutting-plane line, a viewing-plane line, dimension lines, extension lines, leader lines, and a chain line. Can you find all of them?

each view, many features may be hidden, while other features may be curved or symmetrical.

Of utmost importance is the ***visible line.*** This is a thick, continuous line representing all of the edges and surfaces of an object that are visible in the view, Figure 2-3. These lines should be twice as thick as the thin lines of the view. Visible lines give the print reader the shape description of the object. Through the years, many books and drafting teachers have also referred to these as ***object lines.***

Another line that is especially important in the multiview drawing is the ***hidden line.*** This is a line that features thin, short dashes spaced closely together. Earlier standards recommended the dashes be about 1/8″ long and spaced approximately 1/32″ apart. Hidden lines are used to show edges, surfaces, and features not visible in a particular view, Figure 2-3. Hidden lines are used to clarify a drawing. Sometimes, hidden lines are omitted on complex views when the drawing is clear without them. On older drawings, these lines may have been created with a medium weight.

The third type of line especially important in the multiview drawing is the ***center line.*** This is a thin line with alternating long and short dashes used to designate centers of holes, arcs, and other symmetrical features, Figure 2-4. In a circular view, two center lines are used and should form a "plus" in the center of the circle. Some CAD programs do not, by default, show center lines crossing in the center of a circle quite as well as drawings drawn by hand. In these cases, the drafter should adjust the scale of the center line so the short dashes make a plus. Center lines are also used to indicate paths of motion, as shown in Figure 2-2. In addition, on some drawings, only one side of a part is drawn and a symbol is placed on each end of the center line to indicate the other side is identical in dimension and shape. When these symbols are added to the center line, the line can be called a ***symmetry line.***

## Section View Lines

Some lines are used primarily in section view drawings, which are discussed in Unit 6. Section

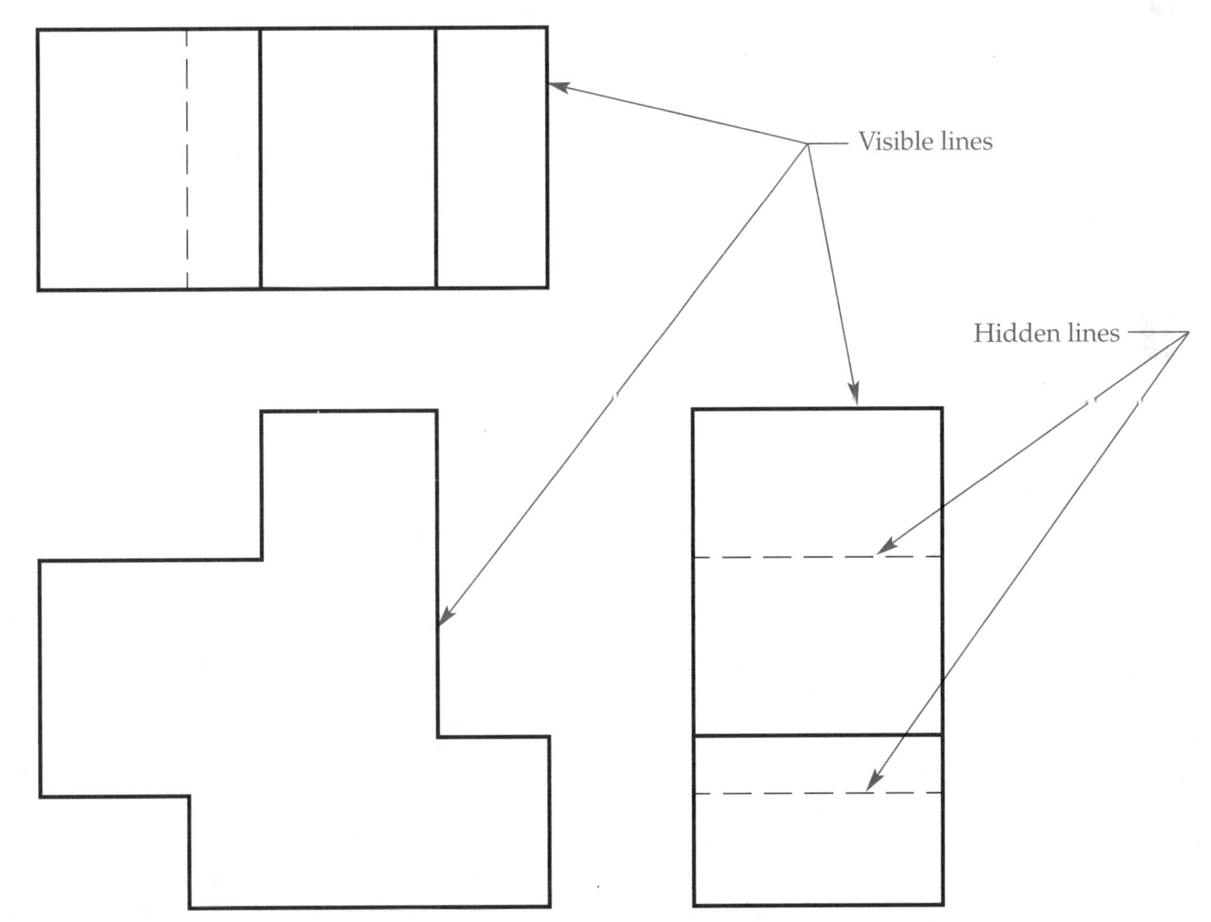


**Figure 2-3.**
A visible line is a thick, continuous line representing all edges and surfaces on an object visible in the view, while hidden lines represent hidden features.

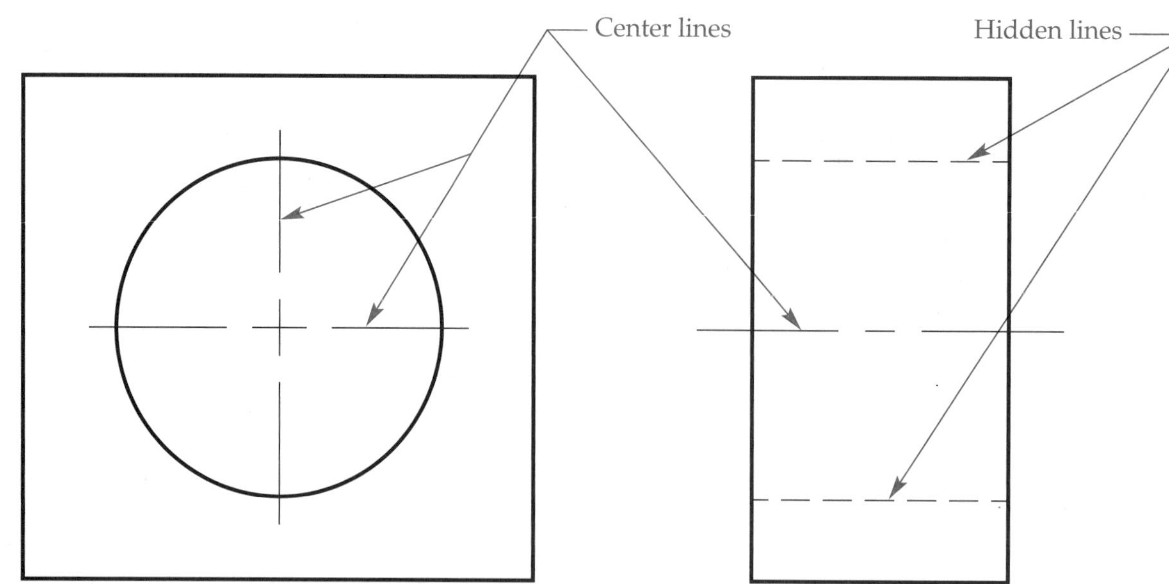


Figure 2-4.
Center lines are thin lines with alternating long and short dashes. They designate centers of holes, arcs, and other symmetrical objects.

views are views that pretend the object has been cut through. Additional conventions are needed to express these views on a drawing.

For surfaces that are assumed to be cut, ***section lines*** are used. These are thin, continuous lines usually drawn at an angle, most commonly 45°. Section lines indicate the surface of an object in a section view that was "cut" by a "cutting plane." Sometimes, section lines have dashes to indicate a particular material. General purpose section lining is the same as the cast iron pattern shown in Figure 2-5. This type of section lining is commonly used for other materials in section unless the drafter or designer wants to indicate the specific material. Some CAD programs refer to a section-lined area as *hatching* or *cross-hatching*.

For section view drawings, a ***cutting-plane line*** can be shown on the view adjacent to the section view to help the print reader know where the "cut" is made. The cutting-plane line should be a thick, dashed line. It usually terminates in a short line at 90° to the cutting plane with arrowheads in the direction of sight for viewing the section. Letters may be used to indicate the section.

There are currently three different choices for the drafter when creating a cutting-plane line. Refer to Figure 2-1. The most common cutting-plane line features a long dash and then two short dashes. Another version represented in the standard is a series of medium-size dashes, about twice as long as hidden line dashes. The most current standard also allows for two "elbows" with identifying labels and arrows. In any case, the lines should be the same thickness as visible lines.

## Dimensioning Lines

While the visible, hidden, and center lines are used to create the shape description of an object, the size description of an object is indicated in annotations known as ***dimensions.*** Dimensions are composed of a variety of lines, all drawn without dashes and all with a thin line weight. Dimensioning is discussed in Unit 9.

The lines that extend the edges of the object out away from the view are called ***extension lines,*** Figure 2-6. Some books, references, and CAD systems refer to these as ***witness lines.*** The purpose of the extension lines is to keep the dimensional annotation away from the shape description. Extension lines begin about 1/16″ away from the object's visible corners and extend about 1/8″ beyond the arrows of the dimension line.

Between the extension lines are ***dimension lines,*** which indicate the extent and direction of part dimensions, Figure 2-6. In most engineering drawings, the dimension line is broken in the middle for the dimensional value. Dimension lines are usually terminated by arrowheads against extension lines. Dimensioning methods are becoming more diverse in industrial applications, including arrowless coordinate dimensioning and

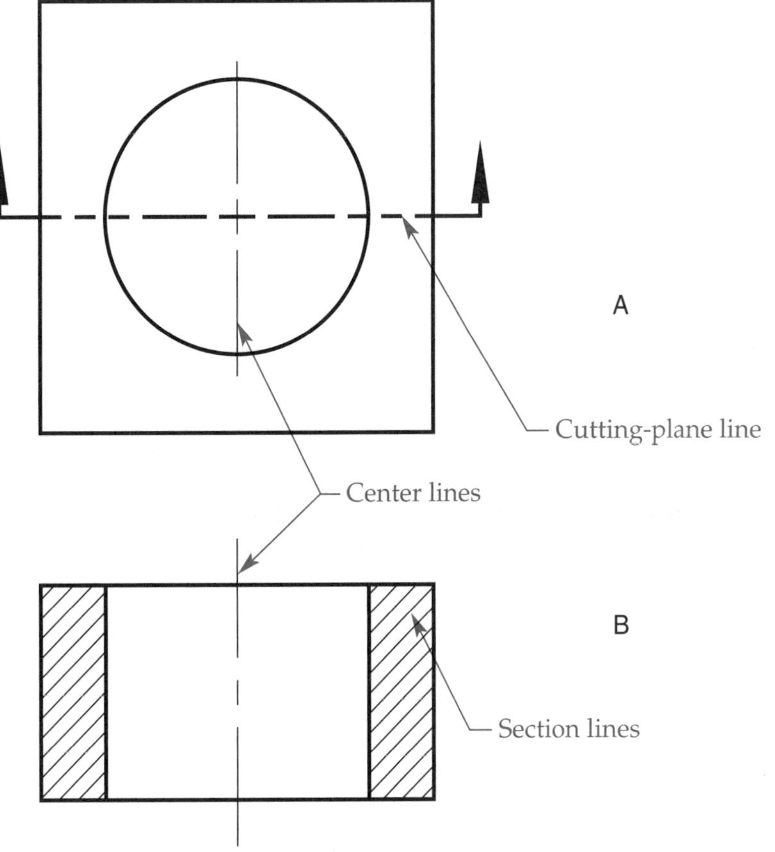


Figure 2-5.
Cutting-plane lines can be made in different ways, but the method indicated in A is common. Section lines, shown in B, are thin, continuous lines usually drawn at 45° and about 1/8″ apart.

geometric dimensioning and tolerancing. These topics are discussed in later units.

In dimensioning, ***leaders*** are used to point to a feature or drawing area to which a local note applies, Figure 2-6. Leaders are also used in conjunction with dimension lines if there is insufficient room for dimensional values within the dimension line. Leaders are usually characterized by an arrow on one end and a shoulder on the other end vertically centered with the beginning or end of the note. They are thin, continuous lines.

## Other Miscellaneous Lines

A few other lines should be discussed in this unit. As you progress through the textbook, you will encounter applications for these lines. These examples will reinforce your ability to discuss the purpose and function of each line, as described here.

On some occasions, it is not practical to view a particular feature of an object in the regular view arrangement, so a viewing direction can be established with a ***viewing-plane line,*** see Figure 2-7. The viewing-plane line is equivalent in appearance to the cutting-plane line, but simply "floats" outside of the object instead of being placed at a hypothetical cutting position.

***Break lines*** are used in drawings to "break out" or "break off" a portion of a view. There are two types, one for short breaks and one for long breaks. The short break line, if drawn by hand, should be freehand and thick. It was even called *freehand short break line* in past standards and references. The short break line is often used in section view drawings if it is desirable for the cutting plane not to go all the way through an object, Figure 2-8. Either of the break lines can be used for shortening the view of an object or creating a partial view. Objects that are constant in detail, yet too long to place on the drawing, like a shovel handle or a long bar, can be shortened by using either type of break line. In some cases, round stock, such as shafts (solid) or pipe (tubular), may need to be broken. In these cases, a conventional "S" break is used, as discussed in Unit 6. When the part to be broken is not round, a short break line is used. These lines are also used to break an edge or surface for clarity of a hidden surface. This line should not be an exaggerated zigzag like

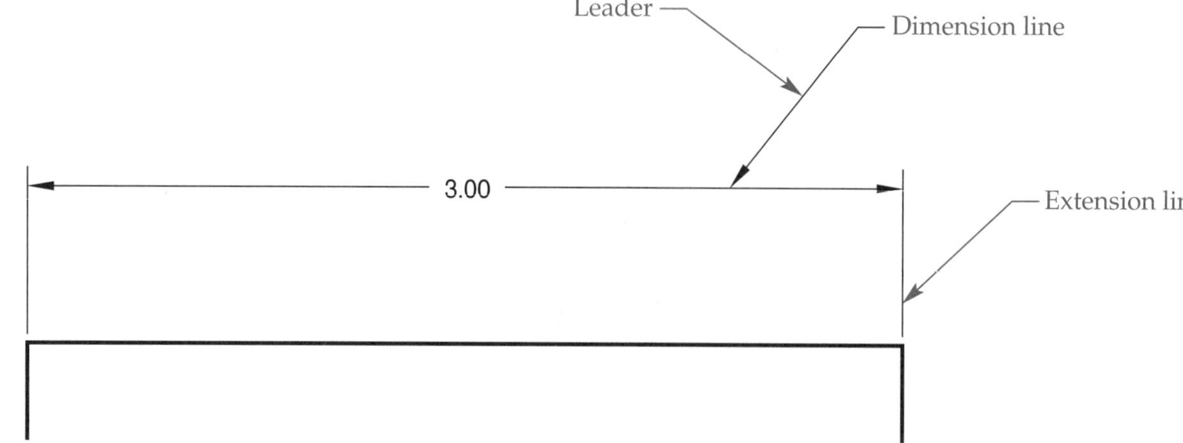


Figure 2-6.
Dimension lines are thin lines used to indicate the extent and direction of part dimensions, while the extension lines extend the object. Leaders are characterized by an arrow on one end and a shoulder on the other end, which is vertically centered with the beginning or end of the note.

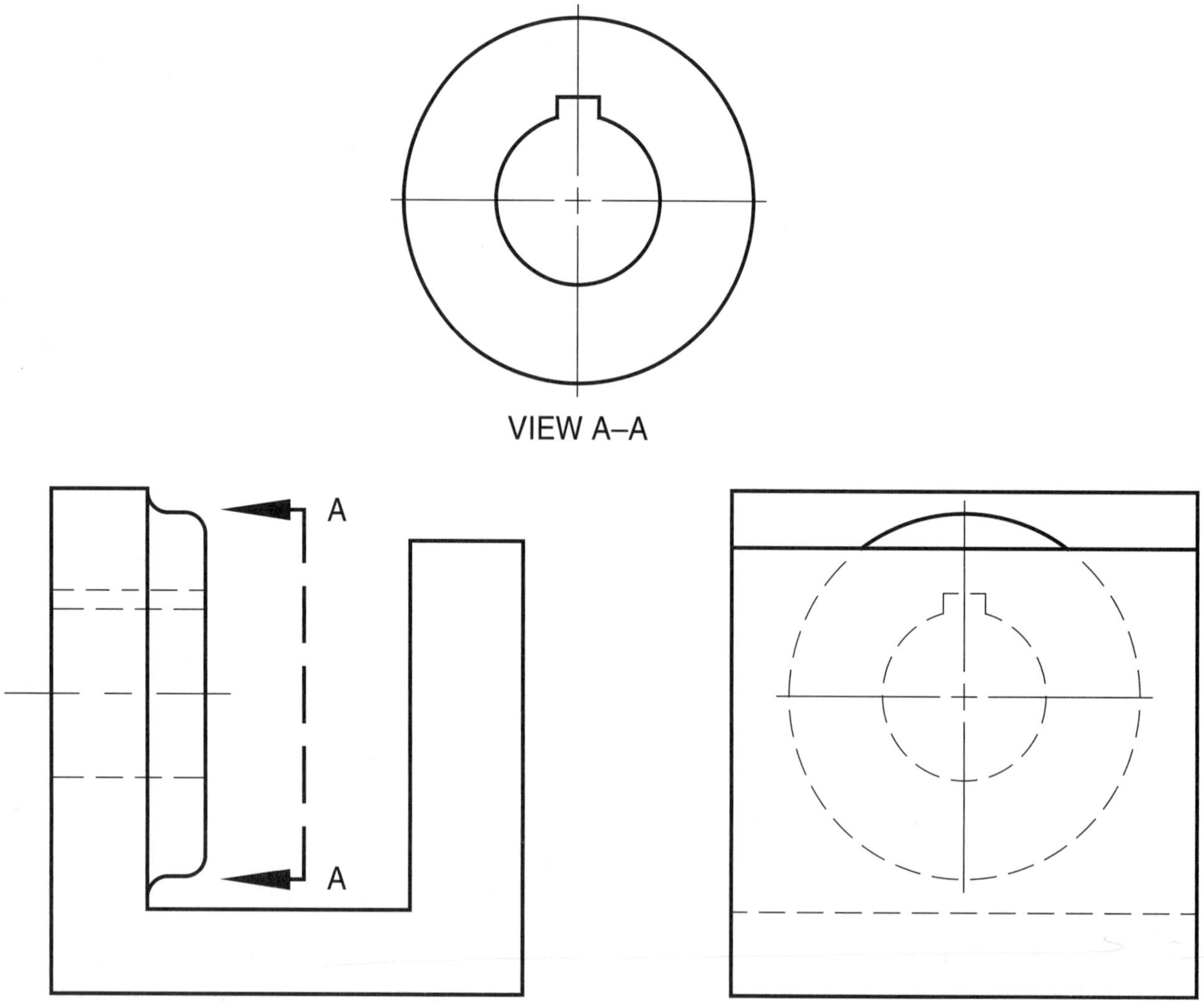

VIEW A–A

A

A

Figure 2-7.
A viewing-plane line for a partial view is drawn the same as a cutting-plane line, although there are three standard options. This viewing-plane line indicates the direction for viewing VIEW A-A.

lightning or splintered wood, but should look more like torn paper. If the part to be broken requires a longer break line, a thin, long break line is used, Figure 2-8.

***Phantom lines*** are thin lines composed of long dashes alternating with pairs of short dashes. The dash pattern is similar to a cutting-plane line, but the line weight should be thin, not thick. Phantom lines are primarily used to indicate alternate positions of moving parts, such as a machine arm, Figure 2-9A; adjacent positions of related parts, such as an existing column, Figure 2-9B; or for repeated detail, Figure 2-9C.

***Stitch lines*** are included in the ASME standard. These lines simply represent the path of a sewing or stitching process. They are comprised of either short dashes with spaces that are the same length or a series of dots approximately 1/8″ mm apart.

The ***chain line*** is used to indicate an area on the drawing wherein something special applies. Notice the NO PAINT note in Figure 2-2. For example, if the last inch of a rod is to be heat treated, a 1″ chain line is drawn next to that area of the part. A local note can point to the chain line. A chain line appears somewhat similar to a center line, with short and long dashes alternating, but is thick instead of thin.

## Standardized Lettering

In engineering drawings, the text and numeric information is referred to as ***lettering.*** Within the context of drafting, *lettering* is not only the letters

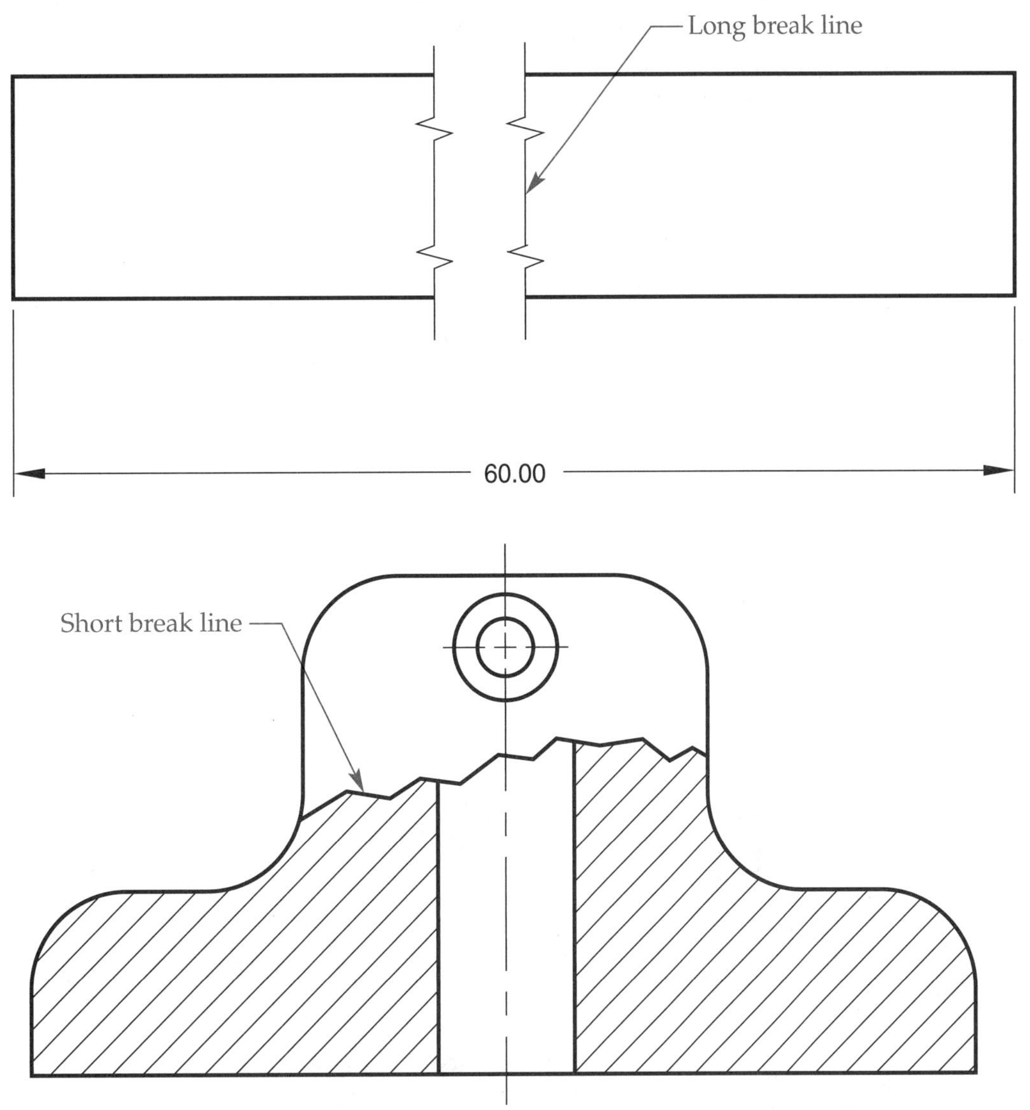


Figure 2-8.
Break lines can be used within views to "break out" sections for clarity or for shortening a view featuring a long and redundant part, such as a handle or tube.

and numerals themselves, but the term also applies to the process of creating those characters. At one time lettering was performed freehand. It was of utmost importance that the lettering be legible, uniform, and standard. Even today, for engineering documents that are created by hand, such as preliminary sketches, it is critical that mistakes are avoided by using good lettering. If you would like to develop skills in creating freehand letters, there is additional discussion of lettering in the appendix.

Within the context of print reading, lettering is covered in this unit to establish the definition and standard expression recommended by the ASME standard. Lettering on an industrial print is to be uppercase lettering, unless lowercase lettering is specifically required. The recommended minimum height for lettering is 3 mm (1/8″), with exceptions for drawing block headings, zone letters and numerals, section and view letters, and other information in the title block, such as part name and number. For additional discussion on title block information, see Unit 3.

By definition, engineering drawings use ***single-stroke Gothic lettering,*** Figure 2-10. The "single-stroke" adjective refers to the freehand technique wherein each letter is made up of a series of single strokes. Each letter is not created with one single stoke. Gothic, as applied to lettering, means simple, sans serif lettering. A Roman style of lettering contains serifs, which are small tails that make letters appear fancier. As CAD technologies and computer fonts have been developed, not everyone has used the same terminology. For example, one popular CAD system has a font called GOTHIC, but it appears like a traditional "old English" font. The same system has a font called ROMAN SIMPLEX that matches the engineering Gothic style rather well, but a ROMAN DUPLEX font that is truly a Roman-style lettering with serifs. This can be confusing when establishing the template drawings and standard CAD management protocols.

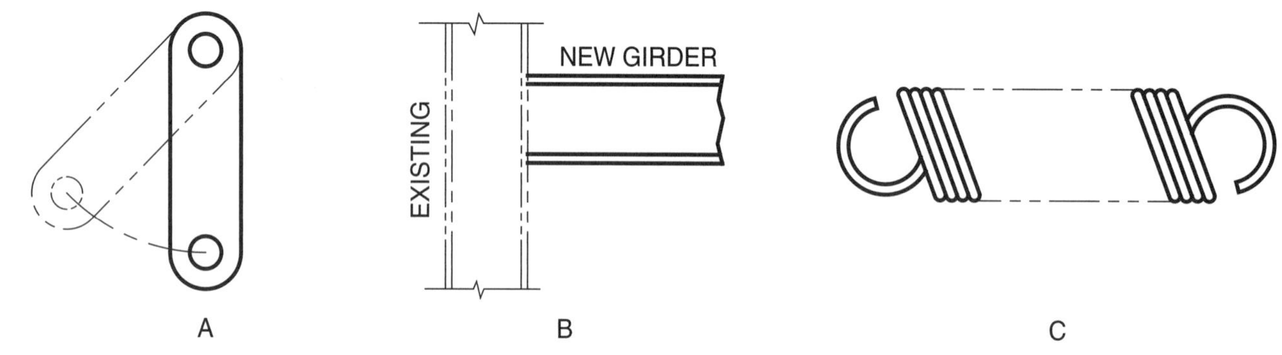


Figure 2-9.
Phantom lines are used to show: A—Alternate positions of moving parts. B—Adjacent positions of related parts. C—Repeated detail.

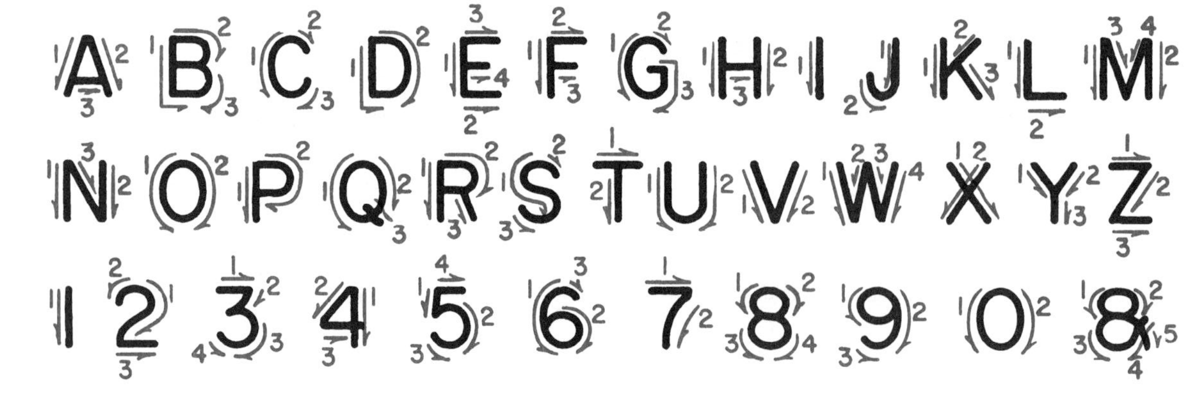


Figure 2-10.
Single-stroke Gothic lettering, usually set in uppercase letters, is the established standard for engineering drawings.

## Review Questions

*Circle the answer of choice, fill in the blank, or write a short answer.*

1. The term for a "generally accepted way of doing things" is ________________.

2. *True or False?* There are only two standard line weights (widths) found in the alphabet of lines indicated in current standards.

3. Which of the following lines does *not* feature any dashes?
   A. General purpose section line.
   B. Hidden line.
   C. Phantom line.
   D. Center line.

4. How many options are there for the style of a cutting-plane line? ________________

   ________________________________

5. List two different lines that feature a repeating pattern of a short dash followed by a long dash:

   ________________________________

   ________________________________

6. Of the following lines, which is *not* drawn thick?
   A. Visible line.
   B. Short break line.
   C. Chain line.
   D. Long break line.

7. A(n) ________________ line is used to point to a feature or drawing area and usually has an arrowhead on only one end.

8. *True or False?* If drawn by hand, the short break line should be drawn with a straightedge.

9. The ________________ line is used to show an alternate position, adjacent related parts, or repeated detail.

10. Which statement listed below regarding characteristics of lettering is false?
    A. Lettering, if done by hand, is composed of a series of single strokes.
    B. Lettering on an engineering drawing should be a Gothic style or font.
    C. Lettering on an engineering drawing has serifs and is very fancy.
    D. Lettering is uppercase, not lowercase.

## Review Activity 2-1

*Match the letter of each illustrated line with the correct name. Items are matched only once.*

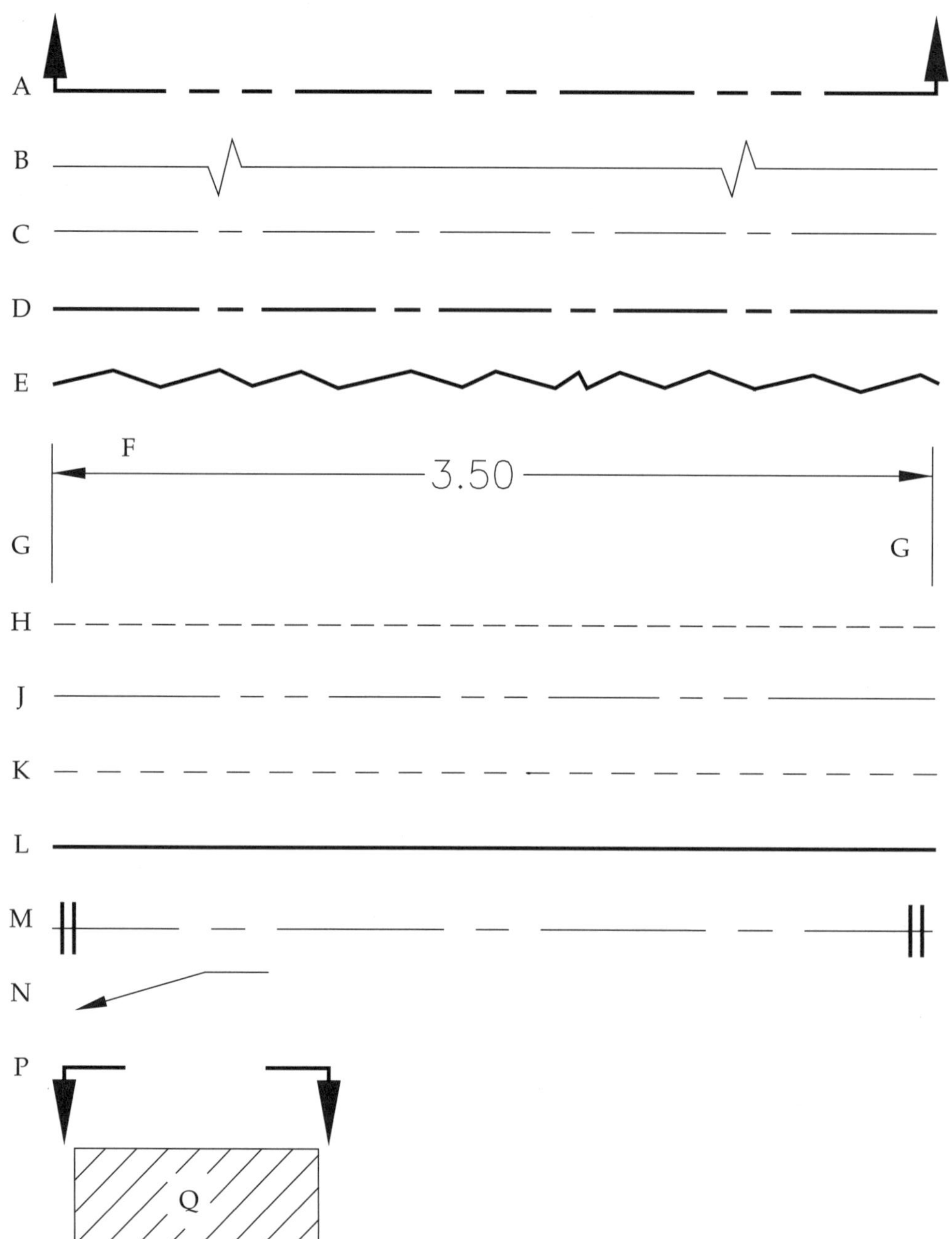


| | | |
|---|---|---|
| _______ 1. Short break line | _______ 6. Leader line | _______ 11. Viewing-plane line |
| _______ 2. Hidden line | _______ 7. Dimension line | _______ 12. Chain line |
| _______ 3. Center line | _______ 8. Extension line | _______ 13. Long break line |
| _______ 4. Phantom line | _______ 9. Cutting-plane line | _______ 14. Stitch line |
| _______ 5. Section line | _______ 10. Visible line | _______ 15. Symmetry line |

## Review Activity 2-2

*Study the drawing below and identify the twelve lines by name in the blanks below.*

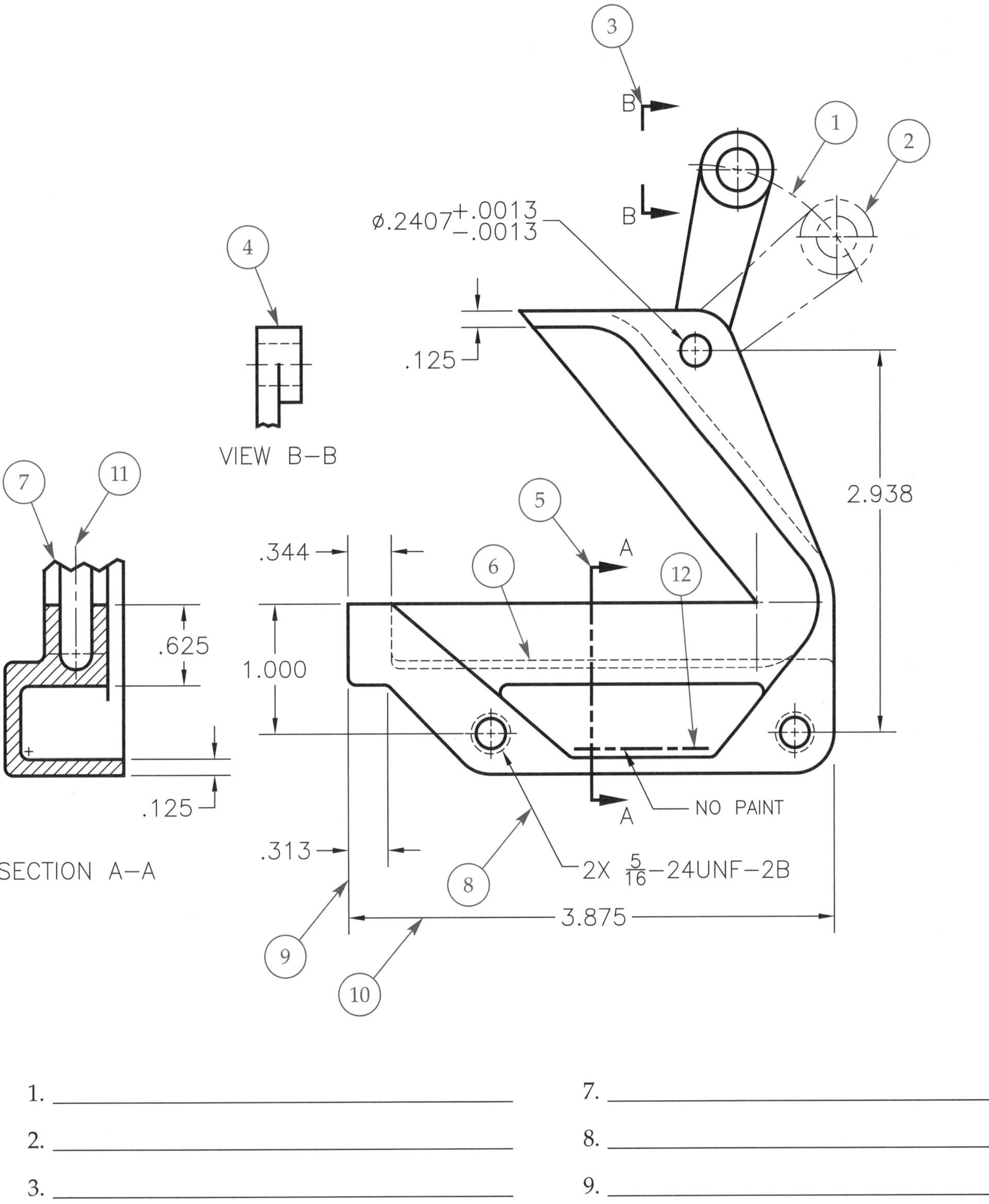


1. ______________________
2. ______________________
3. ______________________
4. ______________________
5. ______________________
6. ______________________
7. ______________________
8. ______________________
9. ______________________
10. ______________________
11. ______________________
12. ______________________

## Industry Print Exercise 2-1

Refer to the print PR 2-1 and answer the questions below.

*Closely examine the print to see which lines from the alphabet of lines are present or absent. For each of the lines listed below, fill in the blank with P for present or A for absent.*

________ 1. Cutting-plane line

________ 2. Dimension line

________ 3. Stitch line

________ 4. Section line

________ 5. Short break line

________ 6. Extension line

________ 7. Leader line

________ 8. Hidden line

________ 9. Chain line

________ 10. Center line

________ 11. Long break line

________ 12. Visible line

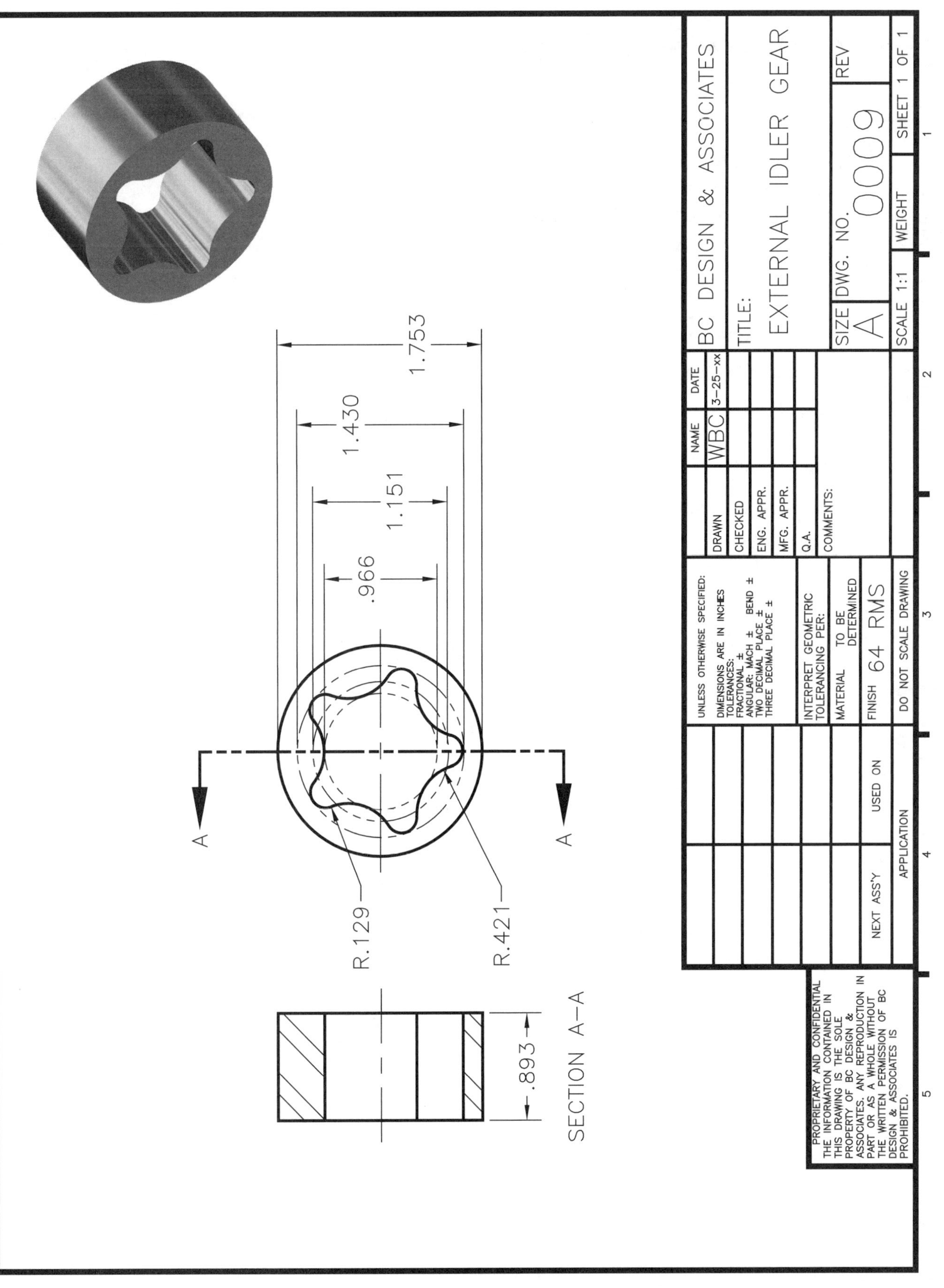


PR 2-1.
Print supplied by BC Design and Associates.

## Industry Print Exercise 2-2

Refer to the print PR 2-2 and answer the questions below.

*Closely examine the print to see which lines from the alphabet of lines are present or absent. For each of the lines listed below, fill in the blank with P for present or A for absent.*

________ 1. Extension line

________ 2. Phantom line

________ 3. Stitch line

________ 4. Visible line

________ 5. Short break line

________ 6. Cutting-plane line

________ 7. Long break line

________ 8. Hidden line

________ 9. Dimension line

________ 10. Center line

________ 11. Leader line

________ 12. Section line

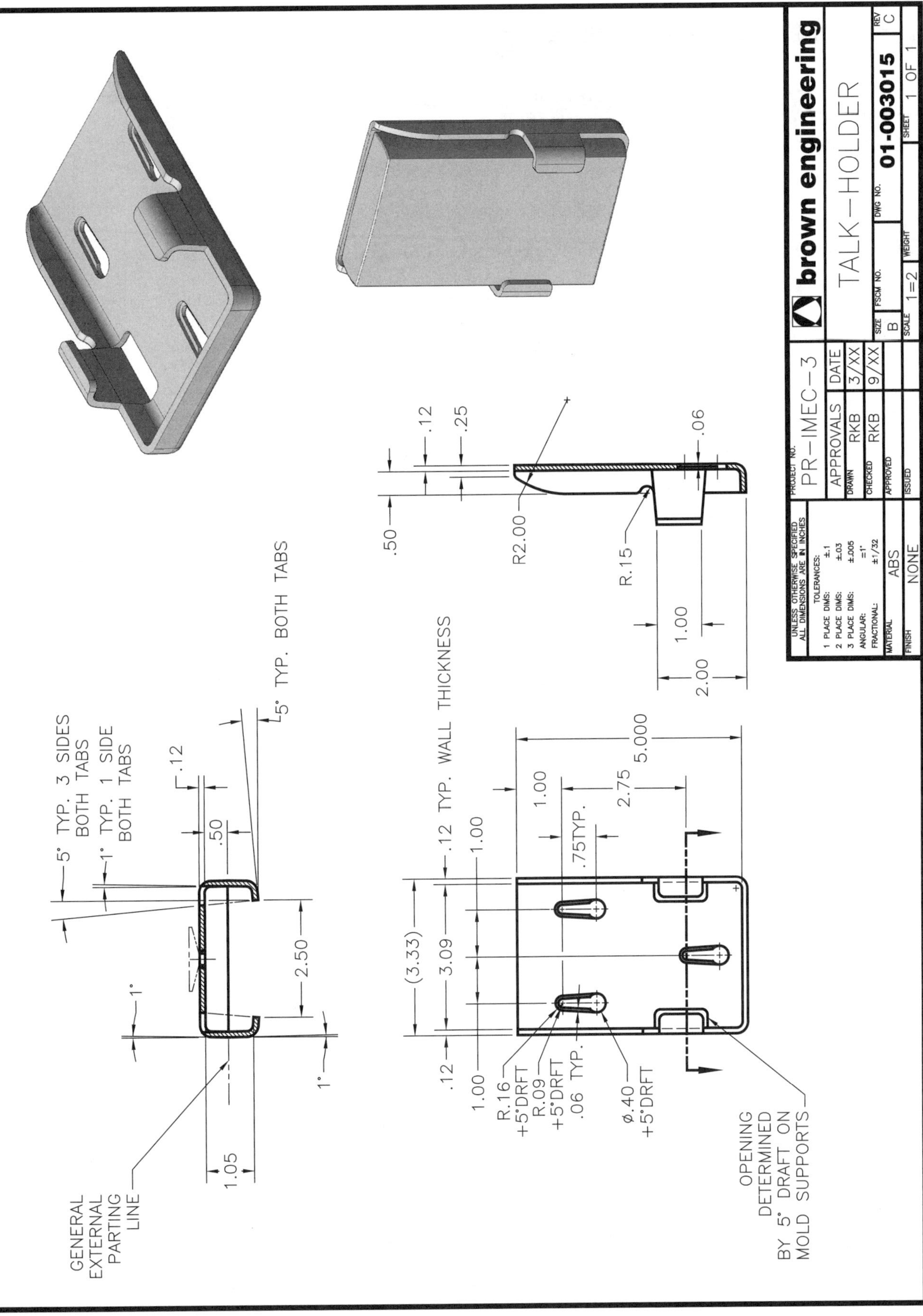


PR 2-2.
Print supplied by Brown Engineering.

## Bonus Print Reading Exercises

The following questions are based on various bonus prints located in the folder at the back of this textbook. Refer to the print indicated, evaluate the print, and answer the question.

**Print AP-001:**

1. Do the visible lines of the object appear to be thicker than the center lines and section lines?
_______________

2. What type of line is used to indicate the circular area being enlarged for View A? _______________
_______________

**Print AP-003:**

3. In how many of the four main views are section lines used? _______________
_______________

4. List the types of lines that appear in Section A-A, including the different types of lines used in dimensioning, if any: _______________
_______________

**Print AP-007:**

5. In the view with one dimension, 8X R.015, what two types of lines are featured besides the leader line? _______________
_______________

**Print AP-008:**

6. What type of line is used to indicate the cylindrical surfaces that are to be heat treated?
_______________

7. What type of line is used to indicate the involute splines that are specified on each end of this shaft?
_______________

**Print AP-012:**

8. What line name is given to the thicker dashed lines with arrows that are labeled A, B, C, etc.?
_______________

**Print AP-015:**

9. Describe the manner (pattern) in which the cutting-plane line is drawn. Remember, there are other options in the ASME standards. _______________
_______________

10. What type of line is used to show the partial view break off? _______________
_______________

11. Name the lines that appear in Section A-A. ____________________

____________________

**Print AP-019:**

12. Name the lines that appear in the enlarged detail A: ____________________

____________________

13. Not counting leader line callouts and arc radius dimensions, how many dimension lines were employed *without* using extension lines? ____________________

____________________

**Print AP-020:**

14. What type of line is used in the upper-left, round view to indicate the remaining gear teeth without drawing all of them? ____________________

____________________

**Print AP-021:**

15. Not counting the lines used for dimensioning, list the types of lines that appear on this print:

____________________

Computers contain many parts. The assembly drawing for the computer will have an extensive parts list.

UNIT 3

# Title Blocks and Part Lists

*After completing this unit, you will be able to:*

**Describe** drawing sheet sizes and formats.

**Identify** marginal information and zoning methods for drawing sheets.

**Identify** the elements of the title block as defined by industry standards.

**Explain** the techniques for identifying parts of an assembly drawing as represented in a basic parts list.

Industrial drawings are placed on a variety of sheet sizes, depending on the complexity or size of the part. The drafter uses both of these factors to determine the physical size of the views that will be needed to clearly show detail and describe the part. All of these factors together determine the final sheet size and main scale for the majority of the views. More complex parts may even require more than one sheet.

ASME Y14.1, entitled *Decimal Inch Drawing Sheet Size and Format*, and ASME Y14.1M, entitled *Metric Drawing Sheet Size and Format*, cover the basics of sheet layout and format, title blocks, and revision blocks. ASME Y14.34, entitled *Associated Lists*, sets forth recommendations for parts lists and data lists.

The primary focus for this unit is on the title block and part lists. Every print reader needs to begin with the title block to get their bearings regarding the task at hand. Within the title block resides information about part name, part number, last time the part was revised, and from what material the part is made. With assembly drawings, the list of material is usually critical to identifying the components that need to be assembled.

## Sheet Size and Format

Drawings are prepared on standard-size sheets, but there are two systems for inch-based paper sizes: engineering and architectural. Figure 3-1 illustrates four of the paper sizes, identified as sizes A through D. As illustrated, engineering sheet sizes are in multiples of 8 1/2″ × 11″, which is commonly referred to as A-size. If D-size paper is cut in half, two C-size sheets results, each 17″ × 22″. Likewise, C-size paper cut in half results in two B-size (11″ × 17″) sheets. Note: often, the A-size sheet for architectural applications is 9″ × 12″. This is because architectural drawings are larger and often based on rolls of paper manufactured in widths of 24″, 36″, or 48″.

In the United States, the decimal-inch system is still predominant, especially with respect to paper sheet sizes. Even if drawings are created with the metric system, smaller printers primarily use trays designed for decimal-inch paper. Roll-size format G, H, J, and K are also available for large parts and assemblies.

Metric sheet sizes are identified as A0, A1, A2, A3, and A4. See Figure 3-2 for a size comparison chart. This unit will not attempt to cover both decimal inch and metric systems, as the content is virtually the same, simply with different units.

## Borders and Zoning

As illustrated in Figure 3-1, the margins for a drawing vary depending on sheet size and available space. The recommended minimum space varies from .25″ to 1.00″. Roll-feed sheets for plotters may require as much as 2″ along one edge. Border lines are technically not in the alphabet of lines. They are usually drawn thick, perhaps even thicker than visible lines.

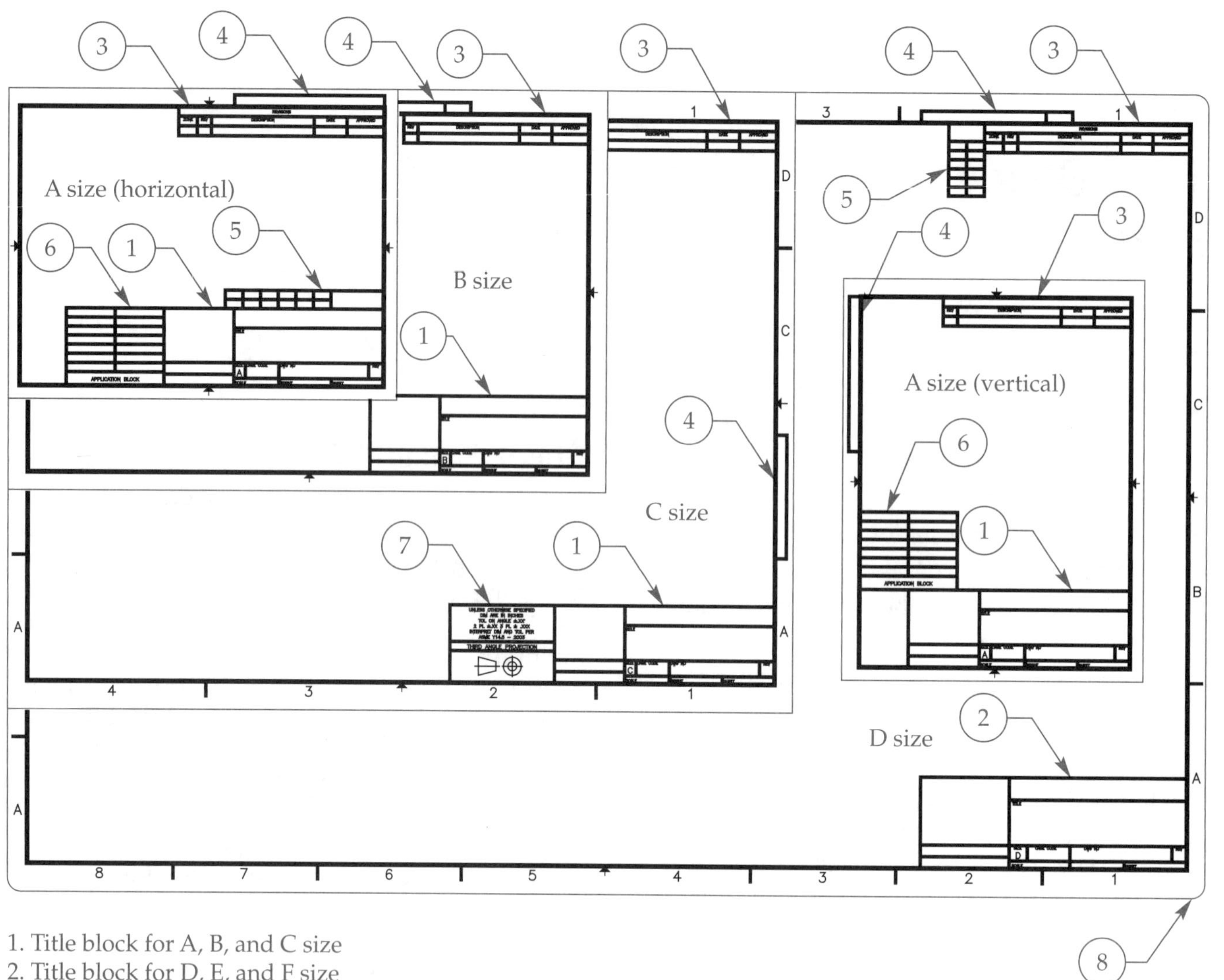


1. Title block for A, B, and C size
2. Title block for D, E, and F size
3. Revision history block
4. Optional margin drawing number block
5. Revision status of sheets block
6. Optional application block
7. Tolerance and projection block
8. Rounded corners optional all sheets

**Figure 3-1.**
Flat-sheet formats are standardized by ASME. Sheet formats A through D in landscape mode and A in portrait mode are shown here.

ASME standards recommend zoning for sheet sizes larger than B-size, although zoning is permissible even for smaller sheets. Outside of the border, ***zoning*** is used to aid in locating details of parts or revision notes, Figure 3-3. The zoning system is similar to that used on highway maps as an aid in locating cities and points of interest. Zones are indicated outside of the border. Numbers are used at intervals from right to left. Letters are used at intervals from bottom to top.

## Basic Title Block Elements

The ***title block*** provides information that aids in identification and filing of the print. The title block also provides supplementary information on the part or assembly. The title block is usually located in the lower, right-hand corner of the print so that it can be seen when the print is correctly folded. This allows for easy reference and filing.

| Letter size | Engineering standard | Architectural standard | Similar to ISO standard |
|---|---|---|---|
| A | 8.5 X 11 | 9 X 12 | A4 |
| B | 11 X 17 | 12 X 18 | A3 |
| C | 17 X 22 | 18 X 24 | A2 |
| D | 22 X 34 | 24 X 36 | A1 |
| E | 34 X 44 | 36 X 48 | A0 |
| E1 | | 30 X 42 | |
| F | 28 X 40 | | |

Figure 3-2.
The engineering standard for paper is different than the architectural standard. Metric paper sizes are also standardized, but American standards are still based on inches.

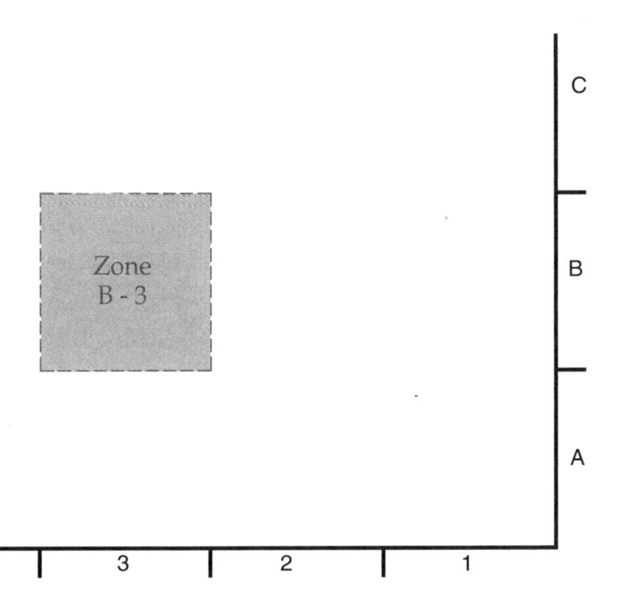


Figure 3-3.
Zoning is used to specify an area of a large print, similar to what is used on highway maps.

While many companies have their own variations of the standard title block, the information found in a title block is similar between companies. Title block information included on plans by most industries is explained in this unit. Understanding the information typically found in a title block will help you properly interpret all title blocks.

The name and address of the company is usually placed in the upper part of the title block, Figure 3-4A. Often, the company logo appears in this space. With CAD systems, it is easy to insert a standard logo into a drawing.

The ***drawing number*** is used to identify and control the print. It is also used to designate the part or assembly shown on the print, Figure 3-4B. The number is usually coded to indicate department, model, group, serial number, and dash numbers. The drawing number may also be shown in the margin of the drawing, usually in a rectangular box. This is designed so the drawing number shows along the top edge when the drawing is folded to A-size, Figure 3-5.

The ***drawing title*** indicates the name of the part. The title should be descriptive, brief, and clearly state an identification of the part or assembly. See Figure 3-4C. The title starts with the name of the part or assembly, followed by descriptive modifiers. When the title is read out loud, the descriptive modifier is read first. For example, the title in Figure 3-6 is read "pressure regulating valve assembly."

The ***sheet area*** of the title block is used for sheet numbering. Sheet numbering is used on multi-sheet prints to indicate the consecutive order and total number of prints. Sheet numbering is often written as SHEET *n* OF *x*, where *n* is the number of the sheet, and *x* is the total number of sheets for that set. See Figure 3-4D.

UNLESS OTHERWISE SPECIFIED
ALL DIMENSIONS ARE IN INCHES
TOLERANCES:
1 PLACE DIMS: +/− .1
2 PLACE DIMS: +/− .03
3 PLACE DIMS: +/− .005
ANGULAR: +/− 1°
FRACTIONAL: +/− 1/32
MATERIAL CRS
FINISH NONE
PROJECT NO. PR−101
APPROVALS DATE
DRAWN (J) RKB 09/01
CHECKED
APPROVED
ISSUED
BROWN ENGINEERING (A)
TITLE PISTON (C)
(E)
SIZE B
CAGE CODE (H)
DWG NO. 5409-3 (B)
REV A (L)
SCALE 1:1 (G)
WEIGHT .234 (F)
SHEET 1 OF 1 (D)

Figure 3-4.
Most title blocks contain similar information. Standard formats are recommended, but not required.

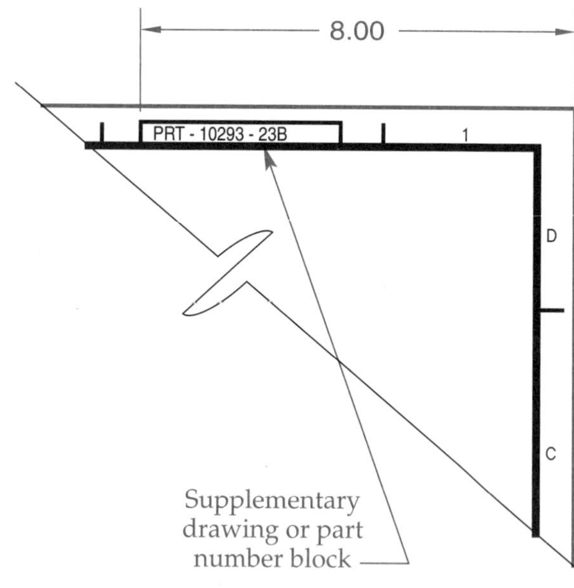


**Figure 3-5.**
Standards allow for the drawing number to be shown in the margin to aid in finding a drawing in the filing drawer.

VALVE ASSY, PRESSURE REGULATING
Part name — Descriptive modifiers

**Figure 3-6.**
A part name often contains descriptive modifiers. When read out loud, the modifiers are read first and then the part name.

The ***size area*** of the title block is used to indicate which sheet size is used for the print. A letter designation such as B, C, or D indicates the standard size paper. Refer to Figure 3-2 for standard paper sizes.

The ***weight area*** of the title block provides the weight of the part, Figure 3-4F. The weight can be either the actual or calculated weight, as indicated. Calculated weight is used during design stages to control the weight of the finished part or assembly. Actual weight is obtained after the part or assembly is actually manufactured.

The ***scale area*** indicates the scale of the drawing. The scale is the ratio between the part as drawn and the actual part. It is usually expressed as *paper* = *real* or *paper*:*real*, where the first number (*paper*) indicates the size measured on the paper and the second value represents the size measured on the actual part. See Figure 3-4G. Typical scale notations are: 1/2″ = 1″ (half size), FULL (actual size), 1:1 (actual size), and 2:1 (twice size). Sometimes 2X or 3X is used to indicate 2:1 or 3:1, respectively. When the scale is shown AS NOTED, several scales are used in the drawing. In this case, each scale is indicated below the particular view to which it pertains. Some drawings, especially diagrams, have no scale. The designation NONE is an appropriate entry in the title block for a drawing with no scale.

It is important to note that even though drawings are usually printed to scale and the scale is indicated in the title block, measurements should *never* be made directly on a print. This is because the print may be reduced in size or stretched. Work from the dimension values given on the print. If you believe these to be in error, report it to the appropriate person in the company.

Some title blocks have an area labeled CAGE CODE, which is a unique identifier given to vendors for the government, Figure 3-4H. CAGE is an acronym for ***Commercial and Government Entity.*** On older drawings, this block may be identified as FSCM (Federal Supply Code for Manufacturers) or NSCM (National Supply Code for Manufacturer). In summary, this area is used to identify part suppliers on drawings. For parts not requiring a CAGE code, this area is left blank or it may be omitted from the title block layout.

On the left side of a title block is an area for the signatures and date of release notation. These areas are not specified in current standards, but may appear as shown in Figure 3-4J. Entries in this area are made by those responsible for making or approving certain facets of the drawing or manufacture of the part. There may be many signatures on a given drawing. Some areas of responsibility are:

- **Drawn.** This area is for the drafter who made the drawing and the date completed.
- **Checked.** This area is for the engineer who checked the drawing for completeness, accuracy, and clarity.
- **Approved.** This area is to record any other required approvals.
- **Issued.** This area indicates the person who finally issues the drawing as available for general use, making it an "official" drawing.

## Intermediate Title Block Elements

There are many other areas of information that will be more clearly understood after some additional units are covered. Within your progression of study, it is important to understand the importance of studying the title block early and often in the process of interpreting the full scope of the product or part. Additional title block information is discussed in future units, but this section is a brief overview of additional information found in some title blocks.

In manufacturing, part dimensions are often within a range. The total amount of the range is the tolerance for that value. Most title blocks include a ***tolerances area,*** which indicates the general tolerance limits for one-, two-, and three-place decimal values and a certain number of degrees for angular dimensions. These limits are to be applied unless the tolerance is otherwise indicated on the drawing. Refer to Unit 13 for a discussion on tolerancing.

The ***finish area*** in a title block indicates general finish requirements, such as paint, chemical, or other. Any specific finish requirements are called out in a local note with the word NOTED in the finish area. Refer to Units 11 and 12 for a discussion on machining specifications and surface texture symbols.

The ***materials area*** in a title block indicates the material used to make the part. Some companies have material numbers assigned to all raw materials. The number of the material is then indicated in the title block. See Figure 3-7.

The ***heat treatment area*** in a title block indicates heat treatment and hardness requirements. See Figure 3-8. The entry may be AS REQUIRED or NOTED, which means the part must conform to the specification block notation or to the callout on the drawing detail. If heat treatment is not required, the word NONE or a diagonal line is entered in the block.

An ***application block*** can be added to the title block. The application block lists the assembly or subassembly in which the part or subassembly is used. See Figure 3-9. Indicating where the part or assembly is used can help determine the effects of a change in the part or assembly. Refer to Unit 16 for a discussion on assembly drawings.

A few companies include information in the title block indicating DRAWING SUPERSEDED BY or DRAWING SUPERSEDES. For example, if PART 111 supersedes PART 110, then PART 111 is the new part that replaced PART 110 in the system. This enables a drawing to report the previous part number or drawing number as a historical record, which helps the print reader locate information connected to the history of the part. This information can also be applied to a drawing that is being discontinued and archived so individuals can find the new drawing and part that superseded the old part.

Other information may appear in a title block. The customer for whom the product is being produced is sometimes listed, as is the contact number. Standards used for inspection may be used to refer to specific sets of inspection standards used by the company. A diagonal line or X appearing in any part of the title block means the item is not required.

MATERIAL
ASTM B124 CA377
BRASS FOR FORGING
NAME
BODY FORGING
(RAW CODE NO. R2803)
SCALE
BY
CHKD
APPD

**Figure 3-7.**
The material area of the title block indicates the material used to make the part.

MATERIAL
CARPENTER 610
HEAT TREAT
HARDEN & DRAW R/C 58 - 60
APPLIED FINISH

**Figure 3-8.**
The heat treatment area in the title block indicates heat treatment and hardness requirements.

| | | | | |
|---|---|---|---|---|
| | | | | |
| | | | | |
| | | | | |
| | | | | |
| | | | | |
| | | | | |
| - 3 | 1 | 1 | K - 139 | 10M3200 |
| PART DASH NO. | NEXT | FINAL | NEXT ASSY. | USED ON |
| | QTY.REQ. PER.ASSY. | | APPLICATION | |

**Figure 3-9.**
The application block, if present, is attached to the title block and lists the assembly or subassembly in which the part is used.

## Revision History Block

A system for reporting the history of changes to a part and its drawing is also necessary to be included on the sheet. Unit 14 discusses revision systems in more detail, but even locating the current revision status for a drawing is an initial step in the print reading process. As a part is revised, the differences are usually recorded on the drawing with alphabetical notations. It is important that the print reader establish which revision the drawing is and to know if a drawing is up-to-date. There are times when a revision is pending.

The ***revision history block*** should be located in the upper-right corner for most sheet layouts, Figure 3-10. This block may include information such as the character applied to a change, what change order number was assigned, the date, and other useful information. Additional columns can also be added as necessary. For multiple-sheet drawings, there may be a revision "status of sheets" block attached near the revision history block.

## Part Lists

Lists of information relative to the drawing often take the form of charts or tables. Recommended layouts are covered in ASME Y14.34, entitled *Associated Lists.* While some of these are beyond the scope of this text, the recommendations for charting a list of parts for an assembly, subassembly, or welding detail drawing are discussed. Additional discussion can be found in Unit 16, Assembly Drawings, and Unit 22, Welding Prints.

| REVISION HISTORY | | | |
|---|---|---|---|
| REV | DATE | DESCRIPTION | APPRVD |
| A | 02/10 | REDRAWN | RKB |
| B | 03/10 | ADDED GD&T | DCW |
| C | 11/10 | REMOVED HOLE AND KEYWAY | MAR |

Figure 3-10.
The revision block contains a record of the changes that have been made to the original drawing.

The name recommended by ASME for a chart listing the parts within a particular drawing is parts list. The ***parts list*** is a tabular chart or form, usually appearing immediately above the title block on a print, Figure 3-11. It provides information with respect to the quantity and description of the parts of a machine, structure, or assembly. The parts list is used primarily for subassembly, assembly, welding, or installation drawings. Throughout the years, the parts list block has also been called a list of materials, materials list, bill of materials, and schedule of parts.

These items are usually included in a basic parts list, as illustrated in Figure 3-11:

- Find number or mark column.
- Quantity required column.
- Part or identifying number (PIN) column.
- Nomenclature or description column.

The find number, or mark, is an optional column if numbers such as 1, 2, 3, etc., have been assigned to a part (in lieu of the actual part number) for the purpose of helping the print reader find the part within the drawing. This find number is often found within a circle, or balloon, which is discussed later in this unit.

The ***quantity required*** column is mandatory and indicates the number of parts required for an assembly. When the letters AR appear in the column in place of a number, it means the number is "as required."

| MARK | QTY | PART NUMBER | DESCRIPTION |
|---|---|---|---|
| 8 | 1 | 8400–356 | INSTRUCTIONS |
| 7 | 1 | 143–5321–150 | LABEL, 100–200 SERIES |
| 6 | 3 | 304–5300–101 | SPRING, GUIDE |
| 5 | 3 | 304–5300–100 | PIN, GUIDE |
| 4 | 1 | 143–5320–410 | BUSHING |
| 3 | 1 | 143–5320–407 | COLLET, SELF–CENTERING |
| 2 | 1 | 143–5321–202 | HUB, 200 SERIES |
| 1 | 1 | 143–5321–201 | BASE, 200 SERIES |
| PARTS LIST | | | |

Figure 3-11.
This basic parts list provides information about the quantity and names of the parts within an assembly.

The ***part or identifying number (PIN)*** column is mandatory, indicating detail part numbers required for the assembly. All company- and subcontractor-manufactured part numbers are listed. Standard parts, such as bearings, bolts, and screws are listed by the company-assigned supply number. Sometimes, special fasteners, such as rivets, might not be listed in the block.

The ***nomenclature*** or ***description*** column is mandatory and provides "real" names for all subassemblies and parts. These "real" names are usually limited to a basic noun description or, if a main assembly has a description listed, an abbreviated name.

Additional columns for the parts list are discussed in Unit 16, Assembly Drawings. Parts lists can become quite involved for assembly drawings that document several different versions of an assembly with one drawing.

## Part Identification

A system for identifying the parts appearing in the parts list is necessary. Most assembly or subassembly drawings use leaders to identify the parts. In addition, sometimes the parts are exploded to help the print reader see the relationship between parts. Many companies use balloons neatly arranged on the drawing to identify the parts. ***Balloons*** are circles containing a number or letter and are usually connected to leader lines, Figure 3-12. Depending on the complexity of the drawing or part names and numbers, detail callouts can appear adjacent to the part on the drawing. See Figure 3-13. A ***detail callout*** includes the part name and number, and may eliminate the need for a parts list block.

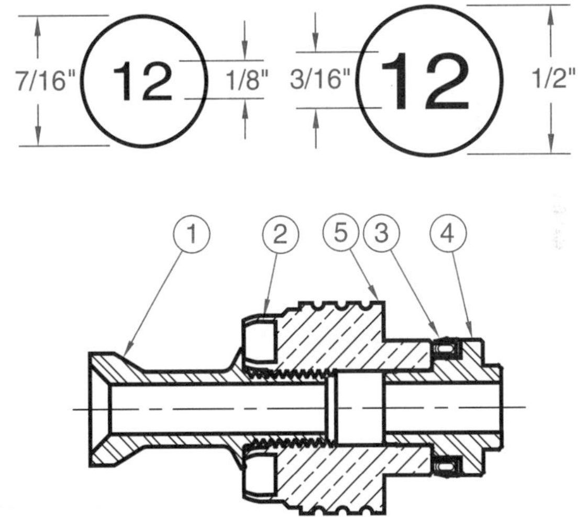


Figure 3-12.
Balloons can be used to identify parts.

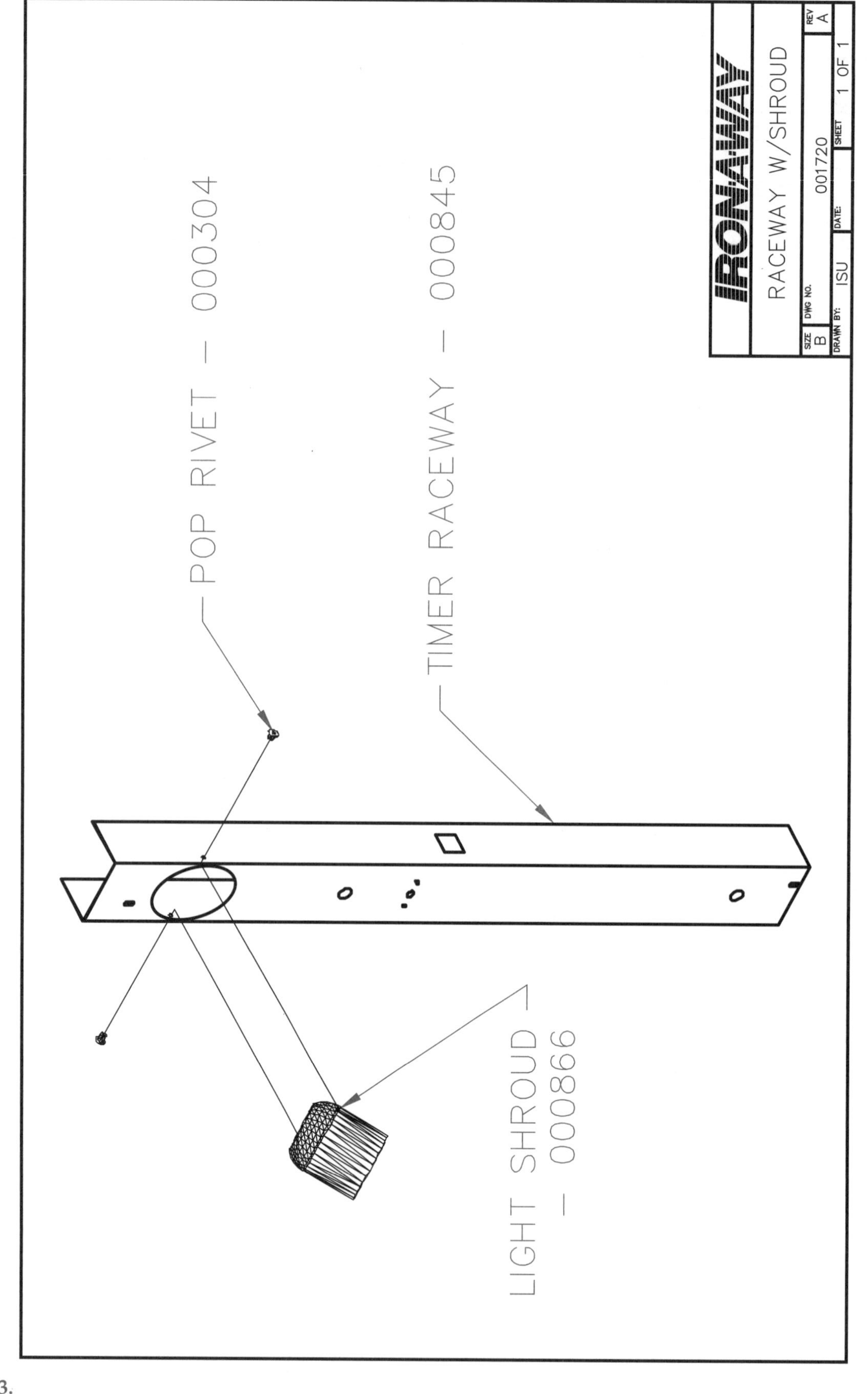


Figure 3-13.
A detail callout includes the part name and number. (Iron-A-Way)

## Review Questions

*Circle the answer of choice, fill in the blank, or write a short answer.*

1. When reading a print, what part of the drawing should be read first? ____________________

____________________________________________

2. *True or False?* A D-size sheet can be cut into four B-size sheets with no waste.

3. The numbers and letters found in the margins of larger drawings are related to ____.
   A. dates
   B. scale
   C. material
   D. zones

4. *True or False?* The company name and address are not placed in the title block for security reasons.

5. Of the following "boxes" in the title block, which is usually the largest?
   A. Drawing number.
   B. Scale.
   C. Sheet size.
   D. CAGE code.

6. The drawing title should be written with the ____________________ last, even though it is read first.

7. *True or False?* A drawing wherein the views are created half size should be indicated as a scale of 2:1 (or 2 = 1) in the title block.

8. In which corner of the drawing should you expect to find the revision history block? ____________

____________________________________________

9. Which of the following is *least* likely to be found in a basic parts list?
   A. Quantity.
   B. CAGE code.
   C. Part name.
   D. Description.

10. Balloons most often contain a(n) ____________________ number associated with the parts list.

## Review Activity 3-1

*Study the illustration on the next page and identify the numbered items by name in the blanks below.*

1. ______________________
2. ______________________
3. ______________________
4. ______________________
5. ______________________
6. ______________________
7. ______________________
8. ______________________
9. ______________________
10. ______________________

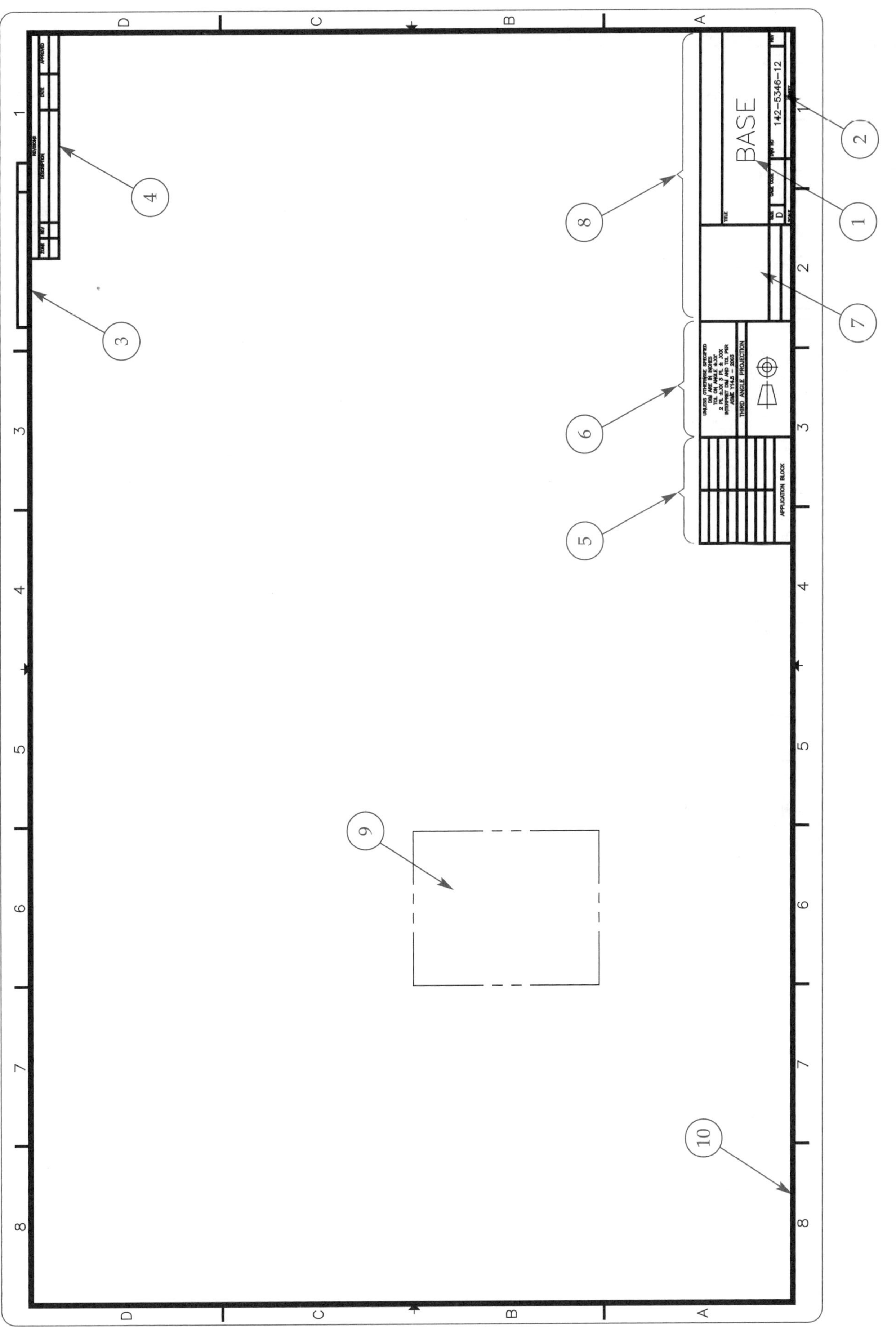
BASE
142-5346-12
D
UNLESS OTHERWISE SPECIFIED
DIM ARE IN INCHES
TOL ON ANGLE ±.X°
2 PL ±.XX 3 PL ±.XXX
INTERPRET DIM AND TOL PER
ASME Y14.5 – 2003
THIRD ANGLE PROJECTION
APPLICATION BLOCK
REVISIONS
1
2
3
4
5
6
7
8
9
10
A
B
C
D

## Review Activity 3-2

*Study the drawing on the next page and answer the questions that follow.*

1. What is the name of the assembly? ____________________

2. How many parts are identified by balloons in this drawing? ____________________

3. How many parts have a quantity of two or more? ____________________

4. Of the parts shown, what is the name of the smallest part? ____________________

5. What is the individual part number of the DRAW ROD? ____________________

6. In the material block, it says, SEE LOM. What does LOM stand for? ____________________

7. For what size sheet is this drawing intended? ____________________

8. Is this drawing part of a multi-sheet set? ____________________

9. What is the name of part number 105-003? ____________________

10. What are the initials of the person who checked this drawing? ____________________

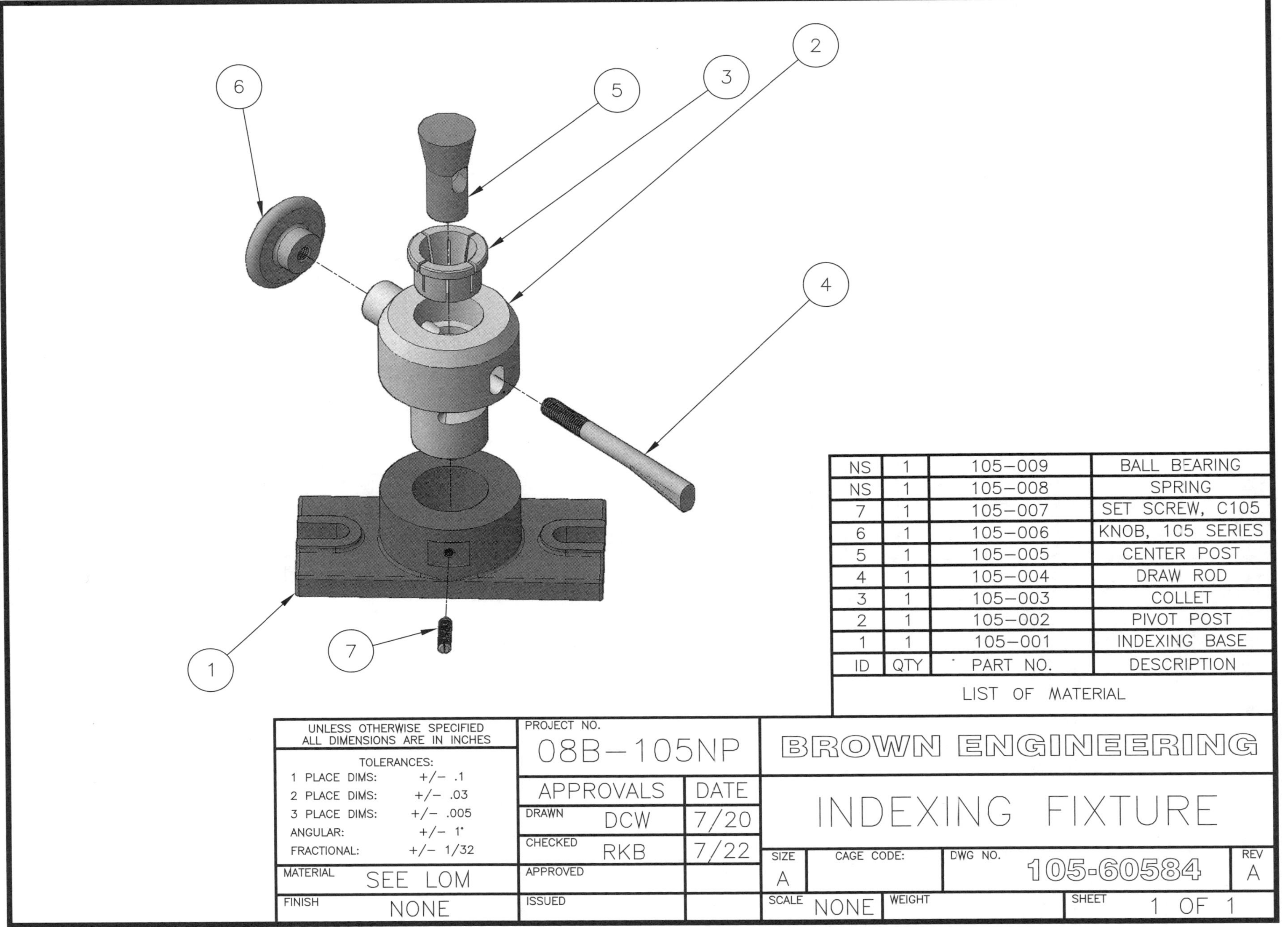


| ID | QTY | PART NO. | DESCRIPTION |
|---|---|---|---|
| NS | 1 | 105−009 | BALL BEARING |
| NS | 1 | 105−008 | SPRING |
| 7 | 1 | 105−007 | SET SCREW, C105 |
| 6 | 1 | 105−006 | KNOB, 105 SERIES |
| 5 | 1 | 105−005 | CENTER POST |
| 4 | 1 | 105−004 | DRAW ROD |
| 3 | 1 | 105−003 | COLLET |
| 2 | 1 | 105−002 | PIVOT POST |
| 1 | 1 | 105−001 | INDEXING BASE |

LIST OF MATERIAL

## Industry Print Exercise 3-1

*Refer to the print in PR 3-1 and answer the questions below.*

1. What is the name of this drawing? ______

2. What is the drawing number? ______

3. How many parts have a quantity of two or more? ______

4. What is the part number of the NUT? ______

5. What is the part number of the KNOB? ______

6. At what scale is this drawing created? ______

7. For what size sheet is this drawing intended? ______

8. In what state is this company located? ______

9. How many balloons are used to locate parts for the parts list? ______

10. What is the last name of the person who checked this drawing? ______

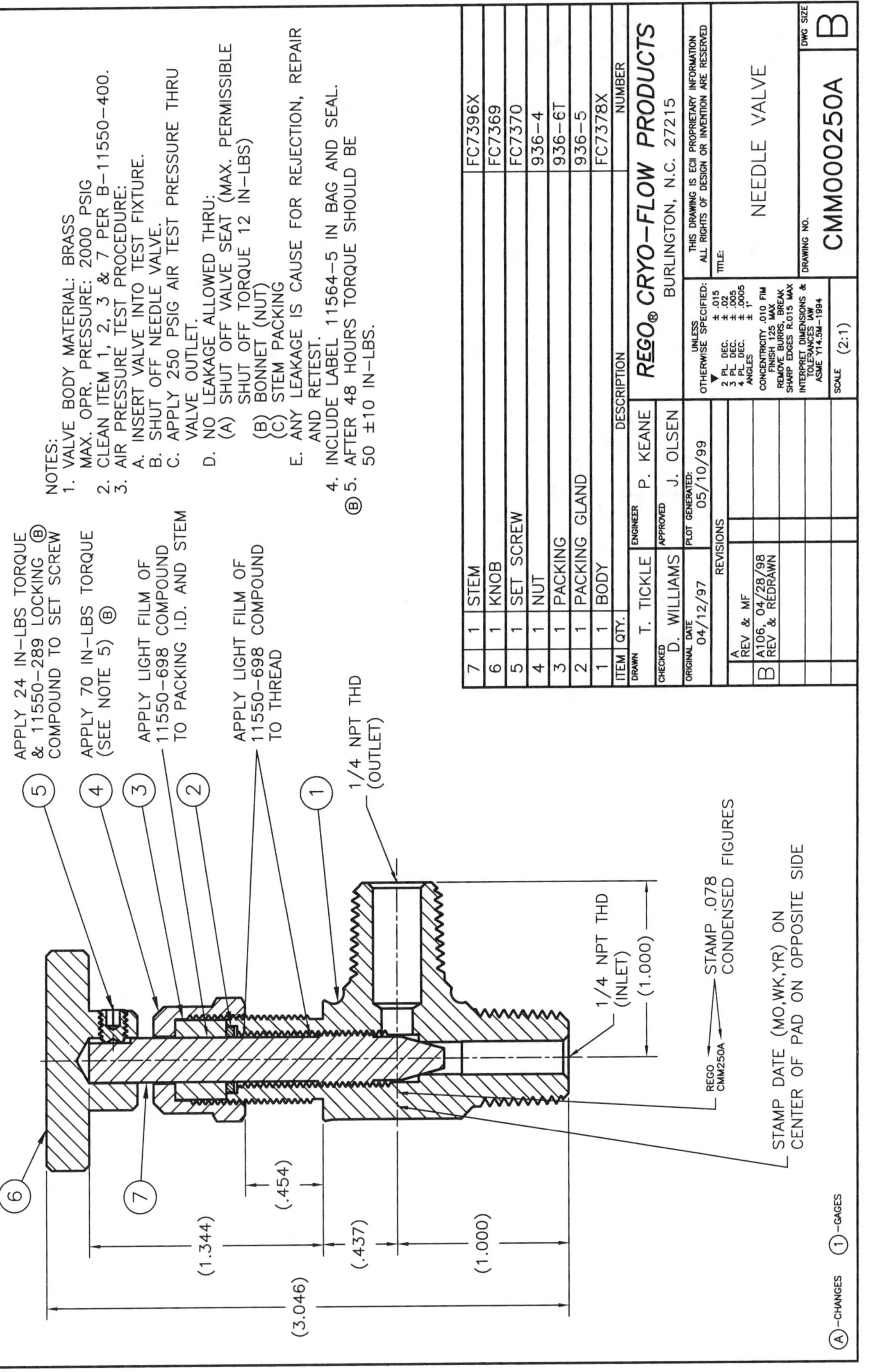


PR 3-1.
Print supplied by RegO Cryo-Flow Products.

## Industry Print Exercise 3-2

*Refer to the print PR 3-2 and answer the questions below.*

1. What is the name of the assembly? ____________

2. How many rows are there for parts in the parts list? ____________

3. What is the name of the part with the greatest quantity needed? ____________

4. What is the name of item #14? ____________

5. What are the initials of the person who drew the drawing? ____________

6. What is the name of the company? ____________

7. For what size sheet is this drawing intended? ____________

8. Is this drawing part of a multi-sheet set? ____________

9. What is the name of part number 000412? ____________

10. What is the individual part number of the WHITE PLASTIC PLUG? ____________

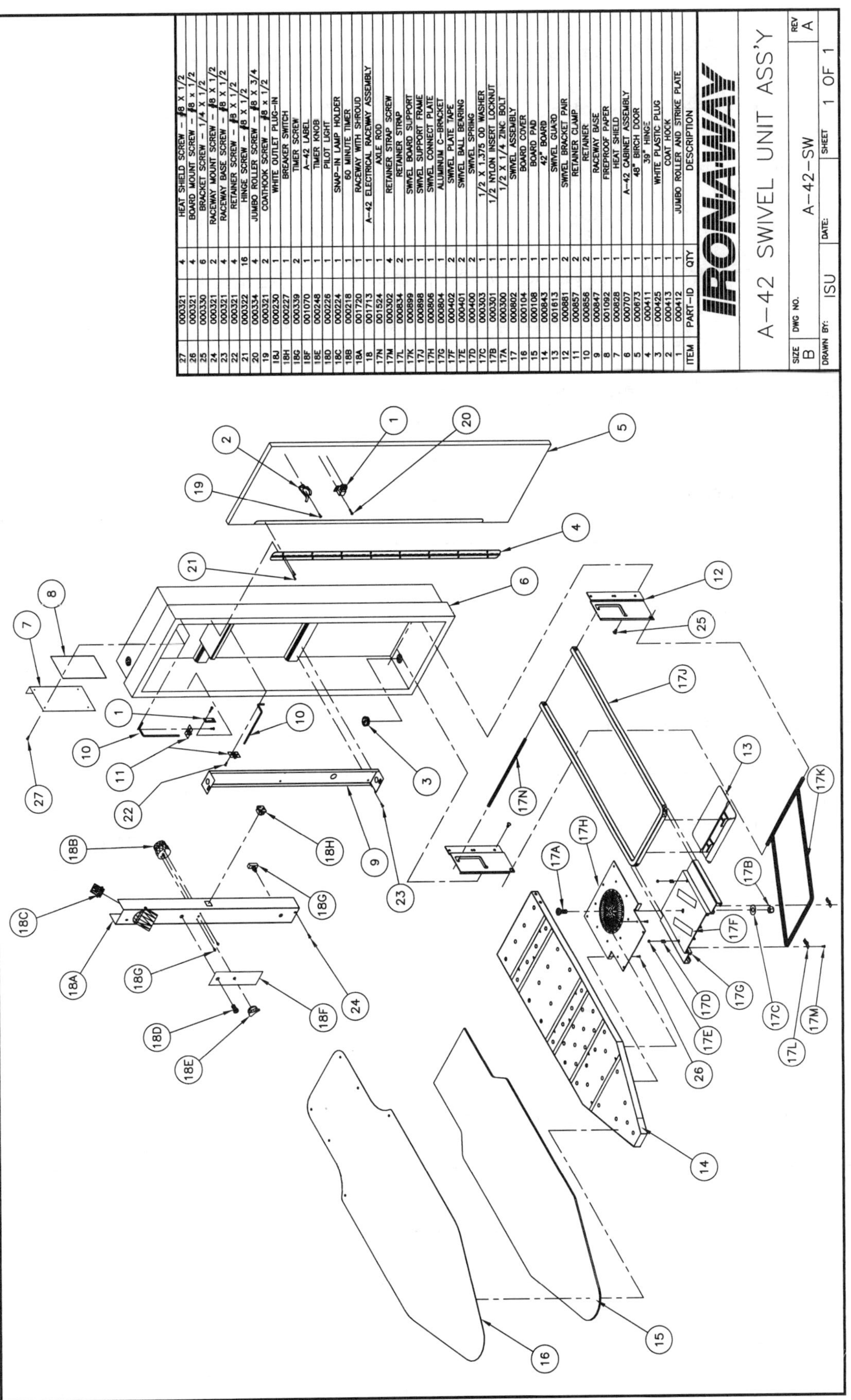


PR 3-2.
Print supplied by Iron-A-Way.

## Bonus Print Reading Exercises

The following questions are based on various bonus prints located in the folder at the back of this textbook. Refer to the print indicated, evaluate the print, and answer the question.

**Print AP-001:**

1. What size paper was used for the original version of this print? ____________________

**Print AP-002:**

2. What size paper was used for the original version of this print? ____________________

3. What is the total quantity of subassemblies required for this assembly? ____________________

**Print AP-003:**

4. For this print, of what significance is the number 81073? ____________________

**Print AP-004:**

5. List the three parts indicated by name in the parts list: ____________________

6. Within what zone is the primary portion of View A? ____________________

**Print AP-005:**

7. This print is sheet ____ of ____ sheets.

**Print AP-006:**

8. What material is specified for this part? ____________________

9. What size paper was used for the original version of this print? ____________________

**Print AP-008:**

10. What title is given to the revision history block on this print? ____________________

**Print AP-009:**

11. This print is sheet ____ of ____ sheets.

12. Within what zone is the 3/4-10 UNC-2B threaded hole ? ____________________

13. Where on the print does it specify the weight? ____________________

____________________

**Print AP-010:**

14. What is the complete name of this subassembly? ____________________

____________________

15. What is the total quantity of parts required for this assembly? ____________________

____________________

**Print AP-011:**

16. According to the title block, at what scale are the main views on the original print? ____________________

____________________

**Print AP-012:**

17. At what scale is the circular view on the original print? ____________________

____________________

18. This welded assembly is primarily made of two parts. List the two parts by name. ____________________

____________________

**Print AP-013:**

19. In general terms, what material is specified for this part? ____________________

____________________

20. In the title block, a special symbol is defined to show locating points that must be smooth and clean. Describe the symbol in basic terms and indicate how many times it appears in the front view.

____________________

**Print AP-016:**

21. The numbers for the parts list on this page increase as they go up the page. Is that a recommended method? ____________________

____________________

22. What is the part number for the label? ____________________

____________________

23. What quantity is required of part number B-9470-12? ____________________

____________________

**Print AP-018:**

24. How many columns are featured in the parts list of this print? ____________________

____________________

25. What size paper was used for the original version of this print? ____________________

____________________

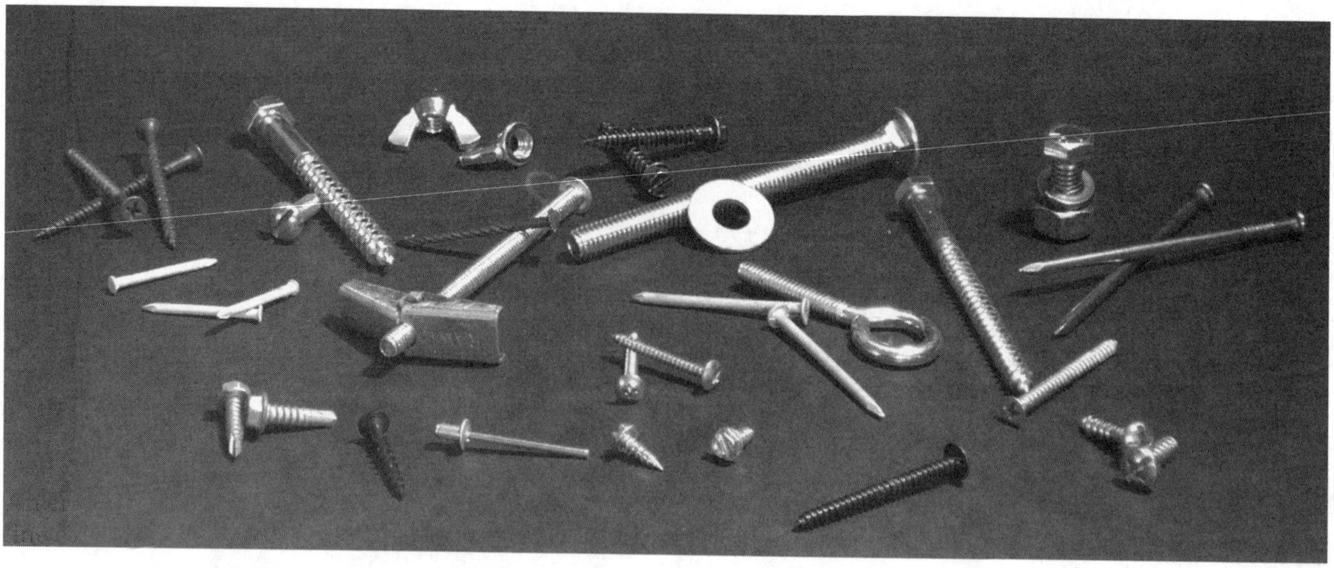

Geometric constructions can be seen everywhere. What geometric shapes can you identify in this group of fasteners?

# UNIT 4
# Geometric Terms and Construction

*After completing this unit, you will be able to:*

**Define** terms related to the geometry of industrial drawings.

**Describe** orientation relationships found within two-dimensional and three-dimensional geometry.

**List** various properties of geometric constructions.

**Identify** two-dimensional geometric shapes.

**Identify** three-dimensional geometric objects.

**Identify** specialized geometric elements that are used in product design.

Lines, arcs, circles, ellipses, splines, and polygons are some of the basic two-dimensional building blocks of everything in life. These two-dimensional elements are joined to create planes, curved surfaces, and sculptured surfaces to form three-dimensional models of design ideas. Traditional "pencil" drafters used mechanical equipment such as triangles, T-squares, compasses, and irregular curves to create various geometric shapes. Drafters have always needed an understanding of geometry to complete designs and drawings, even though CAD programs automate much of the geometric-design process. While someone reading a print does not need to have the same level of understanding as a drafter, knowledge of the terms and geometric constructions is important, especially as the print reader encounters Geometric Dimensioning and Tolerancing, which is discussed in Unit 13.

## Two-Dimensional Shape Terminology

A line segment is one-dimensional. However, much of the geometry in a print is composed of several straight lines to form two-dimensional shapes. Figure 4-1 illustrates many of the shapes with which the print reader needs to be familiar. Line segments can form right angles, acute angles, obtuse angles, complementary angles, or supplementary angles. Parallelograms can be in the form of a square, rectangle, rhombus, rhomboid, trapezoid, or trapezium. A polygon can be in the form of a triangle, square, pentagon, hexagon, heptagon, or octagon. Triangles are three-sided polygons and can be right, equilateral, isosceles, or scalene.

Many terms are associated with lines, circles, and arcs. These are illustrated in Figure 4-2. A ***circle*** is a closed curve where all points on the curve are an equal distance from a given point called the ***center point.*** The ***radius*** is the distance from the center point to a point on the curve. The ***diameter*** is the distance across the circle through the center point. An ***arc*** is a portion of a circle, or an "open" circle. A line segment within a circle connecting two points other than through the center point is referred to as a ***chord.*** A chord is always less than the diameter. A ***sector*** is an area bounded by two radial lines and the included arc. If the included angle is 90°, the sector can be identified as a ***quadrant.*** When a line, circle, or arc is ***bisected,*** it is divided into two halves.

Polygons are created around the outside or inside of a circle, thus the terms ***circumscribed*** and ***inscribed*** are associated with polygons. If the polygon is circumscribed, the "across the flats" measurement is equal to the diameter of the circle. If the polygon is inscribed, the "across the corners" measurement is equal to the diameter of the circle. See Figure 4-3. A common application of this in industry is a hexagon head bolt. The head of the bolt is a hexagon circumscribed about the outside of a circle.

## Three-Dimensional Form Terminology

Two-dimensional shapes can be combined or extended to create three-dimensional forms. With

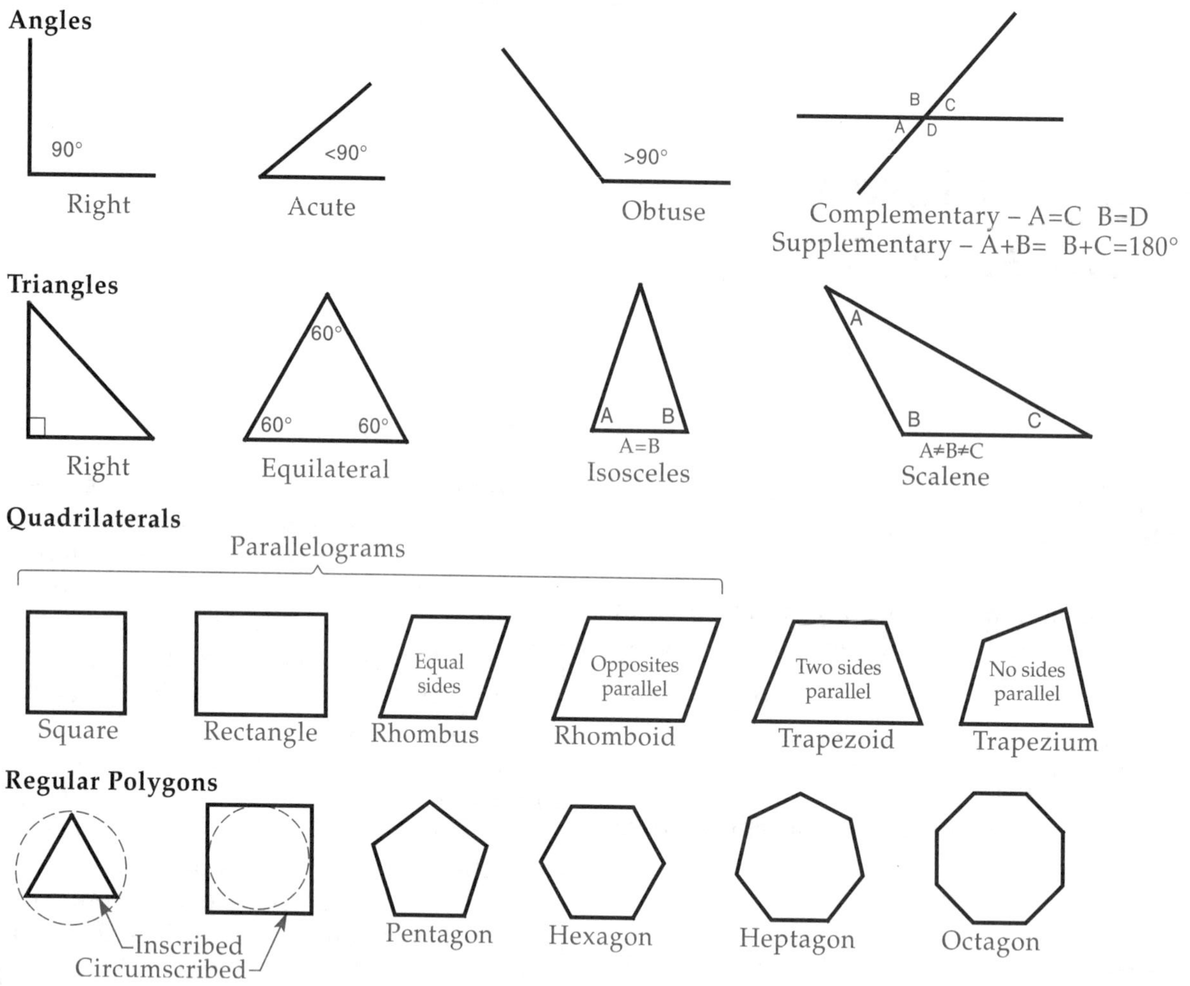

Angles
90°
Right
<90°
Acute
>90°
Obtuse
B
C
A
D
Complementary – A=C B=D
Supplementary – A+B= B+C=180°
Triangles
Right
60°
60°
60°
Equilateral
A
B
A=B
Isosceles
A
B
C
A≠B≠C
Scalene
Quadrilaterals
Parallelograms
Square
Rectangle
Equal sides
Rhombus
Opposites parallel
Rhomboid
Two sides parallel
Trapezoid
No sides parallel
Trapezium
Regular Polygons
Inscribed
Circumscribed
Pentagon
Hexagon
Heptagon
Octagon

**Figure 4-1.**
Much of the geometry in a print is composed of several straight lines that form two-dimensional shapes.

Chord
Diameter
Radius
90°
Tangent
Quadrant
Sector
Arc
Semicircle

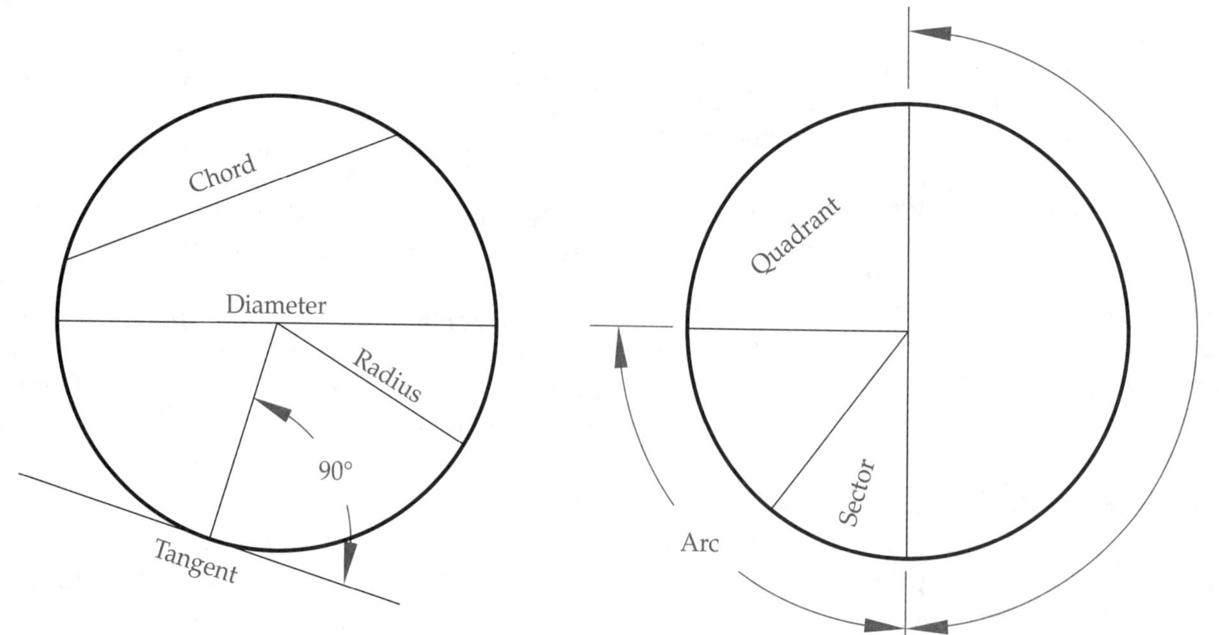

**Figure 4-2.**
Many terms are related to circles and arcs.

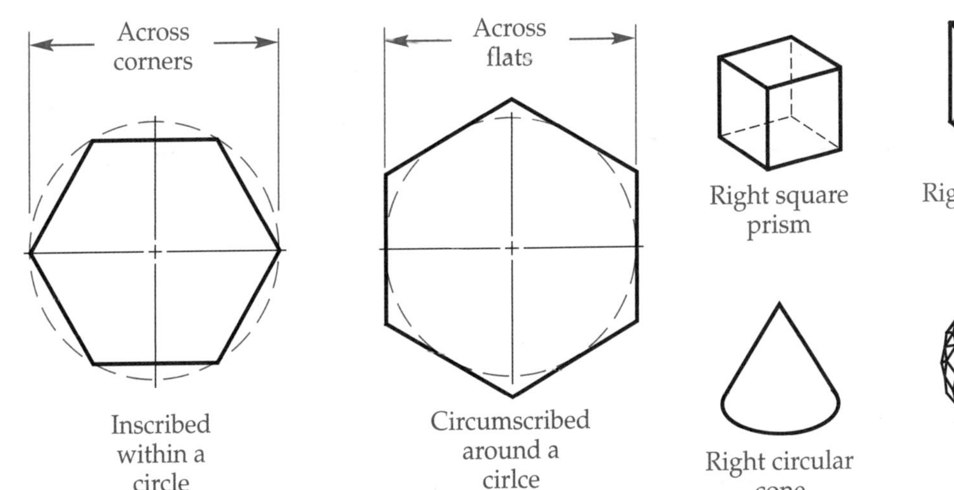


**Figure 4-3.**
Polygons are created around the outside or inside of a circle.

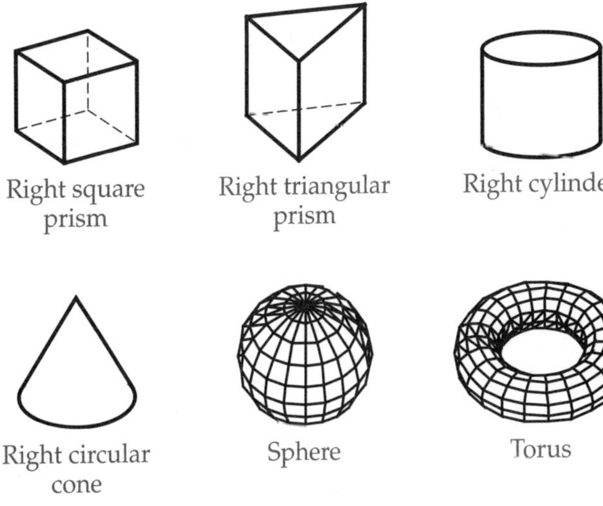


**Figure 4-4.**
Many designs require an understanding of three-dimensional forms.

respect to CAD systems, two-dimensional shapes are often created first, sometimes referred to as "pre-model" shapes. CAD functions such as **EXTRUDE**, **LOFT**, or **REVOLVE** can be used to create models from these shapes.

While the scope of this text is to read prints, more and more prints are created from three-dimensional CAD models, created as either ***surface models*** or ***solid models.*** Therefore, reading prints may involve the study of design solutions that require an understanding of three-dimensional forms. In most CAD systems, the basic forms used to create complex models are referred to as ***primitives.*** Some of the commonly used three-dimensional primitives are prisms, cylinders, pyramids, cones, spheres, and tori (plural for torus), Figure 4-4. The term ***right*** is applied to three-dimensional forms that have an "axis" squarely oriented to the base. The term ***oblique,*** which means slanted, is applied to those forms with an axis somewhat inclined to the base.

## Orientation Terminology

Many geometric terms deal with the orientation of one element of geometry to another. Geometric elements are often parallel, perpendicular, or tangent to each other. These terms can be applied to both two-dimensional and three-dimensional geometry.

***Parallel*** means the elements never intersect, even if extended. See Figure 4-5. Parallel lines run side by side. Parallel surfaces likewise maintain a uniform distance, even if extended. Two-dimensional geometry can also be parallel with three-dimensional geometry. Another way of expressing parallelism is to say that all points of one element are ***equidistant*** (equal distance) from all corresponding points of another element.

***Perpendicular*** means two elements form a 90° angle with each other. This is considered a ***right angle.*** See Figure 4-6. Many geometric principles are based on a right angle. For example, the direction from which you view an object to see the front view is perpendicular to the direction from which you view an object to see the top view. Surfaces can also be perpendicular to each other and two-dimensional lines or axes can be perpendicular to three-dimensional surfaces.

While railroad tracks appear to maintain parallelism as they curve around the bend, there are some different terms for curved elements. Circles and arcs are defined as ***concentric*** if they share a common center point. The three-dimensional extension of concentric is ***coaxial.*** See Figure 4-7. For example, two cylinders that share the same center axis are coaxial. There are several design applications wherein ***coaxiality*** is desired.

Two circles that do not share a center point are said to be ***eccentric.*** Two cylinders with axes that do not align can also be considered eccentric. In the manufacturing department, an analysis of the ***degree of eccentricity*** is sometimes a critical aspect of the quality of the part or assembly.

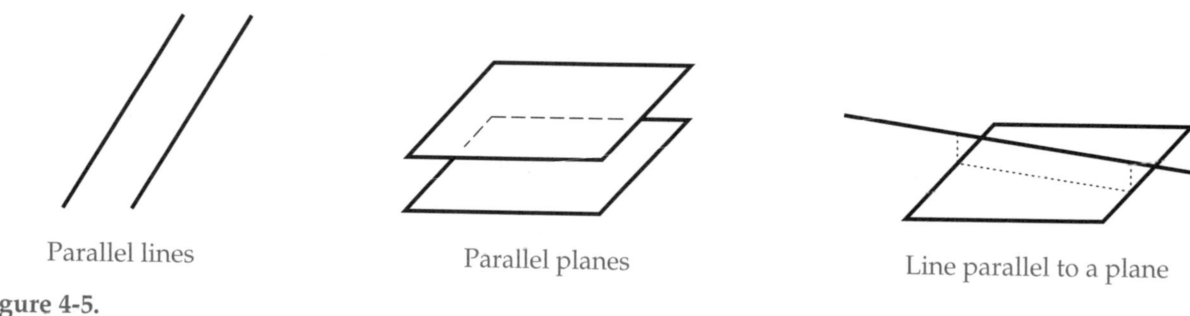


**Figure 4-5.**
Parallel elements never intersect, even if extended. This relationship can exist between 2D and 3D elements.

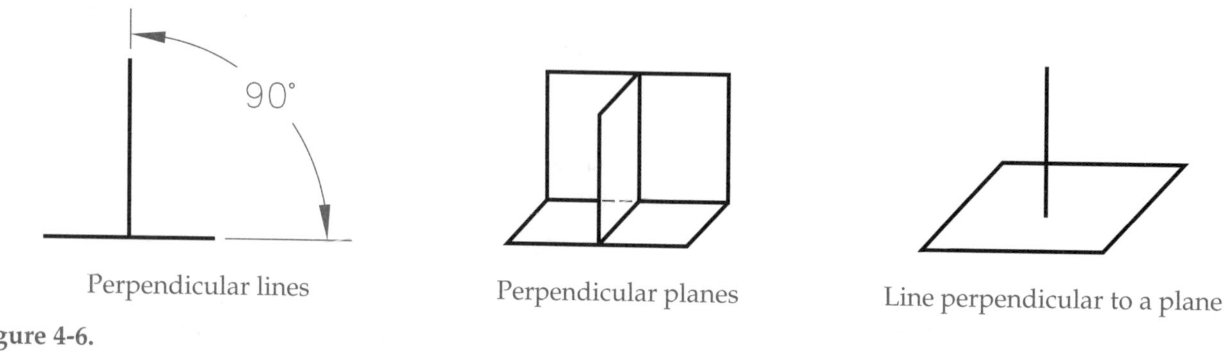


**Figure 4-6.**
Perpendicular elements from a 90° angle to each other.

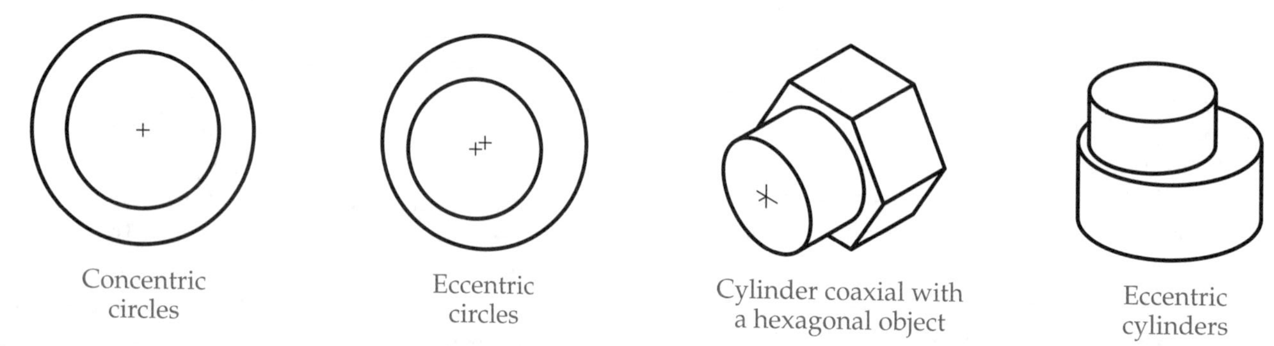


**Figure 4-7.**
Features of a design can exhibit concentric relationships. The extent to which they are not concentric can be called the degree of eccentricity.

***Tangent*** is a relationship between two features best described as "sharing one point." In two-dimensional terms, a straight line is tangent to a circle if it touches the circle at only one point and does not intersect the circle again, even if extended, Figure 4-8. The shared point is called the ***point of tangency.*** For a line and circle, the point of tangency is always found at the end of a line drawn perpendicular to the line from the center point of the arc or circle. Likewise, circles and arcs can be tangent to each other. The point of tangency for two circles or arcs coincides with a segment connecting the center points of the circles or arc. In three-dimensional modeling, flat surfaces can be tangent to curved surfaces or curved surfaces can be tangent to curved surfaces.

## Specialized Geometric Shapes

The term ***spline*** can be defined in various ways. In simple terms, a spline is a curved "line" that smoothly fits through a series of points. In the days of pencil drafting, drafters used irregular-curve devices to form splines. However, CAD tools can quickly calculate splines using complex mathematical functions.

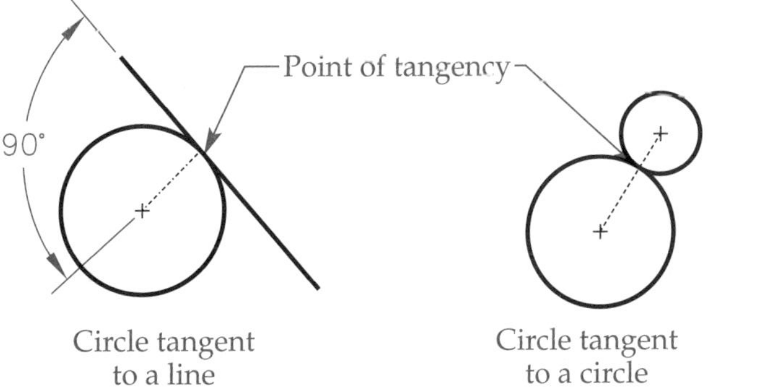


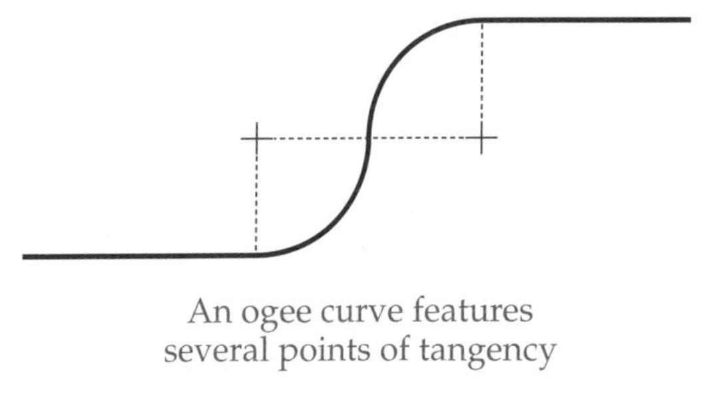


**Figure 4-8.**
Tangent elements share a single point.

Sometimes, splines are referred to as ***B-splines*** or ***Bézier curves.*** In three-dimensional form, B-spline surfaces can be referred to as ***sculptured surfaces*** or ***NURBS surfaces.*** While the explanation of these formula-driven shapes is beyond the scope of this text, it is important to be aware of these terms while studying the geometry of parts and products. In the 21st century, prints are seldom necessary to define these shapes, as the process of designing, manufacturing, and even inspection are all controlled within the computer systems.

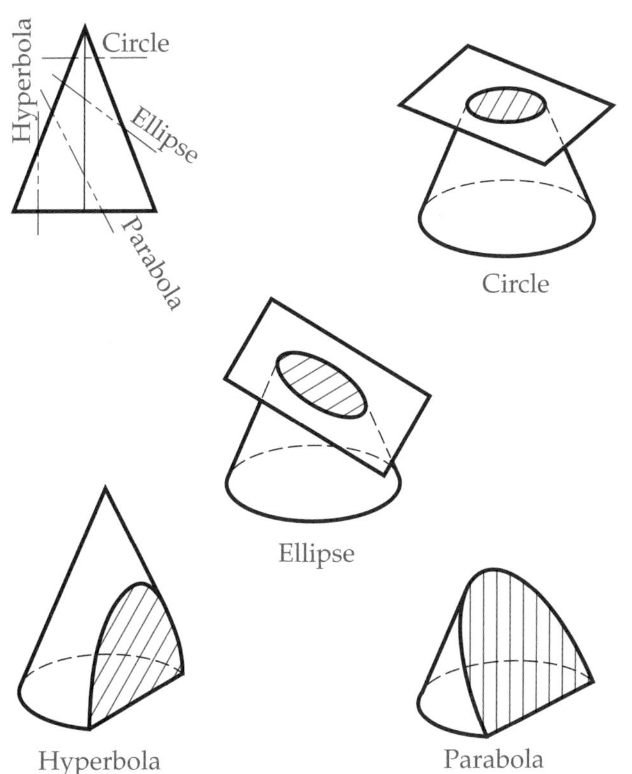


**Figure 4-9.**
Some geometric shapes are based on the way a plane intersects a cone.

Some geometric shapes are based on the way a plane intersects a cone. See Figure 4-9. The resulting shape is called a ***conic section.*** Depending on the angle by which the plane intersects the cone, a circle, ellipse, parabola, or hyperbola is formed. There are many design applications for parabolas, including headlight lenses and bridge arches.

In addition to passing a plane through a cone, an ellipse results when a circle is viewed at an inclined line of sight. For example, a flat, circular disc appears as a circle if the view is perpendicular to the disc. When the disc is inclined away from the viewer, the circle is foreshortened in one direction and appears as an elliptical shape. See Figure 4-10.

A ***helix*** is generated by rotating a point around an axis while moving it along the axis. This may remind you of a barber pole or candy cane stripe. Screw threads are helical ridges formed on the surface of a cylinder. See Figure 4-11. An ***involute*** is the path a point follows if it is attached to the end of a string and the string is unwound from a cylinder. A gear tooth has a surface based on an involute to avoid interference between gear teeth. See Figure 4-12.

## Summary

A fundamental understanding of geometry and geometric constructions is complimentary to the study of print reading. Most drafting texts include a unit that serves as an extensive reference on how to manually construct geometric shapes with instruments. Additional geometric techniques, such as bisecting lines and arcs or dividing a line into equal parts, are also of value to those in technical fields.

Front and side view
perpendicular to a circle

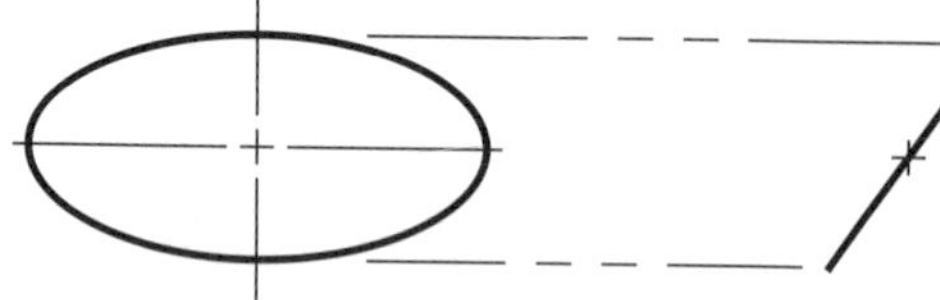

Front inclined to a circle,
side view perpendicular

**Figure 4-10.**
An ellipse results when a circle is viewed at an inclined line of sight.

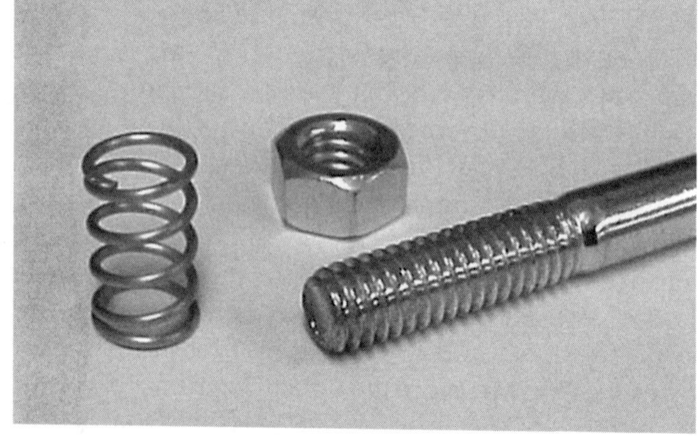

**Figure 4-11.**
Screw threads are helical ridges formed on the surface of a cylinder.

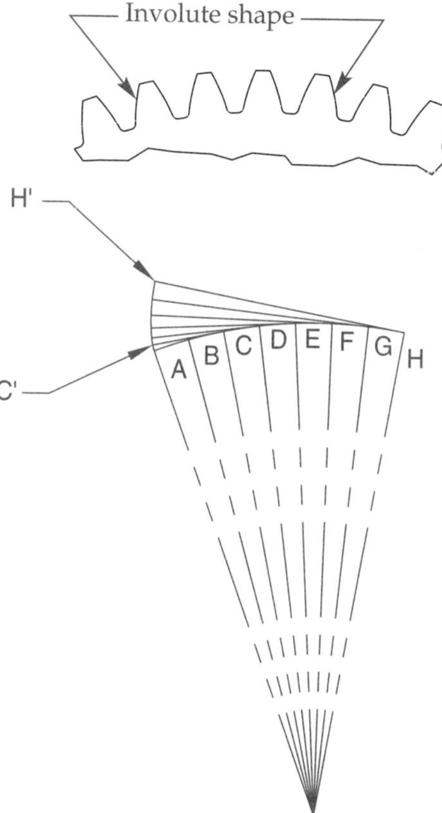


**Figure 4-12.**
A gear tooth has a surface based on an involute.

## Review Questions

*Circle the answer of choice, fill in the blank, or write a short answer.*

1. All of the following terms are two-dimensional *except:*
   A. polygon.
   B. circle.
   C. pyramid.
   D. rhombus.

2. *True or False?* The word *chord* describes a line segment that is less than the diameter of the circle.

3. *Circumscribed or Inscribed?* The "across the flats" measurement of the hexagon is the same as the diameter of the circle that is the basis for the hexagon.

4. With respect to CAD systems, three-dimensional objects that can be used as basic building blocks are called ____________________.

5. For A–C, choose the orientation that describes the given relationship.
   A. *Parallel or Perpendicular?* Orientation of the top surface of a cube to the front surface.
   B. *Parallel or Perpendicular?* Orientation of the top surface of a table to the floor.
   C. *Parallel or Perpendicular?* Orientation of a flagpole to the ground.

6. Given two circles tangent to one another, how do you find the point of tangency? ____________________

   ____________________

   ____________________

7. What name is given to a free-form curve that smoothly fits through points based on a mathematical formula? ____________________

   ____________________

8. *True or False?* The section derived by slicing a cone with a plane will be either elliptical or circular.

9. The screw thread is defined as a ridge that has the shape of a(n) ____________________.

10. Each side of a gear tooth is based on the geometric form known as a(n) ____________________.

## Review Activity 4-1

*Study the drawing on the following page and answer the questions.*

1. Considering four possibilities: parallel, perpendicular, concentric, and tangent, what relationship exists between:
   A. Line A & Line L ______
   B. Line E & Line G ______
   C. Line J & Arc K ______
   D. Line B & Line D ______
   E. Line L and Arc M ______
   F. Circle P & Arc M ______
   G. Line G & Line J ______

2. Fill in the blank with the geometric name (arc, chord, etc.) for each of these letters:
   A. F ______
   B. P ______
   C. N (as related to M) ______
   D. U ______
   E. H ______
   F. V ______

3. If Item V is circumscribed around a circle, what is the radius of the circle? ______

4. How many degrees, measured clockwise along the center line circle, is it from center point to center point?
   A. Center point R to center point S ______
   B. Center point S to center point U ______
   C. Center point U to center point T ______

5. What type of angle is formed between D and E? ______

6. What type of angle is formed between B and A? ______

7. How many degrees are in arc C? ______

8. How many degrees are in arc K? ______

9. How many degrees are in arc F? ______

10. How many degrees are in arc H? ______

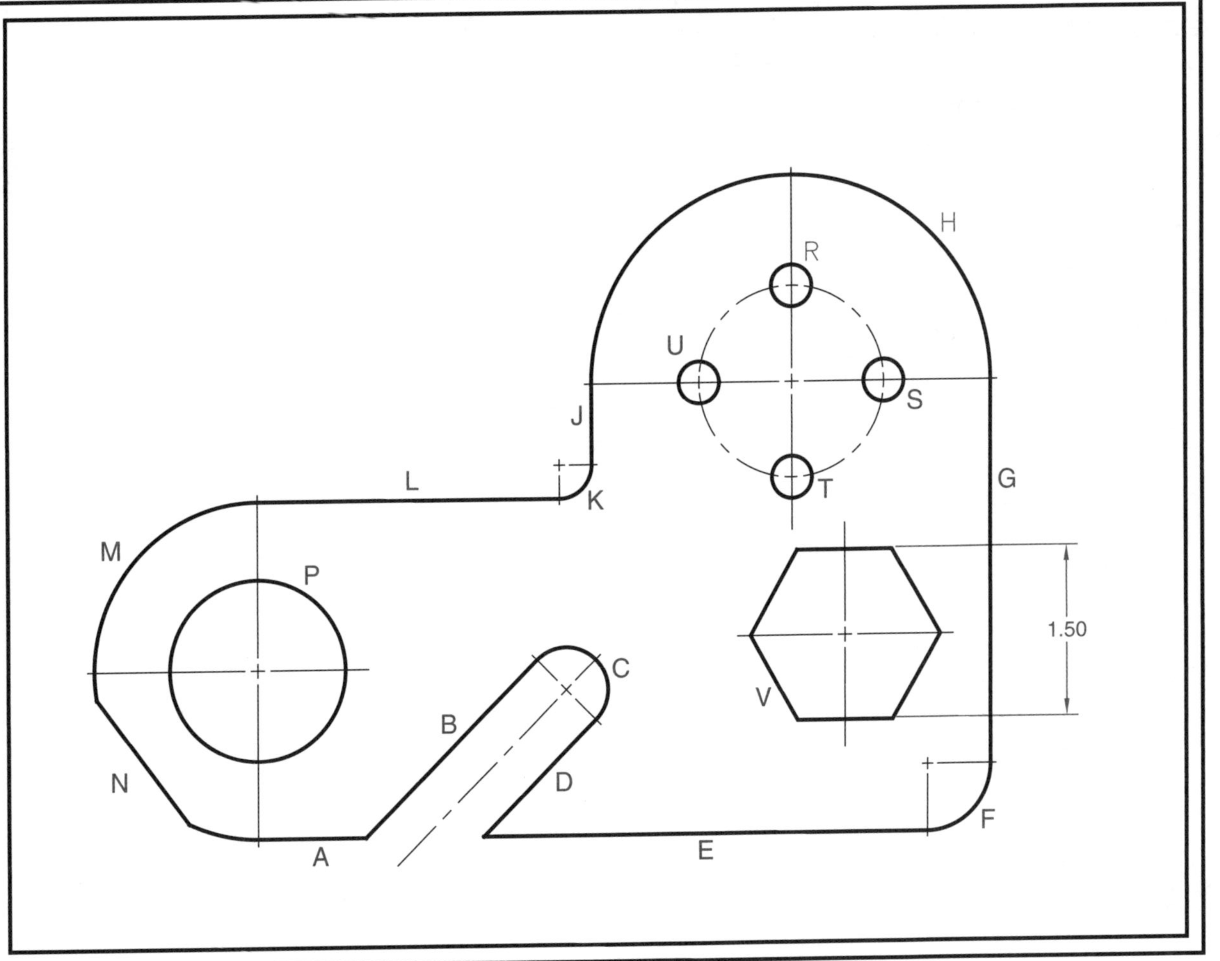
H
R
U
S
J
G
L
K
T
M
P
1.50
C
V
B
N
D
F
A
E

## Review Activity 4-2

*Match the letter of each illustrated shape with the correct name. Items are matched only once.*

_________ 1. Pentagon

_________ 2. Scalene triangle

_________ 3. Equilateral triangle

_________ 4. Acute angle

_________ 5. Trapezoid

_________ 6. Isosceles triangle

_________ 7. Octagon

_________ 8. Square

_________ 9. Right triangle

_________ 10. Hexagon

_________ 11. Rhombus

_________ 12. Rectangle

_________ 13. Obtuse angle

_________ 14. Heptagon

_________ 15. Rhomboid

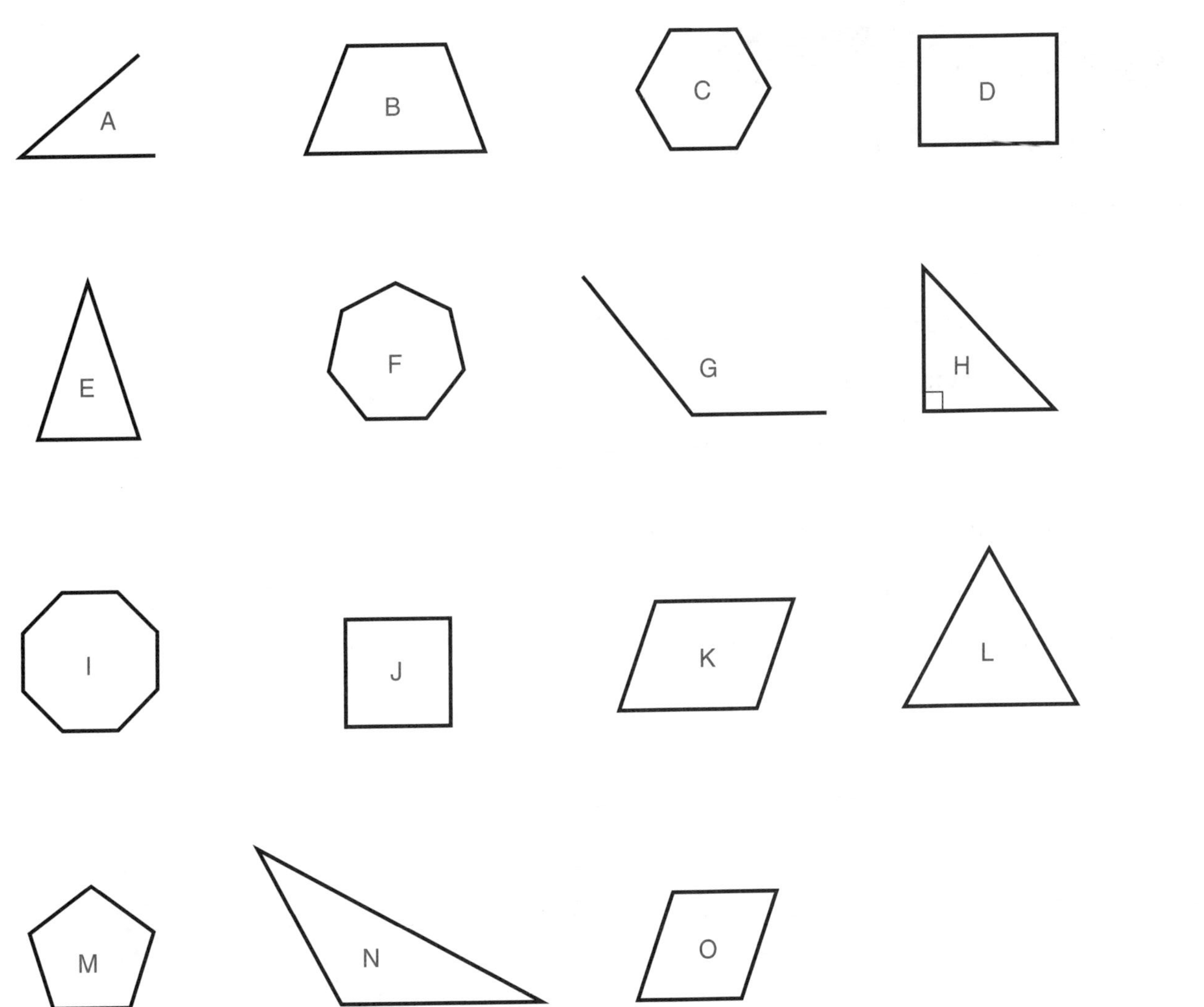
A
B
C
D
E
F
G
H
I
J
K
L
M
N
O

## Industry Print Exercise 4-1

*Refer to the print PR 4-1 and answer the questions below.*

The front view is the view in the lower-left corner. In that view…

1. How many circles are there? ____________________
2. How many arcs are there? ____________________
3. How many points of tangency are there? ____________________
4. How many sets of concentric circles are there? ____________________

The top view is the view in the upper-left corner. In that view…

5. How many circles are there? ____________________
6. How many hidden (dashed) lines are there parallel to each other? ____________________
7. How many lines, if any, are tangent with the circle? ____________________

The right-side view is in the lower-right corner. In that view…

8. How many vertical visible lines are there? ____________________
9. What geometric term describes the general shape of that view? ____________________
10. What relationship does the dimension line have with its extension line? ____________________

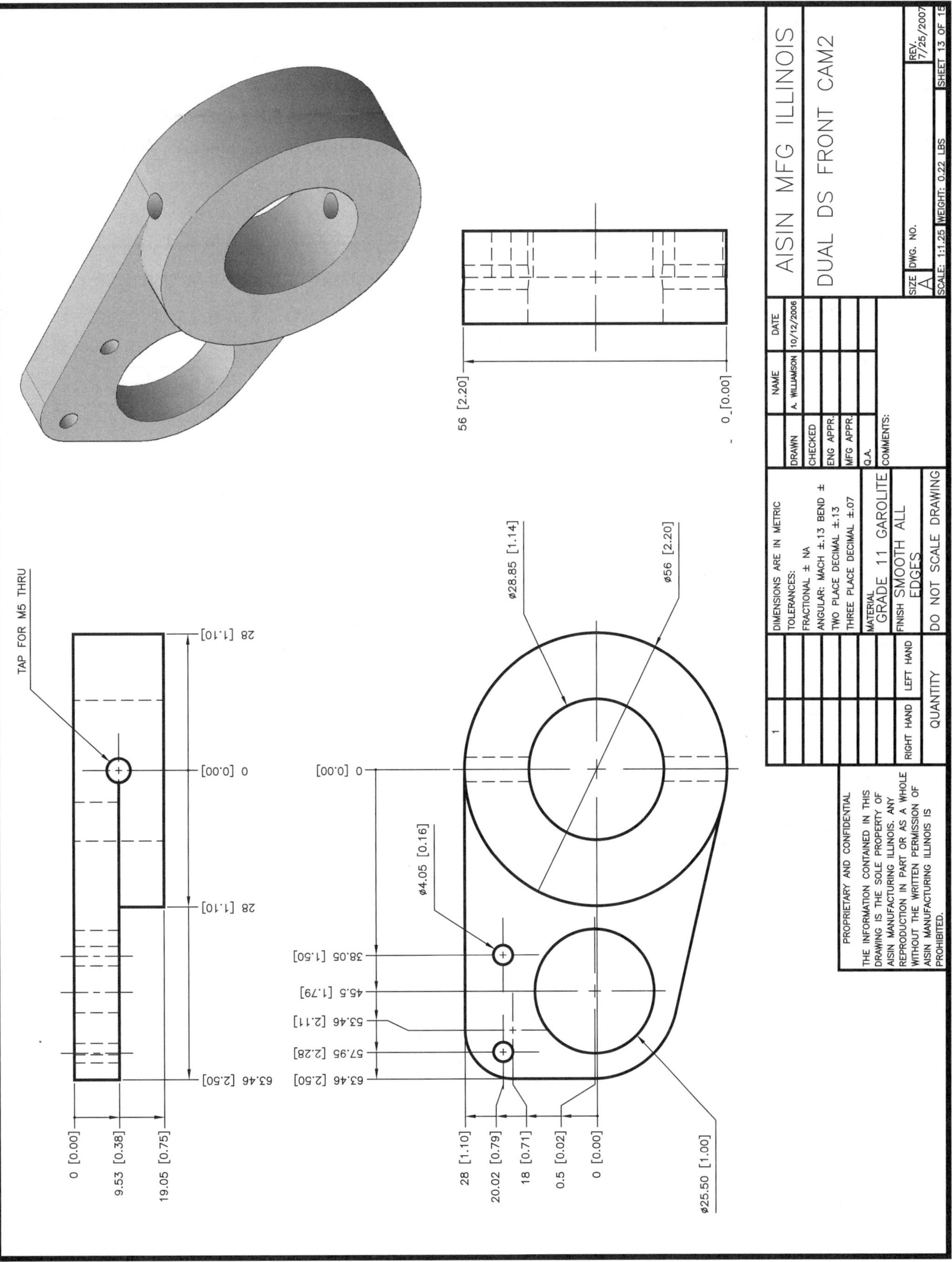


PR 4-1.
Print supplied by AISIN Manufacturing Illinois.

## Industry Print Exercise 4-2

*Refer to the print PR 4-2 and answer the questions below.*

This drawing has two part drawings on one sheet, with one part slightly different than the other part. Multiview drawings are discussed in the next unit. For questions 1–5, look only at the geometric shape of the "top view" of either part. This print features inch values with metric equivalents in brackets. Also, in some dimensions, a comma is used as the decimal marker instead of a decimal point.

1. What term can be applied to relationship between the center of the semicircle and the center of the large circle? ______________________________

2. How many sets of parallel lines are there? ______________________________

3. How many points of tangency can be found in the basic shape of either object? ______________

4. As drawn, the functional center line of the shape is not vertical, but rather is rotated ______ degrees.

5. There is an arc that is *not* tangent to other surfaces. What geometric term can be applied to a straight line that connects its two endpoints? ______________________________

6. In summary, what is the key difference between the two parts (Part A on the left and Part B on the right)? ______________________________

7. For the part with the two small circles, those represent holes that go through the part from top to bottom. What 3D relationship do the hole axes have with each other? ______________

8. It can be assumed that the designer intended the two small holes and the one larger hole to be ______________________________ to the top and bottom surfaces.

9. For each drawing, below the top view is a smaller view that shows the part thickness. What geometric shape can be used to describe the overall shape of that smaller view? ______________

10. For Part B on the right, there is one dimension for 125 mm and another dimension for 22.45 mm. What geometric orientation relationship exists between these two measurement "directions"?

______________________________

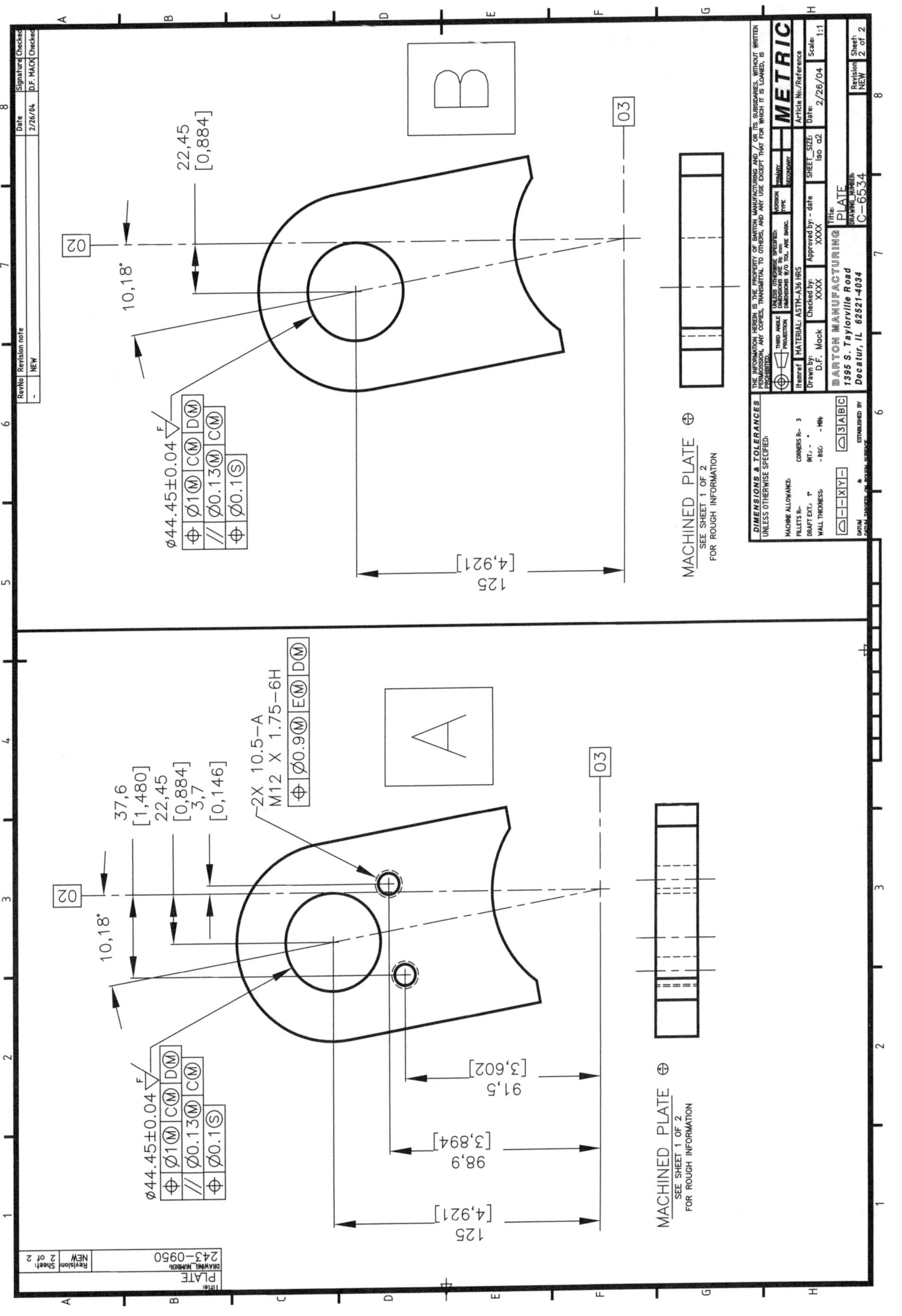


PR 4-2.
Print supplied by Barton Manufacturing.

## Bonus Print Reading Exercises

The following questions are based on various bonus prints located in the folder at the back of this textbook. Refer to the print indicated, evaluate the print, and answer the question.

**Print AP-002:**

1. What term describes the relationship of the axis of the 1/2″ pipe plug with the main, longer axis of the bottom assembly? ____________________

____________________

2. With respect to the part of subassembly 1 that attaches to subassembly 2, which word best describes the shape of the material at the connection: round tube, hexagon tube, or square tube. ____________

____________________

3. The top assembly featured in this drawing appears to have two tall stems, each about 14.5″ tall. What relationship exists between the two axes of the stems? ____________________

____________________

**Print AP-003:**

4. This part features some concave arcs that are equally spaced. There are some larger arcs (.072 radius) spaced ________ degrees apart and some smaller arcs (.030 radius) spaced ________ degrees apart.

**Print AP-007:**

5. This part features four slots. How many degrees are there between the center plane of each slot?

____________________

6. One of the views features four circles and a set of center lines. What terms apply to the relationship the four circles have with each other? ____________________

____________________

**Print AP-010:**

7. What geometric term applies in a general way to the main objects of this assembly, which is a primary reason only one view is required? ____________________

____________________

**Print AP-011:**

8. This part features a V-groove for a belt. What term applies to the included angle of the V-groove: acute, obtuse, right, or oblique? ____________________

____________________

9. Note 7 is specifying what type of relationship between the belt groove and the main axis as determined by datum D? ____________________

____________________

10. Note 5 specifies some shallow balance holes can be drilled 180° away from what other feature?

______________________________

11. How many degrees are there between each of the M8 × 1.5 threaded holes? ______________________________

**Print AP-012:**

12. How many degrees are there between each of the lobes on this part? ______________________________

13. How many degrees are there between each of the mounting holes on this part? ______________________________

**Print AP-014:**

14. This metric part has two holes specified as .1094″ DR THRU TO CROSSHOLE. What geometric relationship is implied between the axes of these holes and the hole that is 500 mm deep?

______________________________

15. What geometric term applies to the groove of this ball screw? It uniformly progresses along the central axis as it revolves around the axis. ______________________________

**Print AP-015:**

16. What relationship exists between the surface of the pedal pad and the pedal arm, as shown in the auxiliary view in the upper-right corner of the drawing? ______________________________

17. The pedal arm on this part curves at a radius of 11.875. What term is given to the relationship between the 11.875 radius curve and the 3.000 radius curve? ______________________________

**Print AP-018:**

18. In the top view of this assembly, one of the inserts is shown. The other inserts are not drawn, but rather a location for each is indicated by a center line. How many degrees separate each of these insert positions? ______________________________

19. The cylindrical body part of this assembly tapers from a larger diameter to a smaller diameter. What geometric term applies to the tapered portion of the body? ______________________________

**Print AP-019:**

20. What geometric term is used to describe the manner in which the worm gear ridge is wrapped around the shaft, as shown on the left end of this part? ______________________________

21. What geometric term appears in the Fine Pitch Spur Gear Data table to describe the base circle diameter of 1.17462? ________________________________________

________________________________________

**Print AP-024:**

22. What geometric term applies to the opening that is centered approximately 4.310 from the left, assuming the corners were sharp rather than rounded? ________________________________________

________________________________________

23. What geometric term applies to the curve at each end of the 11 long slots on the right side?

________________________________________

**Print AP-025:**

24. What geometric term applies to the relationship the adjacent sides of this part have to each other after the bends are made? ________________________________________

________________________________________

25. Are any of the holes in this part square (equal sides)? ________________________________________

________________________________________

# UNIT 5
# Multiview Drawings

*After completing this unit, you will be able to:*

**Explain** the relationship between an orthographic projection and a multiview drawing.

**Identify** and **define** the three dimensions of an object.

**Define** the three regular views.

**Identify** three principal planes of projection.

**Explain** three visualization principles for multiview drawings.

**Identify** the three types of flat surfaces.

**Explain** characteristics of cylindrical surfaces.

**Explain** characteristics of fillets, rounds, and runouts.

**Identify** differences between third-angle and first-angle projection.

The purpose of a drawing is to show the size and shape of the object. A drawing can also provide certain information on how an object is to be made. Various methods of presentation are available to the designer or drafter. However, the best way to show every feature of an object in its true size and shape is to use an arrangement of more than one view known as a ***multiview drawing.*** Multiview drawings are created using the principles of orthographic projection. Many drafting and print reading texts use the terms *orthographic projection* and *multiview drawing* interchangeably.

## Orthographic Projections of Views

Any view of an object drawn by a drafter can be explained as the projection of an object's features onto a two-dimensional plane. For a drawing, the plane is often a piece of paper or computer screen. Simply defined, ***orthographic projection*** is a system wherein parallel lines, called ***projectors,*** are used to project the object onto a projection plane. The projectors are perpendicular to the projection plane, thus resulting in an exact and very precise view. If more than one projection plane is used, the result is a ***multiview projection.*** In general terms, a multiview drawing is a drawing *based* on the principles of orthographic projection.

The different views of a multiview drawing are systematically arranged. This allows anyone "reading" the drawing to "connect" the views together, thus forming a mental picture, Figure 5-1.

## Projection Explained

A skilled technician reading a print must be able to ***visualize*** the object as a whole. This means they must be able to look at the views in a drawing and interpret those into a mental picture. Understanding how views are projected and arranged will help you later in the visualization process.

One way to help you understand the multiview system is to observe how a cardboard box unfolds. Each side of the box is oriented similar to orthographic projection views. The sides are at right angles to each other and have a definite relationship. See Figure 5-2. If the front of the box remains in position, the four adjoining sides unfold similar to how the views of a multiview drawing are arranged.

Now think of the cardboard box as made out of glass. Place an object inside of the glass box and imagine that the points of the object are projected onto the glass planes as views. See Figure 5-3. Imaginary projection lines are used to bring the separate views to each projection plane. If the glass box is unfolded like the cardboard box, six views are shown in an orthographic arrangement. Once you have perfected mentally projecting orthographic views using the glass box, you will be able to see orthographic views looking at any object, Figure 5-4.

Figure 5-1.
The views of a multiview drawing are systematically arranged so anyone can visualize the object.

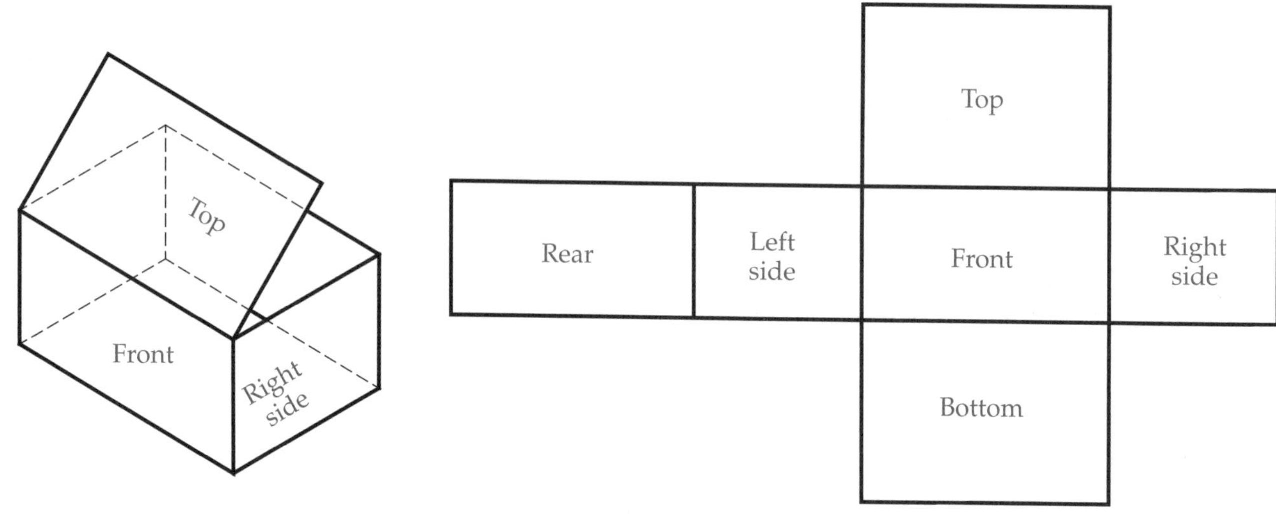


Figure 5-2.
Each side of an unfolded cardboard box is oriented similar to orthographic projection views.

## Selection of Views

An orthographic projection can result in six ***normal*** views, as previously illustrated. Since there are three directions in space, an object can be viewed from the right or left, top or bottom, and front or back. However, only those views necessary to clearly describe the object need to appear in the drawing. Seldom is an object so complex that all six normal views are required. Usually, the necessary details can be shown in two or three views. Three views will almost always fully describe an object, but more views can be used if there is a lot of detail on the opposite sides of an object. Directions other than the six normal directions are defined as auxiliary views, which are discussed in Unit 7.

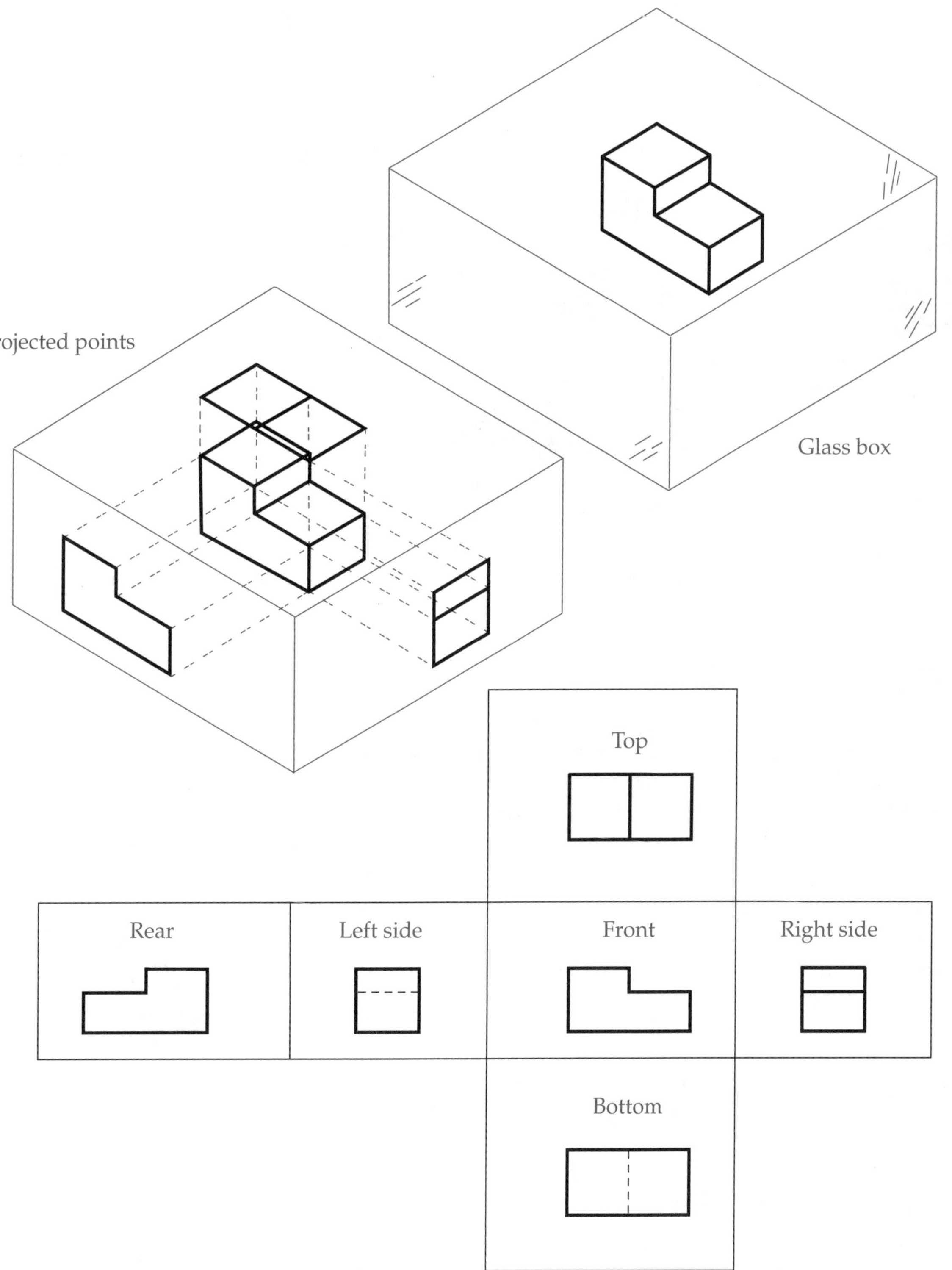


**Figure 5-3.**
If an object is placed inside of a glass box and projected onto each side, when the glass box is unfolded, six views are shown in an orthographic arrangement.

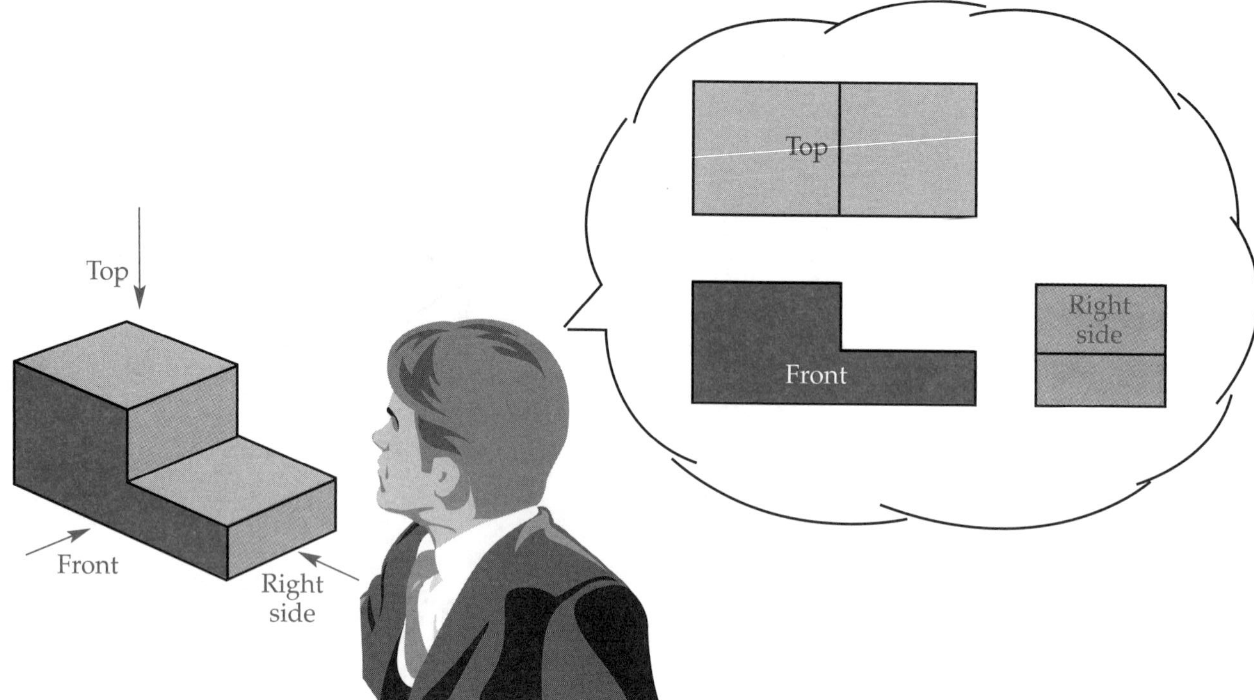


**Figure 5-4.**
Looking at a three-dimensional object, you must be able to see the orthographic views of the object.

In educational settings, the ***front, top,*** and ***right side*** views are typically used to describe an object. This type of drawing can be referred to as a three-view drawing. Yet, many objects made of flat sheet metal require only one view, while cylindrical objects may only require two views. The following rules will help in the selection of views.

- Only views clearly describing the shape of the object should be drawn.
- Select views containing the fewest hidden lines. For reference, compare the two side views in Figure 5-3.
- If practical, draw the object in its functioning (operating) position.
- If practical, draw the view best describing the shape of the object as the front view.

## Dimensions of an Object

One of the keys to reading a multiview drawing is familiarity with the terminology used for the dimensions of an object. It is critical to discuss the scientific explanation of projections using standard terms. As you have probably noticed, each projected view is two dimensional, even though the object is three dimensional.

Throughout the field of drafting, three terms predominantly used for the three dimensions of an object are height, width, and depth. The height and width of the object are shown in the front view, but not the depth. The width and depth of the object are shown in the top view, but not the height. The height and depth of an object are shown in a side view, but not the width.

***Height*** is defined as the top-to-bottom measurement for an object, as you look at the front view. Remember, the front of the object is not always selected as the front view! ***Width*** is defined as the left-to-right measurement of an object, as you look at the front view. ***Depth*** is defined as the front-to-back distance of the object, as based on the orientation of the object for the views. Standardizing the terms used for the measurements is critical to studying print reading and for completing the exercises in this unit.

## Visualization of Objects

Another necessary step on the road to successful print reading is understanding how views are created by projection. You will examine how flat, or ***planar,*** surfaces are oriented in space and how they appear in a multiview drawing. At this stage, three principal planes of projection will be established and only three regular views used. The front view will project onto a plane called the ***frontal plane,*** the top view will project onto a plane called the ***horizontal plane,*** and the right side view will project onto a plane called the ***profile plane.*** See Figure 5-5.

With these three planes of projection in mind, here are some principles to help you read the print.

- **Principle One.** A flat surface is oriented perpendicular, parallel, or inclined to a plane of projection. See Figure 5-6.
- **Principle Two.** As a result of Principle One, all flat surfaces appear in a multiview drawing as:
  - A) a line, if oriented perpendicular.
  - B) true size and shape, if oriented parallel.
  - C) foreshortened, if oriented inclined. See Figure 5-7.
- **Principle Three.** All surfaces appear in *every* view of a multiview drawing, even if only as a line and even if represented by a hidden line.

## Three Types of Surfaces

There are three basic types of flat surfaces in an orthographic projection. A ***normal surface*** is parallel to one of the three projection planes and, therefore, perpendicular to the other two. For example, each surface of a cube is normal and the top flat surface of a cylinder is normal. If normal surfaces are examined with respect to the three principles:

- A normal surface appears true size and shape in only one view.
- The normal surface appears as a line in two of the three regular views.

Think about this for a while. Very often, when you look at a line in a multiview drawing, you are looking at the edge view of a surface. When you look at the front view of a cube, you see the top and right side surfaces as lines. Study Figure 5-8A. With respect to the top surface of the cube in the front view, if you only "see" the front edge along the front surface, you are still thinking in two dimensions. As your visualization ability increases, you will see these lines as surfaces that extend back.

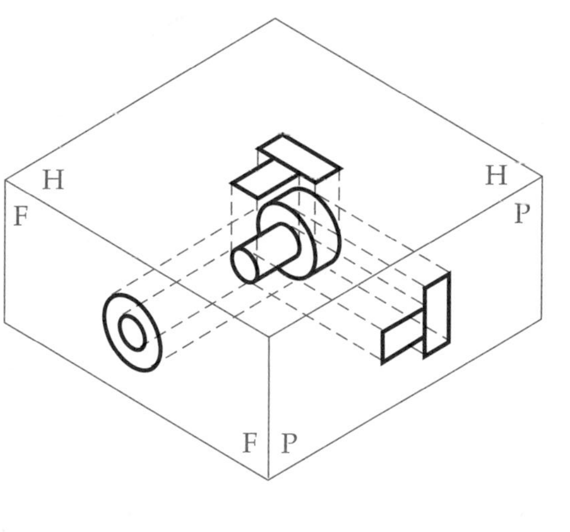


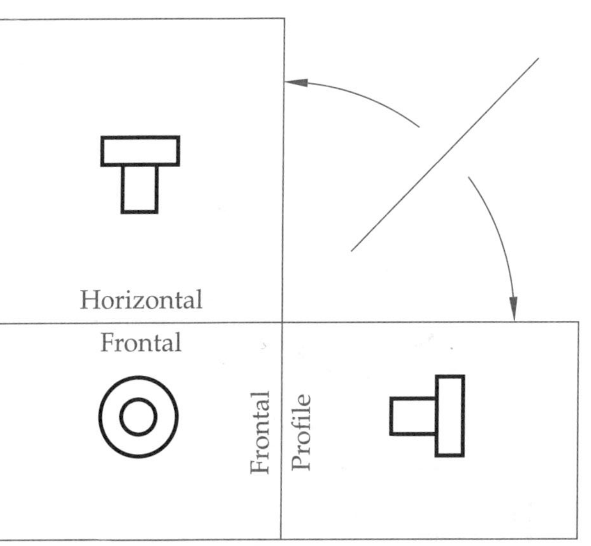


**Figure 5-5.**
Three basic projection planes are used to explain multiview drawings.

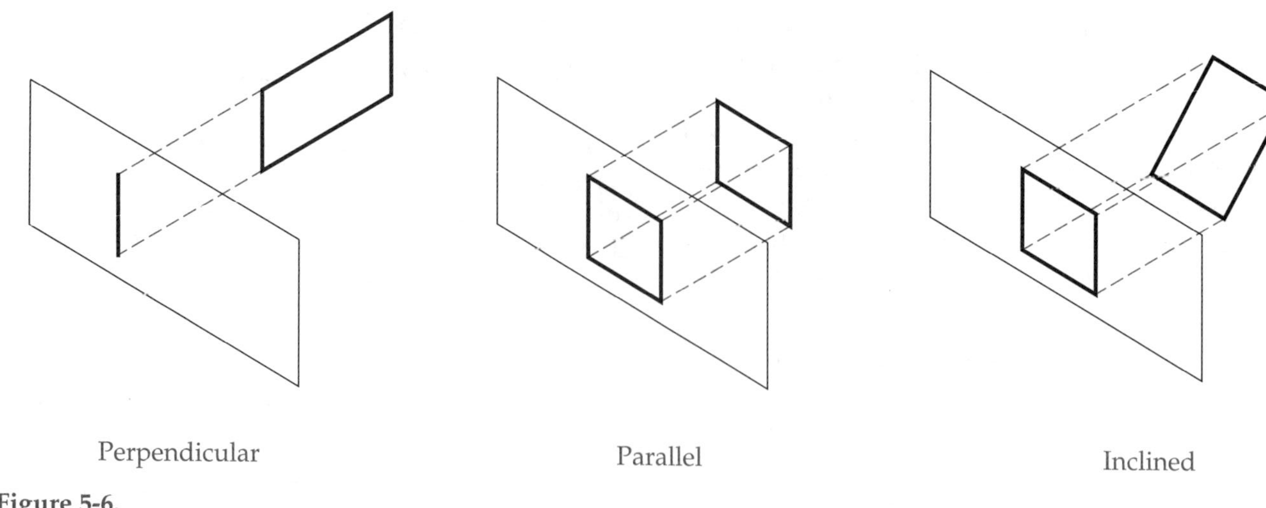


**Figure 5-6.**
A flat surface is oriented perpendicular, parallel, or inclined to a plane of projection.

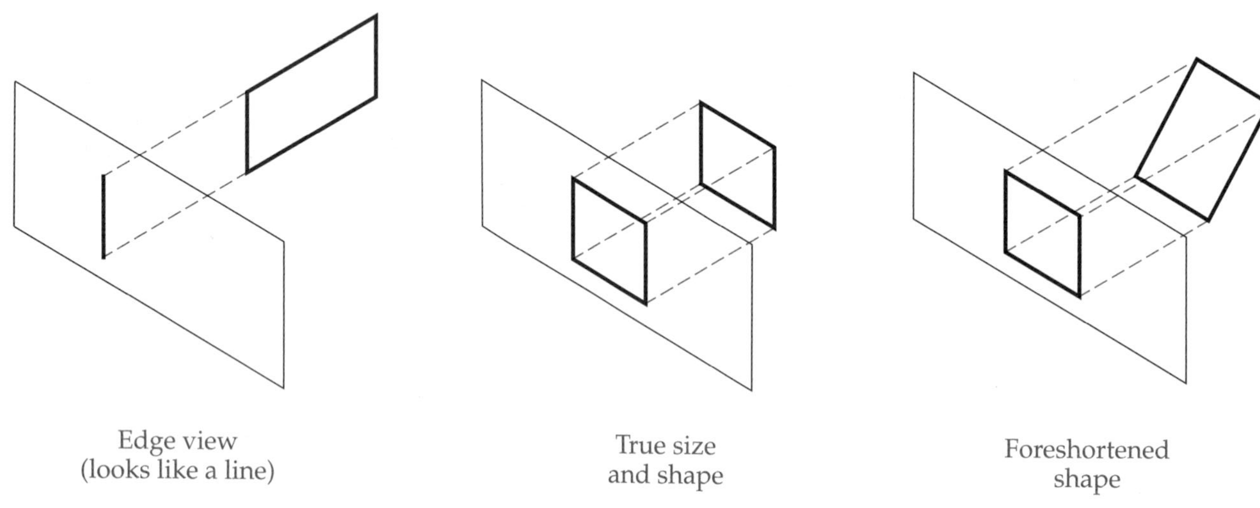


**Figure 5-7.**
A flat surface can appear in a projection as an edge, in true size and shape, or foreshortened.

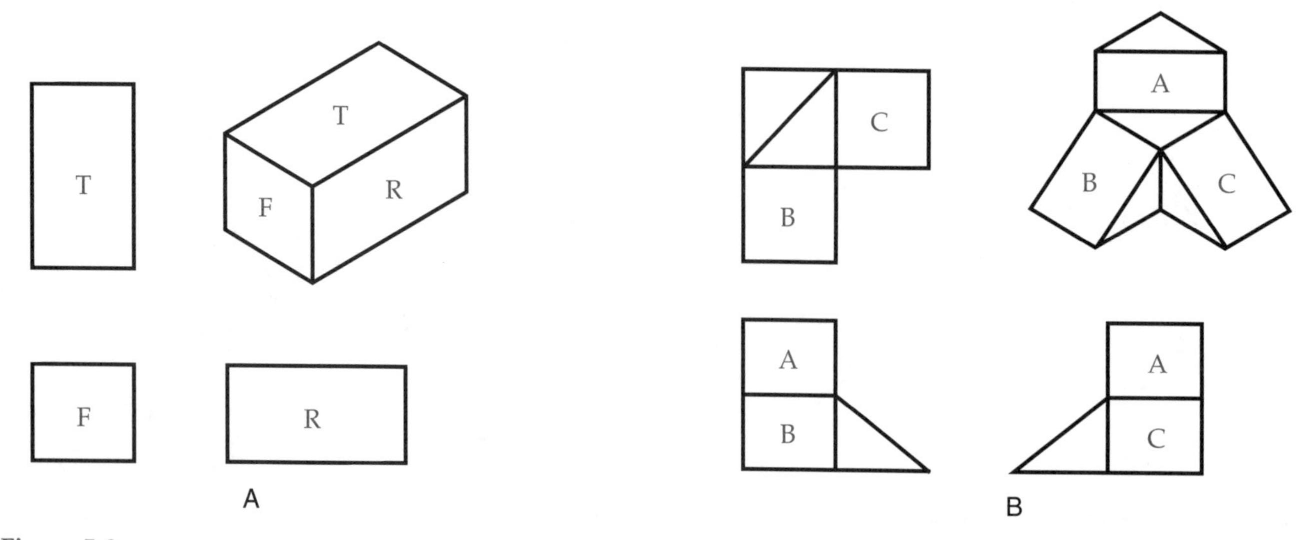


**Figure 5-8.**
A—This object has normal surfaces. B—This object has normal and inclined surfaces.

A second type of flat surface is the inclined surface. An ***inclined surface*** is perpendicular to one plane of projection, but inclined to the other two planes of projection. If inclined surfaces are examined with respect to the three principles:

- An inclined surface appears as a line in only one of the three regular views.
- An inclined surface appears as a foreshortened shape in two of the three regular views.

Study Figure 5-8B. Surface C is perpendicular to the frontal plane, so it appears as a line in the front view. However, it is inclined to the horizontal and profile planes, so it appears as a foreshortened shape in those two views. Analyze surfaces A and B in the same way. In summary, the shape of an inclined surface appears twice in three regular views, but a normal surface shape only appears once! The normal surface is true size and shape, but the inclined surface is never true size and shape in a regular view.

A third type of planar surface is the oblique surface. An ***oblique surface*** is not only inclined, but rotated. Therefore, it is inclined to all three planes of projection. It is not true shape and size in any view. In fact, it may appear a little distorted due to the projection angle it forms with the projection plane. It also does not appear as a line in any view. See Figure 5-9.

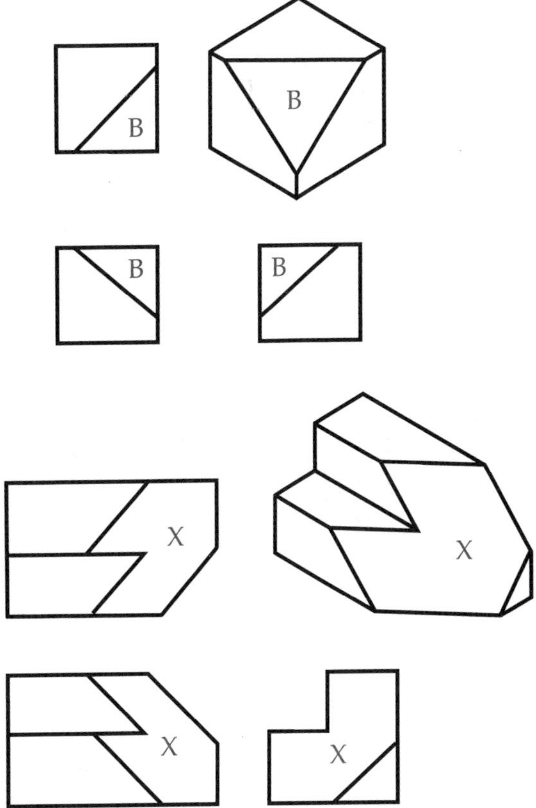


**Figure 5-9.**
An oblique surface is not only inclined, but rotated. Surfaces B and X are oblique surfaces.

## Cylindrical and Curved Surfaces

Cylindrical surfaces present another set of visual challenges to the print reader. Technically, cylindrical surfaces are made of thousands of "elements" that form a curved surface about an axis, Figure 5-10. The designer has often planned it so that a flat surface is tangent to a curved surface, thus making a smooth transition between the curve and the flat. See objects B and D in Figure 5-10. In these cases, no lines are shown at the element of tangency. Be aware, however, some CAD programs that automatically generate the views from the model show these elements of tangency. Also, when flat surfaces form intersections and cutouts with cylindrical surfaces, the projections can be tricky. Study Figure 5-11 to help you visualize how cylindrical surfaces are projected in multiview drawings.

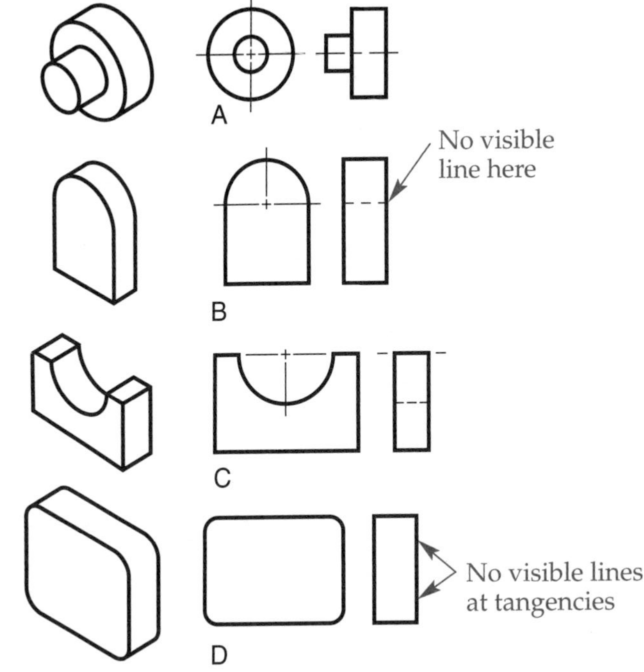


**Figure 5-10.**
Cylindrical surfaces. A flat surface is often tangent to a curved surface, as shown in B and D.

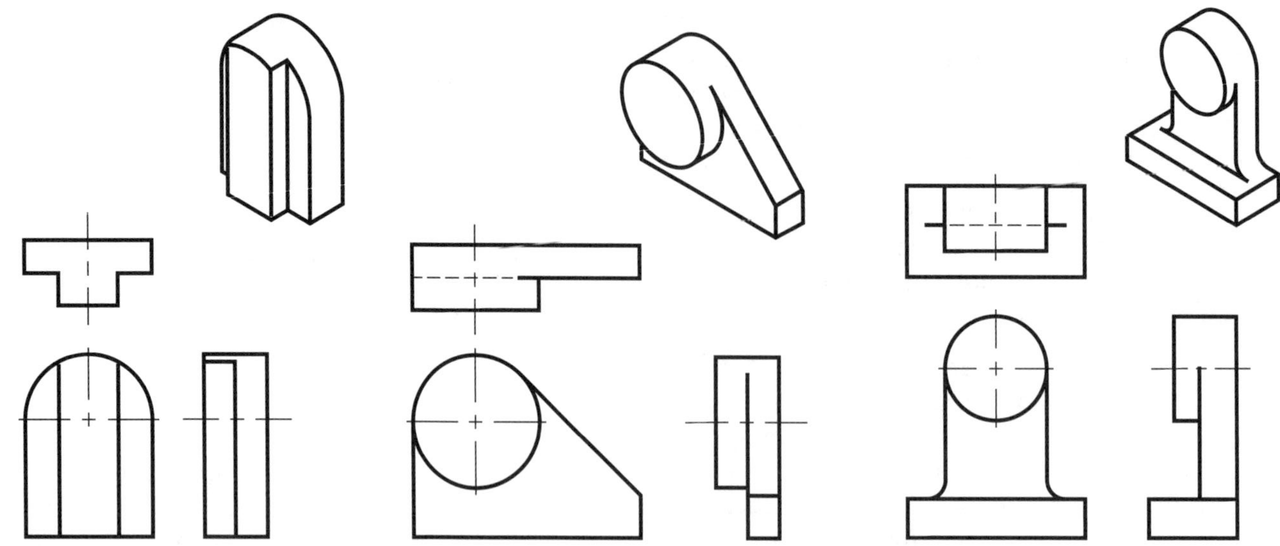

Figure 5-11.
This figure helps illustrate how cylindrical surfaces are projected in multiview drawings.

## Fillets, Rounds, and Runouts

Molded and cast objects have rounded edges called fillets and rounds. See Figure 5-12. ***Fillets*** are interior rounded edges. ***Rounds*** are exterior rounded edges. Most CAD programs have a command called **FILLET** to perform this task of rounding corners, both interior and exterior.

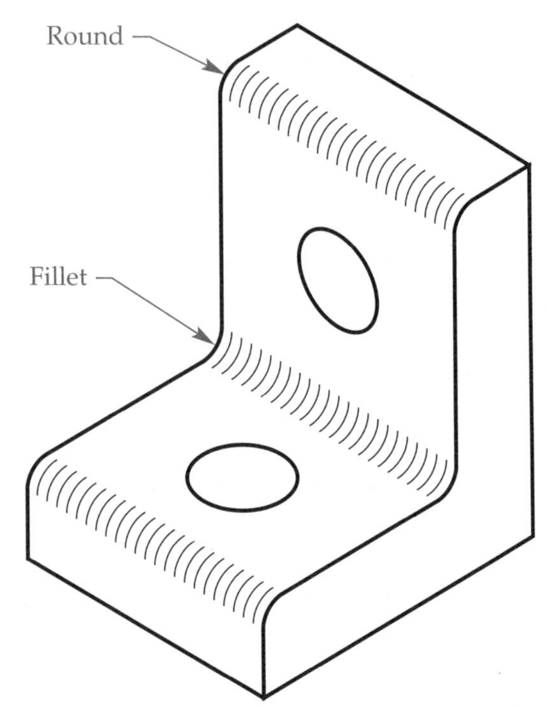


Figure 5-12.
Molded and cast objects have rounded edges called fillets and rounds.

These rounded corners also create situations where the drafter has to decide whether or not to project the edge. Conventional practice calls for the lines to be shown if needed for clarity in describing the object. This is sometimes a judgment call. If the edge is shown, the line is projected as if the corner is sharp or square. When a rounded corner intersects a curved surface, the edge fades, or "tails" out. This is called a ***runout.*** Note, however, this is different than *runout* as defined in Unit 13, Geometric Dimensioning and Tolerancing. See Figure 5-13 for an example of how runouts may be drawn. This is another case wherein views automatically created from a 3D model by the CAD program may differ from views created by a CAD operator using two-dimensional CAD lines and conventional practice.

## Meanings of a Multiview Line

In summary, there are many different perimeter shapes that a flat surface can have, from circular, to square, to polygonal. Yet, if the surface is perpendicular to the projection plane, it simply projects as a line. Therefore, many lines in a multiview drawing represent the edge view of planar surfaces. Curved surfaces also are seen as edge views if the curved surface is perpendicular to the projection plane. See Figure 5-14.

A second meaning a multiview line can have is simply an intersection—an edge where two surfaces meet and nothing more. A third meaning a multiview line can have is the maximum contour element of a curved surface. In some views, this will be represented by a line. These are the three

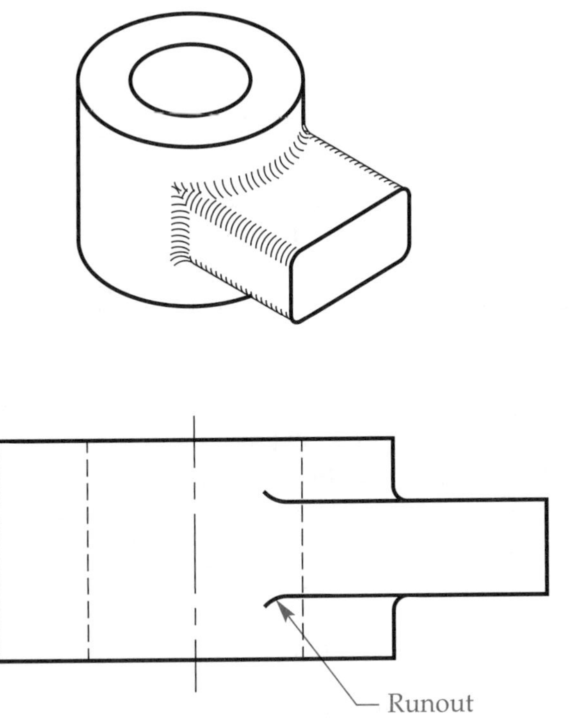


Figure 5-13.
When rounded edges intersect curved surfaces, a runout is used to show how the edge tails out. Computer-generated views from a CAD model may show this differently.

meanings a line in a multiview drawing can have, whether it is visible or hidden. As you learn how to read prints, keep these three meanings in mind.

## First-Angle and Third-Angle Projection

The system of projection explained earlier in the unit is common throughout the United States. The projection of a view is basically what the viewer sees when looking at the object through the glass projection plane. The glass box is unfolded in such a way as to place the top view above the front view. This is known as ***third-angle projection.*** In many countries, especially in Europe, a slightly different projection system is used that results in the views being located differently.

The two projection systems used in industrial drawings are identified as third-angle projection and first-angle projection. These two types result from a theoretical division of space into four quadrants by vertical and horizontal planes, Figure 5-15. The viewer of the four quadrants is considered to be in front of the frontal plane and above the horizontal plane. The views are arranged by folding the two planes into one by collapsing the second and fourth quadrants and the views are then seen from the front. As a result of this, there are no second or fourth angle projections. If there were, the views would overlap.

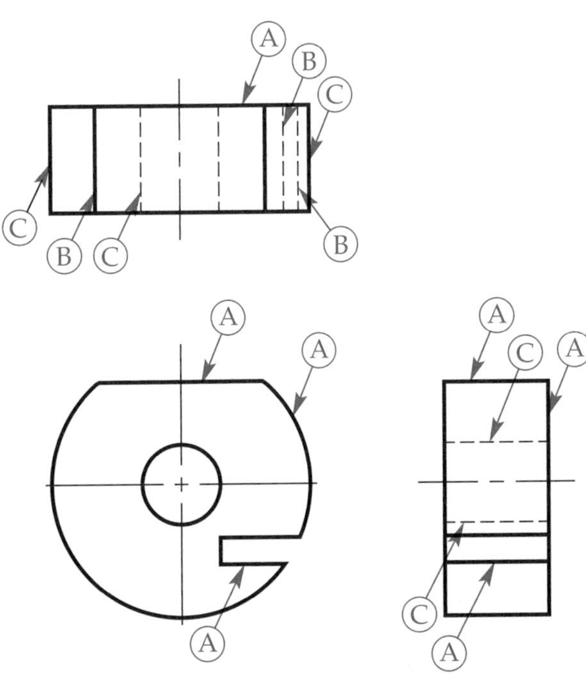


A = Edge view of a flat or curved surface

B = Intersection of two surfaces only (just an edge)

C = Maximum contour of a curved surface

Figure 5-14.
This figure shows how to interpret several lines on a multiview drawing.

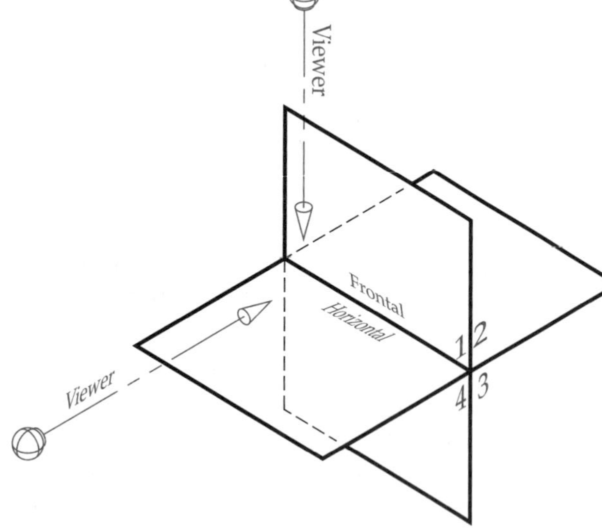


**Figure 5-15.**
The two types of projection are based on a theoretical division of space into four quadrants. Quadrants two and four are not used.

As presented earlier in the unit, in third-angle projection, the object resides in the third angle of space. So, the projection planes are considered to be between the viewer and the object. The views are projected toward the viewer onto the planes. See Figure 5-16. When quadrants two and four are collapsed, the top view appears above the front view.

In contrast, in ***first-angle projection,*** the object resides in the first angle of space. So, the projection planes are on the opposite sides of the object as the viewer. In other words, the object is between the viewer and the projection planes. See Figure 5-17. The individual views are the same as those obtained in third-angle projection, but their arrangement on the drawing is different. In essence, the top view ends up below the front view and a right side view is on the left side of the front view.

In summary, the individual views are the same for both angles of projection. The only difference between the two types is the arrangement of views on the drawing. The ASME and ISO standard symbols to indicate first-angle and third-angle projection are shown in Figure 5-18. One of the two versions of the symbol should be included in the title block for drawings that are read within the international community.

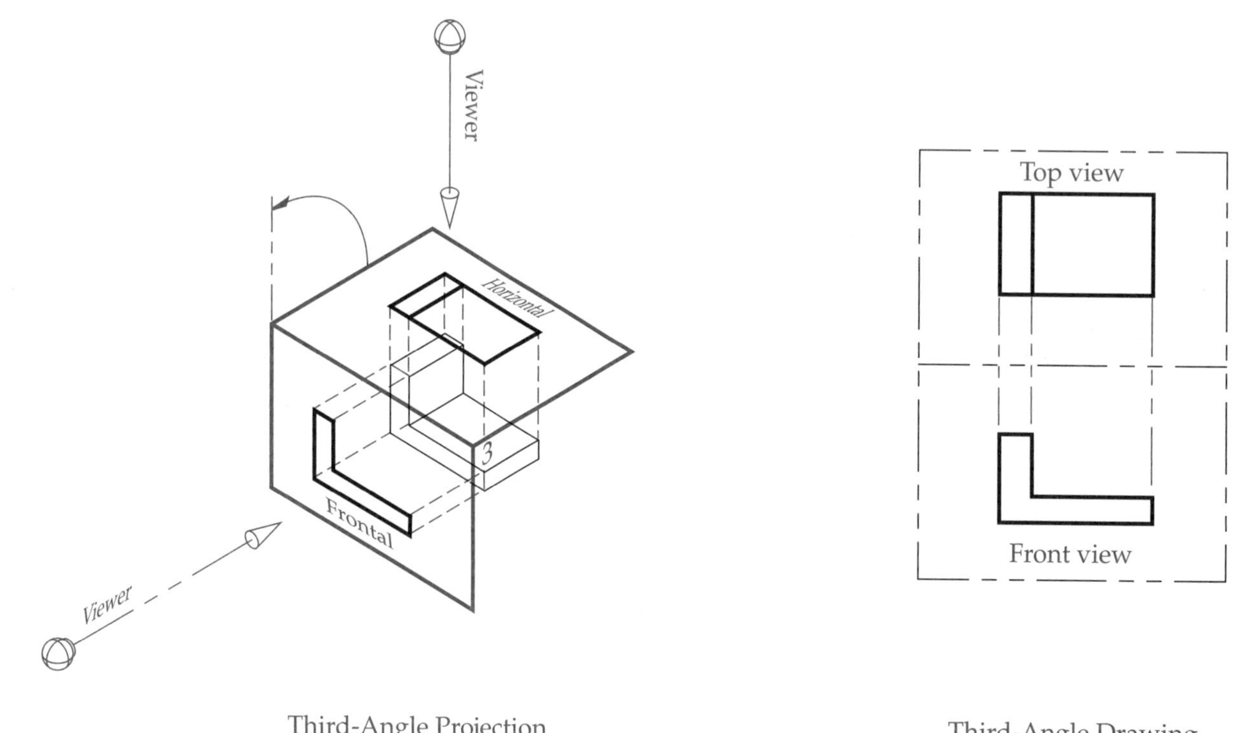


**Figure 5-16.**
In third-angle projection, the projection plane is considered to be between the viewer and the object and the views are projected toward the viewer onto the plane.

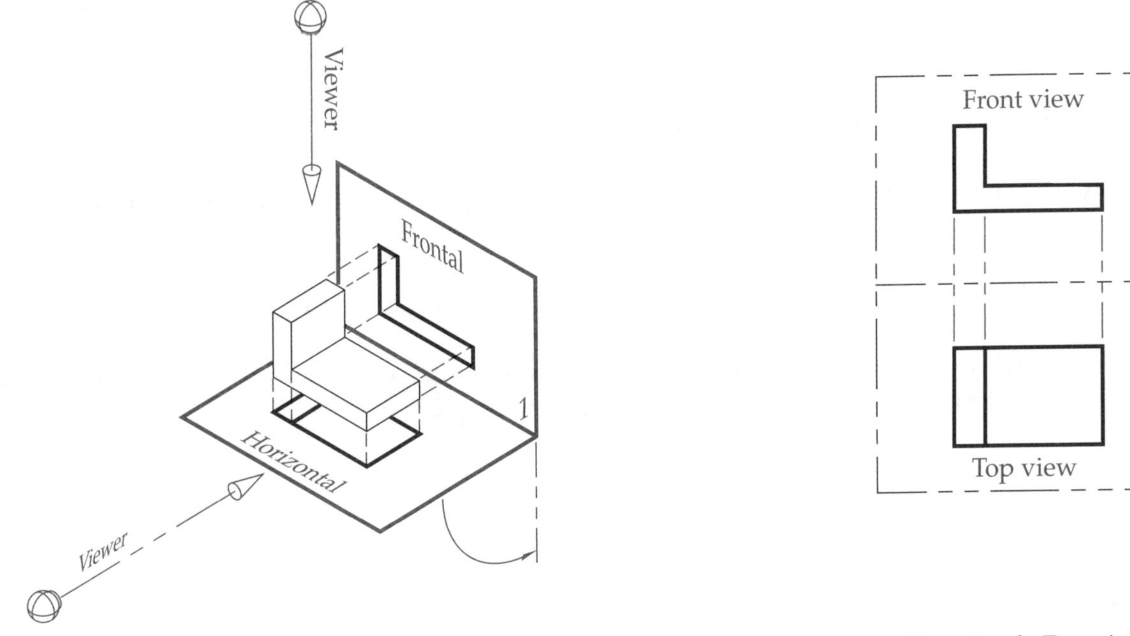


**Figure 5-17.**
In first-angle projection, the projection plane is on the opposite side of the object as the viewer and the views are projected onto the plane on the far side of the object.

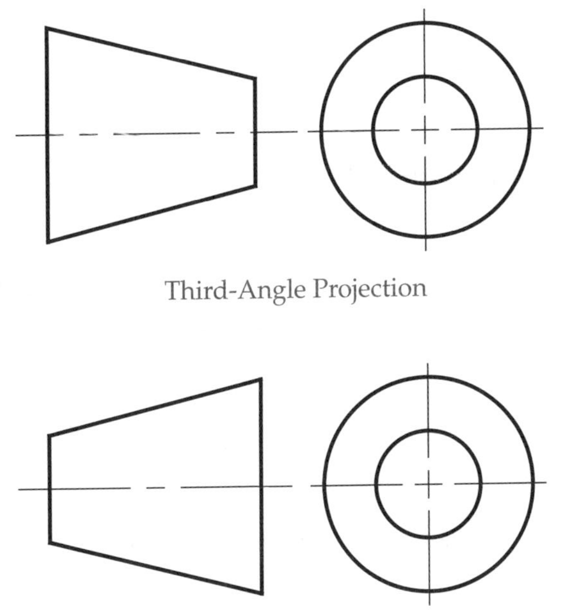


**Figure 5-18.**
The type of projection for a drawing is indicated by one of these two symbols appearing in the title block.

## Review Questions

*Circle the answer of choice, fill in the blank, or write a short answer.*

1. Multiview drawings are created using the principles of ______________ projection.
2. Any view of an object can scientifically be explained as the projection on a object's features on to a(n) ______________.
3. *True or False?* In orthographic projection, the projectors are parallel with each other.
4. *True or False?* While multiview drawings are often arranged in a certain manner, it does not matter how they are arranged on the paper as long as they are correctly labeled.
5. There are ____ normal ways to view an object.
   A. two
   B. four
   C. six
   D. eight
6. List the three predominate terms used to define the three dimensions of an object: ______________
   ______________
7. List the names given to the three principal planes of projection: ______________
   ______________
8. *True or False?* A flat surface is oriented to a plane of projection in one of three ways: perpendicular, parallel, or inclined.
9. If a surface is perpendicular to a plane of projection, it projects on to that plane as a(n) ______________.
10. Flat surfaces are defined by how they are oriented to the three principal planes of projection. List the three names given to flat surfaces: ______________
    ______________
11. An inclined surface will appear as a(n) ______________ in one view and as a foreshortened shape in ______________ views.
12. *True or False?* When a flat surface transitions into a cylindrical surface, a line is shown at the element of tangency.
13. *True or False?* Interior rounded corners are called rounds and exterior rounded corners are called fillets.
14. Of the following, select the one statement that is *not* a meaning a line in a multiview drawing can have:
    A. Edge view of a flat surface.
    B. Maximum contour element of a curved surface.
    C. Intersection edge where two surfaces meet, nothing more.
    D. Tangency element between a curved surface and a flat surface.
15. In the United States, industrial prints use the ______________-angle projection system, whereas many other countries use ______________-angle projection.

## Review Activity 5-1

**Normal Surfaces**

Study each pictorial (3D) drawing and the identification letters placed on, or pointing to, the normal surfaces. Match the ID letter to the corresponding number for each of the multiview (orthographic) callouts. Note: The letter I is not used.

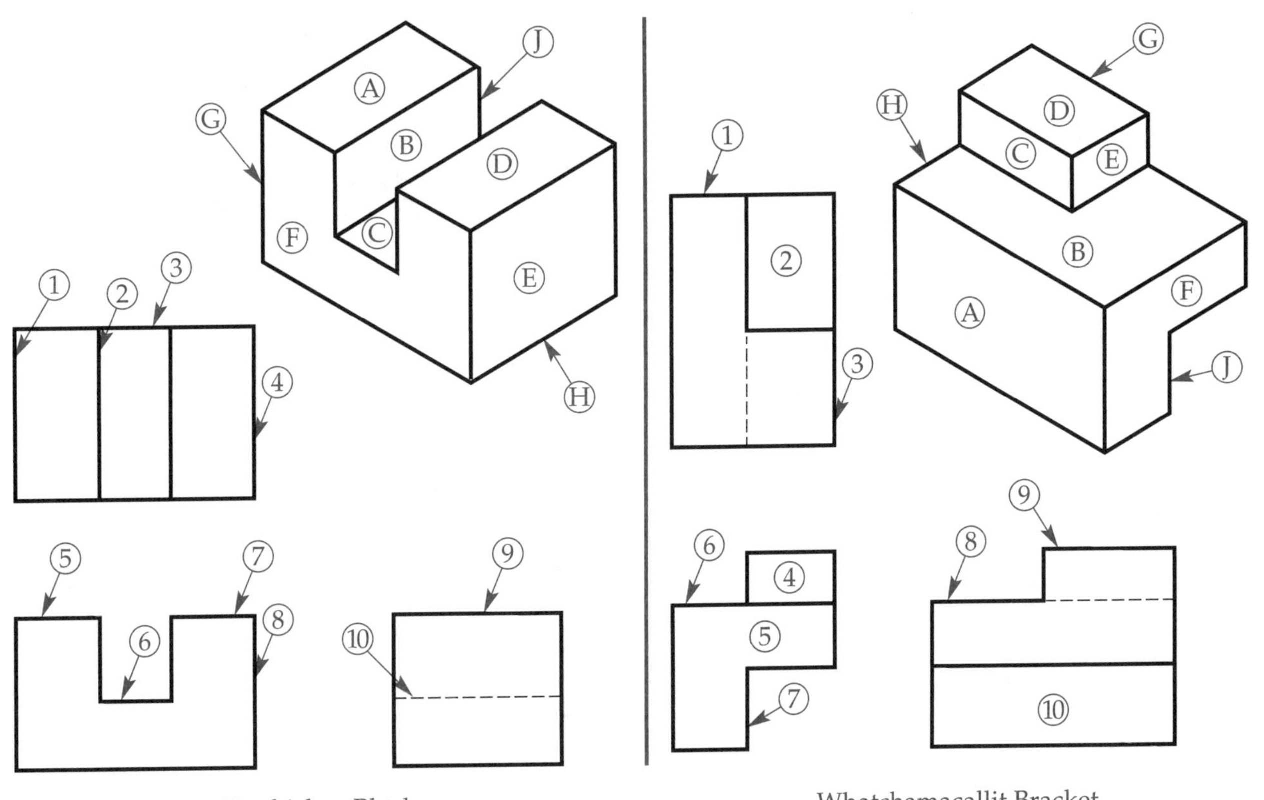


Doohickey Block

Whatchamacallit Bracket

Note: In the pictorial view, arrows pointing directly to a line are referencing a surface that is not visible, rather "around the back" of the object.

**Doohicky Block**

| | |
|---|---|
| _______ 1 | _______ 6 |
| _______ 2 | _______ 7 |
| _______ 3 | _______ 8 |
| _______ 4 | _______ 9 |
| _______ 5 | _______ 10 |

**Whatchamacallit Bracket**

| | |
|---|---|
| _______ 1 | _______ 6 |
| _______ 2 | _______ 7 |
| _______ 3 | _______ 8 |
| _______ 4 | _______ 9 |
| _______ 5 | _______ 10 |

## Review Activity 5-2

**Normal and Inclined Surfaces**

Study each pictorial (3D) drawing and the identification letters placed on, or pointing to, the normal or inclined surfaces. Match the ID letter to the corresponding number for each of the multiview (orthographic) callouts. Note: The letter I is not used.

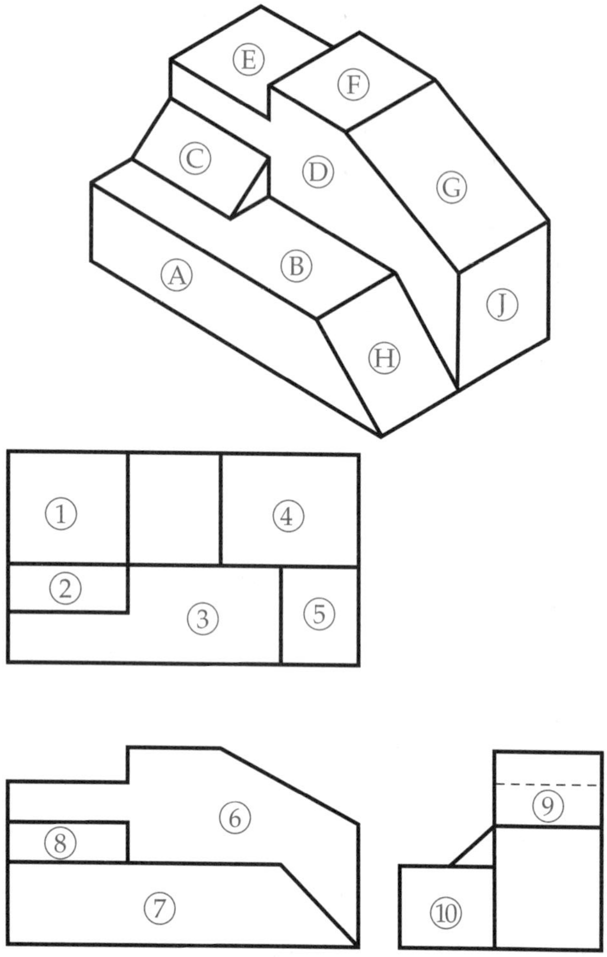

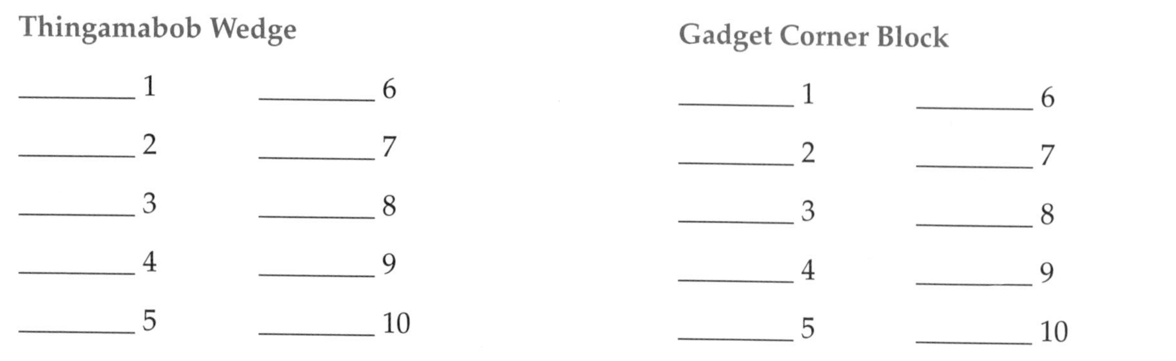

Thingamabob Wedge

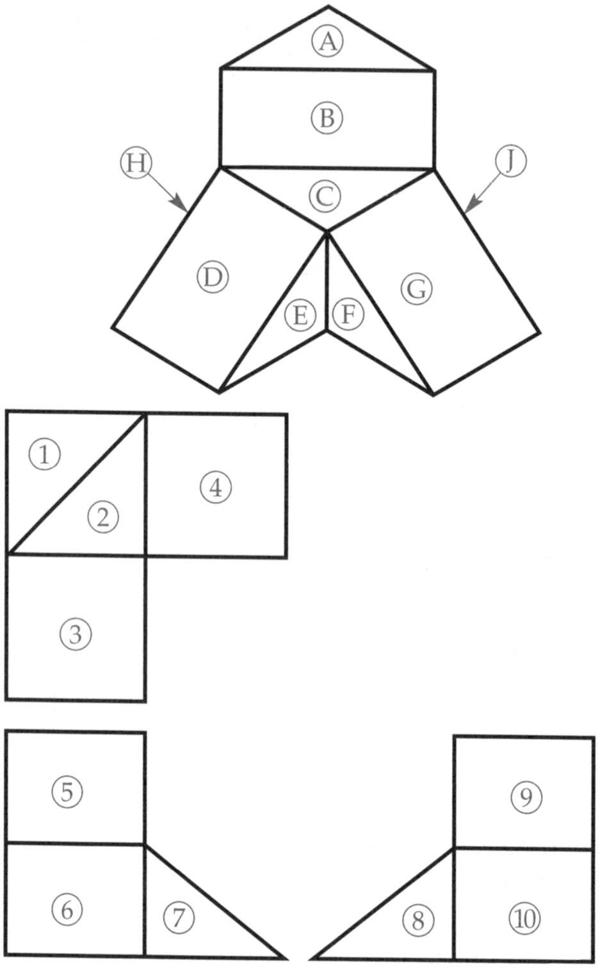

Gadget Corner Block

Note: In the pictorial view, arrows pointing directly to a line are referencing a surface that is not visible, rather "around the back" of the object.

**Thingamabob Wedge**

_______ 1 _______ 6

_______ 2 _______ 7

_______ 3 _______ 8

_______ 4 _______ 9

_______ 5 _______ 10

**Gadget Corner Block**

_______ 1 _______ 6

_______ 2 _______ 7

_______ 3 _______ 8

_______ 4 _______ 9

_______ 5 _______ 10

## Review Activity 5-3

**Sketching Missing Lines**

Sketch the missing line(s) in the following multiview drawings. The final drawing should present a clear view of the object and all views must agree. Include all visible lines, hidden lines, and center lines.

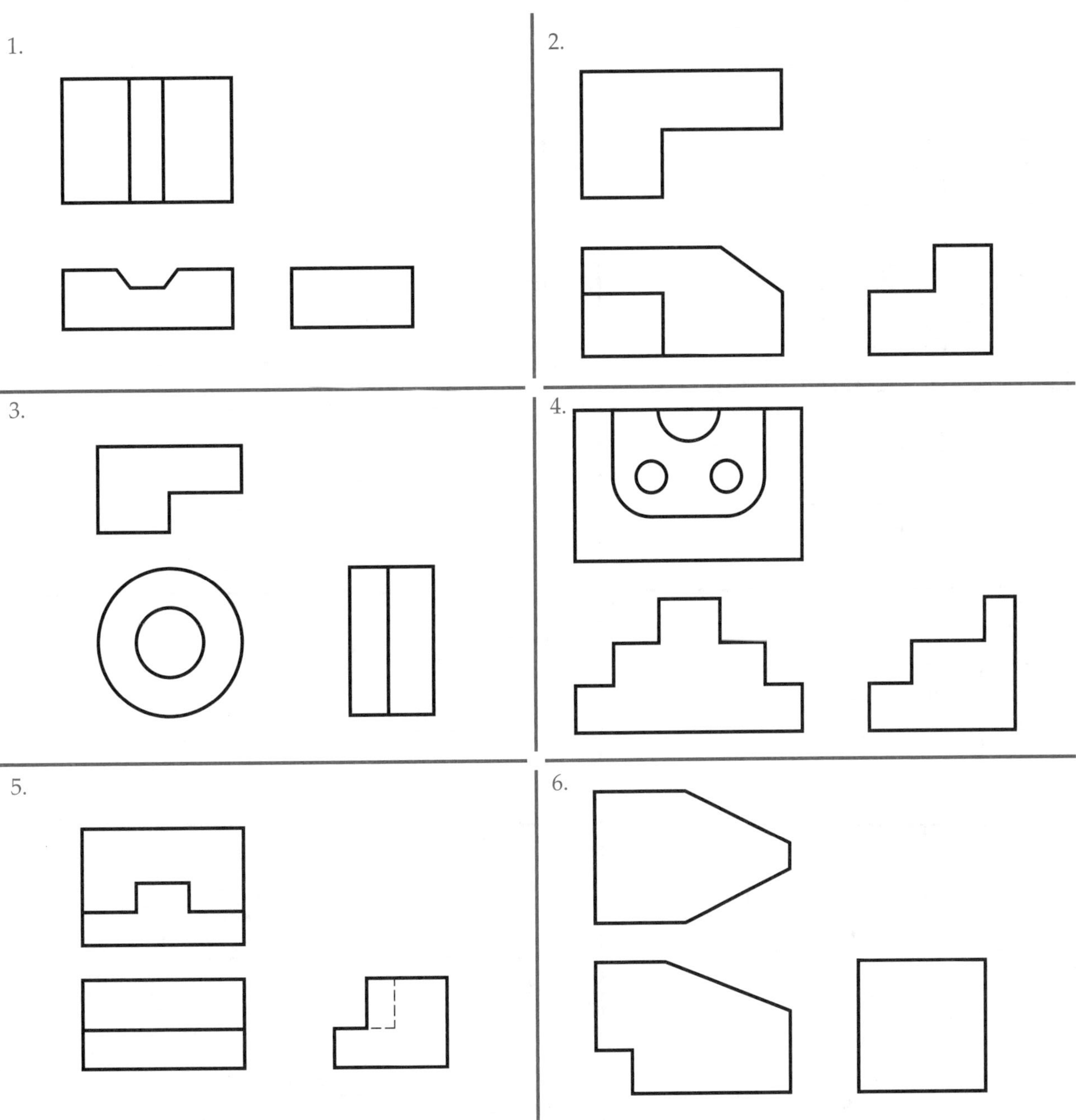

# Review Activity 5-4

## True Size and Shape Identification

Analyze each of the multiview drawings. Place a T for *true size and shape* or an F for *foreshortened size and shape* in the blanks below each multiview. Remember, normal surfaces appear true size and shape in only one view. Inclined surfaces appear foreshortened in shape and size in two views. The first problem is done for you as an example.

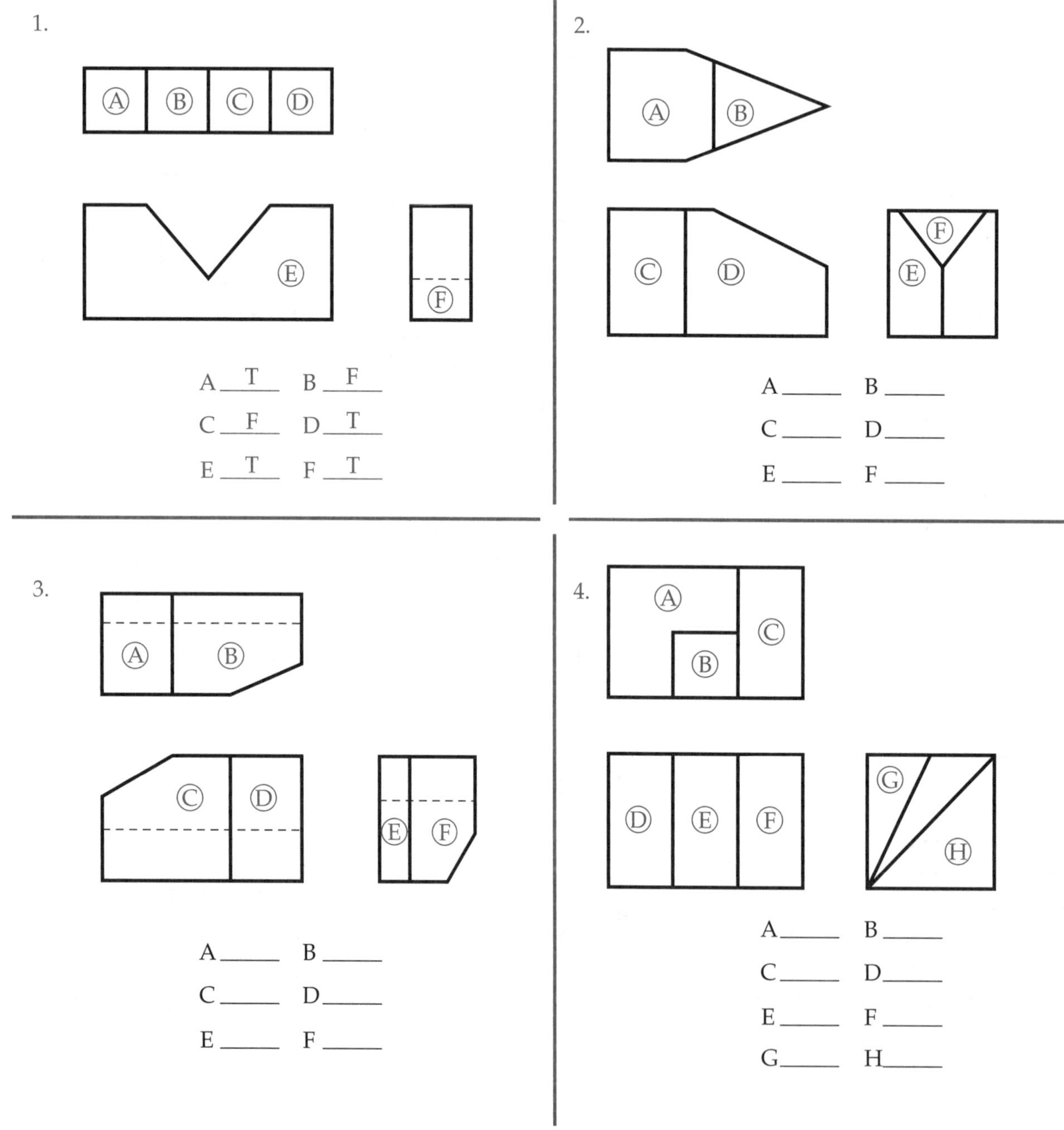

# Review Activity 5-5

## Projection Plane Orientation and Surface Type Identification

For each of the four pictorial drawings, imagine a frontal projection plane in front of the object, a horizontal plane of projection above the object, and a profile projection plane to the right, as described in this unit. In the charts below, fill in the number of surfaces oriented to each of the three projection planes as indicated for each column (parallel, perpendicular, or inclined). Include *all* surfaces, even the bottom, back, and left-side surfaces. The first object has been done for you as an example. Also, enter the number of each type of surface in the second chart. Note: The total for each row of the first chart should be the same and it should also match the total of the second chart, as shown in the first problem.

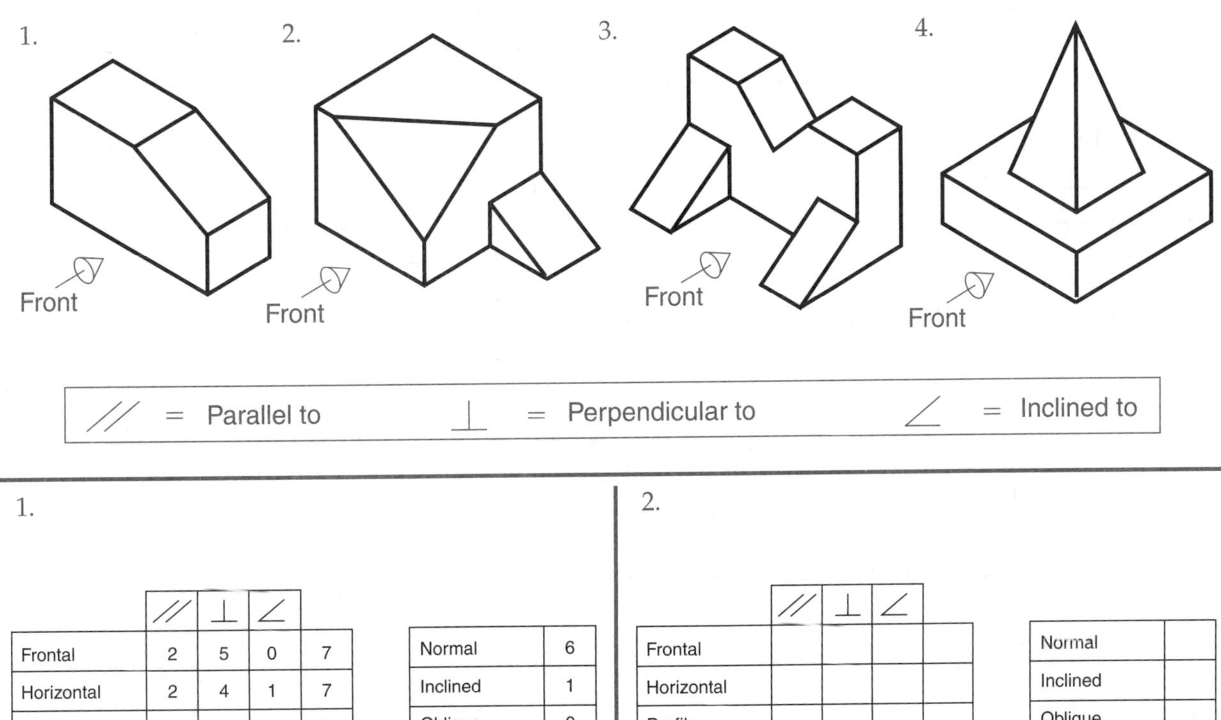


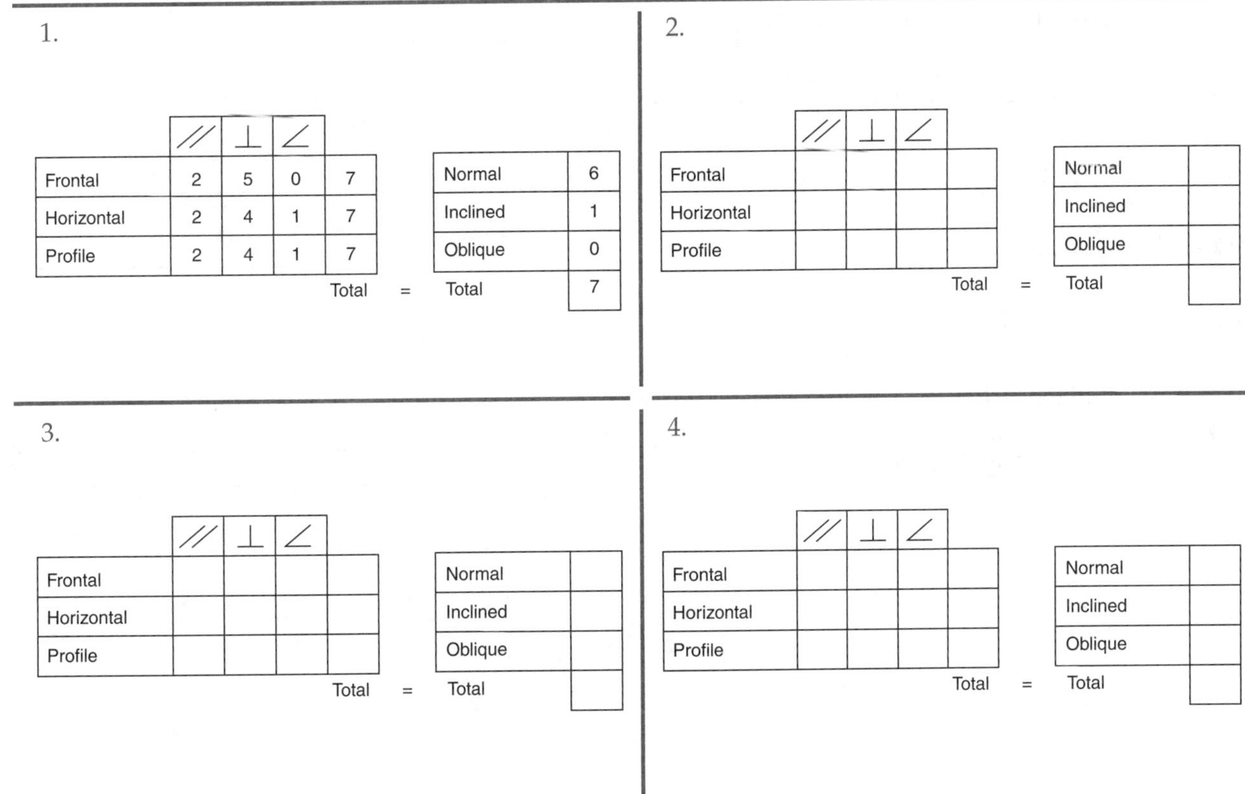

1.

| | // | ⊥ | ∠ | Total |
|---|---|---|---|---|
| Frontal | 2 | 5 | 0 | 7 |
| Horizontal | 2 | 4 | 1 | 7 |
| Profile | 2 | 4 | 1 | 7 |

Total =

| | |
|---|---|
| Normal | 6 |
| Inclined | 1 |
| Oblique | 0 |
| Total | 7 |

2.

| | // | ⊥ | ∠ | Total |
|---|---|---|---|---|
| Frontal | | | | |
| Horizontal | | | | |
| Profile | | | | |

Total =

| | |
|---|---|
| Normal | |
| Inclined | |
| Oblique | |
| Total | |

3.

| | // | ⊥ | ∠ | Total |
|---|---|---|---|---|
| Frontal | | | | |
| Horizontal | | | | |
| Profile | | | | |

Total =

| | |
|---|---|
| Normal | |
| Inclined | |
| Oblique | |
| Total | |

4.

| | // | ⊥ | ∠ | Total |
|---|---|---|---|---|
| Frontal | | | | |
| Horizontal | | | | |
| Profile | | | | |

Total =

| | |
|---|---|
| Normal | |
| Inclined | |
| Oblique | |
| Total | |

## Review Activity 5-6

### Piston

Study the print and answer the questions.

1. Which view is represented by:
   A. ______
   B. ______
   C. ______

2. How many visible lines are in view:
   A. ______
   B. ______
   C. ______

3. How many hidden lines are in view:
   A. ______
   B. ______
   C. ______

4. What are the three overall dimensions of the object:
   Height ______
   Width ______
   Depth ______

5. How many normal surfaces does this object have? ______

6. How many inclined surfaces does this object have? ______

7. How is the depth of surface X determined? ______

8. How wide is surface X? ______

9. How wide is surface Z? ______

10. Given three choices (edge view, intersection, contour):
    A. What does line 10A mean? ______
    B. What does line 10B mean? ______

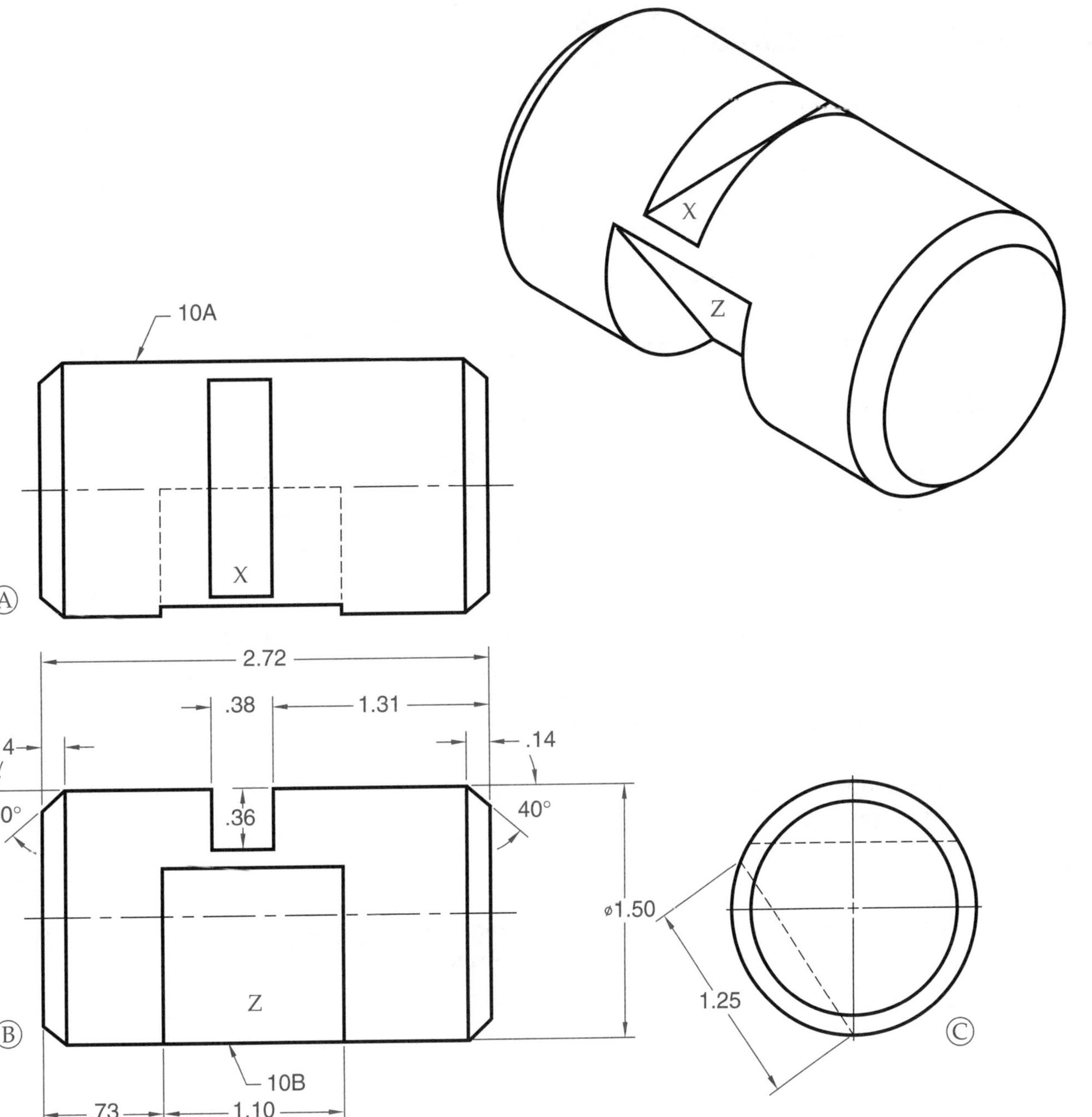
X
Z
10A
X
A
2.72
.38
1.31
.14
.14
40°
.36
40°
Z
B
10B
.73
1.10
⌀1.50
1.25
C

## Industry Print Exercise 5-1

*Refer to the print PR 5-1 and answer the questions below.*

1. This drawing has the three regular views. What are they? ____________________
2. What term is applied to the dimension value of 2 1/8″? ____________________
3. The circular feature in the right side view is ____________________ in the front view.
4. What is the height of this part? ____________________
5. The top view shows the true size and shape of a surface that encompasses four holes. What type of surface is it? ____________________
6. What is the total width of this part? ____________________
7. Which letter of the alphabet, M, H, or L, best describes the general shape of the normal surface facing the frontal plane? ____________________
8. How many hidden lines are there in the right side view? ____________________
9. How does the bottom, flat surface of the object appear in the top view: true size and shape, foreshortened shape, or edge view? ____________________
10. In the top view, the right-most vertical line has what meaning: edge view of a surface, intersection of two surfaces only, or maximum contour of a curved surface? ____________________

*Review questions based on previous units:*

11. What is the name of this part? ____________________
12. What is the drawing number or part number? ____________________
13. In what state is this company located? ____________________
14. What is the five-digit material code specified for this part? ____________________
15. What is the last name of the person who drew this drawing? ____________________

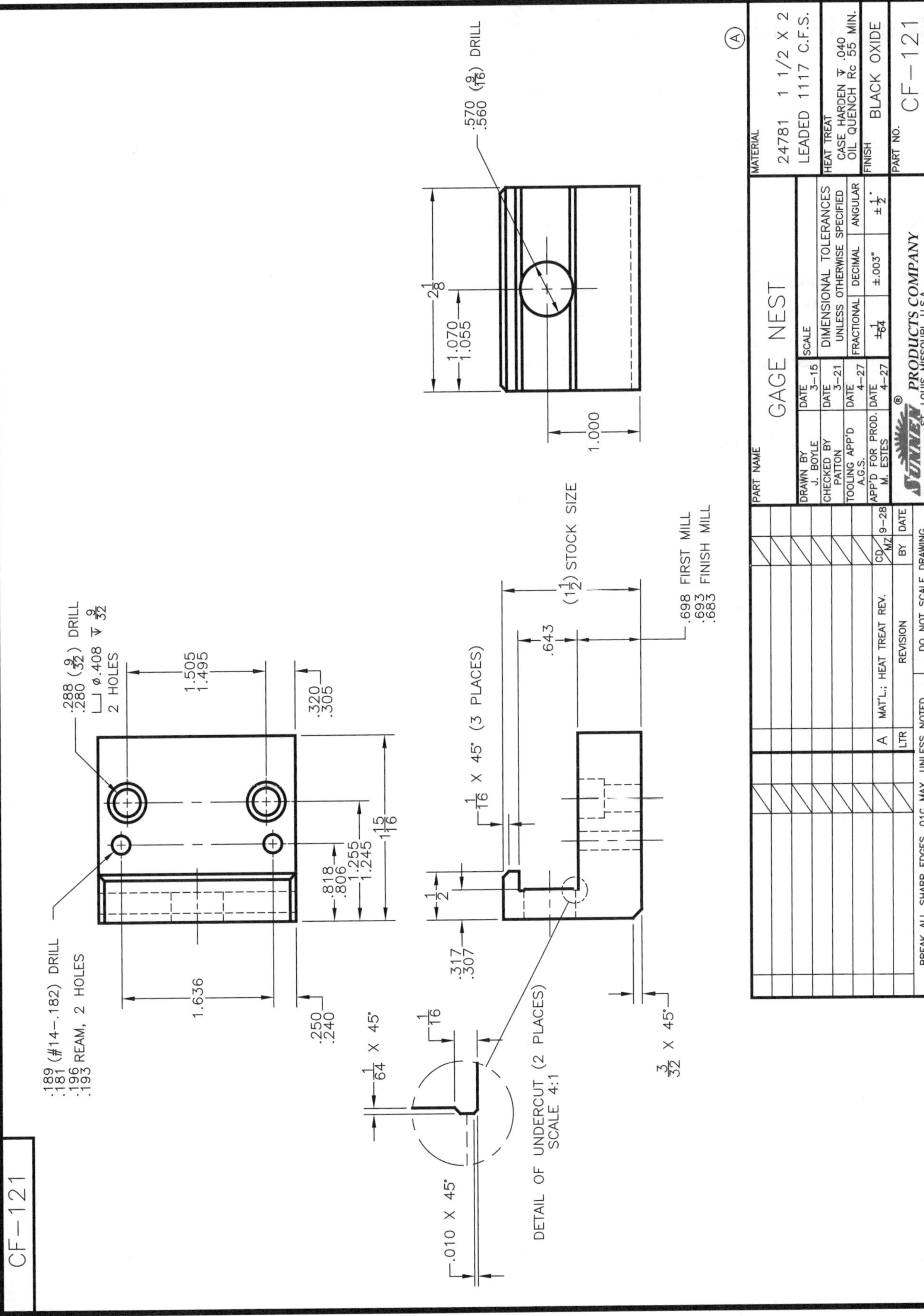


PR 5-1.
Print provided by Sunnen Products Company.

## Industry Print Exercise 5-2

*Refer to the print PR 5-2 and answer the questions below.*

1. Not counting the removed view and pictorial views, how many regular views does this object have? ____________
2. How many of the regular views feature height? ____________
3. If the view that is closest to the center of the sheet is the front view, what is the name of the view directly below it? ____________
4. The flat surface that shows true size and shape in the top view is parallel to the ____________ plane of projection.
5. In which view is the viewing-plane line shown for the removed view? ____________
6. What two dimensions (names) are featured in the left side view? ____________
7. Does this object appear to have any inclined surfaces? ____________
8. Which dimension is greater, width or depth? ____________
9. What is the name of the removed view? ____________
10. Are hidden lines shown in the top view? ____________

*Review questions based on previous units:*

11. Which view features cutting-plane lines? ____________
12. What is the drawing number or part number? ____________
13. How many views feature section lines? ____________
14. What is the most recent date associated with this drawing? ____________
15. The annotation on the left side view features a couple of ____________ lines.

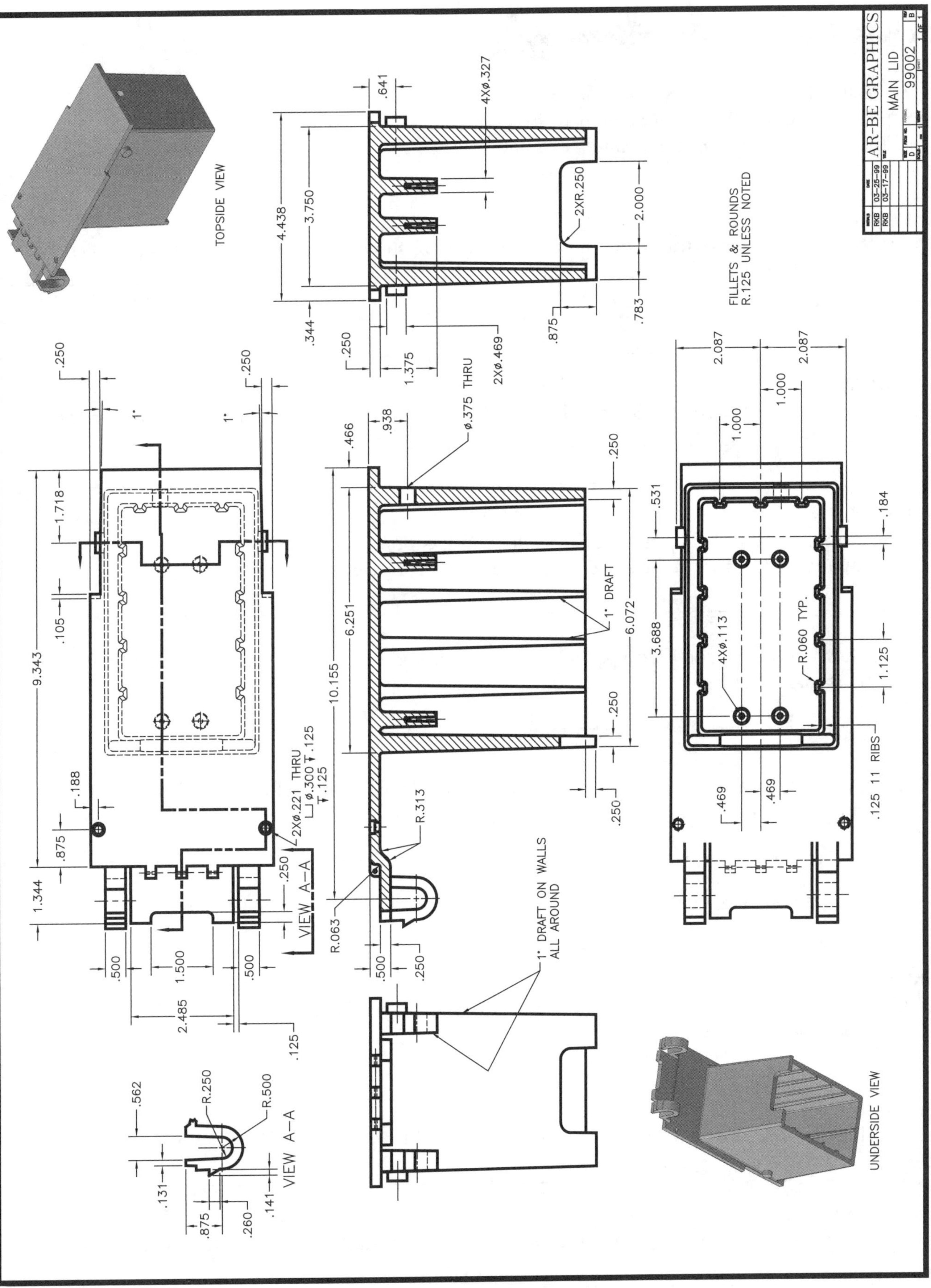


PR 5-2.
Print provided by Ar-Be Graphics.

## Bonus Print Reading Exercises

The following questions are based on various bonus prints located in the folder at the back of this textbook. Refer to the print indicated, evaluate the print, and answer the question.

**Print AP-001:**

1. On the left side view of this print, there is a horizontal, straight-line segment that does not connect to other lines. In basic terms, explain why. ______________________

**Print AP-004:**

2. If the view labeled Section B-B is a right side view, what name would be applied to the other four main views? ______________________

**Print AP-005:**

3. This print features what appears to be a side view (Section G-G) that does not have an accompanying front view. Briefly explain where the front view is located. ______________________

4. The view labeled Section J-J is a partial view, as indicated by the short break line. Does it appear to be an orthographic view? ______________________

**Print AP-006:**

5. The two views near the bottom of the print are called detail views. Which one of the views above (left side, front, or right side) shows the same geometry, but at the normal 1:1 scale? ______________________

6. In general terms, what name is applied to the two groove views near the bottom of the print? ______________________

**Print AP-007:**

7. Carefully examine the drawing. With respect to the enlarged Detail C, how many times larger is it than the other multiview views? ______________________

**Print AP-009:**

8. For this drawing, are there other views besides View E-E and View F-F? If so, where are they? ______________________

**Print AP-012:**

9. What name is given on this print to the view that is not in multiview orientation or arrangement? ______________________

**Print AP-015:**

10. Is the left side view on this print a complete view or a partial view? ____________________

**Print AP-016:**

11. Are the views of this assembly drawing arranged orthographically in proper third angle projection?

____________________

**Print AP-021:**

12. What is the most likely reason the upper-right view features short break lines? ____________________

**Print AP-022:**

13. One view features the depth measurements .300, .250, and .040. How many other main views (not including detail views) on this print feature the depth direction? ____________________

**Print AP-024:**

14. If the lower view is considered to be a front view, what label would apply to the view located on the upper-right side of the print? ____________________

**Print AP-025:**

15. Is the view labeled FORMED VIEW a normal view defined by multiview practice? ____________________

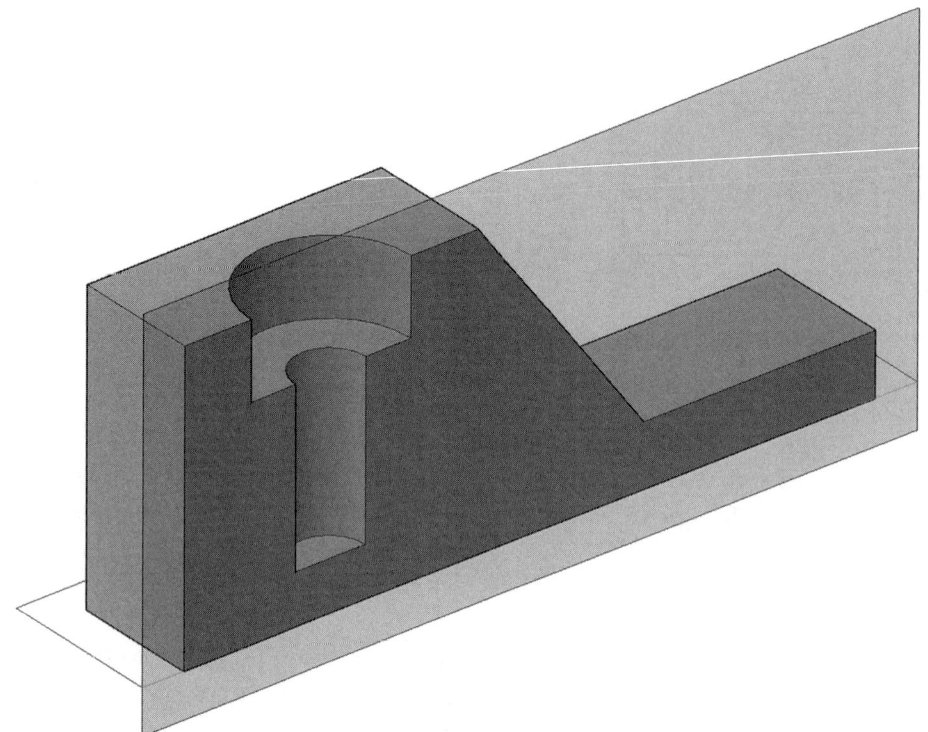

CAD software typically has a tool that allows a section view to be quickly created from a solid model.

# UNIT 6
# Section Views

*After completing this unit, you will be able to:*

**Explain** the characteristics of a drawing that features a full section view.

**Describe** the characteristics of a half section view.

**Describe** the characteristics of an offset section view.

**Identify** the characteristics of a sectional view that features aligned features.

**Identify** the characteristics of a view that features a broken-out section.

**Compare** revolved sections and removed sections and **identify** the characteristics of these sections.

**Explain** conventional practices applied to section views.

**Identify** the proper representation of partial sections and outline sections.

**Identify** the proper representation of separate parts in an assembly section view.

Sometimes the regular views of a drawing cannot clearly show the interior features of an object, so a type of view known as the section view is used to help show this interior detail. A ***section view*** is created by passing an imaginary cutting plane through the object and removing the part nearest to the observer. This allows a direct view of the interior details. Section views are also called sectional views, cross sections, or simply sections. See Figure 6-1. The standard practices for sectional views are covered in ASME Y14.3, entitled *Multiview and Sectional View Drawings.*

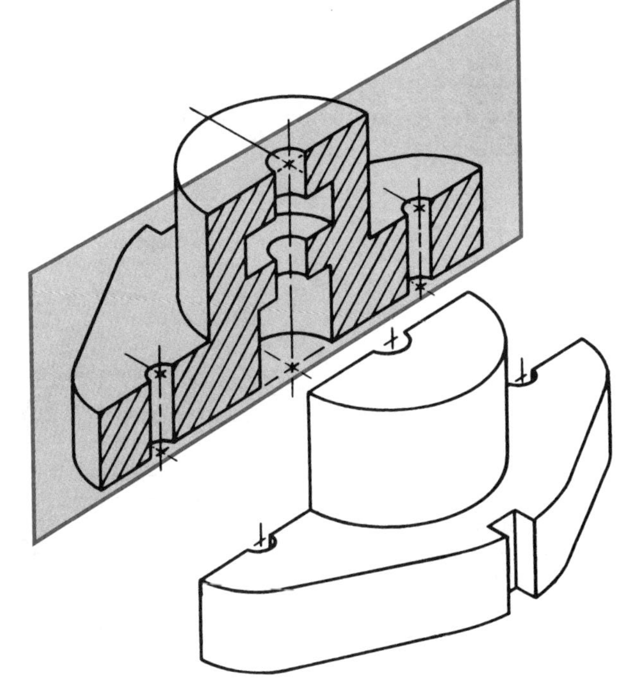

Figure 6-1.
A section view allows a direct view of interior detail.

## Section View Lines and Principles

The alphabet of lines contains two line types used for section views. These are section lines and cutting-plane lines. ***Section lines*** shade the area of material cut by the imaginary cutting plane. The ***cutting-plane line*** represents the edge view of the cutting plane. It is placed in the adjacent view to indicate the location of the cut.

Section lines, also called *crosshatch* or *hatching,* can be created in a variety of patterns. Different patterns can be used to indicate the type of material used for the part, Figure 6-2. In practice, however, section lines are often drawn with continuous lines about 1/8″ apart at a 45° angle. These 45° lines, like those shown in Figure 6-3, indicate cast iron or malleable iron, but can be used for general purpose section lines. The section lines in a section view are not to be the only method designating the material for the object. The material specification should also be listed in the title block, materials block, or in a note on the drawing.

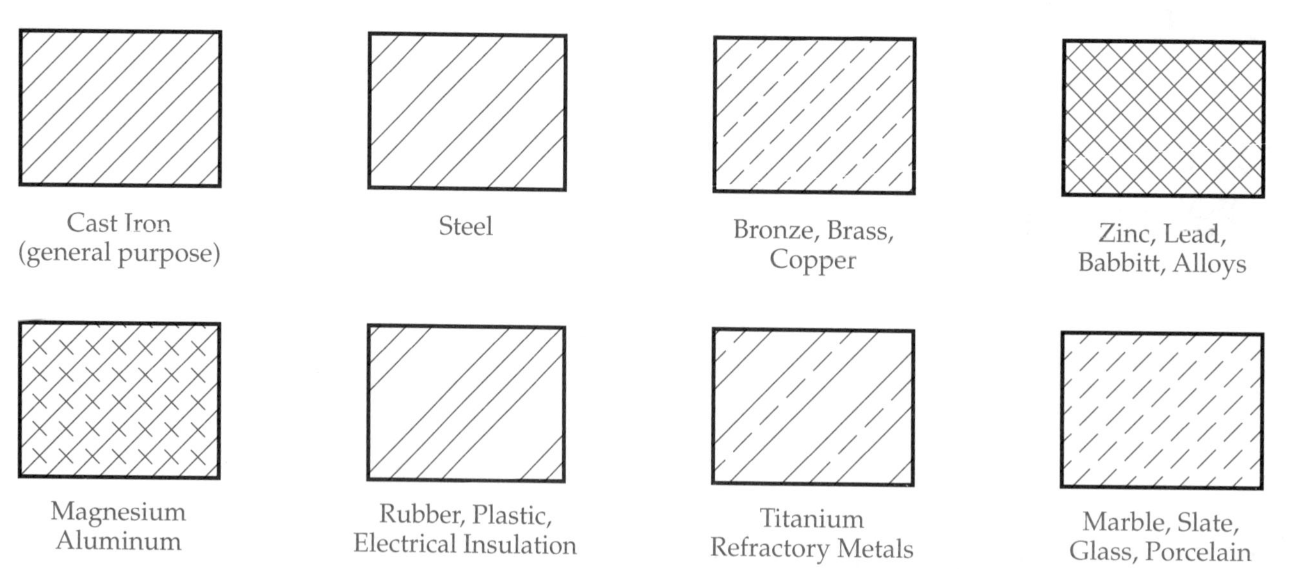


**Figure 6-2.**
Section line patterns are standardized for different types of materials. Most of these are available for industry CAD systems, but general purpose section lining is still common.

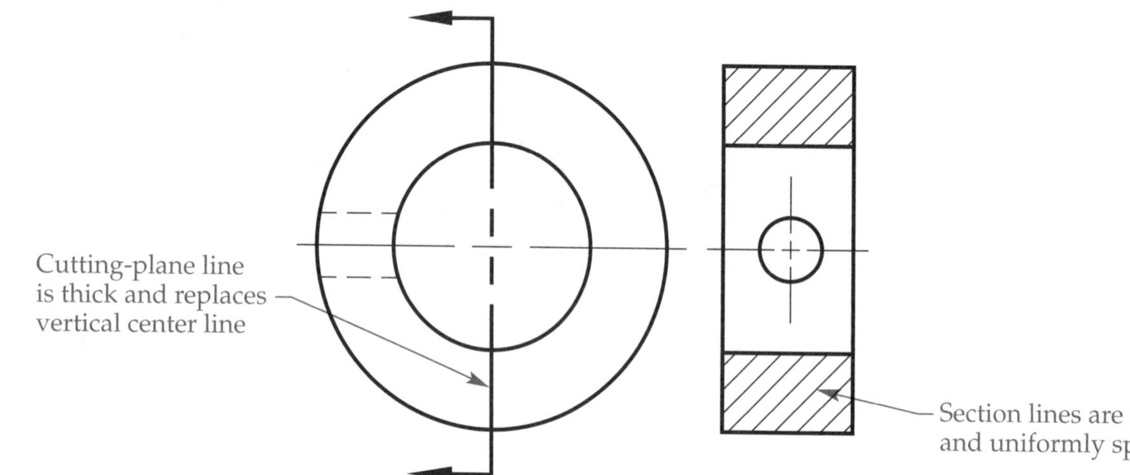


**Figure 6-3.**
Cutting-plane lines and section lines are used in section view drawings.

While section lines are usually drawn at 45°, they can be drawn at any angle, as long as they are not parallel or perpendicular to one of the major visible lines. The angle should be the same throughout the section view of a single part. A section view may serve as one of the regular views on the drawing or it may appear as an additional view.

Cutting-plane lines are thick and may be drawn as a series of medium dashes or as a series of two short and one long dashes. An additional option is to simply show the cutting-plane line with thick "elbows." Arrowheads are placed on each end of the cutting-plane line to indicate the viewing direction for the section view. Most often, capital letters are also placed near each arrowhead, such as A, B, C, etc., to help identify the sectional views on the page with the corresponding cutting plane. The cutting-plane line can be omitted on a simple symmetrical object, but the standards recommend the cutting-plane line be drawn if the cutting plane is bent or offset or if the section view that results from the cut is not symmetrical.

When possible, the section view should be arranged in orthographic projection with the other views. Scale and sheet size are two factors that are considered when deciding where to place the section view. For multi-sheet drawings, it is highly desirable to keep the section view on the same sheet as the view that shows the cutting plane, although section names and zoning can assist in cross-referencing sections to other sheets if required.

## Section View Types

There are several types of sections. Each is designed for a specific purpose. A ***full section*** is created when the cutting plane passes entirely through the object. For all practical purposes, the object is "cut in half." See Figure 6-4. Notice how much more clearly the internal details appear in the section view than in the right side view, where the internal details are shown as a hidden line. Lines visible in the section view are shown, but hidden lines usually are omitted. Even if there are hidden features behind the sectioned area, showing hidden lines reduces the clarity of the section view. For complex parts, showing a few hidden lines on a section view is permitted if it eliminates the need to draw another view.

A symmetrical object is the same on both sides of the centerline. The ***half section*** is created when a symmetrical object is drawn with one-half of the view as a section and the other half as a regular view. See Figure 6-5. In essence, the half section cuts "halfway" through the object. In your mind, think of this as if one-quarter of the object is removed. The cutting-plane line is shown on the adjacent view indicating where the part has been cut and the direction from which the section is viewed. Notice there is only one arrowhead on the cutting-plane line for the half section view. Also note the preferred method of showing the section view with a center line dividing the sectioned half from the regular half, *not* a visible line. After all, the object is really not cut. The resulting section is really a "double exposure" with two half views as one.

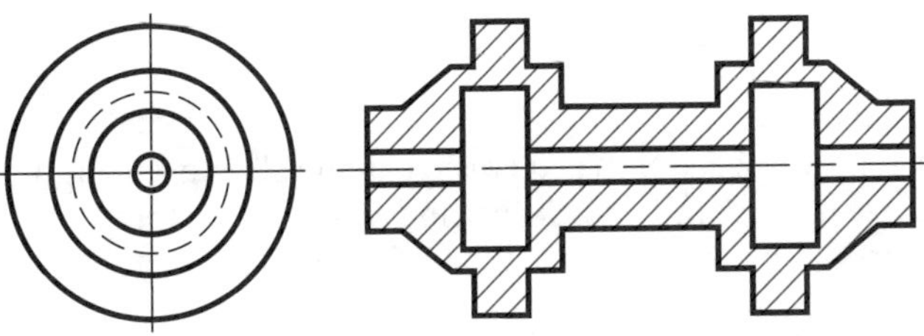

**Figure 6-4.**
A full section is created when the cutting plane passes entirely through the object. The cutting-plane line can be omitted on basic, symmetrical full sections such as this one.

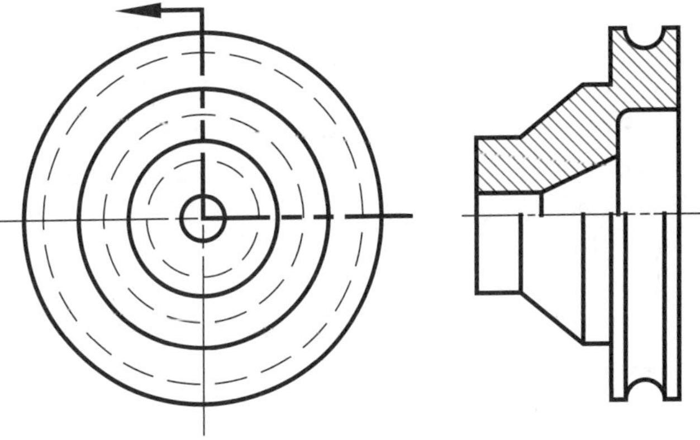

**Figure 6-5.**
The half section is created when a symmetrical object is drawn as a blended view with one-half as a section view and the other half as a regular view.

An ***offset section*** is drawn when the essential internal details do not appear on one flat plane through the object. The cutting-plane line is drawn offset through the object to include the desired features. Those features are then shown on one plane in the section view. See Figure 6-6. The section view itself does *not* show the bends in the cutting plane and appears as a flat plane.

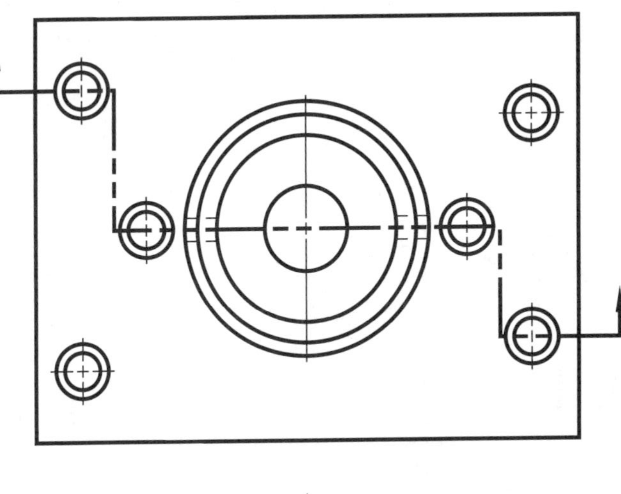

**Figure 6-6.**
An offset section is drawn with a cutting-plane line offset through the object to include the desired features.

An ***aligned section*** is usually drawn for a cylindrical object with an odd number of features. The cutting-plane line is offset through the features such that they can be "rotated" to a normal vertical or horizontal plane and projected to the section view. See Figure 6-7. The result is equivalent to the feature being "aligned" with a normal full section cutting plane.

A ***broken-out section*** is created when a small portion of a part is exposed to show the interior construction. This is like starting to cut the object with a plane, but then breaking off a piece of the object, leaving the rest of the object shown in a regular way. A cutting-plane line in an adjacent view is unnecessary. The sectioned portion of the view is separated by a short break line, discussed in a previous unit. See Figure 6-8. This type of section view can also be used to show exterior and interior details on the same view, similar in fashion to a half section. As in other cases, the hidden lines in the non-sectioned portion of the view may be drawn if needed to explain other interior details or may be omitted.

A ***revolved section*** is used for a feature such as a wheel spoke or long steel bar. The cutting plane slices through the feature parallel to the line of sight, but the "cut" shape is revolved 90° directly on the regular view as if it were an overlay, as shown in Figure 6-9 at A. To further clarify the view, the part may be broken on each side of the section, as shown in Figure 6-9 at B.

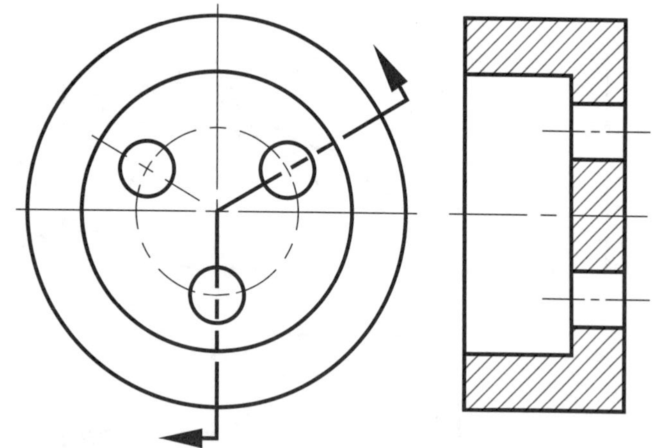

**Figure 6-7.**
The cutting-plane line of an aligned section is offset through the features so they can be "rotated" to a normal vertical or horizontal plane and projected to the section view.

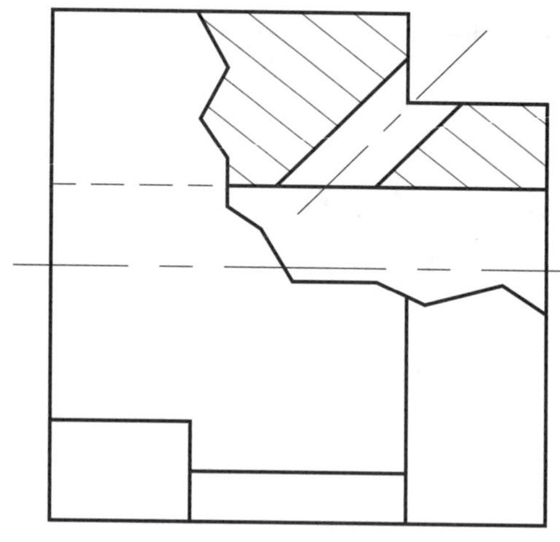

**Figure 6-8.**
A broken-out section is created when a small portion of a part is exposed to show the interior construction.

A ***removed section*** is similar to a revolved section, but the section view is shown in another place on the drawing. The removed section is usually "out of projection" with the other views. Removed sections are frequently used as detail sections. They may even be shown at a different scale so the view is enlarged to clarify detail, Figure 6-10.

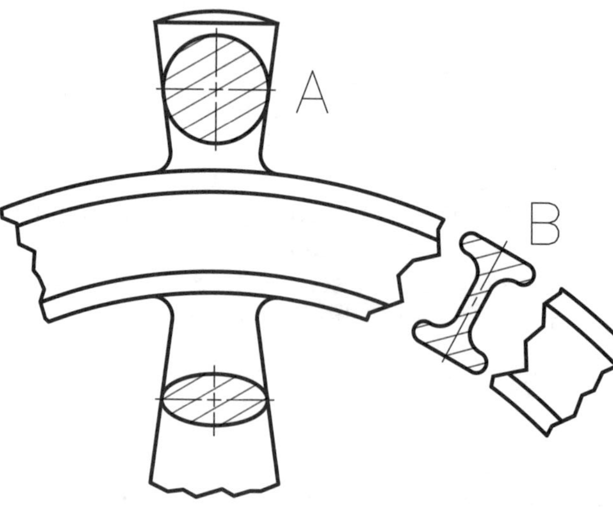


**Figure 6-9.**
A revolved section has a cutting plane slicing through the feature parallel with the line of sight, but the "cut" shape is rotated 90° directly on the regular view.

**Figure 6-10.**
A removed section is shown "out of place" on the drawing.

## Conventional Practice in Section Views

Section views present many opportunities for conventional practice. A ***conventional practice*** is when drafters break the principles or rules of projection theory for the sake of clarity, but standard references make everyone aware of the rules for that particular exception. One example of a conventional practice can be called the "rib rule." Many sectioned objects have thin walls, webs, or ribs that help support a feature of the object, as in Figure 6-11. To avoid a false impression of thickness, it is a *conventional practice* to leave the web or rib without section lines. Also, the rib is outlined, even though the cutting plane passes through it. This rule applies when the cutting plane is parallel to the thin material, not when the cutting plane passes across it in a perpendicular fashion. In some cases, it may be clearer to show the rib sectioned with alternating lines to distinguish it from the rest of the sectioned area.

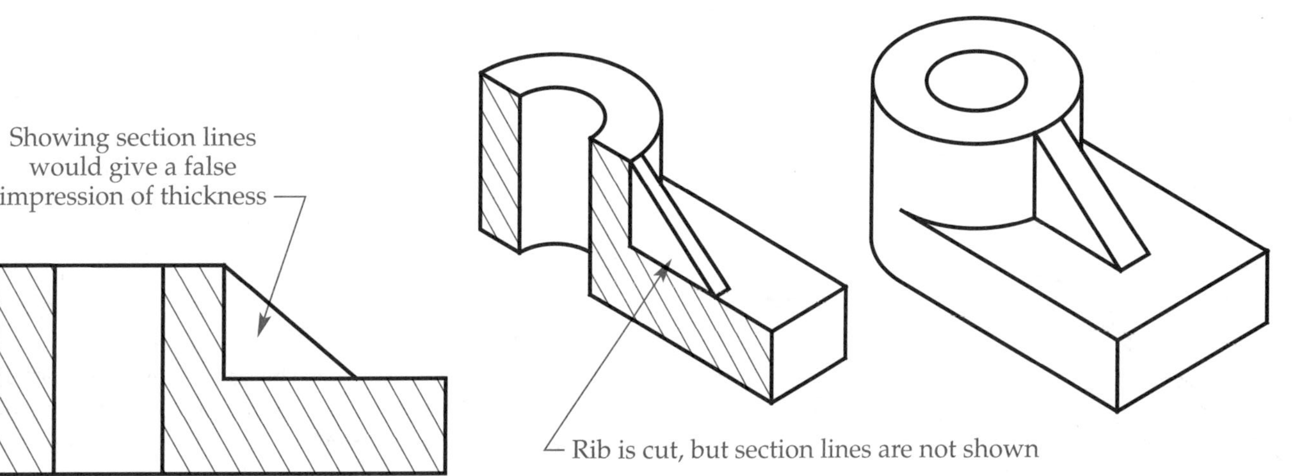


**Figure 6-11.**
It is conventional practice to leave webs or ribs without section lines.

The aligned section illustrated in Figure 6-12 also shows the conventional practice for spokes of a wheel. As with the rib rule, spokes and arms in a section view are not section lined to distinguish between these and a solid web.

## Other Section View Practices

There are a few other practices specific to section views that should be explained in this unit. A section can be a ***partial section.*** This practice allows for a removed section to show details of an object without drawing complete views. In Figure 6-13, the partial section is labeled SECTION A-A and is removed to the side. It could have been drawn at a different scale to enlarge the view, if desired.

If a drawing is clear without complete section lines, the section lines can be segments drawn only along the visible object lines. This is called an ***outline section*** and is only used on large views with a lot of surface area to be section lined. See Figure 6-14. Large parts are often drawn with outline sections. The fewer lines makes the drawing more easy to read.

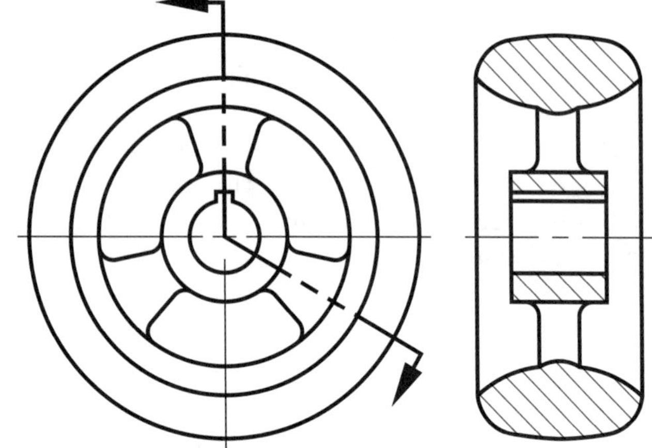

**Figure 6-12.**
It is conventional practice to leave spokes without section lines.

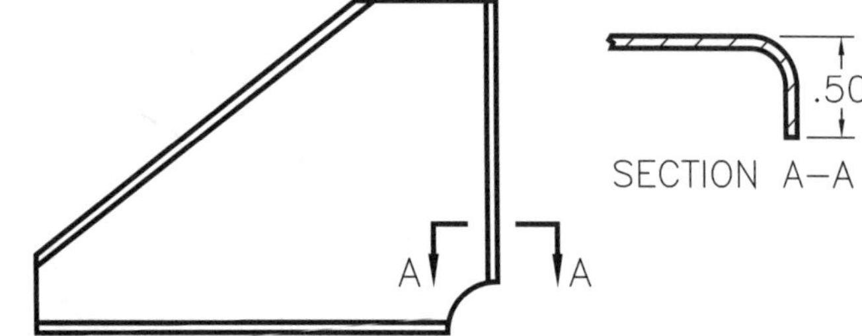


**Figure 6-13.**
A partial section view shows details of an object without drawing complete views.

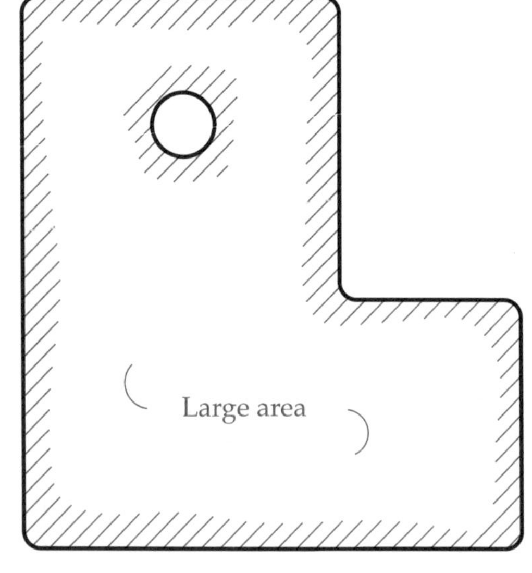


**Figure 6-14.**
The lines of an outline section are drawn only along the visible lines.

Auxiliary views are projected in directions other than the six normal directions. These views are discussed in the next unit. The principles of section views can logically be applied to auxiliary views when the cutting plane is oriented in an auxiliary view direction, Figure 6-15. Any type of section found on regular views—full, half, broken out, etc.—may also be found on auxiliary views. When an auxiliary view is sectioned and the portion of the view extending beyond the section is not fully shown, the view should follow the guidelines for any partial view of an object.

## Sections in Assembly Drawings

When multiple parts are shown assembled in a drawing, the drawing is called an assembly drawing. Assembly drawings are discussed in Unit 16. With respect to assembly drawings of multiple parts, the different parts are distinguished by orienting the section lines at different angles. The largest part may feature section lines that slope to the right, the second part may feature section lines that slope to the left, and then additional parts may have other angles or closer spacing, Figure 6-16.

Also note in Figure 6-16 the conventional practice of leaving fasteners and shafts uncut as the cutting plane passes through the assembly. Fasteners and shafts are usually more recognizable by their exterior features.

Figure 6-15.
An auxiliary view can also be a section view.

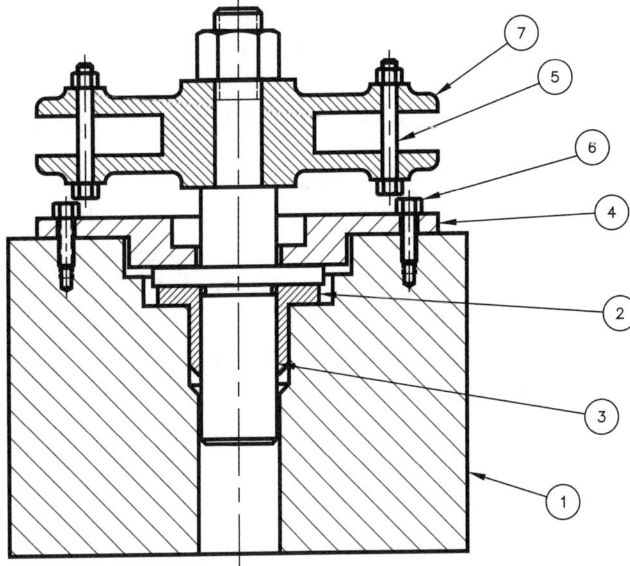


Figure 6-16.
Fasteners and shafts are not sectioned in assembly drawings. Each part should have a unique appearance to the section lines.

In assembly drawings, thin parts such as rubber inserts, gaskets, and thin plates, are often shown differently in section than other parts. As the linework in thin shapes is so close together, the shape and thickness of thin parts may simply appear shaded solid, Figure 6-17.

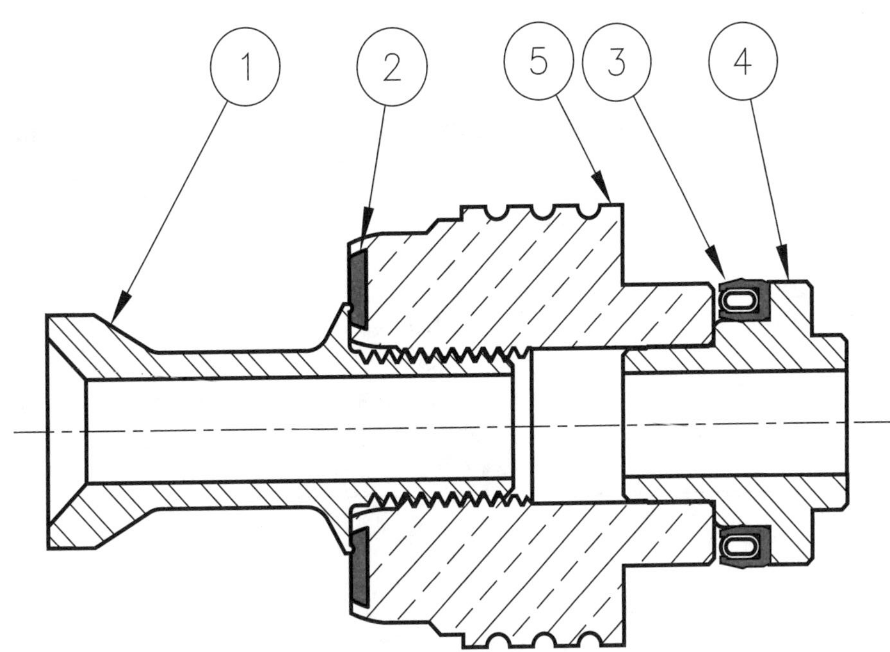


Figure 6-17.
The shape and thickness of thin parts in an assembly section view are simply shaded solid.

## Review Questions

*Circle the answer of choice, fill in the blank, or write a short answer.*

1. What two lines from the alphabet of lines are used exclusively in section view drawings?

_______________________________________________

2. *True or False?* Although section lines may be drawn in different patterns for different materials, they should always slope up to the right at 45°.

3. Of the following, which is *not* a type of section view featured in this unit?
   A. Full
   B. Half
   C. Quarter
   D. Offset

4. A half section view features a sectional view in combination with a regular view. What type of line should be used to separate these two views? _______________

_______________________________________________

5. An object with an odd number of features (spokes, holes, etc.) often has a(n) _______________ section so the odd features can be rotated around the axis into a vertical cutting plane, resulting in a view that is easier to read and more informative.

6. Which type of section view appears like an overlay directly on the regular view? _______________

_______________________________________________

7. The type of section view where the section view is no longer aligned in projection with other views is:
   A. removed.
   B. revolved.
   C. broken-out.
   D. offset.

8. List two specific features within a single object that, although cut by the cutting plane, usually will *not* be section lined, shaded, or hatched in the section view: _______________

_______________________________________________

9. *True or False?* If a section view has a large area of section lining, the section lines can just be drawn around the edges, which is known as an edge section.

10. List two specific items within a section view of several assembled objects that, although centered on the cutting plane, will be left *whole* in the section view: _______________

_______________________________________________

## Review Activity 6-1

### Section View Identification

Analyze each of the section views shown below. Identify the type of section view.

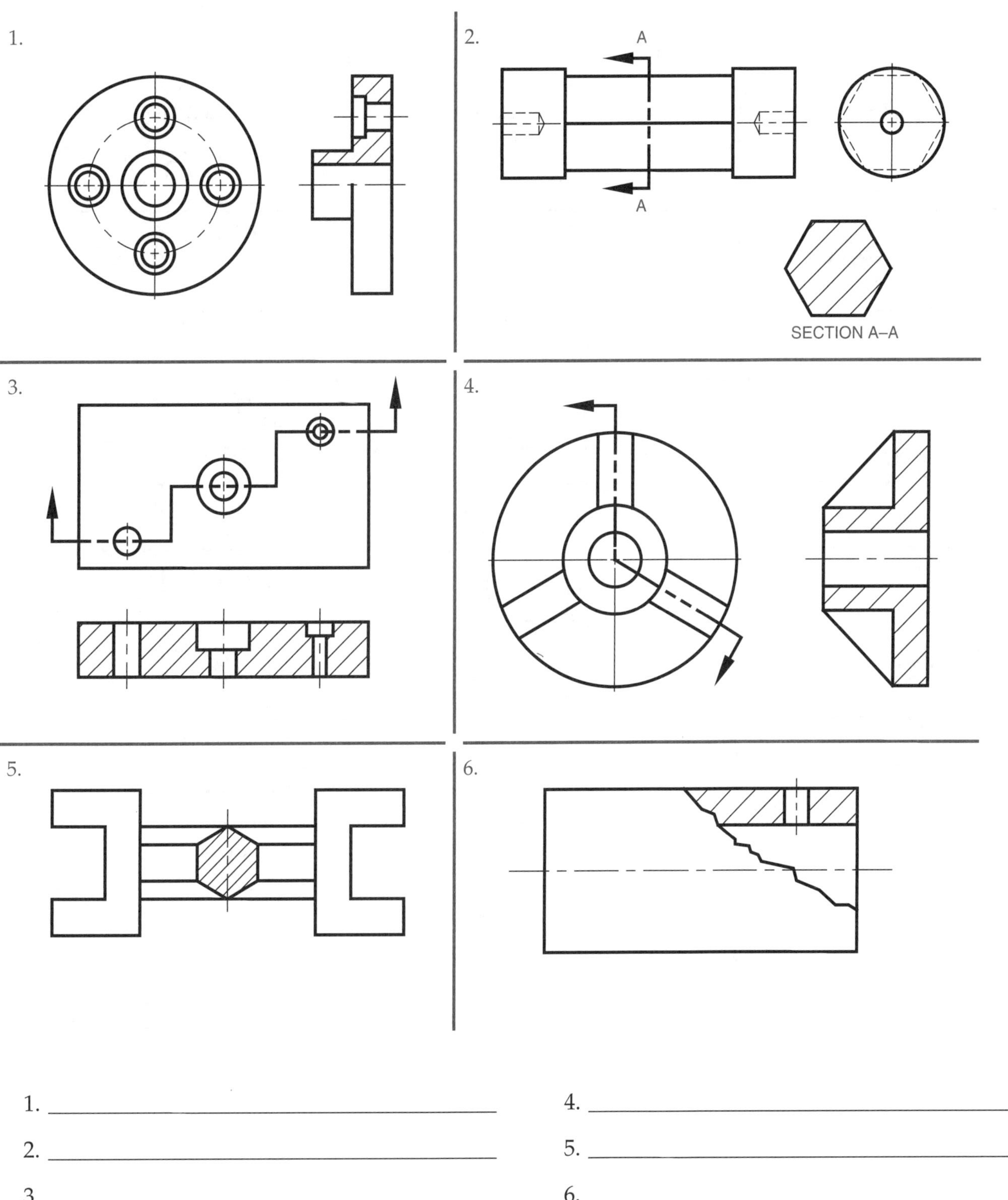


1. ______________________
2. ______________________
3. ______________________
4. ______________________
5. ______________________
6. ______________________

## Review Activity 6-2

### Section Views

For each of the numbers below, select the correctly drawn section view (A, B, C, or D). Place the letter of the correct section view in the blank provided.

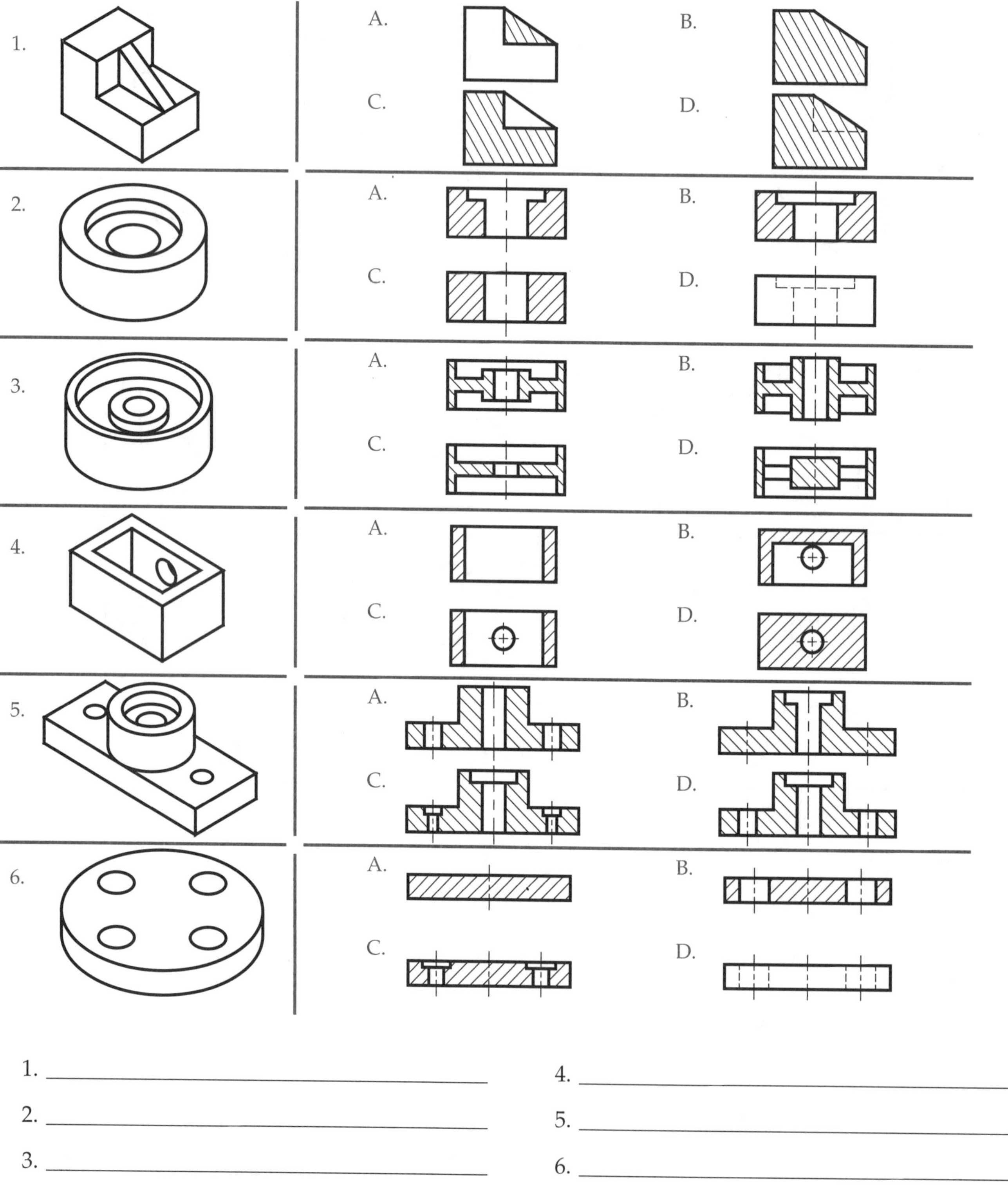


1. ______________________
2. ______________________
3. ______________________
4. ______________________
5. ______________________
6. ______________________

## Review Activity 6-3

### Section Views

For each of the numbers below, a front section view is missing. Place the letter of the correct section view in the blank provided. Note: To make these problems a little more challenging, center lines have been omitted!

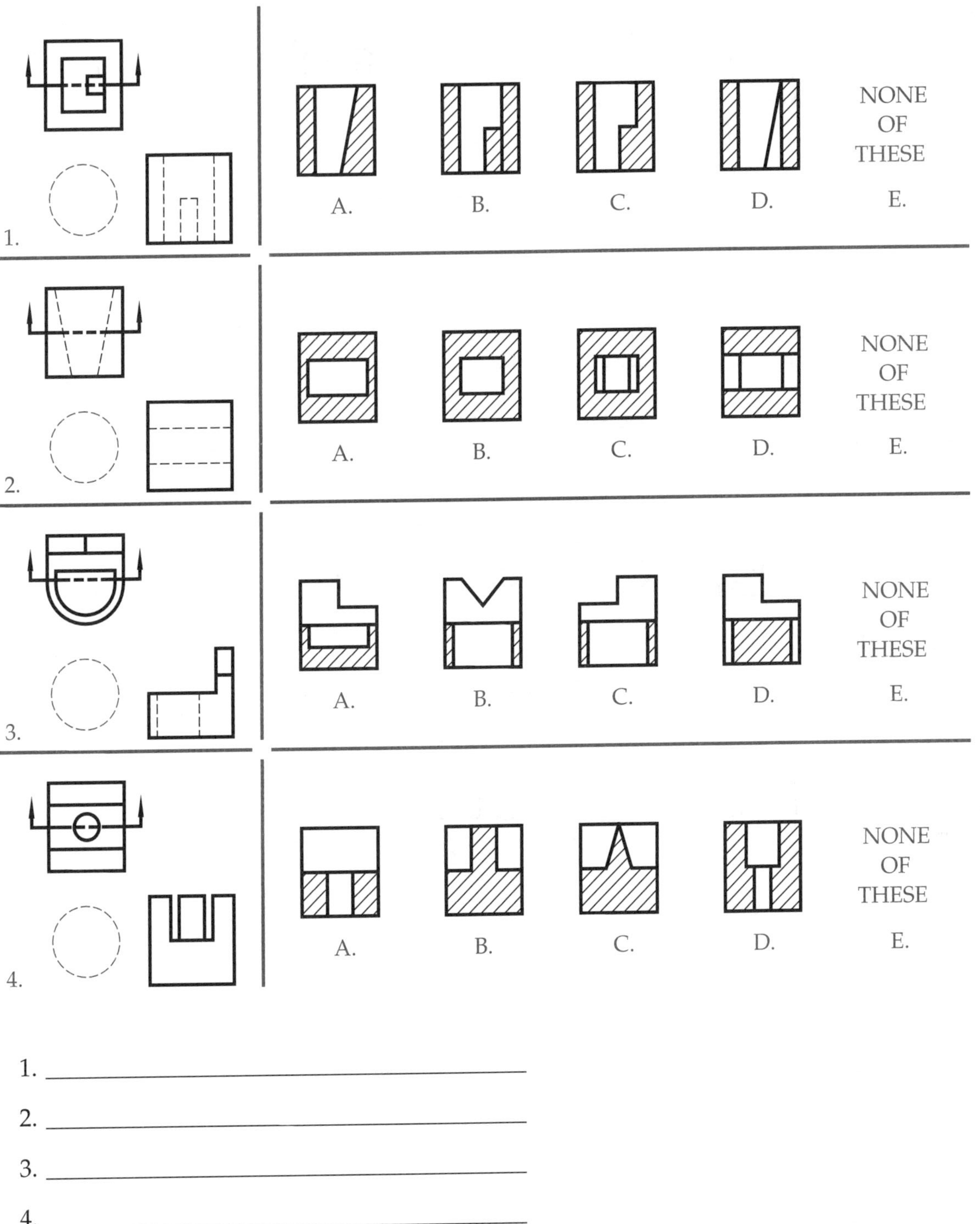


1. ______________________

2. ______________________

3. ______________________

4. ______________________

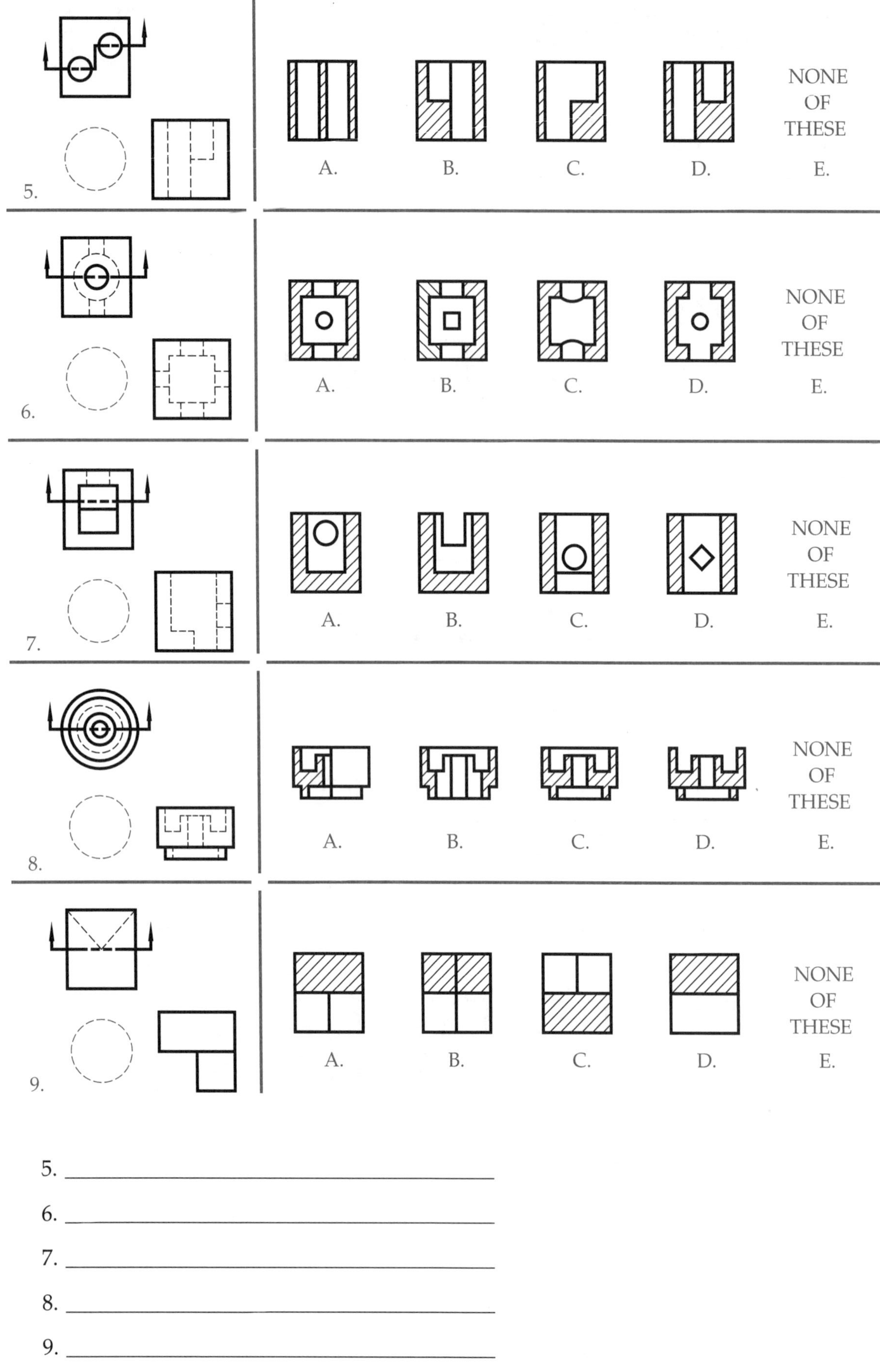
5.
A.
B.
C.
D.
NONE OF THESE
E.
6.
A.
B.
C.
D.
NONE OF THESE
E.
7.
A.
B.
C.
D.
NONE OF THESE
E.
8.
A.
B.
C.
D.
NONE OF THESE
E.
9.
A.
B.
C.
D.
NONE OF THESE
E.

5. ______________________

6. ______________________

7. ______________________

8. ______________________

9. ______________________

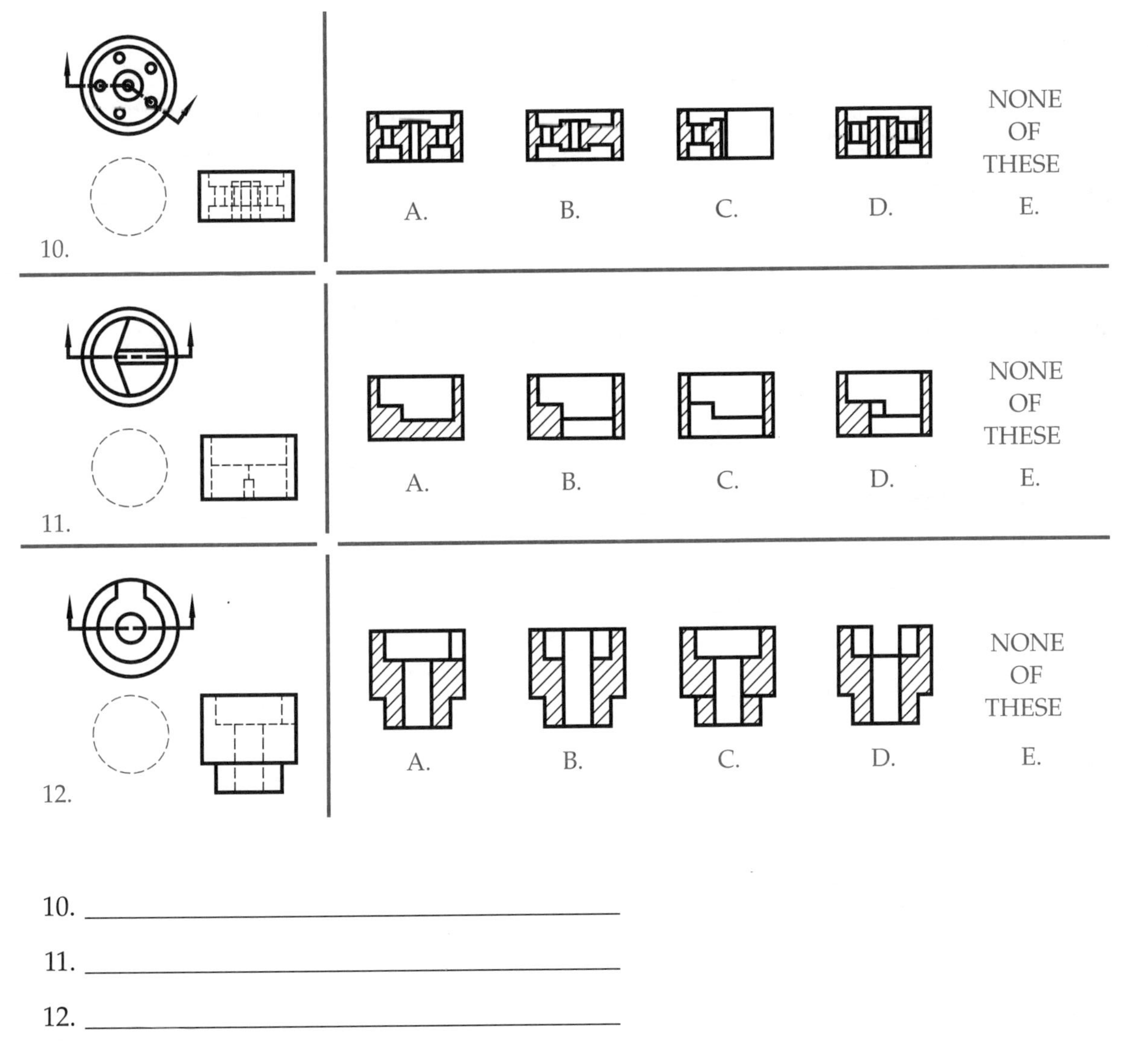
10.
A.
B.
C.
D.
NONE
OF
THESE
E.
11.
A.
B.
C.
D.
NONE
OF
THESE
E.
12.
A.
B.
C.
D.
NONE
OF
THESE
E.

10. ______________________________

11. ______________________________

12. ______________________________

## Industry Print Exercise 6-1

*Refer to the print PR 6-1 and answer the questions below.*

1. What type of section view is shown on this print? ____________
2. Is the cutting-plane line shown? ____________
3. Are there any hidden lines shown in the section view? ____________
4. The section line pattern indicates cast iron. Is this the material for the part? If not, what is the material? ____________
5. Does this drawing feature a top view? ____________

*Review questions based on previous units:*

6. What is the drawing number or part number? ____________
7. What is the name of this part? ____________
8. Was this drawing checked by someone? ____________
9. Does this drawing have center lines? ____________
10. According to the note in zone E6, in what state is the supplier located? ____________
11. On what size paper was the original drawing created? ____________
12. On what part number(s) will this part be used? ____________
13. At what scale was the original drawing created? ____________
14. In the section view, the leftmost vertical object line has what meaning? ____________
15. In the section view, what numerical dimension value is given for the overall depth? ____________

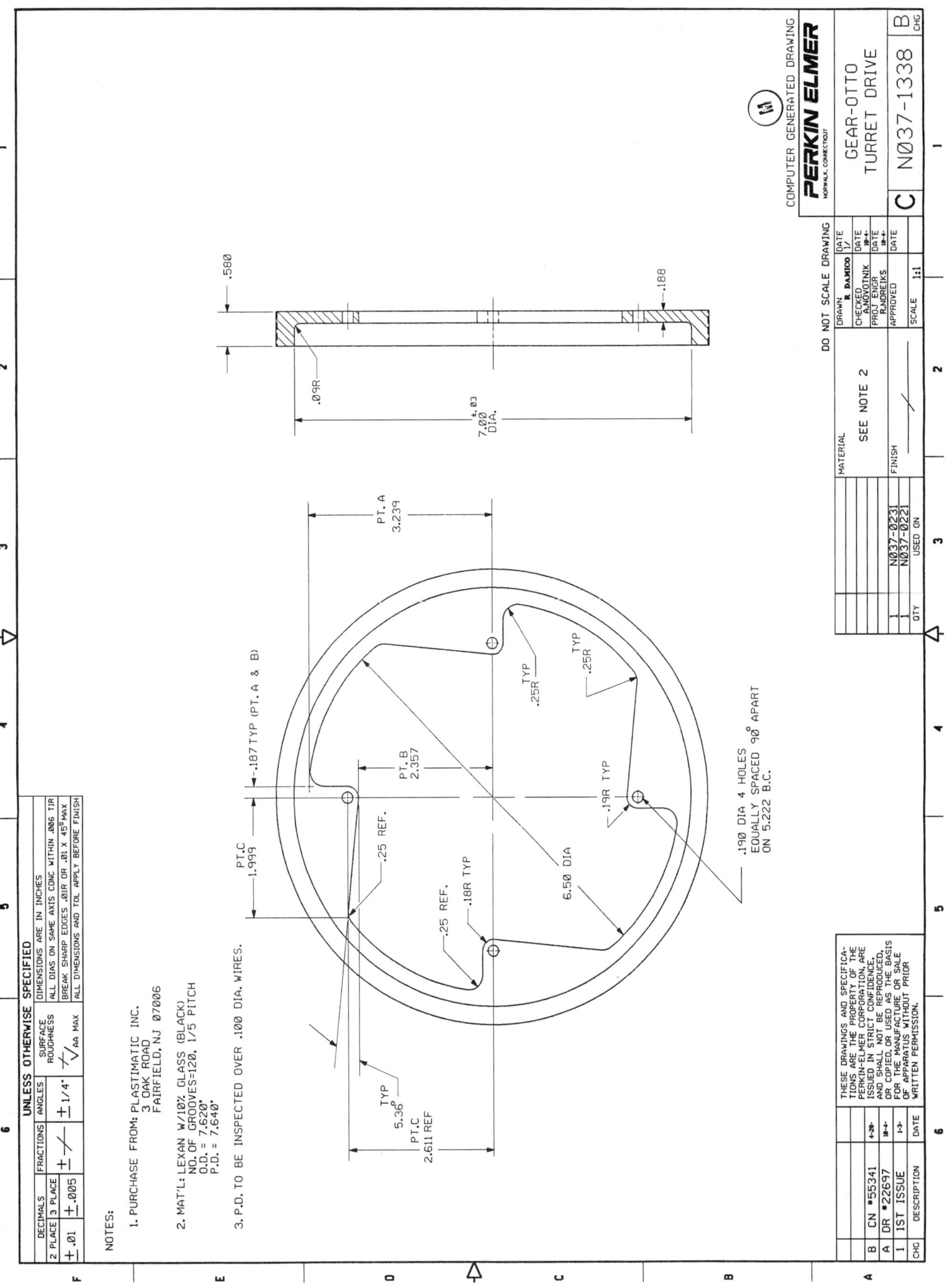


PR 6-1.
Print supplied by Perkin Elmer.

## Industry Print Exercise 6-2

*Refer to the print PR 6-2 and answer the questions below.*

1. If the main section view is considered the front view, what term would describe the round view on the left side of the drawing? ______
2. Is the cutting-plane line shown? ______
3. Does the section line pattern shown in the section view match the type of material specified in the title block? Refer to Figure 6-2 in the unit. ______
4. What term could be applied to the section in Detail A? ______
5. What type of section view is represented by this drawing, full or half? ______

*Review questions based on previous units:*

6. What is the drawing number or part number? ______
7. What is the name of this part? ______
8. Was this drawing checked by someone? ______
9. Does this drawing have center lines? ______
10. In what state is the company that owns this print? ______
11. What is the name of the jagged line in the view labeled Detail A? ______
12. Besides a cylinder, what other 3D geometric form is featured in the body shape of this object? ______
13. At what scale was the original drawing created? ______
14. In the section view, the uppermost horizontal object line has what meaning? ______
15. In the round view, is the smallest visible circle an internal feature (hole) or an external feature (post)? ______

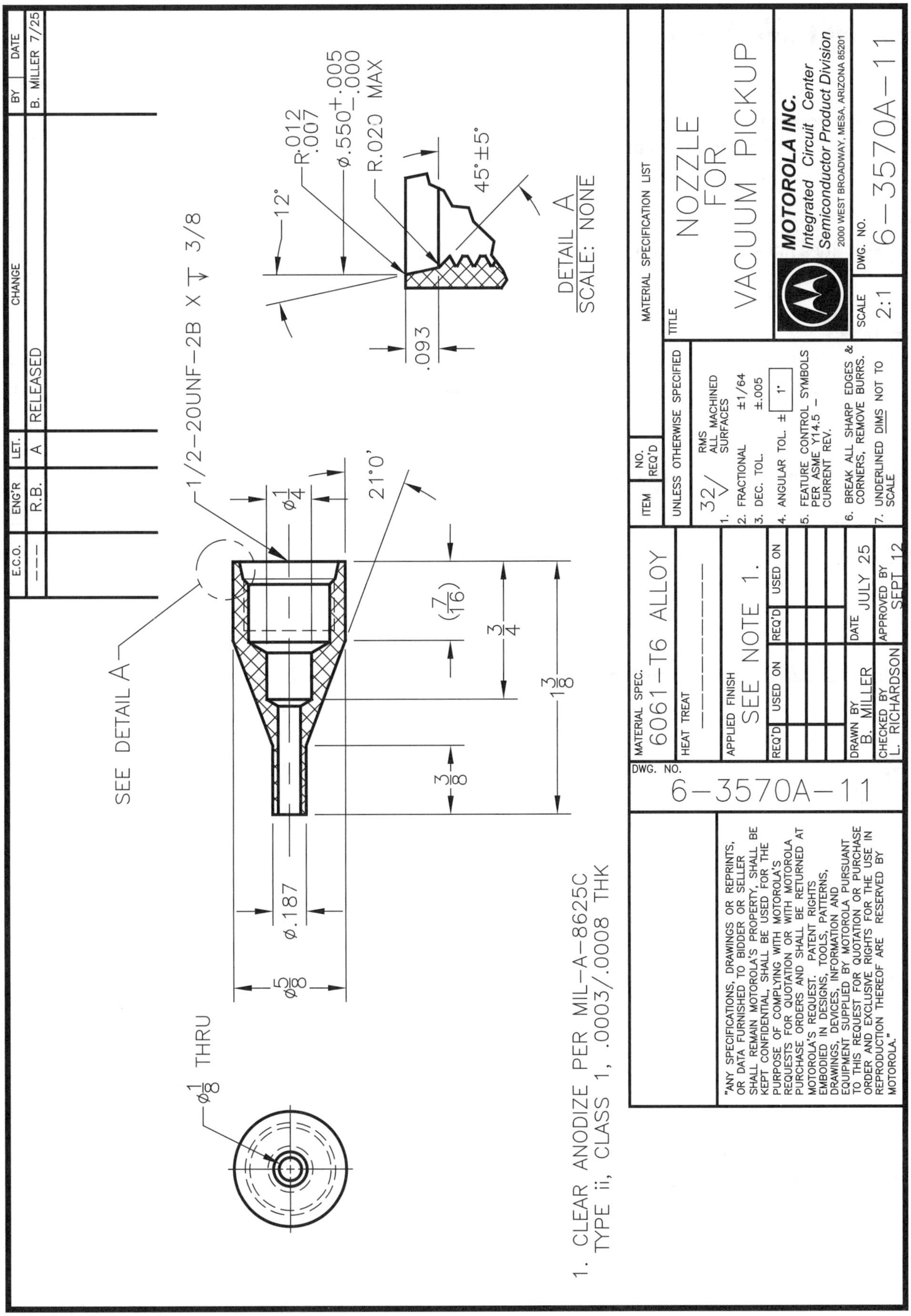

| E.C.O. | ENG'R | LET. | CHANGE | BY | DATE |
|---|---|---|---|---|---|
| --- | R.B. | A | RELEASED | B. MILLER | 7/25 |

SEE DETAIL A

1/2–20UNF–2B X ⟂ 3/8

ϕ1/4

21°0'

(7/16)

3/4

1 3/8

3/8

ϕ.187

ϕ5/8

ϕ1/8 THRU

12°

R.012/.007

ϕ.550+.005/–.000

R.020 MAX

.093

45°±5°

DETAIL A
SCALE: NONE

1. CLEAR ANODIZE PER MIL–A–8625C
TYPE ii, CLASS 1, .0003/.0008 THK

| ITEM | NO. REQ'D | MATERIAL SPECIFICATION LIST |
|---|---|---|

MATERIAL SPEC.: 6061–T6 ALLOY

HEAT TREAT: ———————

APPLIED FINISH: SEE NOTE 1.

| REQ'D | USED ON | REQ'D | USED ON |
|---|---|---|---|
| | | | |

DRAWN BY: B. MILLER
DATE: JULY 25
CHECKED BY: L. RICHARDSON
APPROVED BY: SEPT 12

UNLESS OTHERWISE SPECIFIED
1. 32 RMS ALL MACHINED SURFACES
2. FRACTIONAL ±1/64
3. DEC. TOL. ±.005
4. ANGULAR TOL. ± 1°
5. FEATURE CONTROL SYMBOLS PER ASME Y14.5 – CURRENT REV.
6. BREAK ALL SHARP EDGES & CORNERS, REMOVE BURRS.
7. UNDERLINED DIMS NOT TO SCALE

TITLE: NOZZLE FOR VACUUM PICKUP

MOTOROLA INC.
Integrated Circuit Center
Semiconductor Product Division
2000 WEST BROADWAY, MESA, ARIZONA 85201

SCALE: 2:1
DWG. NO.: 6–3570A–11

DWG. NO. 6–3570A–11

"ANY SPECIFICATIONS, DRAWINGS OR REPRINTS, OR DATA FURNISHED TO BIDDER OR SELLER SHALL REMAIN MOTOROLA'S PROPERTY, SHALL BE KEPT CONFIDENTIAL, SHALL BE USED FOR THE PURPOSE OF COMPLYING WITH MOTOROLA'S REQUESTS FOR QUOTATION OR WITH MOTOROLA PURCHASE ORDERS AND SHALL BE RETURNED AT MOTOROLA'S REQUEST. PATENT RIGHTS EMBODIED IN DESIGNS, TOOLS, PATTERNS, DRAWINGS, DEVICES, INFORMATION AND EQUIPMENT SUPPLIED BY MOTOROLA PURSUANT TO THIS REQUEST FOR QUOTATION OR PURCHASE ORDER AND EXCLUSIVE RIGHTS FOR THE USE IN REPRODUCTION THEREOF ARE RESERVED BY MOTOROLA."

**PR 6-2.**
Print supplied by Motorola, Inc.

## Bonus Print Reading Exercises

The following questions are based on various bonus prints located in the folder at the back of this textbook. Refer to the print indicated, evaluate the print, and answer the question.

**Print AP-001:**

1. This print does not show a cutting-plane line. Is this acceptable practice or an error? ______

**Print AP-002:**

2. When looking at Section A-A, how does the print reader know that this is several parts or pieces and not just one part? ______
3. Most of the sections on this print are full sections. What other method of sectioning is used? ______

**Print AP-003:**

4. What type of section is shown in Section B? ______
5. Section B uses a center line between the left half and the right half. Is this recommended practice? ______

**Print AP-004:**

6. Based on the shape of the cutting-plane line, what type of section is Section B-B? ______
7. Except for Section B-B, what other type of section view is used on the main views? ______

**Print AP-007:**

8. What material is being specified by this section line pattern? (Refer to Figure 6-2.) ______
9. How many cutting-plane lines are shown on this print? ______

**Print AP-010:**

10. What type of section describes this one-view drawing? ______

**Print AP-011:**

11. This print features a section view labeled Section A-A. In this particular case, is this label necessary?

_______________

**Print AP-012:**

12. How many removed sections would you expect to find on the other sheets of this detail drawing?

_______________

**Print AP-013:**

13. What type of view is Section D-D? _______________

_______________

**Print AP-015:**

14. What type of section is featured on the front view of this object? _______________

_______________

**Print AP-016:**

15. Which two parts located on the central cutting plane did *not* get sectioned? _______________

_______________

16. Are there any section lines on this part that are material specific? _______________

_______________

**Print AP-019:**

17. What type of section is featured to show the hole on the worm gear end of this part? _______________

_______________

**Print AP-021:**

18. Besides a full section, what other type of section is used on this print? _______________

_______________

This object has a surface that does not align with any of the standard orthographic views. An auxiliary view must be drawn to represent that surface in true size and shape.

# UNIT 7
# Auxiliary Views

*After completing this unit, you will be able to:*

**Explain** the purposes of auxiliary views.

**Identify** auxiliary views and **explain** their relationship to regular views.

**Discuss** terms related to auxiliary view projection.

**Read** prints that incorporate auxiliary views.

Drafters sometimes find it necessary to use auxiliary views to fully describe an object. ***Auxiliary views*** are projected in an auxiliary direction to show the true shape and size of features that cannot be shown in the "normal" views. In a previous unit, six "normal" views were identified: the front, top, right side, left side, rear, and bottom. Simply put, any other view is an auxiliary view.

## Auxiliary View Principles and Purposes

As explained in Unit 5, an inclined surface will not project true size and shape in any of the normal views. The block in Figure 7-1 features an inclined surface 1-2-3-4. In the front view, the edge view of the surface is featured because the surface is perpendicular to the *frontal* plane. The two other

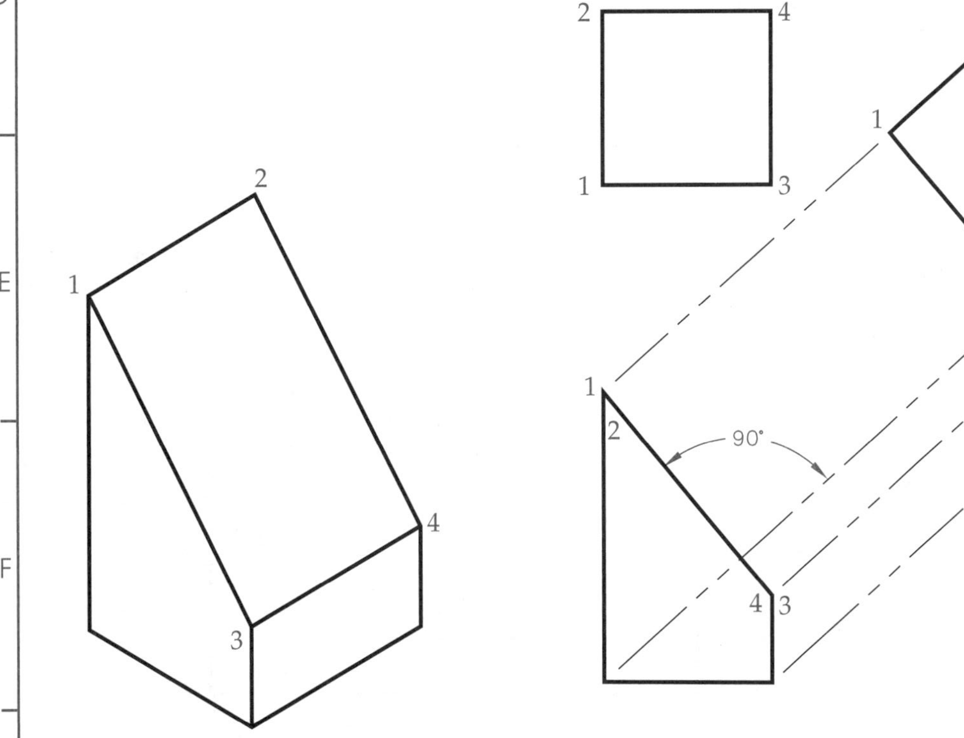


**Figure 7-1.**
In this case, an auxiliary view is needed to show the true shape and size of the inclined surface.

regular views show a foreshortened shape of surface 1-2-3-4 because the surface is inclined to the *horizontal* and *profile* planes.

An auxiliary view of the surface can be projected onto an auxiliary projection plane strategically oriented to create a view that will show the true size and shape of the surface. Applying orthographic principles, if a plane of projection is parallel to the inclined surface and the surface is projected with parallel projectors perpendicular to the projection plane, the result will be true size and shape. This is a common use of the auxiliary view.

As illustrated in Figure 7-2, the auxiliary plane is "hinged" to the frontal plane of projection and swung around into alignment with the other planes. While not standard practice, the folding lines between the views may be present as reference lines. If so, they should be phantom lines and may be marked with projection plane labels. Reference lines such as these are more common when reading the standards or drafting texts, Figure 7-3. Accordingly, it is helpful in visualizing the spatial relationship between two views if you think of the phantom reference line as a 90° bend between two adjacent planes.

Another purpose for an auxiliary view would be to create a view that shows the true angle between two surfaces, known more technically as the ***dihedral*** angle. The dihedral angle between two planes is shown true only in a view that features the point view of the edge of intersection between the two surfaces. In cases where a V-groove runs in a normal direction, there is likely a normal view that shows the true angle. In Figure 7-4, the V-groove is inclined to the normal directions of sight, so an auxiliary view is required to show the true angle between the two surfaces of the groove.

## Visualizing Auxiliary Views

Visualizing auxiliary views should not be that much different than visualizing regular views, but the angles do seem to play tricks on the mind. As with regular views, learning how to visualize the views comes with practice. One technique is to picture the object being turned while your line of sight remains fixed. Another technique is to picture the object remaining fixed, while you move yourself around the object to view it from different vantage points. With auxiliary views, you can imagine that your eye is located in such a way as to squarely look at the inclined surface or to look squarely at the point view of an inclined V-groove.

As described above and in earlier units, views of an object are related to each other and should be lined up with each other in ways that help the print reader visualize the part. Remember, any view adjacent to another view is like turning the object 90°. In the case of auxiliary views, the turn axis is not vertical or horizontal, but rather at an angle. To help with this, sometimes a print reader can simply rotate the print when reading an auxiliary

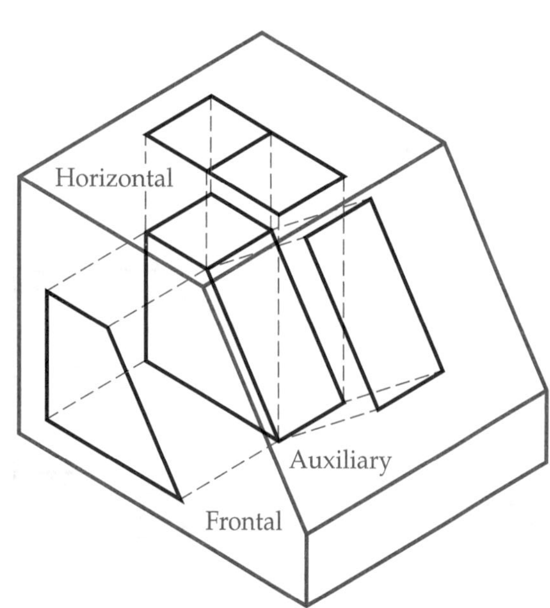


**Figure 7-2.**
If a plane of projection is placed parallel to an inclined surface, the projection onto that plane is true size and shape.

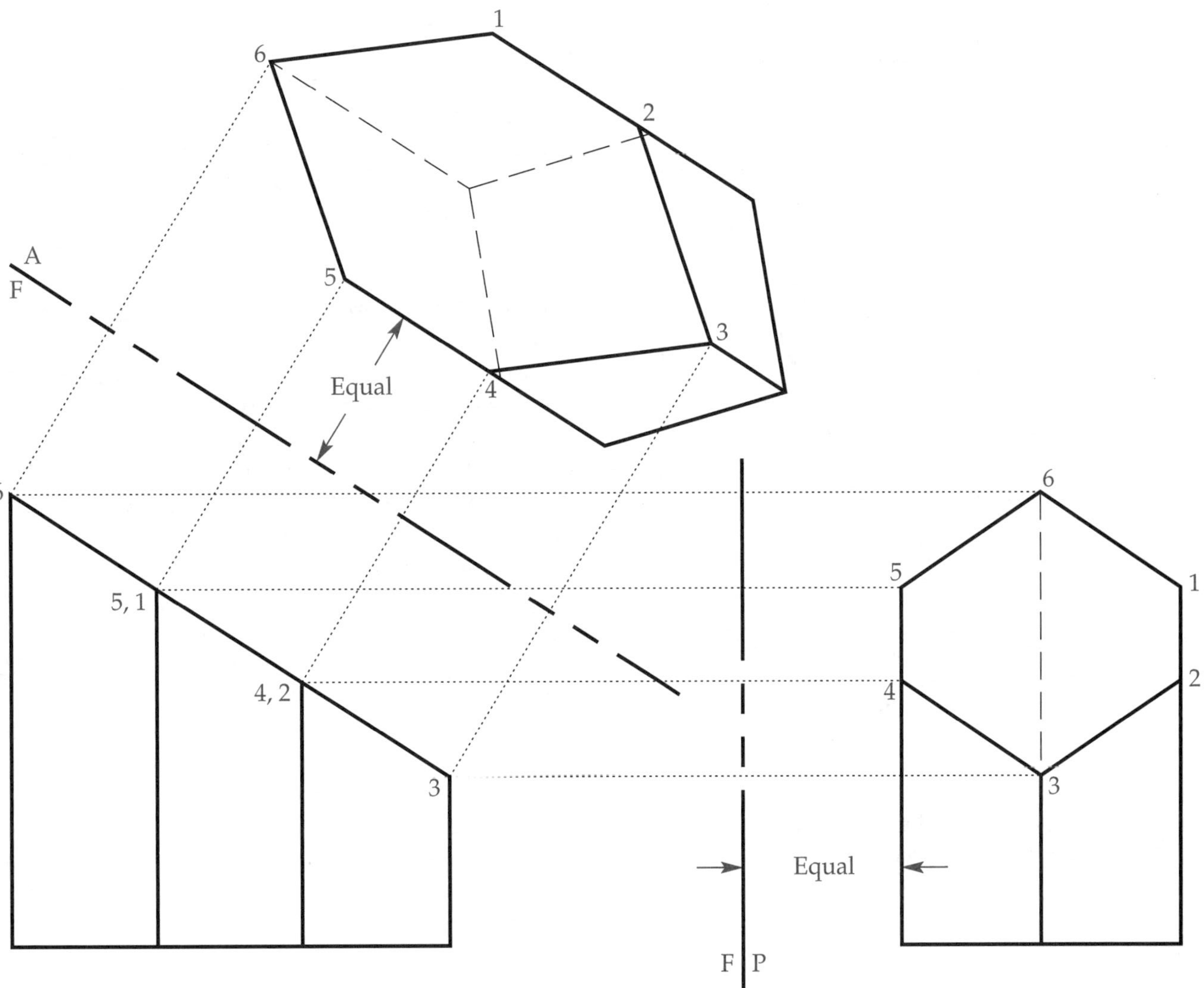


**Figure 7-3.**
Reference lines may be indicated in some textbooks or standards. These are not usually part of an industry print. A = auxiliary, F = frontal, and P = profile. These lines represent the hinge line of the projection planes. Distances from the reference lines to the views are equal as indicated.

view so the lines of projection between the normal view and auxiliary view are vertical or horizontal, Figure 7-5. This technique is helpful for those who are accustomed to reading regular views.

In the case of inclined surfaces, the auxiliary views may be projected from any view in which the inclined surface appears as a line. Depending on the angle, auxiliary views can be difficult to locate on the sheet. The complexity of the part, the scale of the drawing, and available paper space all present challenges in locating the auxiliary view. As with regular views, viewing-plane lines can also be used. Auxiliary views can be placed out of projection, but still turned and tilted in the auxiliary view orientation dictated by the viewing-plane line, Figure 7-6.

## Auxiliary View Types

Auxiliary views should be projected and aligned with regular views. They are not labeled as auxiliary views, just as front, top, and right side views are not labeled. In drafting textbooks, and perhaps in industry standard manuals, there are terms used to help explain the auxiliary views and the system of projection that allows them to be properly constructed.

With that in mind, an auxiliary view projected from a regular view is called a ***primary*** auxiliary view. In some cases, only the inclined or special feature is shown, which can be achieved with a partial primary auxiliary view. Figure 7-7 shows a partial primary auxiliary view projected from the front view to provide a true representation of

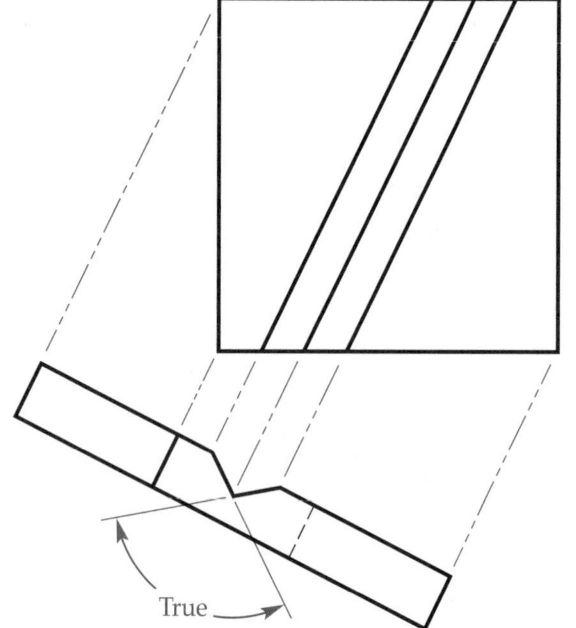


**Figure 7-4.**
An auxiliary view projected to show the point view of the intersection edge is needed to show the true dihedral angle of the groove.

the inclined, rectangular feature of the object. The location dimensions for the hole pattern and the size of this inclined surface are foreshortened in the regular top and right side views. The partial primary auxiliary view provides not only a true size and shape description of that portion of the object, but provides the best view in which to locate the dimensions.

Some textbooks define auxiliary views by the primary dimension that is featured in the auxiliary view. For example, in multiview projection, the front view of an object does not feature or show the depth dimension of the object, only width and height. By principles of projection, *any* view projected from the front view will show depth (front-to-back dimension) because it is a 90° turn of that view. So, just as the right side view and top view feature the depth dimension of an object, any auxiliary view projected from a front view also features the true depth of the object. These auxiliary views could be identified as ***depth auxiliary views.*** When manually constructing depth auxiliary views, one dimension is constructed by projection from the front view, but the depth measurement is obtained from some other view projected from the front view, such as the top or a side view.

Under this system of defining auxiliary views, an auxiliary view projected from a top view is called a ***height auxiliary view.*** Any auxiliary view projected from a right side view is called a ***width auxiliary view***. In summary, these terms are not likely to be used on a print, but do help describe

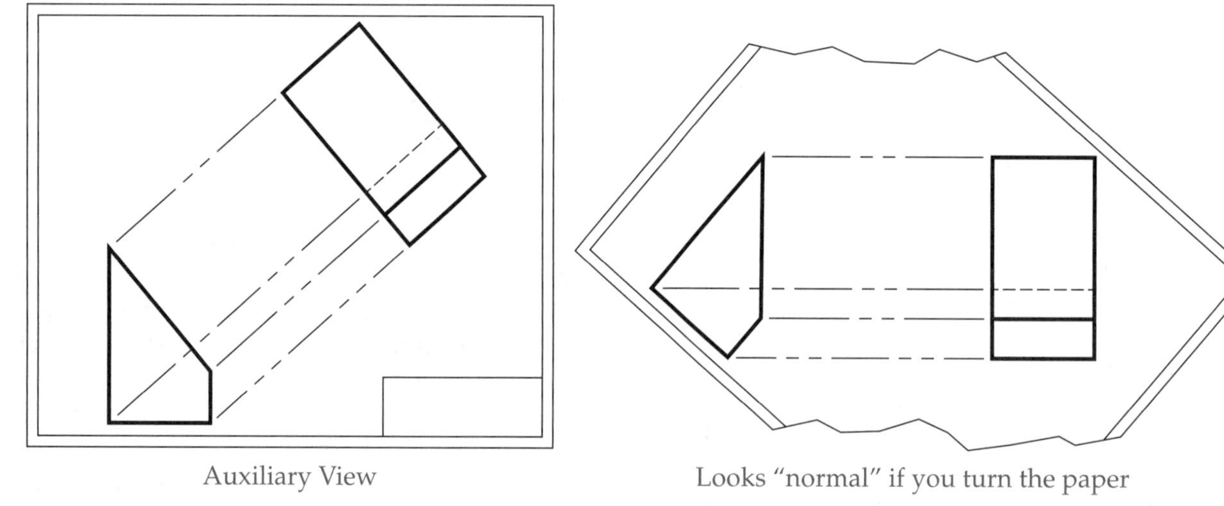


**Figure 7-5.**
Rotating the print when reading an auxiliary view can help the reader visualize the view.

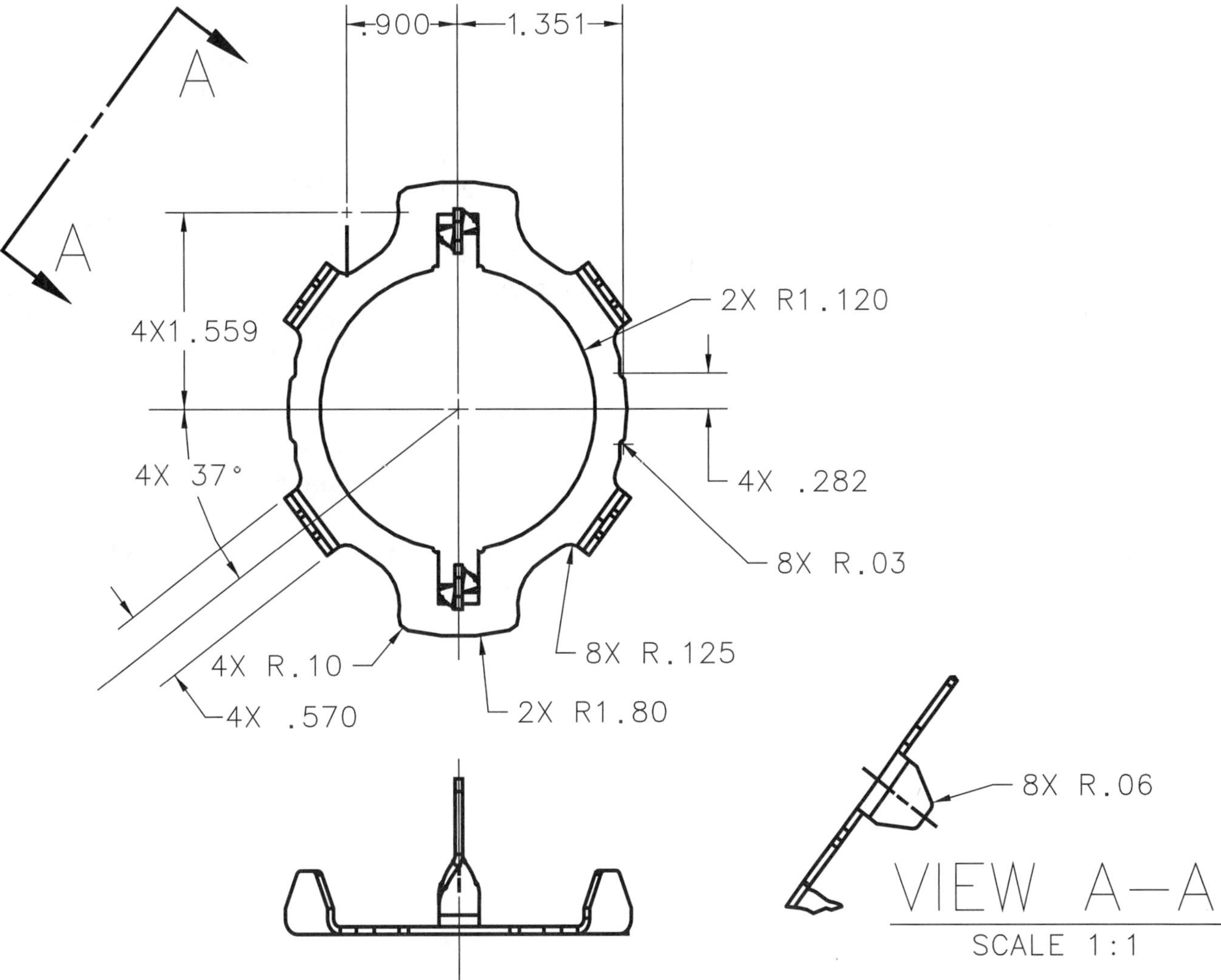


**Figure 7-6.**
This auxiliary view was moved out of projection, but still maintains the auxiliary angle represented by the viewing-plane line.

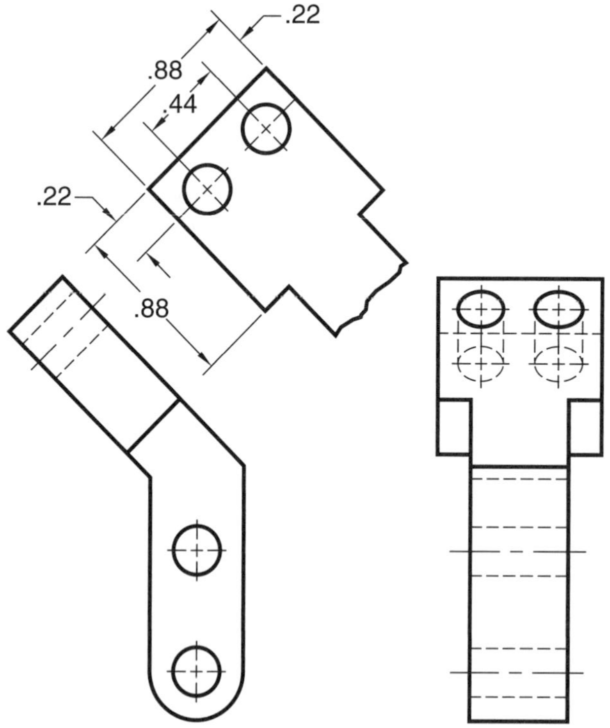

Figure 7-7.
The partial auxiliary view provides a true representation of the inclined feature of this object.

the system of auxiliary view projection. An understanding of these terms may help the print reader determine the dimensional values and how they apply to the object.

A primary auxiliary view is projected from a normal view that shows the edge view of an inclined surface. If the surface is oblique, then not only is it inclined, but rotated. Creating a view that shows the true size and shape of an oblique surface first requires the drafter to create a primary auxiliary view that shows the oblique surface as a line, or edge view. After this, a ***secondary auxiliary view*** can be projected from the primary auxiliary view that shows the true size and shape of the oblique surface. See Figure 7-8.

In recent years, CAD systems offer the ability to create views from any direction directly from the 3D model. While this requires less visualization and drafting skill for the person creating the drawing than in the days of manual drafting, it is still critical that the views are properly arranged and the drawing or print still follows standard practices. As new technologies evolve, be ready to see drawings that might be impacted by automation. As 3D modeling becomes more and more prevalent and 3D documents such as PDF and DWF become more sophisticated, the definition of an auxiliary view may change.

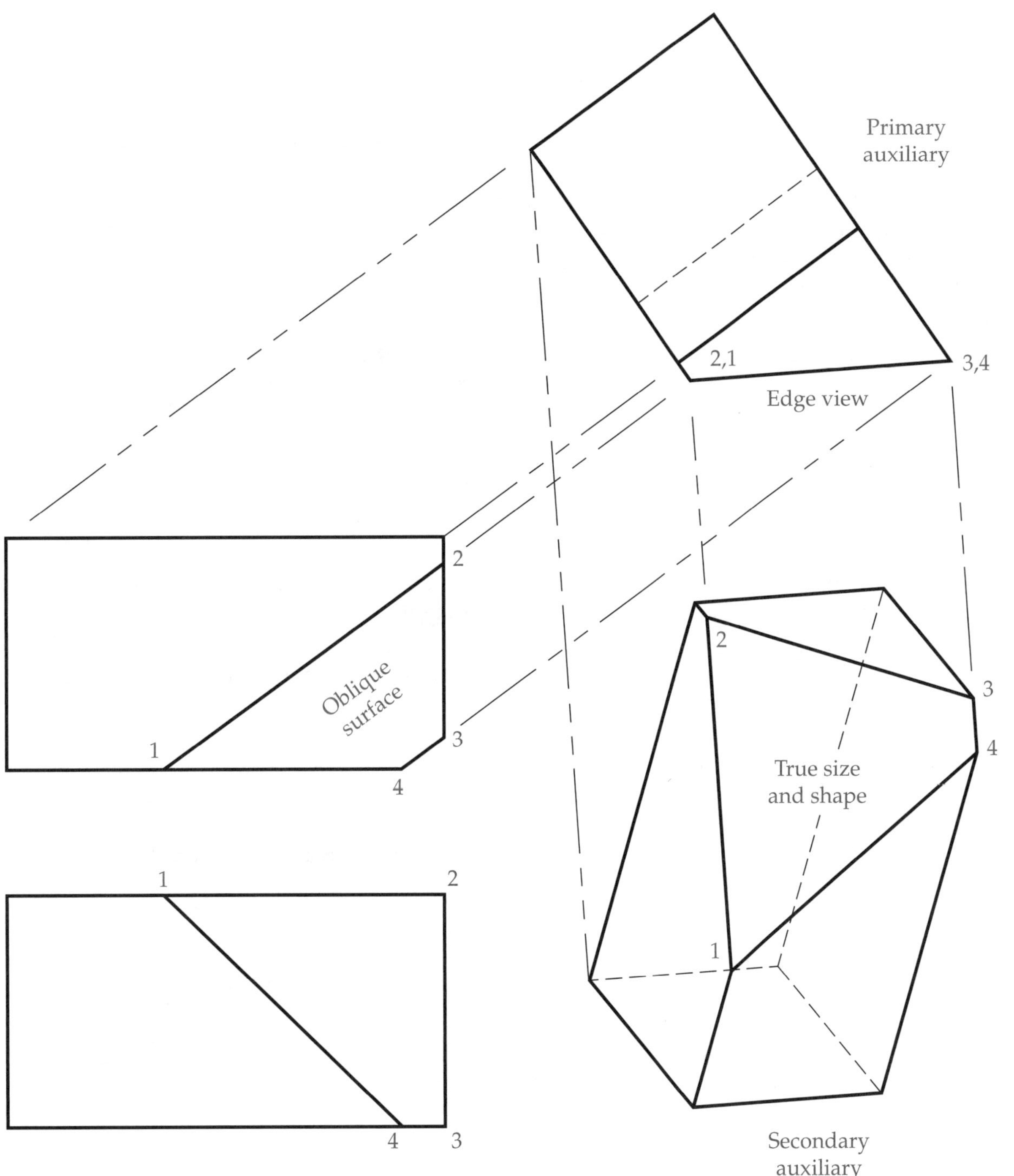


**Figure 7-8.**
Secondary auxiliary views are necessary to show the true size and shape of an oblique surface.

## Review Questions

*Circle the answer of choice, fill in the blank, or write a short answer.*

1. What are the six normal views of an object? ______________________________

______________________________________________

2. *True or False?* There is an unlimited number of possible directions for an auxiliary view, depending on the angle of features within an object.

3. Of the following, which is a common reason to use an auxiliary view?
   A. To show a turned and tilted view of a normal surface.
   B. To show the true size and shape of an inclined surface.
   C. To show the object as if it has been cut in half.
   D. When the title block gets in the way of a normal right side view.

4. A technical document explaining auxiliary views might feature a drawing that shows the ____________ line between an auxiliary plane and a frontal plane. The appearance of this reference line is the same as a phantom line.

5. What term is given to the true angle between two surfaces? ____________________

______________________________________________

6. When a drafter projects an auxiliary view from a normal view in such a way as to show the true size and shape of the inclined surface, the projection lines are ____________ to the inclined surface.

7. If there is not room to show an auxiliary view projected in alignment with a regular view, then:
   A. use a normal view.
   B. use a secondary auxiliary.
   C. use a section view instead.
   D. a viewing-plane line can be used.

8. What term is applied to an auxiliary view projected directly from a normal view? ____________

______________________________________________

9. *True or False?* All views projected from the front view feature the width dimension, so auxiliary views projected from the front view are sometimes referred to as width auxiliary views.

10. To have a view that features the true size and shape of an oblique surface, a(n) ____________ auxiliary view must be projected from a primary auxiliary view.

## Review Activity 7-1

*Examine the case study that follows. Given a complete front view, a partial top view, and two partial auxiliary views, answer the following questions.*

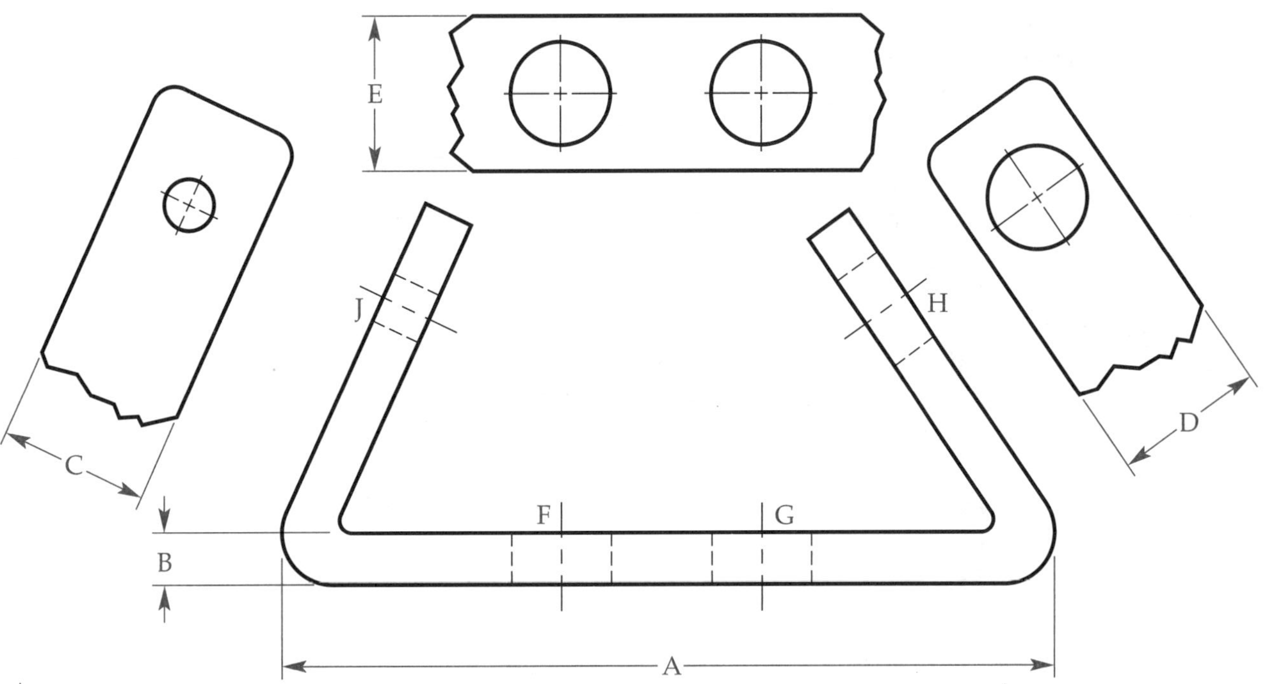


1. Which of the three dimension terms—height, width, or depth—are represented by each of the following dimensions?
   A. ______________
   B. ______________
   C. ______________
   D. ______________
   E. ______________

2. If a normal right side view would have been shown to the right of the front view, what geometric shape would describe the appearance of hole H? ______________

*For questions 3 through 10, place a check next to T for top, F for front, and/or A for auxiliary to indicate which view(s) meets the stated criteria.*

3. Shows the true height of hole F:
   T ____ F ____ A ____

4. "Looks" straight through hole H:
   T ____ F ____ A ____

5. "Looks" straight through hole G:
   T ____ F ____ A ____

6. Shows the axis-to-axis distance from hole F to hole G:
   T ____ F ____ A ____

7. Shows a "depth" location for hole J:
   T ____ F ____ A ____

8. Shows the true shape of the inclined surfaces:
   T ____ F ____ A ____

9. Shows the true angle between the inclined surfaces and the horizontal surfaces:
   T ____ F ____ A ____

10. Shows the "depth" location difference between hole F and hole G, if any:
    T ____ F ____ A ____

## Review Activity 7-2

*Examine the case study that follows. This drawing features some threaded holes and machining specifications that have not yet been covered, but is a good example for this unit. Answer the following questions.*

1. If the section view of this drawing is the front view, what other three views are shown? Include the word *partial* if applicable. ______________________________

______________________________

2. *True or False?* There are some hidden lines that were omitted from the right side view.

3. Of the following, which is a true statement about the auxiliary view?
   A. The view was projected from the right side view.
   B. The view shows the true size and shape of an inclined surface.
   C. The main purpose of the view is to show a true dihedral angle.
   D. The 3.125″ dimension is *not* a true dimension.

4. If the auxiliary view were to be labeled by the dimension it features that is *not* featured in the front view, what dimension is it? ______________________________

______________________________

5. How does the shape of the inclined hole appear in the right side view? ______________

______________________________

6. The auxiliary view is a partial view, as indicated by the ______________ line that is jagged and irregular.

7. If the drawing page is turned in such a way so the longest center line of the auxiliary view is horizontal, the front view is still the front, but the auxiliary view could be called the right side view. In this case, what would the view formerly the right side view be called?
   A. Secondary auxiliary view.
   B. Primary auxiliary view.
   C. Left side view.
   D. Partial width auxiliary view.

8. What term is given to an auxiliary view projected directly from a normal view? ______________

______________________________

9. *True or False?* The auxiliary view is oriented to provide the viewer with a "point view" of the axis of the .453″ diameter hole.

10. List the different lines from the alphabet of lines featured in the auxiliary view: ______________

______________________________

______________________________

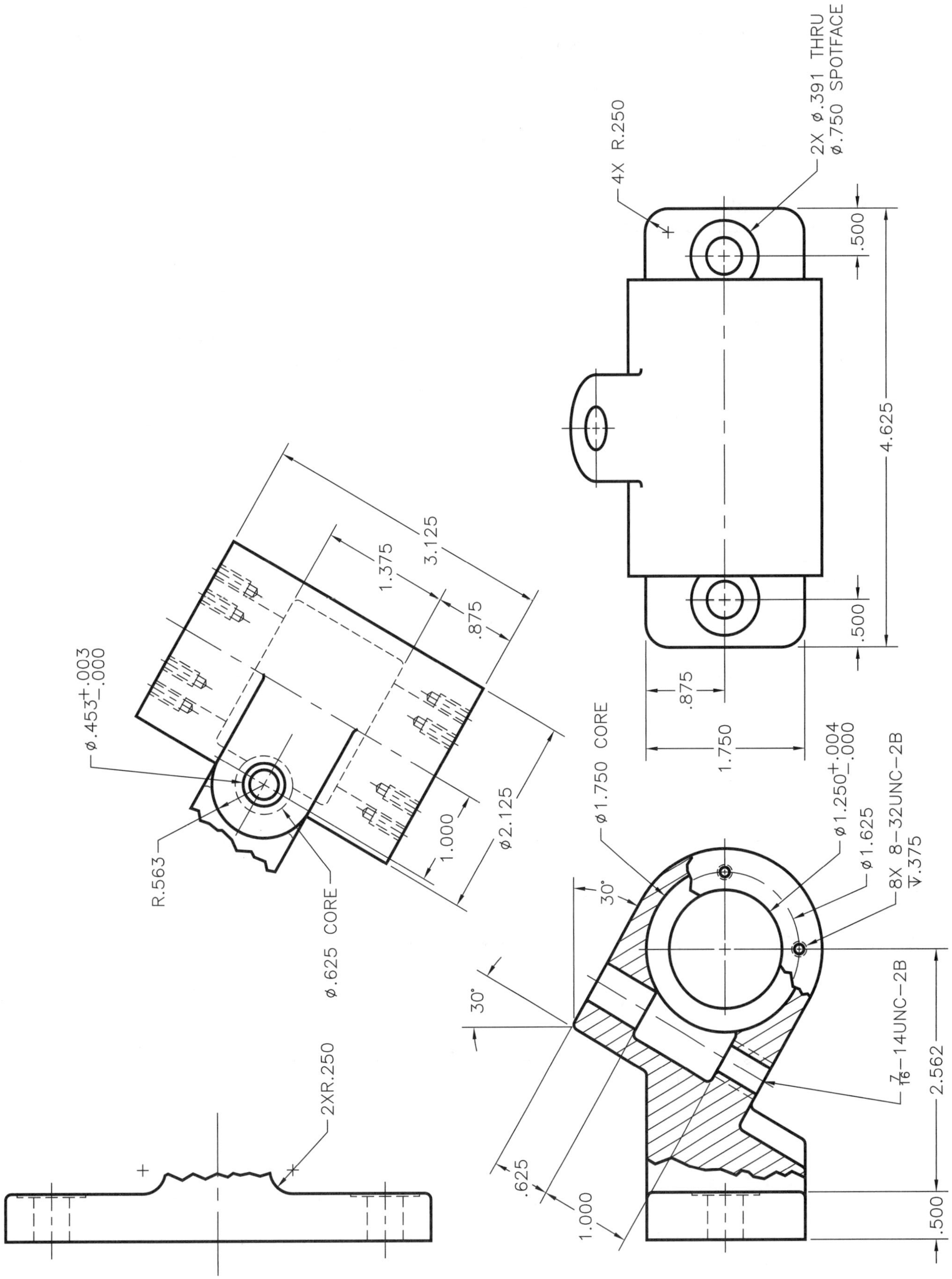

ø.453 +.003 −.000
R.563
2XR.250
ø.625 CORE
1.375
3.125
1.000
.875
ø2.125
30°
.625
1.000
30°
ø1.750 CORE
4X R.250
.875
1.750
2X ø.391 THRU
ø.750 SPOTFACE
ø1.250 +.004 −.000
ø1.625
$\frac{7}{16}$−14UNC−2B
8X 8−32UNC−2B
↧.375
.500
.500
4.625
.500
2.562

## Industry Print Exercise 7-1

*Refer to the print PR 7-1 and answer the questions below.*

1. For this print, is SECTION A-A an auxiliary view? ____________________
2. If the print has a regular front view and regular right side view, how many auxiliary views are projected from the right side view? ____________________
3. Are there any hidden lines shown in the section view? ____________________
4. Analyze the lower-left auxiliary view. If this represents a depth dimension, which value shown in view is the depth direction, the .63 measurement or the .57 measurement? ____________________
5. Detail A is an enlarged partial view. Is it considered to be an auxiliary view also? ____________________

*Review questions based on previous units:*

6. What is the drawing number or part number? ____________________
7. What is the name of this part? ____________________
8. Is there a cutting-plane line on this drawing? ____________________
9. Is the lower-left view a section view? ____________________
10. At what scale is the enlarged detail A? ____________________
11. On what size paper was the original drawing created? ____________________
12. Do the visible lines of this drawing appear to be 2X thicker than the other lines? ____________________
13. According to the title block, what ASME standard is important to this drawing? ____________________
14. As you visualize this part, does it have any holes in it? ____________________
15. In this drawing, the material is specified outside of the title block as a note. In inches, how thick is the material used to make this part? ____________________

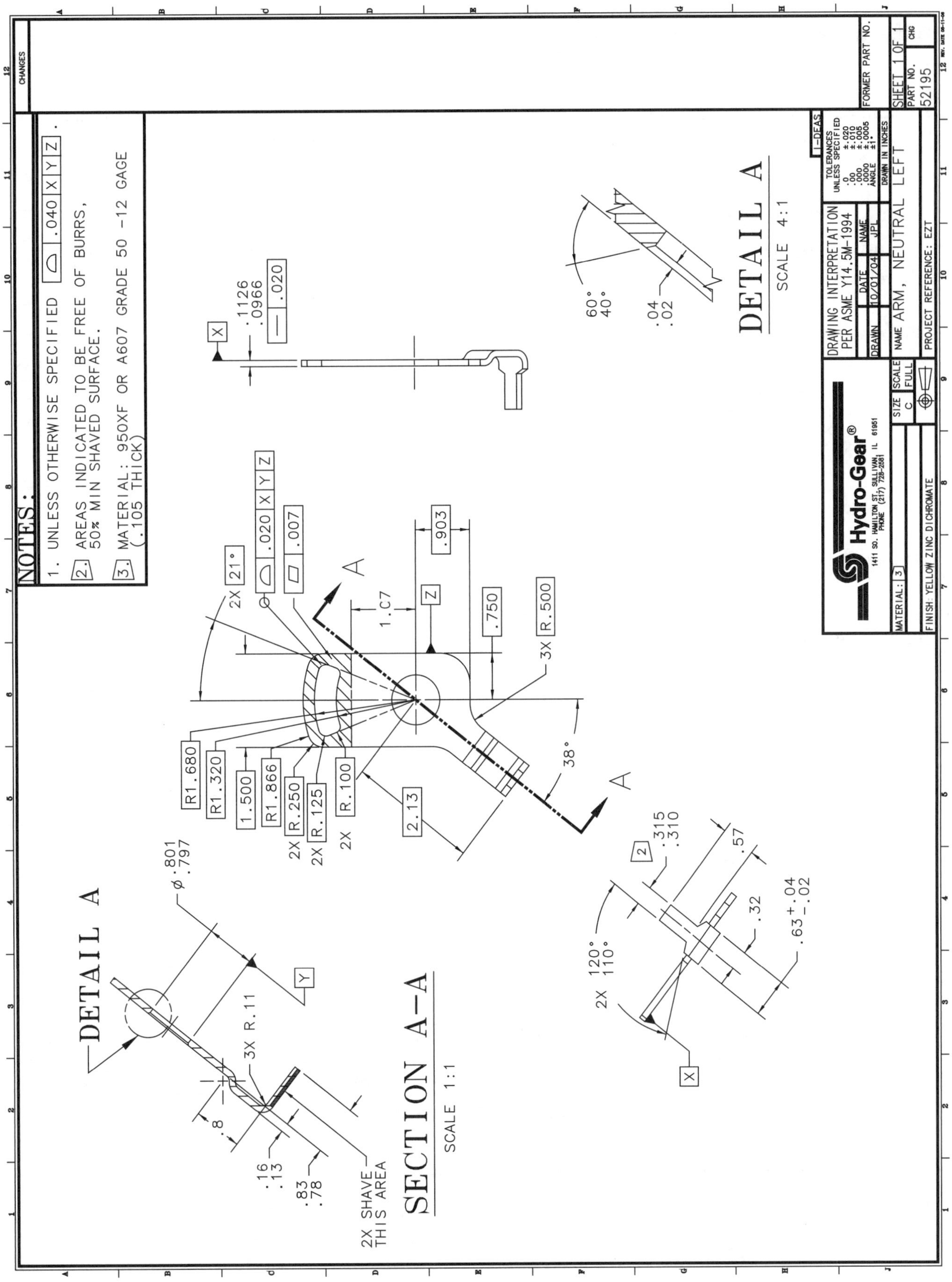

NOTES:

1. UNLESS OTHERWISE SPECIFIED [⌓ | .040 | X | Y | Z].
2. AREAS INDICATED TO BE FREE OF BURRS, 50% MIN SHAVED SURFACE.
3. MATERIAL: 950XF OR A607 GRADE 50 -12 GAGE (.105 THICK)

DETAIL A
SCALE 4:1

SECTION A–A
SCALE 1:1

DETAIL A

Hydro-Gear®
1411 SO. HAMILTON ST., SULLIVAN, IL 61951
PHONE (217) 728-2581

DRAWING INTERPRETATION PER ASME Y14.5M-1994

| DRAWN | DATE | NAME |
|---|---|---|
| | 10/01/04 | JPL |

TOLERANCES UNLESS SPECIFIED: .0 ±.020; .00 ±.010; .000 ±.005; .0000 ±.0005; ANGLE ±1°
DRAWN IN INCHES

I-DEAS

SIZE C | SCALE FULL

NAME ARM, NEUTRAL LEFT

SHEET 1 OF 1
PART NO. 52195
FORMER PART NO.

MATERIAL: 3
FINISH: YELLOW ZINC DICHROMATE
PROJECT REFERENCE: EZT

CHANGES

PR 7-1.
Print supplied by Hydro-Gear.

## Industry Print Exercise 7-2

*Refer to the print PR 7-2 and answer the questions below.*

1. For this print, which is the auxiliary view, SECTION A-A or SECTION B-B? ____________

2. How many auxiliary views are there? ____________

3. For this print, the view with the cutting-plane lines is the front view. The lowest view on the page, directly below it, is then called a(n) ____________ view.

4. Is the view in the upper-left corner a complete view or a partial view? ____________

5. The view in the upper-left corner is projected to create a view that shows the ____________ size and shape of a "tab".

*Review questions based on previous units:*

6. Does this drawing have extension lines? ____________

7. Does the style of lettering for the large SECTION A-A label follow ASME guidelines for Gothic lettering? ____________
____________

8. What material is specified in the title block? ____________

9. This drawing layout has a place for changes (revisions) to be recorded; are there any changes recorded on this sheet? ____________

10. What is the name of the company that owns this print? ____________

11. Was this drawing checked by someone? ____________

12. What is the name of this part? ____________

13. What is the drawing number or part number? ____________

14. In SECTION A-A, there are two "tabs" that are section-lined and two tabs that are not. Are the two tabs without section lining showing true size and shape in this section view? ____________

15. How many enlarged detail views are there? ____________

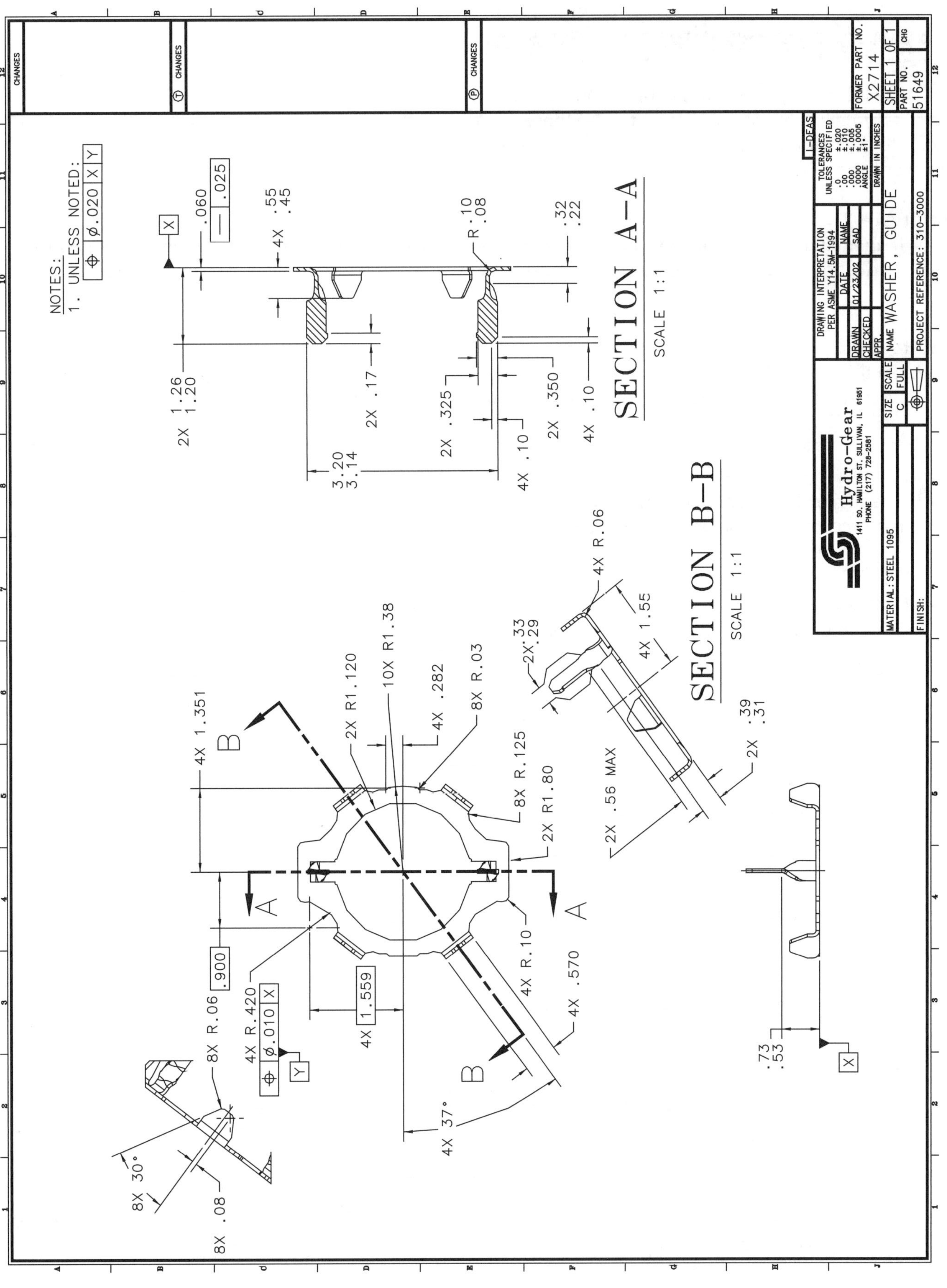


PR 7-2.
Print supplied by Hydro-Gear.

## Bonus Print Reading Exercises

The following questions are based on various bonus prints located at the back of this textbook. Refer to the print indicated, evaluate the print, and answer the question.

**Print AP-001:**

1. If an auxiliary view were used to show a view looking into the hole that is inclined 15°, would the auxiliary view be projected above the left side view, above the section view, or below the section view?

**Print AP-002:**

2. On this print, there is a stem and knob assembly at an angle to the main view. If an auxiliary view were used to show the round knob as a true circle, could it be projected off of the view?

**Print AP-006:**

3. The major diameter (100 mm) of this part is interrupted by a flat surface on top. Is an auxiliary view required to show the true size and shape of that flat surface?

**Print AP-007:**

4. How many auxiliary views are found on this print?
5. Is the SECTION B-B auxiliary view of this print in projection or removed out of projection?
6. At what apparent angle (relative to a horizontal line) is this auxiliary view projected?

**Print AP-013:**

7. What is the name of the auxiliary view on this print?
8. Is the auxiliary view of this print in projection or removed out of projection?

**Print AP-015:**

9. In relation to the part, what is the name of the 2″ portion of this part featured in the upper-right auxiliary view?
10. One of the auxiliary views in this print is jogged out of position so it can fit on the page. What type of line is used to show the jogged alignment?

11. In the two auxiliary views, the break lines indicate the views are also ____________________ views.

12. What is the specification for the hole that is shown in a true circular position in one of the auxiliary views? ______________________________

______________________________

**Print AP-026:**

13. This print does not have auxiliary views, but if an auxiliary view were to be used to show the true size and shape of part T150619, from which view would it be projected: front, top, or right side?

______________________________

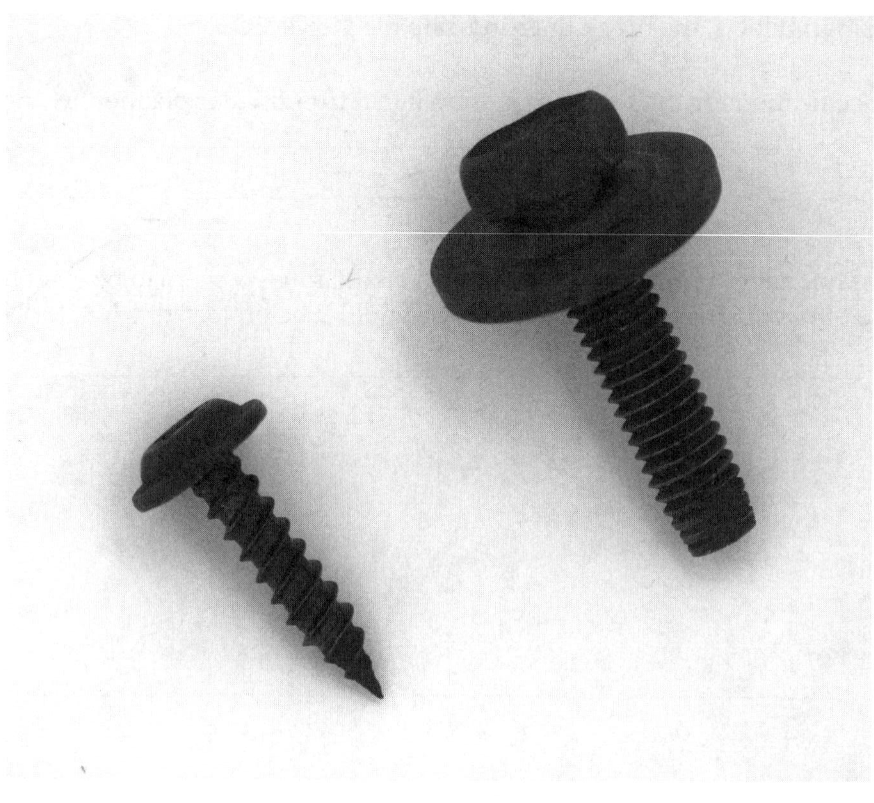

Screw threads are common on fasteners. There are many types of screw threads.

UNIT 8

# Screw Thread Representation

*After completing this unit, you will be able to:*

**Define** terms related to screw threads and fasteners.

**Identify** common screw thread forms.

**Describe** three methods for representing screw threads.

**Discuss** screw thread characteristics that may affect the specification.

**Explain** the different parts of a screw thread specification or callout.

**Discuss** the differences between metric threads and inch threads.

**Identify** standard pipe thread representation and designations.

A ***screw thread*** or ***thread*** is a ridge of a particular shape, often a V, that follows a helical path around the surface of a cylindrical surface. A helix can be seen in common shapes such as a barber pole or candy cane. A thread can also be compared to wrapping a rope around a cylinder, keeping each coil snug against the cylinder and the previous coil. By definition, a helix is formed by a point uniformly progressing along a path with a constant orientation to the axis of a cone or cylinder. Threads can be formed by hand, ground or rolled on a lathe, or formed using thread-cutting machines.

Threads are a very important feature of many industrial machine parts. Threads such as the American National or the Unified Standard (V-type) are used for bolts, screws, and nuts and for fastening machine parts together. Other thread forms are designed for special purposes, such as transmitting power along an axis, as in the lead screw of a machine lathe. Pipe threads are also machined according to standards and are designed for particular functions.

## Thread Terms and Definitions

ASME B1.7, entitled *Screw Threads: Nomenclature, Definitions, and Letter Symbols*, is the best resource for screw thread terms and definitions. A few of the terms are given here. Refer to Figure 8-1 as you read the following definitions.

- An ***external thread*** is a thread on the outside of a shaft. The threads on a bolt are external threads.
- An ***internal thread*** is a thread on the inside of a hole. The threads in a nut are internal threads.
- The ***crest*** is the top edge or surface of a ridge.
- The ***root*** is the bottom or "bottomland" between two ridges.
- The ***major diameter*** is the largest diameter of a thread, measured crest-to-crest across the diameter of the cylinder.
- The ***minor diameter*** is the smallest diameter of a thread, measured root-to-root across the diameter of the cylinder.
- The ***pitch diameter*** is an imaginary diameter where the width of the thread ridge is equal to the width of the groove between ridges. This is important for designing the fit of threaded parts.
- The ***pitch*** is the distance from one point on a screw thread to the same point on the next thread. American inch threads are specified by ***number of threads per inch***. If there are 16 threads per inch, the pitch is 1/16″. The pitch on metric threads is specified in the thread note.
- The ***lead*** is how far a thread advances when the threaded part is rotated one complete revolution. For most threads, the lead and pitch are equivalent. For multiple threads, the lead is double the pitch for double threads or triple the pitch for triple threads. Multiple threads are covered later in this unit.

Root
Crest
Axis
Major dia.
Minor dia.
Pitch dia.
Pitch
Thread angle
External Thread
Root
Crest
Internal Thread

**Figure 8-1.**
A few of the terms associated with screw threads are illustrated here.

## Screw Thread Forms

The ***thread form*** describes the shape of the ridges (or grooves) that form the threads, Figure 8-2. Most thread forms used in industry today are based on approved American standards. There are over a dozen ASME standards related to threads. ASME B1.1 is the recognized standard in the United States for straight threads used on screws, bolts, nuts, and other threaded parts. These threads are called Unified threads and are based on inch increments. Inch threads are still used in some countries outside of the U.S., such as Canada and Great Britain. Unified threads are similar to the older American National thread form, but are easier to manufacture. Unified threads have the general appearance of a "sharp V" and appear to be similar to metric threads, which are also based on a V-form. While metric threads have different parameters for thread depth and amount of bottomland allowed, the appearance on the print is identical to inch threads.

There are various uses for threads other than for fasteners. Thread forms in use for purposes other than fasteners include Acme, stub Acme, buttress, knuckle, and square. Acme and buttress threads are used in applications that exert force, such as vices and jacks. The knuckle thread form is common to the household incandescent lightbulb and other various international applications. Each of the thread forms may or may not have an American standard. For example, there are no standards for square threads, while knuckle thread standards are available in Germany. The following table shows applicable ASME standards for some common threads.

| Screw Thread Form | ASME Standard |
|---|---|
| Unified inch threads (UN & UNR) | ASME B1.1 |
| Acme screw threads | ASME B1.5 |
| Stub Acme screw threads | ASME B1.8 |
| Buttress inch screw threads | ASME B1.9 |
| Metric screw threads (M profile) | ASME B1.13M |
| Pipe threads, general purpose (inch) | ASME 1.20.1 |
| Dryseal pipe threads (inch) | ASME 1.20.3 |

## Thread Representation

A true projection view of a screw thread is very complex due to the helical shape, which projects as an irregular curve. Therefore, threads are represented on a print in conventional ways. In earlier units, *conventional practice* is defined as a standard way of doing something, even if this method breaks the scientific principles on which views are based. The three conventional methods in which threads may be represented on drawings are known as detailed, schematic, and simplified. See Figure 8-3. While the standards shown in the above table set forth the guidelines for manufacturing threads, ASME Y14.6 establishes the standards for representing threads in industrial drawings and prints.

The ***detailed*** method most closely represents screw threads as they actually appear. This convention is sometimes used to show the geometry of a thread form as a portion of a greatly enlarged

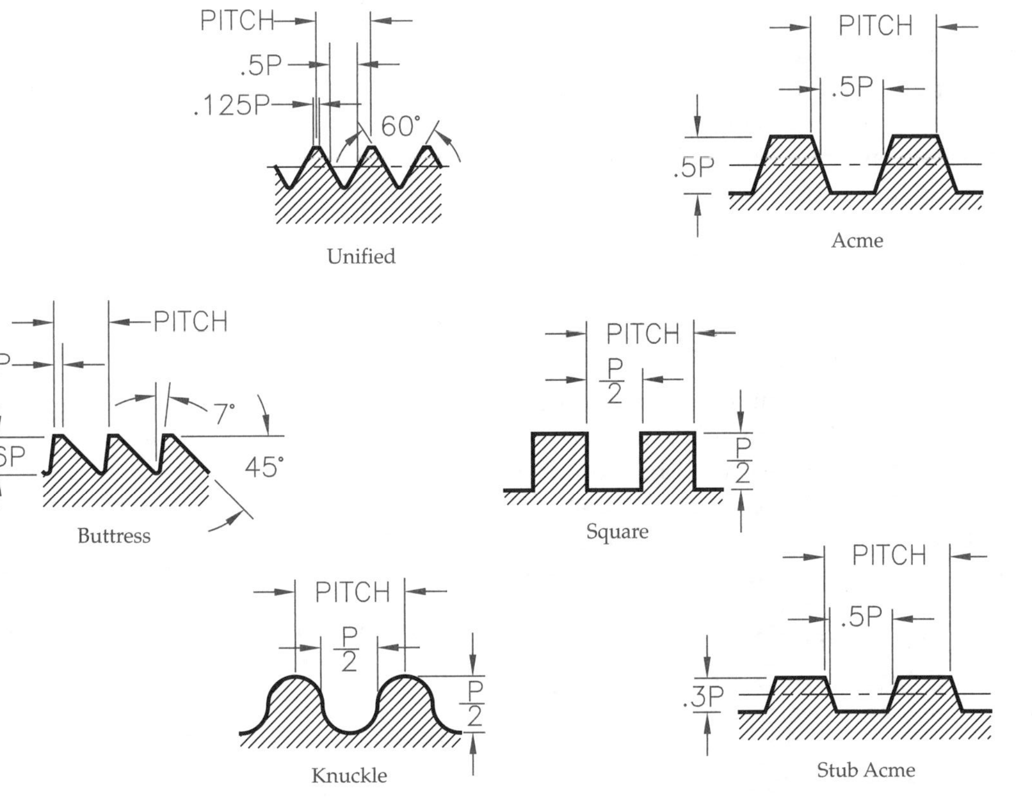


**Figure 8-2.**
The thread form describes the shape of the ridge or groove that forms the threads.

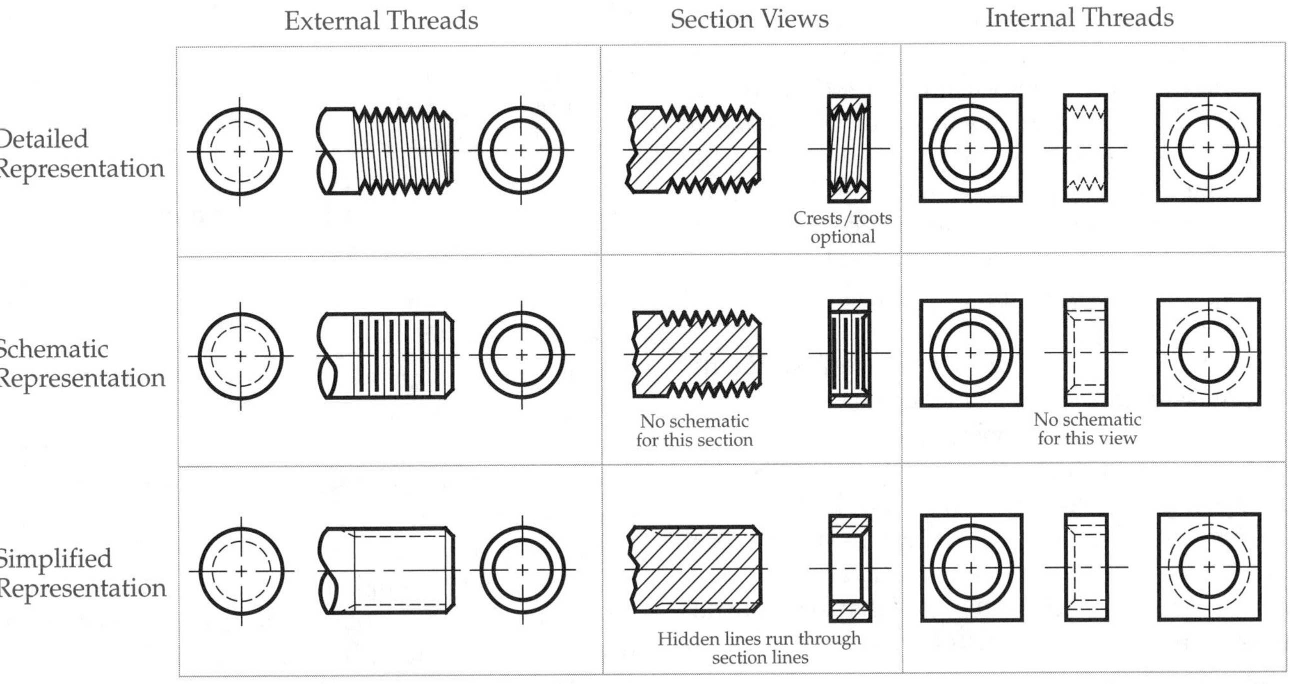


**Figure 8-3.**
The three conventional methods for representing threads are detailed, schematic, and simplified.

detail on a drawing. A rule of thumb for drafters is to draw detailed threads if the major diameter appears larger than 1″ on the drawing. Under this rule, if a 3″ diameter is drawn quarter scale, the actual diameter on the drawing is .75″, so detailed representation would usually not be applied. If the 3″ diameter thread is placed on the drawing at half scale, a detailed representation might be a better choice.

For smaller diameters and reduced-scale views of threaded features, the ***schematic*** method and ***simplified*** method are representations more quickly and easily drawn, yet they are easy to read and help describe the threaded feature. The simplified convention is perhaps the easier of these two conventions to use and has been more common in recent years. It is also recommended for assembly drawings. One disadvantage of the schematic representation is that the spacing looks good at one particular scale, but the spacing may require adjustment if the view is enlarged. With simplified representation, this is not an issue. The ASME standard recommends that simplified and schematic representations not be mixed in the same drawing unless there is a pressing reason to do so.

Certain thread elements can be confusing on a drawing. For example, if the end of an external thread is chamfered, the resulting visible circle may coincide with the minor diameter. As a result, a visible circle will be shown instead of a hidden circle. On threaded holes that are also countersunk, two visible circles may result. If the countersink diameter is close, but not exactly the same as the major diameter, the hidden circle may be omitted for the major diameter. For special conditions, an enlarged detail may be appropriate. Study Figure 8-3 again to see the various ways the three thread representations appear.

In real life, threads do not just stop suddenly. In recent standards, the "fadeout" or "runout" of the threads beyond the last fully cut thread may also be shown so dimensions can be added. On the drawing, dimensions can be applied to show the amount of external or internal surface that must have a complete, fully cut thread. Likewise, dimensions can be added to show that even partial threads must not infringe on a certain area or feature. See Figure 8-4.

## Thread Characteristics

There are several other characteristics besides form associated with threads. These include the thread series, tolerance (class), whether or not the thread is a multiple thread, and the direction of the thread. The following sections describe these additional characteristics.

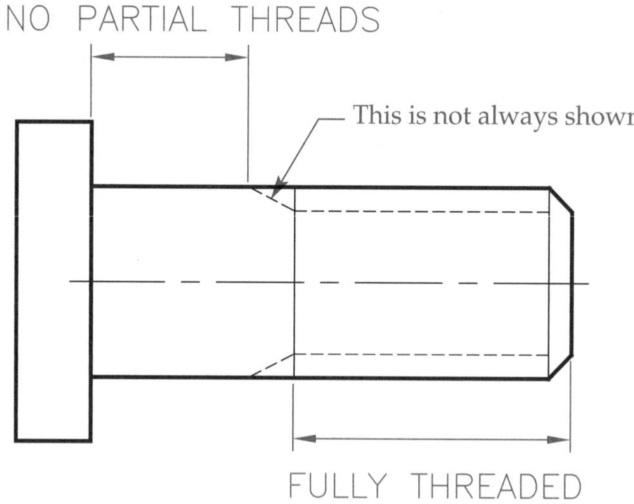


**Figure 8-4.**
Partial threads may run beyond the design length needed for full threads. Dimensions may be added to clarify the boundaries of full and partial thread requirements.

Schematic and simplified thread representations do not reflect the thread form, pitch, thread direction, or multiple-start status. In other words, the drawing looks the same regardless of the characteristics discussed in the next sections. In addition, the spacing for schematic representation is usually just a factor of the major diameter and nothing more. Larger diameters can have more generous spacing. Some drafters may attempt to space the schematic lines based on the pitch, but this is not required, nor practical in many cases.

### Thread Series

There are four main series of Unified screw threads—coarse, fine, extra fine, and constant pitch. See Figure 8-5 for standard values such as basic diameter, threads per inch, tap drill size, and screw numbers for diameters smaller than 1/4″. Each series has its own designated number of threads per inch.

The ***Unified coarse*** thread series is designated UNC. It is used for nuts, bolts, screws, and general uses where fine threads are not required.

The ***Unified fine*** thread series is designated UNF. It is used where the length of the threaded engagement is short and where a small lead angle is desired.

| Nominal Diameter | Basic Diameter | Coarse NC & UNC | | Fine NF & UNF | | Extra Fine NEF & UNEF | |
|---|---|---|---|---|---|---|---|
| | | Thds per in | Tap drill dia | Thds per in | Tap drill dia | Thds per in | Tap drill dia |
| 0 | .060 | | | 80 | .0469 | | |
| 1 | .073 | 64 | No.53 | 72 | No.53 | | |
| 2 | .086 | 56 | No.50 | 64 | No.50 | | |
| 3 | .099 | 48 | No.47 | 56 | No.45 | | |
| 4 | .112 | 40 | No.43 | 48 | No.42 | | |
| 5 | .125 | 40 | No.38 | 44 | No.37 | | |
| 6 | .138 | 32 | No.36 | 40 | No.33 | | |
| 8 | .164 | 32 | No.29 | 36 | No.29 | | |
| 10 | .190 | 24 | No.25 | 32 | No.21 | | |
| 12 | .216 | 24 | No.16 | 28 | No.14 | 32 | No.13 |
| 1/4 | .250 | 20 | No.7 | 28 | No.3 | 32 | .2189 |
| 5/16 | .3125 | 18 | F | 24 | I | 32 | .2813 |
| 3/8 | .375 | 16 | .3125 | 24 | Q | 32 | .3438 |
| 7/16 | .4375 | 14 | U | 20 | .3906 | 28 | .4062 |
| 1/2 | .500 | 13 | .4219 | 20 | .4531 | 28 | .4688 |
| 9/16 | .5625 | 12 | .4844 | 18 | .5156 | 24 | .5156 |
| 5/8 | .625 | 11 | .5313 | 18 | .5781 | 24 | .5781 |
| 11/16 | .6875 | ... | ... | ... | ... | 24 | .6406 |
| 3/4 | .750 | 10 | .6563 | 16 | .6875 | 20 | .7031 |
| 13/16 | .8125 | ... | ... | ... | ... | 20 | .7656 |
| 7/8 | .875 | 9 | .7656 | 14 | .8125 | 20 | .8281 |
| 15/16 | .9375 | ... | ... | ... | ... | 20 | .8906 |

| Nominal Diameter | Basic Diameter | Coarse NC & UNC | | Fine NF & UNF | | Extra Fine NEF & UNEF | |
|---|---|---|---|---|---|---|---|
| | | Thds per in | Tap drill dia | Thds per in | Tap drill dia | Thds per in | Tap drill dia |
| 1 | 1.000 | 8 | .875 | 12 | .922 | 20 | .953 |
| 1–1/16 | 1.063 | ... | ... | ... | ... | 18 | 1.000 |
| 1–1/8 | 1.125 | 7 | .904 | 12 | 1.046 | 18 | 1.070 |
| 1–3/16 | 1.188 | ... | ... | ... | ... | 18 | 1.141 |
| 1–1/4 | 1.250 | 7 | 1.109 | 12 | 1.172 | 18 | 1.188 |
| 1–5/16 | 1.313 | ... | ... | ... | ... | 18 | 1.266 |
| 1–3/8 | 1.375 | 6 | 1.219 | 12 | 1.297 | 18 | 1.313 |
| 1–7/16 | 1.438 | ... | ... | ... | ... | 18 | 1.375 |
| 1–1/2 | 1.500 | 6 | 1.344 | 12 | 1.422 | 18 | 1.438 |
| 1–9/16 | 1.563 | ... | ... | ... | ... | 18 | 1.500 |
| 1–5/8 | 1.625 | ... | ... | ... | ... | 18 | 1.563 |
| 1–11/16 | 1.688 | ... | ... | ... | ... | 18 | 1.625 |
| 1–3/4 | 1.750 | 5 | 1.563 | ... | ... | ... | ... |
| 2 | 2.000 | 4.5 | 1.781 | ... | ... | ... | ... |
| 2–1/4 | 2.250 | 4.5 | 2.031 | ... | ... | ... | ... |
| 2–1/2 | 2.500 | 4 | 2.250 | ... | ... | ... | ... |
| 2–3/4 | 2.750 | 4 | 2.500 | ... | ... | ... | ... |
| 3 | 3.000 | 4 | 2.750 | ... | ... | ... | ... |
| 3–1/4 | 3.250 | 4 | ... | ... | ... | ... | ... |
| 3–1/2 | 3.500 | 4 | ... | ... | ... | ... | ... |
| 3–3/4 | 3.750 | 4 | ... | ... | ... | ... | ... |
| 4 | 4.000 | 4 | ... | ... | ... | ... | ... |

Figure 8-5.
Standard thread specifications for Unified coarse, fine, and extra-fine threads are illustrated in this table.

The ***Unified extra fine*** series is designated UNEF. It is used for very short lengths of thread engagement and for thin-wall tubes, nuts, ferrules, and couplings.

The ***Unified constant-pitch*** series is designated UN with the number of threads per inch preceding the designation, such as 8UN. This series of threads is for special purposes, such as high-pressure applications. Constant-pitch threads are also used for large diameters where the other thread series do not meet the requirements.

Additional Unified threads have joined the standards in recent years. Technically, Unified threads are defined with a "flat-root contour", but thread tool wear results in some rounding. A rounded root actually results in a stronger part. For these reasons, there are a couple of Unified designations that specify the nature of the rounded root. The UNR designation indicates a root radius of 0.10825*P*, but this can only be applied to an external thread. The UNJ is another rounded root form originating in the military standards. It uses a slightly larger radius (0.15011*P*) as the minimum mandatory radius. In addition to these, a designation of UNS indicates a Unified form that is special or non-standard.

## Thread Classes

After classification by form and series, threads are further classified by manufacturing tolerance. These classes are 1A, 2A, and 3A for external threads and 1B, 2B, and 3B for internal threads. On some older drawings, classes 2 and 3 may appear without a letter designation.

Classes 1A and 1B replace the older American Standard class 1. They are used in applications requiring minimum binding. The tolerance of these classes allows for frequent and quick assembly or disassembly of parts.

Classes 2A and 2B are threads with tighter tolerances. They are used for general purposes such as for nuts, bolts, screws, and normal applications by mass production industries.

Class 3A and 3B are threads with very stringent and close tolerances. They are used for applications in industries requiring tighter tolerances than the preceding classes of 1A and 1B or 2A and 2B.

## Multiple Threads

In addition to the single-ridge screw thread, threads can also be manufactured with two or three ridges running side-by-side. These have been commonly referred to as double threads or triple threads, but more recently are referred to as ***multiple-start threads.*** Double threads can be called ***two-start threads*** and triple threads can be called ***three-start threads.*** Due to the increased slope angle of the thread, the holding power of a double or triple thread is less.

Conceptually, it may help to think of a rope wrapped around a pole. If one rope is wrapped, it is like a single thread. If two ropes are wrapped, they are like double threads. Three ropes wrapped around the pole are like triple threads. Looking at three ropes wrapped around a pole, it may appear as one rope, but the "angle" of progression is steeper than it would be if one rope is wrapped around the pole. This steeper angle corresponds to a different lead.

The print reader should not expect to have a visual indication of multiple threads. If the drafter properly applies a detailed representation, the slope of the crest line is inclined 1/2 pitch on the side view of a thread. On the "back side" of the cylinder, the crest slopes the other 1/2 pitch. Study the hidden lines in Figure 8-6 that represent the crest angle on the "back side" of the cylinder. A detailed double thread should indicate a crest directly above a crest. The slope of the crest line is then one pitch. On the "back side" of the cylinder, the crest line slopes an additional pitch. Remember, the *lead* of a double thread, or two-start thread, is twice the pitch. One revolution of the threaded cylinder advances two pitches because there are two ridges. Likewise, a triple thread, or three-start thread, advances three pitches.

With schematic and simplified representation, there is no difference in the graphic symbolization, regardless of the threads. In summary, the print reader should rely solely on the thread specification note. Do *not* rely on the visual representation as a means of determining if the thread is single, double, or triple.

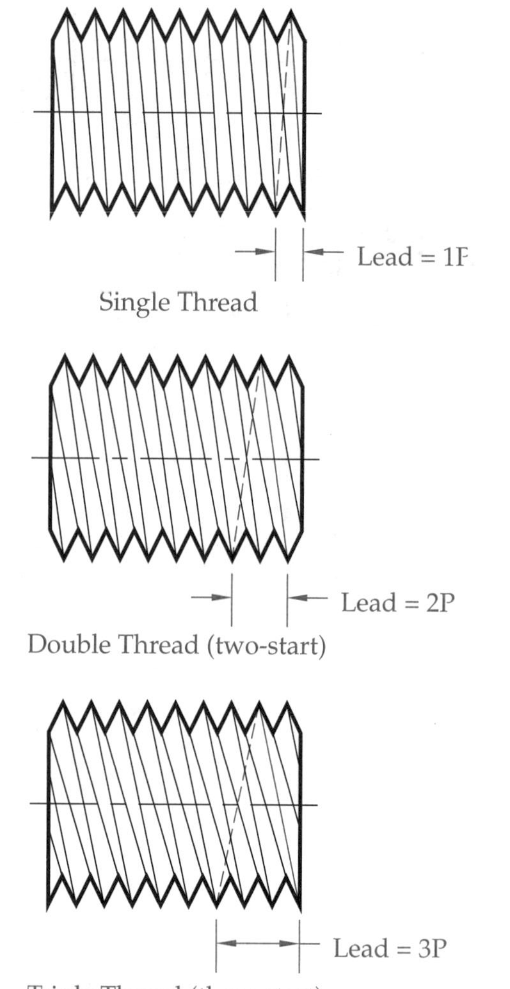


**Figure 8-6.**
If drawn in detailed representations, multiple-start threads should show the crest lines sloping accordingly. The hidden lines in the illustration represent the slope on the "back side." The double thread crest slopes one pitch on the near side and an additional pitch on the back side, which explains the lead being equal to twice the pitch.

### Left-Hand Threads

Threads are most commonly created so that clockwise revolutions advance the threaded parts together. Common nuts and bolts tighten with clockwise turns. These threads are called ***right-hand threads.*** However, some applications require threads that tighten with counterclockwise movement. These threads are referred to as ***left-hand threads*** or ***reverse threads.***

Left-hand threads are usually found on items where right-hand threads may produce unwanted loosening, perhaps due to motion. Examples include threads on bicycle pedals, threads on a toilet tank flush handle, the arbor nut on a table saw, and similar applications. Left-hand threads are indicated in a print using a callout. Detailed representations should also reflect a slope angle opposite that of right-hand threads, Figure 8-7.

## Specification of Screw Threads

A screw thread and all of its characteristics are specified on a drawing using a standard note with a leader and an arrow pointing to the thread. The note contains information describing the specific

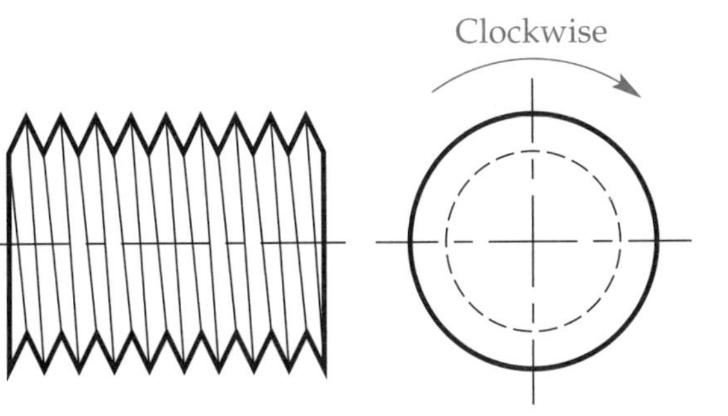


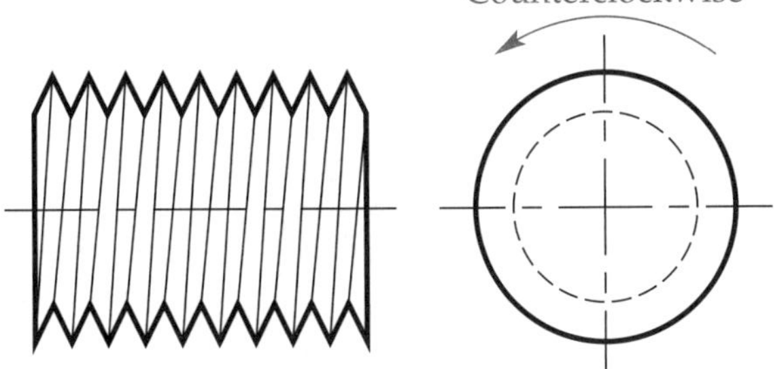


Figure 8-7.
Right-hand and left-hand threads.

thread and may be composed of a variety of details. In Figure 8-8, the note on the right is interpreted as:

1) Nominal size (major diameter or screw number for smaller diameters) is 1″.
2) Number of threads per inch is five (5) threads per inch.
3) Thread form is Acme.
4) Thread class number or symbol and internal/external designation is 2A, which is external threads.
5) The threads are left hand (LH). This designation is only included for left-hand threads; no designation is used for right-hand threads.

Since fractional numbers are widely used, they are acceptable for the nominal size, even though the rest of the drawing is using decimal values. If decimal values are used for the nominal size, they should be shown to four places (unless the fourth place is a zero). For common numbered screw threads, usually those smaller than 1/4″, the decimal equivalent should be shown in parentheses to three places, as in this example: #10(.190)-32 UNF-2A.

In newer standards, additional elements of the thread note include the designation MOD to indicate there is a special modification or qualifying information. If MOD is placed in the thread note, another line is added to explain the modification. For inspection, there are standard thread gauging systems guided by ASME B1.3 that determine the acceptability of screw threads. System 21, for example, allows the inspector to accept screw threads with a *go* or *no-go* ring-and-plug system. This gauging system number can also be added in parentheses at the very end of the thread note.

**Example 1:**

$\frac{3}{8}$-24 UNF-3A MOD (21)

MAJOR DIA .3648-.3720 MOD

**Example 2:**

$1\frac{1}{2}$-10 UNS-3B MOD (21)

MINOR DIA 1.398-1.409 MOD

PD 1.4350-1.4413

MAJOR DIA 1.500 MIN

In Example 1, the modification is described in line two. In this case, the print specifies the major diameter should be between .3648 and .3720, a specification that requires a slight modification to the crests of the thread. Thread gauging system 21 is also specified. In Example 2, the Unified special threads callout means a special diameter and pitch diameter will be specified in the following rows, but gauging system 21 is still the method of inspection.

Unless otherwise specified, threads are right hand and single start. The letters LH after the class symbol specify the thread as left hand. In former practice, the word DOUBLE may also have been shown to indicate a double thread or the word TRIPLE to indicate a triple thread. Single threads and right-hand threads have never required RH or SINGLE to be specified.

Newer standards have implemented some additional techniques for multiple-start threads. Multiple threads can be described by replacing the number of threads per inch with the pitch and lead given in the note. Following are two examples of a thread note for a standard Unified multiple-start thread.

**Former practice**

.750-16 UNF-2A TRIPLE

**Current practice**

.750-.0625*P*-.1875*L*(3 STARTS)UNF-2A

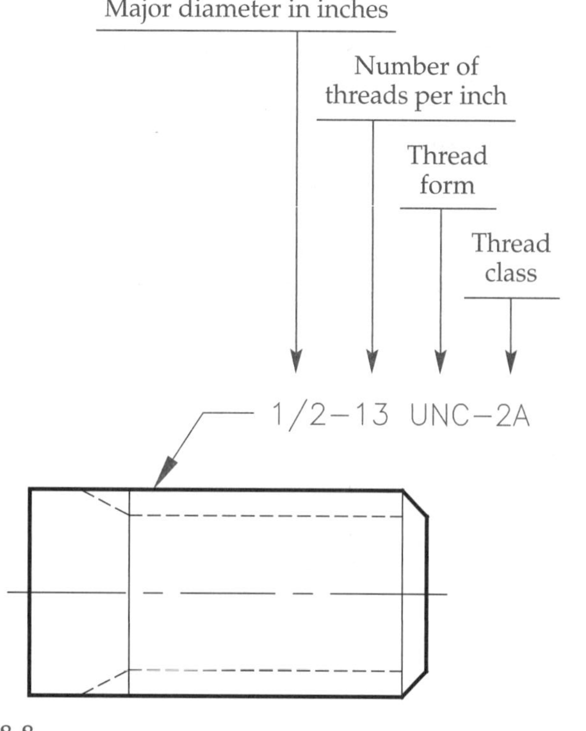


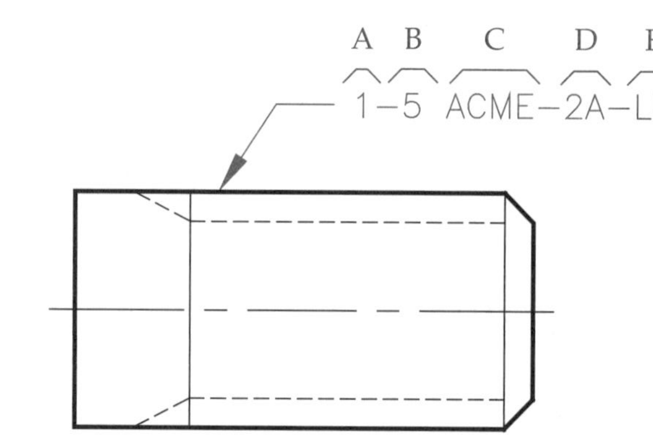


**Figure 8-8.**
A screw thread is specified in a local note that may be composed of several components.

Other specifications for threads may be given. These additional specifications may include thread length, hole size (for internal threads), or chamfer. Giving the pitch diameter below the thread note is a common option adopted by many companies. See Figure 8-9. Additional manufacturing specifications are discussed in Unit 11, Machining Specifications and Drawing Notes.

If the tolerance, or range of size, for the thread pitch diameter is given, it should be placed below the note. However, on some prints, the tolerance may appear on the same line as the thread note.

$\frac{3}{4}$-10 UNC-2B
PD .6850 TO .6927

The specification for a constant-pitch series thread with a tolerance for the pitch diameter is listed as follows.

$2\frac{1}{4}$-8 UN-3A
PD 2.1688 TO 2.1611

Additional information about tolerances is covered in Unit 10, Tolerancing.

## Metric Threads

Metric threads are graphically represented in the same manner as Unified threads, but the format of the specification note is different. In a note for metric threads, the letter M is followed by "diameter × pitch." The M designates the thread as a metric series. In former practice, if the pitch was not indicated, the print reader assumed the standard coarse thread pitch applied. Metric fine threads always had the pitch designated. Today, the preferred standard is to always specify the thread pitch. In Figure 8-10, the M8 represents a nominal thread diameter of 8 mm with the standardized fine thread pitch of 1.0 mm. For an 8 mm diameter, the standard coarse thread pitch is 1.25 mm. If the thread note were specified simply M8 (older practice), then the standard coarse pitch could be assumed. Current standards recommend the diameter, pitch, tolerance class, and gauging system be specified. However, in actual practice, you may find many prints do not indicate the tolerance and gauging system.

The specification for metric series threads does not indicate the number of threads per unit of length, such as threads per millimeter or threads per centimeter. Instead, the pitch is simply stated in the specification.

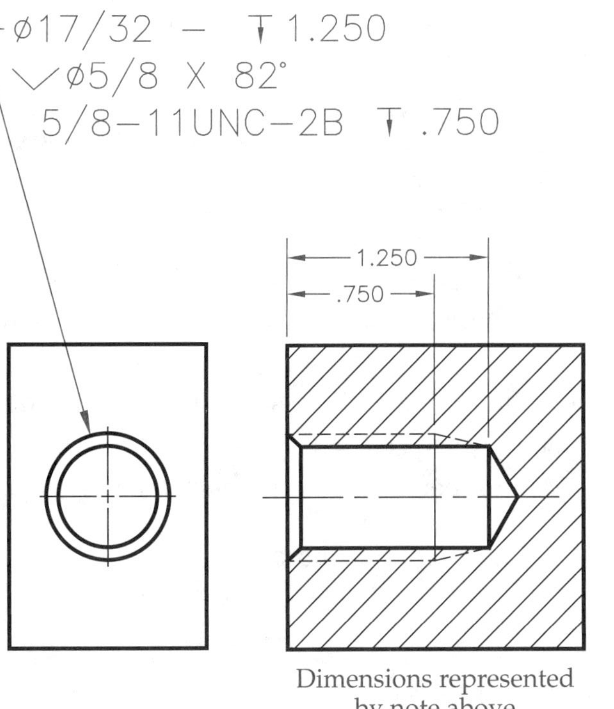


**Figure 8-9.**
Additional specifications for a thread may include thread length, hole size and depth, and chamfer specification.

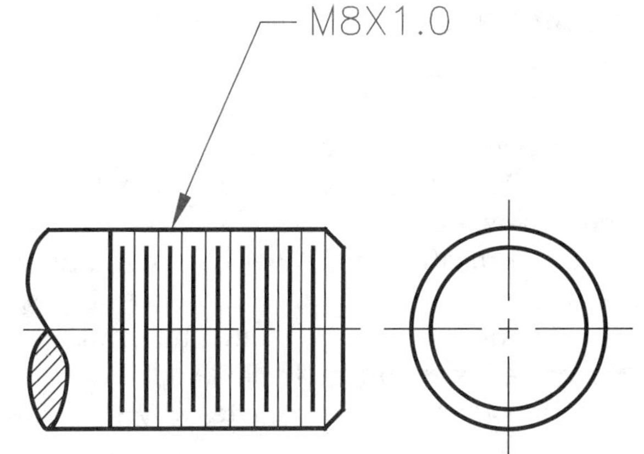


**Figure 8-10.**
Metric threads are specified differently from other threads.

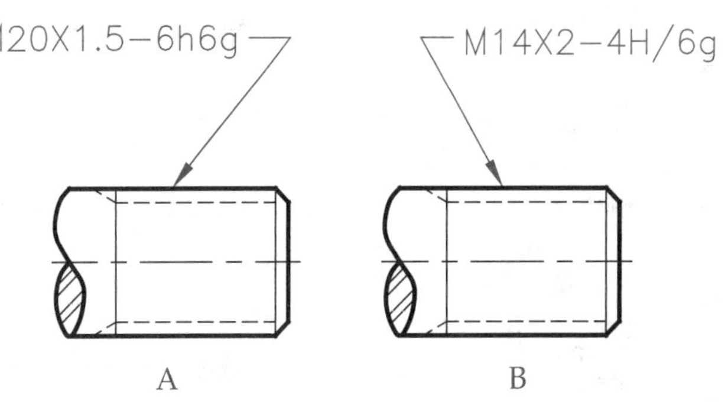


**Figure 8-11.**
The tolerance class for metric threads can be specified in different ways. A—The tolerance class for the pitch diameter is a little more precise than the major diameter. B—The internal thread is specified more precisely than the external thread.

The tolerance and class of fit for metric threads are designated by adding numbers and letters in a certain sequence to the callout. The thread designation in Figure 8-11A calls for a 20 mm diameter fine thread with a pitch of 1.5 mm. In addition, the thread note specifies a tolerance class of 6h6g. The first part, 6h, is a tolerance grade applied to the pitch diameter. The second part, 6g, is a tolerance grade applied to the crest diameter tolerance. If these two diameters have the same tolerance class, then only one is given. The numeral choices range from 3 to 9 for external threads and 4 to 8 for internal threads. Alphabetical characters "g" or "h" (external) or "G" or "H" (internal) are used in the United States for the "position" of the tolerance. An H specification is more precise with zero allowance and a G specification has a small allowance. In brief, the 2A/2B classification for inch threads is approximately equivalent to 6g/6H for metric threads.

If the note also covers the expected class of fit between two threaded parts, the note can specify both internal and external tolerance class. In Figure 8-11B, the thread note specifies a tolerance class specification of 4H/6g. This indicates the nut (internal thread) for the threaded feature should be a precise tolerance class 4H fit, while the pitch diameter and major crest diameter of the external threads both have a medium grade fit (6g). The scope of this textbook does not allow for an in-depth explanation of metric thread tolerancing, but there are many excellent resources available for those who work with metric thread applications.

## Pipe Threads

The three forms of pipe threads used in industry are regular, aeronautical, and Dryseal pipe threads. The ***regular pipe thread*** is the standard for the plumbing trade. The ***aeronautical pipe thread*** is the standard in the aerospace industry. Regular and aeronautical pipe thread forms must be filled with a lute or sealer to prevent leakage in the joint. The ***Dryseal pipe thread*** is the standard for automotive, refrigeration, and hydraulic tube/pipe fittings. Dryseal pipe threads do not allow leakage, even without the use of sealer. This is due to the metal-to-metal contact at the crest and root of the threads.

## Representation and Specification

Straight and taper pipe threads are graphically represented in a similar manner as other screw threads. However, the major diameter of pipe threads is shown with a taper angle of approximately 3° to the center axis, Figure 8-12.

The specifications for American Standard Pipe Threads are listed in sequence of nominal size, number of threads per inch, and the symbols for the thread series and form. For example, the thread specification 1/2-14 NPT designates a 1/2″ nominal size, 14 threads per inch, American Standard taper pipe thread.

## Pipe Thread Designations

The following pipe series symbols used for American Standard pipe threads.

- **NPT.** American Standard taper pipe thread for general use
- **NPTR.** American Standard taper pipe thread for railing joints
- **NPSC.** American Standard pipe thread for couplings
- **NPSM.** American Standard pipe thread for free-fitting mechanical joints
- **NPSL.** American Standard pipe thread for loose-fitting mechanical joints with locknuts
- **NPSH.** American Standard pipe thread for hose couplings

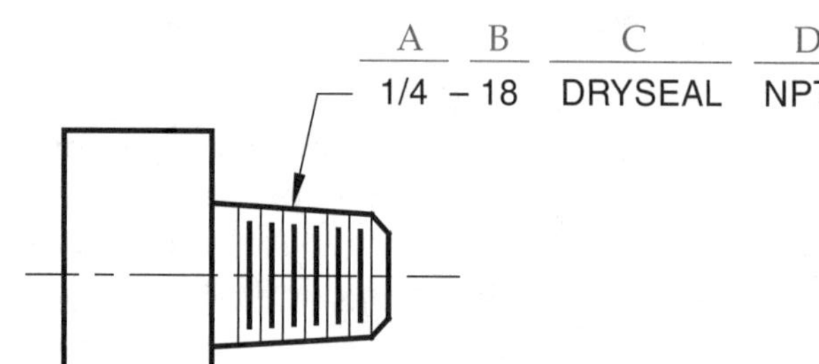


Figure 8-12.
Pipe threads are drawn similar to normal threads, but the major diameter is tapered at a slight angle to the center axis.

Pipe series symbols used for Dryseal pipe threads are designated as follows.

- **NPTF.** Dryseal American Standard pipe thread
- **PTF-SAE SHORT.** Dryseal SAE short taper pipe thread
- **NPSF.** Dryseal American Standard fuel internal straight pipe thread
- **NPSI.** Dryseal American Standard intermediate internal straight pipe thread

A typical specification for Dryseal pipe thread is shown in Figure 8-12. This specification includes:

1) Nominal size (i.e., 1/4″).
2) Number of threads per inch (i.e., 18).
3) Form (if Dryseal) .
4) Pipe series symbol (i.e., NPTF = Dryseal American Standard pipe thread).

## Summary

The representation and specification of screw threads is multifaceted. In this unit, you have examined a few of the most common characteristics of the graphical illustrations that will be found on the print and some of the thread notes the print reader may encounter. Some of the notes require an in-depth knowledge of the manufacturing processes involved in making or inspecting the threads. ASME Y14.6 recommends that the name and thread standard be referenced on all drawings so misunderstandings will be avoided. Having these references, as well as other resources such as a machinist's handbook, can be invaluable to the print reader when threads are represented on a print.

## Review Questions

*Circle the answer of choice, fill in the blank, or write a short answer.*

1. What geometric shape describes the path of a screw thread? ____________________

2. *True or False?* Threads have one purpose, which is holding parts together.

3. For each of the following, give the term that matches the definition.
   A. The bottom between ridges.
   ____________________
   B. Distance from one point of a thread to the same point on the next thread.
   ____________________
   C. Smallest diameter of a threaded feature.
   ____________________
   D. Distance of advance when a threaded feature turns 360°.
   ____________________

4. A thread form that has a 7° slope on one side of the ridge and a 45° slope on the other side is the ____________________ thread form.

5. On industrial prints, which of the following is *not* a way that screw threads usually appear?
   A. Detailed
   B. Schematic
   C. Simplified
   D. True projection

6. Of the choices for representing threads, which method is the most realistic? ____________________

7. For each of the notations below, write out the full name of the thread series.
   A. UN
   ____________________
   B. UNF
   ____________________
   C. UNC
   ____________________
   D. UNEF
   ____________________

8. *True or False?* Inch screw threads are classified by how precise they are made, with 2 being most common, 1 being more loose, and 3 being more precise.

9. A two-start thread is when there are two ridges that form the screw thread, resulting in a(n) ____________________ equal to twice the pitch.

10. What term describes threads that are made so that counterclockwise motion advances the thread instead of the normal clockwise motion? ______________________________

______________________________

11. *True or False?* Even if the drawing or print uses decimal numbers throughout, fractions can still be used in thread notes because these sizes are still commonly expressed that way.

12. What is indicated by the number 21 in parentheses at the end of a thread note? ______________

______________________________

13. For a thread note that states M24 X 3, which of the following is false?
    A. This is a metric thread.
    B. The pitch is 3 mm.
    C. The minor diameter is 24 mm.
    D. The thread is right-handed.

14. One key visual clue that a thread is a pipe thread is the ______________ angle of the major diameter lines.

15. What proper name is applied to pipe threads designed to be leakproof without sealers? __________

______________________________

## Review Activity 8-1

*Examine the chart that follows. Fill in the blanks for the sample threads shown in views A–D. Either read the thread specification note or use the paper scale provided at the bottom to measure or transfer distances with a separate scrap sheet of paper. The guidelines are to assist you with neat capital lettering. Also, answer question 10 related to the metric thread specification.*

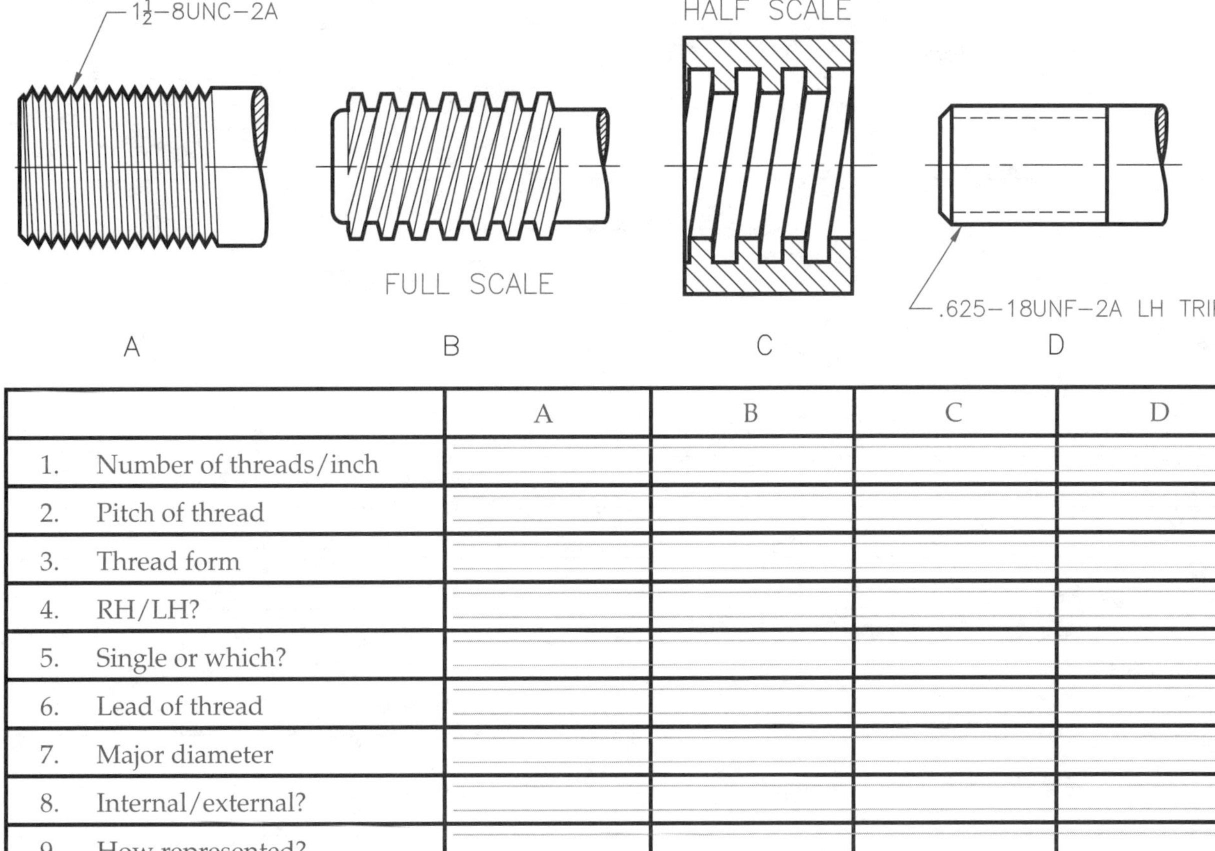


| | A | B | C | D |
|---|---|---|---|---|
| 1. Number of threads/inch | | | | |
| 2. Pitch of thread | | | | |
| 3. Thread form | | | | |
| 4. RH/LH? | | | | |
| 5. Single or which? | | | | |
| 6. Lead of thread | | | | |
| 7. Major diameter | | | | |
| 8. Internal/external? | | | | |
| 9. How represented? | | | | |

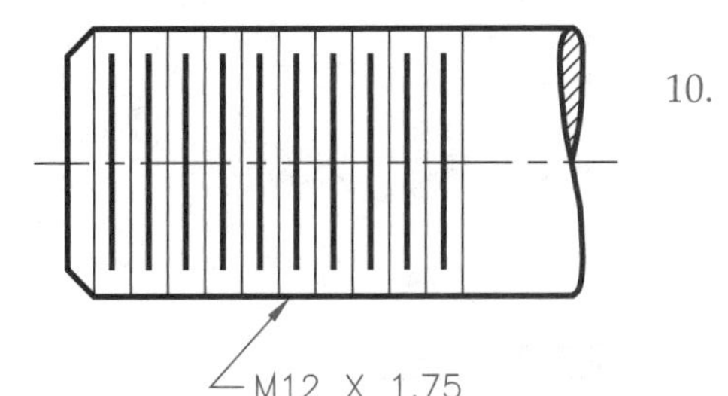


10. For the metric thread:
    A. What is the major diameter? ________________
    B. What is the pitch? ________________
    C. How is it represented? ________________

0 1 2

Copy or transfer distances to scale to answer questions about B and C above

## Review Activity 8-2

*For each of the thread notes below, write out the description using complete words. The first one has been done as an example.*

1. 3/8-16 UNC-2A
   3/8″ major diameter, 16 threads per inch, Unified coarse thread form, with a Class 2 medium fit, external thread

2. 1-20 UNEF-3B

3. .625-8 ACME-LH

4. M10 X 1.5-6H/5g6g

5. 3/4-14 NPT

*For each of the thread descriptions below, create a standard thread note specification.*

6. 1-1/2″ major diameter, 12 threads per inch, Unified fine thread form, with a Class 2 fit, internal thread, left-handed ______________________________

______________________________

7. #5 or 1/8″ major diameter, 40 threads per inch, Unified coarse thread form, with a Class 2 fit, external thread ______________________________

______________________________

8. 1/2″ major diameter, 10 threads per inch, Stub Acme thread form, internal thread, right-hand

______________________________

9. Metric thread, 20 mm major diameter, 2.5 mm pitch, and "6g" grade for external pitch diameter and external major diameter ______________________________

______________________________

10. 1/4″ pipe size, 18 threads per inch, Dryseal American standard pipe thread ______________

______________________________

## Industry Print Exercise 8-1

*Refer to the print PR 8-1 and answer the questions below.*

1. On the left end of this object, the thread note indicates 10UNC. What is the pitch for this thread? ________

2. Is either of the threads on this part left-handed? ________

3. What type of thread representation is used on this print? ________

4. On the right end of the object, what is the major diameter of the thread? ________

5. What thread form is used for the threads on this sheet and is it coarse, fine, or extra fine? ________

6. If a nut were screwed onto the right end of this part, one 360° turn of the nut would advance the nut what distance? ________

7. Does there appear to be a chamfer on one or both ends of this part?

*Review questions based on previous units:*

8. Is there a cutting-plane line on this print? ________

9. Are there hidden lines featured on this print? ________

10. What is the part number for this drawing? ________

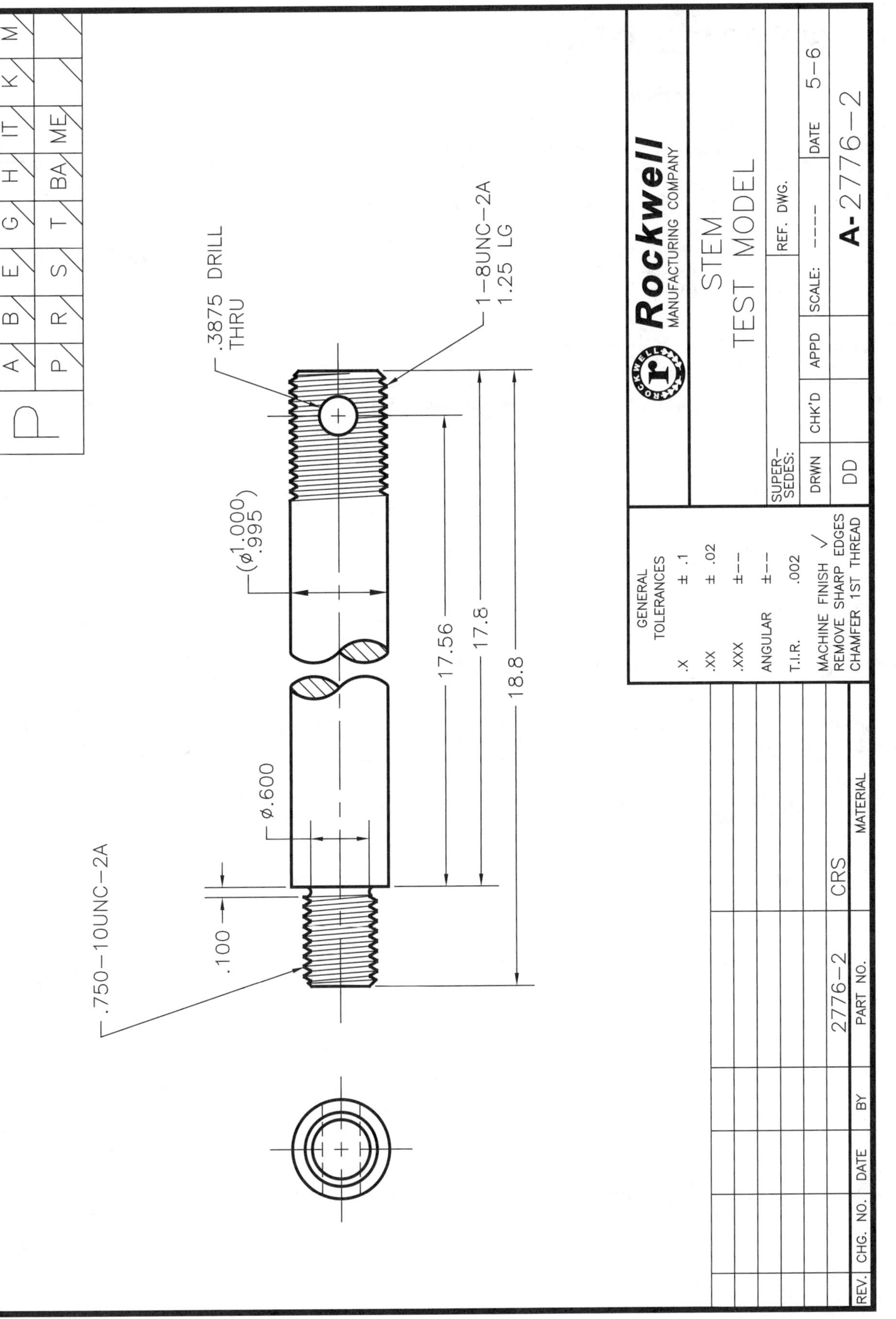


PR 8-1.
Print supplied by Rockwell Manufacturing Company.

## Industry Print Exercise 8-2

*Refer to the print PR 8-2 and answer the questions below.*

1. How many threaded holes are there for the part on this print? ____________________
2. What method of representation is used to indicate the threaded holes in the section view? ____________________
3. What is the pitch of the threads for this part? ____________________
4. What is the major diameter of the threads for this part? ____________________
5. Use Figure 8-5 in this unit to determine if these threads are coarse, fine, or extra fine. ____________________

*Review questions based on previous units:*

6. What paper size is the original version of this print? ____________________
7. What is the material for this part? ____________________
8. What is the name of this part? ____________________
9. Examine the cutting-plane line and the right-side section view. What name is given to the type of section view featured on this print? ____________________
10. Does this drawing feature an auxiliary view? ____________________

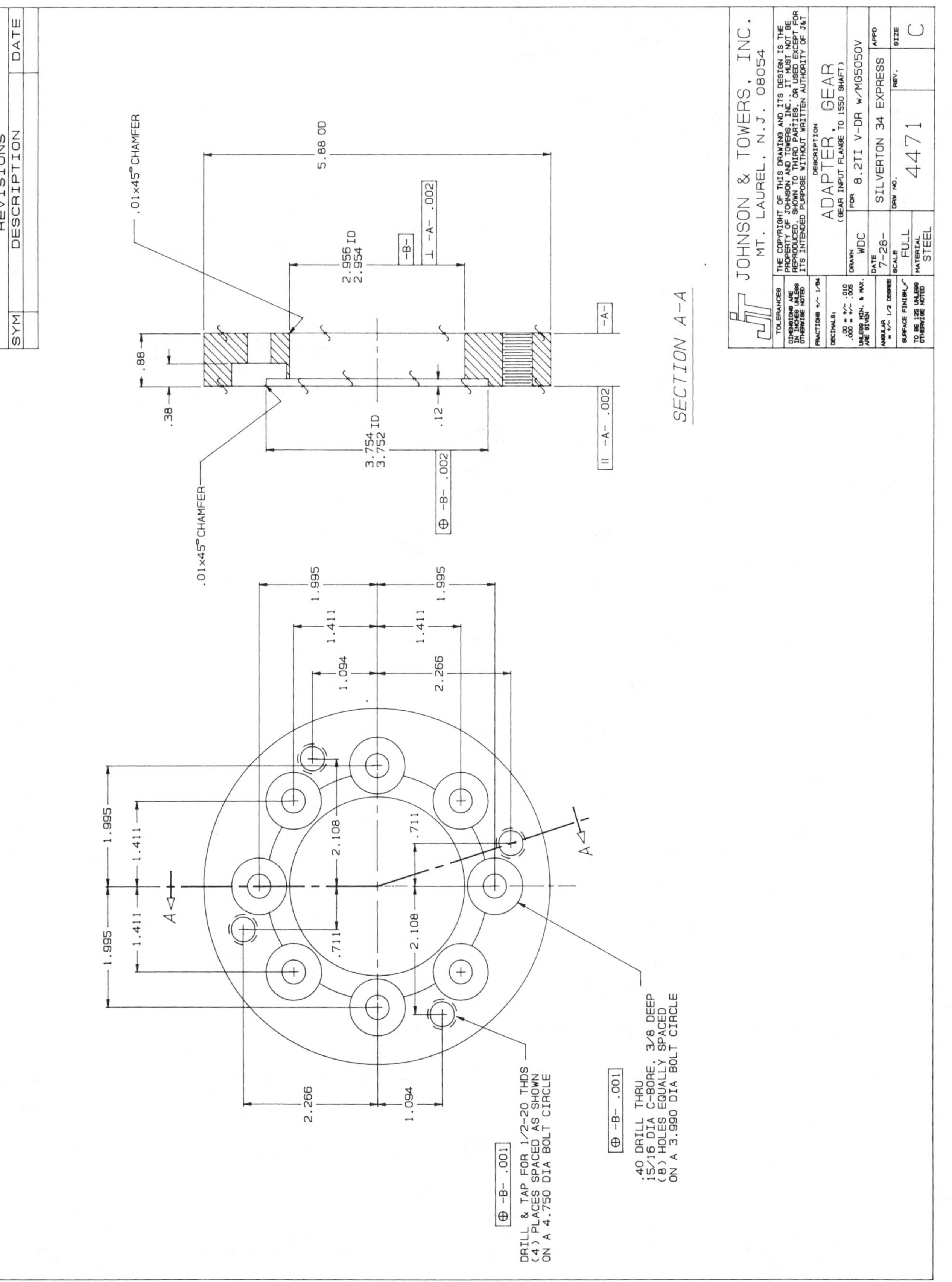


PR 8-2.
Print supplied by Johnson and Towers, Inc.

## Bonus Print Reading Exercises

The following questions are based on various bonus prints located in the folder at the back of this textbook. Refer to the print indicated, evaluate the print, and answer the question.

**Print AP-001:**

1. Is the National Gas Outlet thread shown on this print right handed or left handed? ______________
2. What is the major diameter of the National Gas Tapered thread shown on this print? ______________
3. What is the smallest thread pitch specified on this print? ______________
4. Give the diameter of the countersink leading into the largest threaded hole. ______________

**Print AP-003:**

5. What does UNEF stand for in the thread note on this print? ______________
6. In the thread note, what does 2A specify? ______________
7. The major diameter for this thread is 3/8, but the thread is shown detailed because the scale of the drawing is 4:1 and, therefore, the diameter on the print is drawn at a size of ______________.

**Print AP-004:**

8. This assembly drawing includes a threaded stud. By what method are the threads represented? ______________

**Print AP-005:**

9. What type of thread is specified in the view labeled Section J-J? ______________

**Print AP-006:**

10. What is the major diameter of the threaded feature on this print? ______________

**Print AP-009:**

11. In inches, how much farther is the hole drilled (at maximum) beyond the minimum-full-thread depth? ______________

**Print AP-011:**

12. How many threaded holes does this part feature? ____________________

13. What method of thread representation is used on this print? ____________________

**Print AP-013:**

14. How many threaded holes are specified on this part? ____________________

**Print AP-014:**

15. Even though this part is drawn and dimensioned as metric, what is the thread note given in inches?

____________________

16. In the circular right side view, what thread is specified for the hole at 12 o'clock? ____________________

17. For the threads that are specified as M60 × 2, what does this 2 represent? ____________________

**Print AP-015:**

18. What is the thread note specified on this print? ____________________

There are many standards available to industry. Basic dimensioning practices are covered in ASME Y14.5.

# UNIT 9
# Dimensioning

*After completing this unit, you will be able to:*

**Identify** terms and measurements associated with dimensioning mechanics.

**Identify** symbols that have been standardized for dimensioning notations.

**Explain** the choice and placement rules drafters use.

**Identify** and **discuss** various systems and methods for dimensioning.

**Describe** additional types of dimensions used in special ways.

Print reading requires the reader to understand the shape description of the object. This is called visualization. Previous units primarily focus on describing the shape of an object or part with different view systems, sectional techniques, and special symbols for screw threads. The second major component of print reading is the size description. The reader must understand the size of a part as specified on the print. The size description can be called ***dimensioning.*** Dimensions are annotations to the views describing the size or location of each feature. ASME Y14.5, entitled *Dimensioning and Tolerancing,* is the standard that covers basic dimensioning practices. Unit 10, Tolerancing, and Unit 13, Geometric Dimensioning and Tolerancing, are also based on this standard.

Dimensioning can be a very difficult area of drafting to learn. The drafter must not only know the *mechanics* of how to create all the dimension and extension lines, including sizes and spacing, but also how to choose *which* dimension to place and *where.* Sometimes, this can be a real challenge. Unfortunately, many drafters and engineers have not had sufficient training in how to properly dimension a drawing. Therefore, some prints may not follow standard recommendations for dimensioning as well as they should, making them hard to read.

This unit explains some of the rules a drafter should use to choose and place dimensions. This will assist the print reader in knowing which dimensions to expect and where to look for a dimension. As a print reader, you can be a valuable team member by identifying ways in which to improve a drawing with more concise or appropriate dimensioning.

## Dimensioning Mechanics

***Dimensioning mechanics*** can be defined as the instructions or guidelines for what size and spacing to use for all of the components of the dimension. For example, arrowhead appearance, lettering size, and spacing for extension lines are all part of dimensioning mechanics. Even CAD systems that automatically create dimensions need to be set up and managed so that the automatic dimensioning conforms to standards. Simply put, mechanics is the part of dimensioning concerned with what dimensions look like, rather than where dimensions are placed or whether or not they are even needed.

### *Lines Used in Dimensioning*

The standard lines used in dimensioning are presented in Unit 2, which include the extension line, dimension line, and leader line. Figure 9-1 not only illustrates these lines, but illustrates spacing and size guidelines for gaps and extensions.

Dimension lines are always parallel to the distance dimensioned with an arrowhead on each end. They are usually broken in the middle for the dimensional value.

Extension lines extend from the part or feature being dimensioned. The line extends approximately 1/8″ beyond the arrowhead. About 1/16″ should be allowed between the extension line and the object to clearly separate the shape description from the size description. Center lines can be used as extension

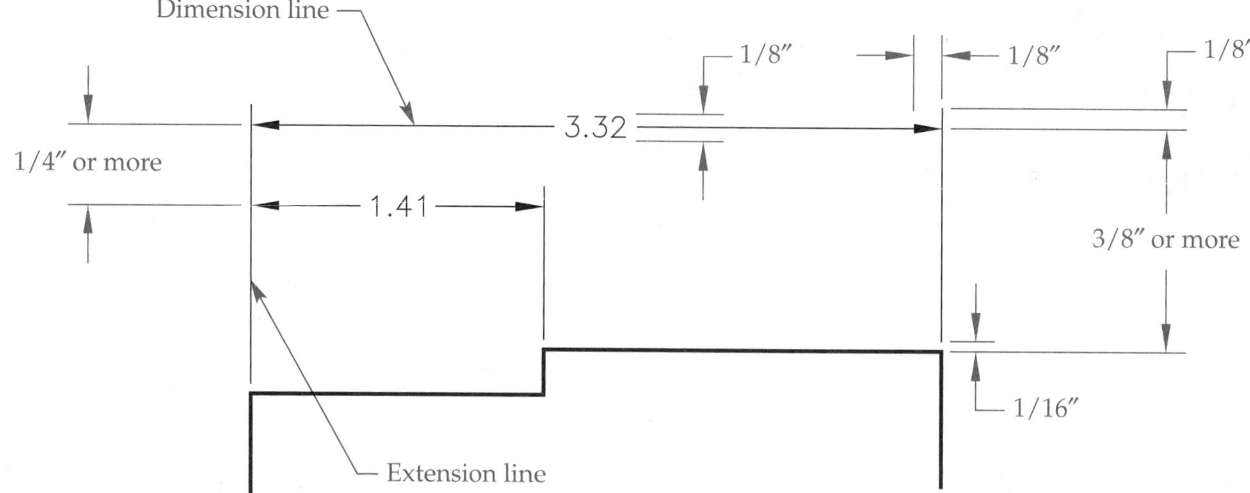


**Figure 9-1.**
This illustration describes some of the dimensioning mechanics the drafter uses to place dimension lines and extension lines in a clear and organized manner.

lines, in which case they cross visible lines without a gap. The first dimension should be at least 3/8″ away from the view. Subsequent dimensions should be another 1/4″ or more beyond the previous one.

Arrowheads should be about 1/8″ long, drawn sleek and slender, sharply formed, and precisely terminated at the extension line. Lettering is usually 1/8″ tall with 1/16″ gaps on each side to allow clearance around the number value.

Leader lines are used for local notes that point to a particular feature on the drawing, Figure 9-2. Leader lines are commonly used to point to a hole or screw thread. In the case of a hole, the leader should line up with the center of the hole and be at an angle. The leader line typically has a shoulder of about 1/4″ at the end leading into the beginning or end of a note.

## *Standard Dimensioning Symbols*

Dimensional values often require symbols, Figure 9-3. In previous practices, it may have been common to express the machining process for a hole, for example, as 1/2″ DRILL or .750″ REAM. In another example, it may have been common to simply state the word RADIUS or DIAMETER. In more recent standards, however, it is recommended the dimensioning callouts use symbols for radius and diameter and

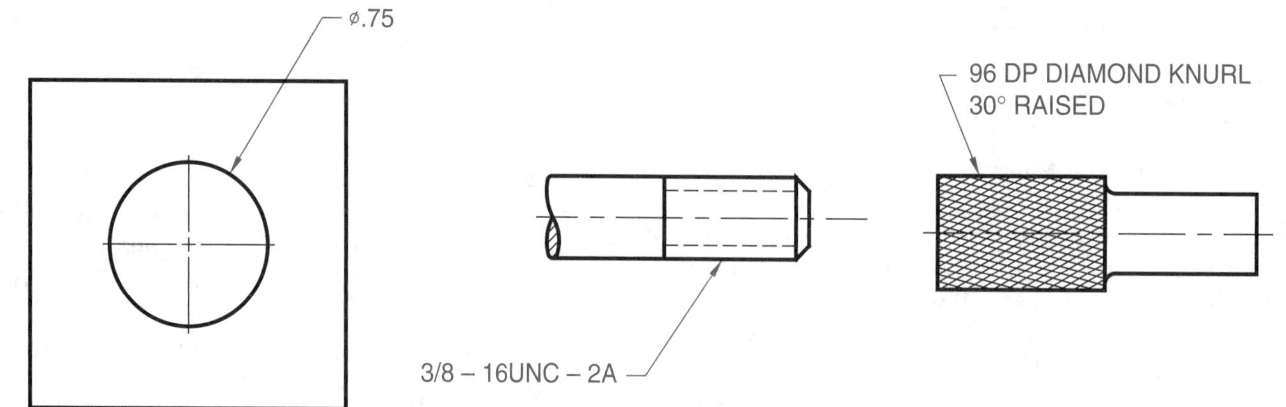


**Figure 9-2.**
Leader lines are used for local notes that point to a particular feature on the drawing.

| Symbol | Meaning |
|---|---|
| Ø | Diameter |
| R | Radius |
| SR | Spherical radius |
| SØ | Spherical diameter |
| CR | Controlled radius |
| X | Places or by |
| ( ) | Reference |
| ⌒ | Arc length |
|  | Slope |
|  | Square |
|  | Counterbore or spotface |
| ∨ | Countersink |
|  | Depth (or deep) |

**Figure 9-3.**
ASME recommends standard symbols that avoid the use of words when possible.

stating the machining process should be avoided unless there is a pressing reason to do so. As well, use of English words and acronyms is discouraged in a manufacturing setting where drawings may be used by non-English-speaking organizations.

## *Placement of Dimensional Values*

There are two methods of placing dimensional values on a drawing, print, or sketch. These are the unidirectional and aligned methods. ***Unidirectional dimensions*** are dimensions placed so they read from the bottom of the drawing. ***Aligned dimensions*** are dimensions placed so they read either from the bottom or right-hand side of the drawing with the number guidelines aligned with the dimension line, Figure 9-4. Unidirectional dimensions are most common in industrial prints. The aligned method is more common on architectural drawings, woodworking drawings, and other drawings that use fractional expressions.

The drafter or designer creating the print has one global rule that supersedes all suggestions:

> Dimension the drawing so it is clear and not confusing to the reader.

With this in mind, the following recommendations are also part of the dimensioning mechanics.

- It is permissible for extension lines to cross each other.
- It is desirable that dimension lines do *not* cross each other.
- Line sub-dimensions up with each other.
- Avoid crossing dimension lines with leader lines.
- Small dimensions create the need to break the dimension line into two pieces, Figure 9-5.

Angular dimensions are used on prints to indicate the size of angles in degrees (°) and fractional parts of a degree. Fractional parts of a degree are minutes (′) and seconds (″). A complete circle contains 360°. One degree contains 60′ (minutes). One minute contains 60″ (seconds). For an angular dimension, the dimension line is a curve with its center at the vertex of the dimensioned angle, Figure 9-6.

With respect to the dimensional values, in the U.S. customary system of measurements, linear dimensions are usually given in inches for measurements of 72″ or less. For measurements of greater than 72″, a combination of feet and inches is used. Where the print calls for accurate machining with close tolerances, the dimensions are usually given as decimal inches. This is the general practice in aerospace, automotive, electrical and electronic, machine tool, sheet metal, and similar industries. In the cabinetmaking, construction, and structural industries, dimensions are given in feet, inches, and fractions of an inch.

## Dimensioning Rules of Choice and Placement

In addition to the mechanics of dimensioning, the drafter or designer must then set about choosing which dimensions to use and then select the best view for those dimensions. In industrial mass production, parts are dimensioned in such a way that there is only one way to find the location or size of a feature.

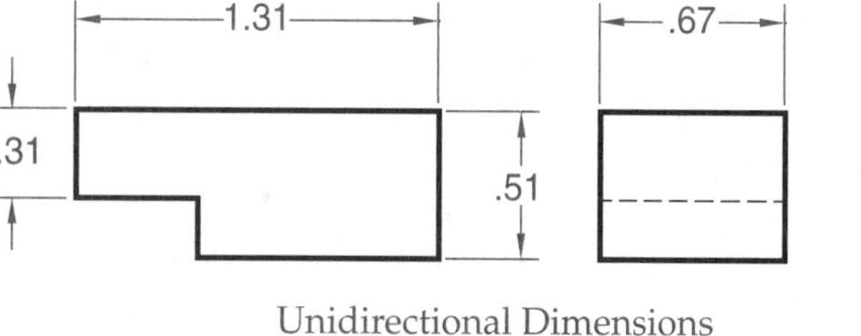


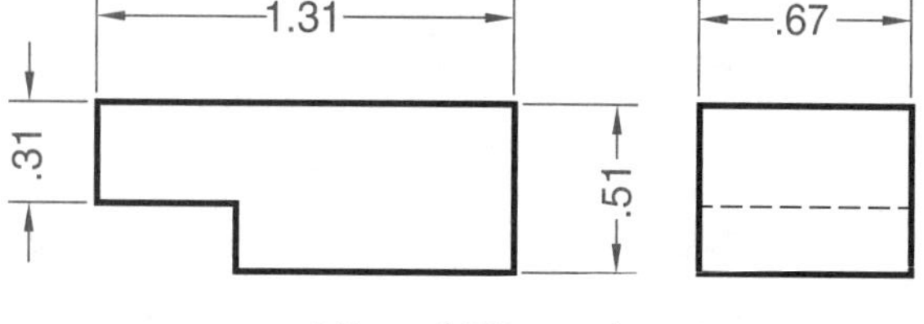


**Figure 9-4.**
There are two methods of placing dimensions on a drawing: unidirectional and aligned.

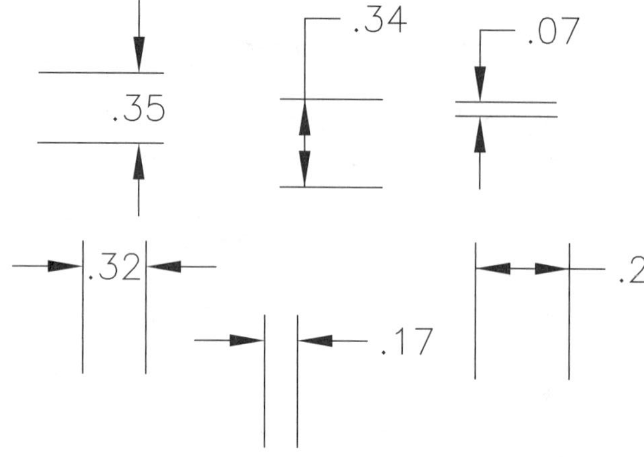


**Figure 9-5.**
The dimension line must be broken and separated in various ways for small dimensional values.

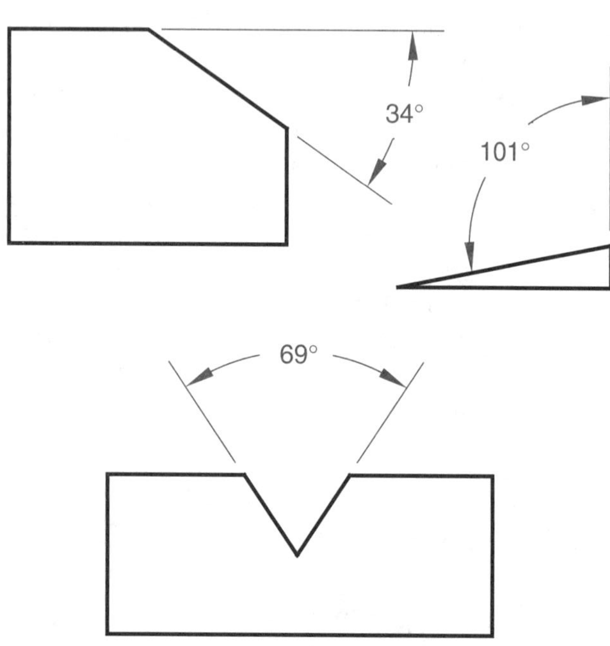


**Figure 9-6.**
For an angular dimension, the dimension line is a curve with its center at the vertex of the dimensioned angle.

The elements of *choice* and *placement* are the real difficult aspects of dimensioning. Knowing the guidelines for choice and placement may help the print reader know where to look for dimensions on a print. However, the function of the part, the purpose of a feature, and/or the machining and inspection processes may dictate dimension locations that do not follow these guidelines. In summary, the guidelines for choice and placement are just that—guidelines.

In keeping with the global rule "to clearly express the size," dimensions should be located off of the views, but still central to the drawing. Therefore, you should find most dimensions between the views. Notes and leader lines, however, are likely around the perimeter of the view arrangement.

A feature should be dimensioned in the view that is most descriptive of the feature's shape. This is known as ***contour dimensioning.*** A feature such as a slot or notch has its shape, or contour, shown in one particular view. Closely related to this principle is a drafter's rule to avoid dimensioning to a hidden line or the center line of a hidden hole. Figure 9-7 shows two versions of a drawing with contour dimensioning, one properly constructed and one poorly constructed.

For objects made of external cylindrical features, the diameter of the cylinder should be dimensioned in the "rectangular" view. See Figure 9-8. Even if a cylindrical feature is milled flat across one or more sides, the diameter of the cylinder is usually given. Older standards called for the diameter symbol to be omitted if there is a "round" view. However, current standards request the symbol to be used in all cases. Leader lines should only be used to point to a circle if it is a hole. Holes located around a bolt circle, or "circle of centers," can be located in several ways. Some common methods are shown in Figure 9-9.

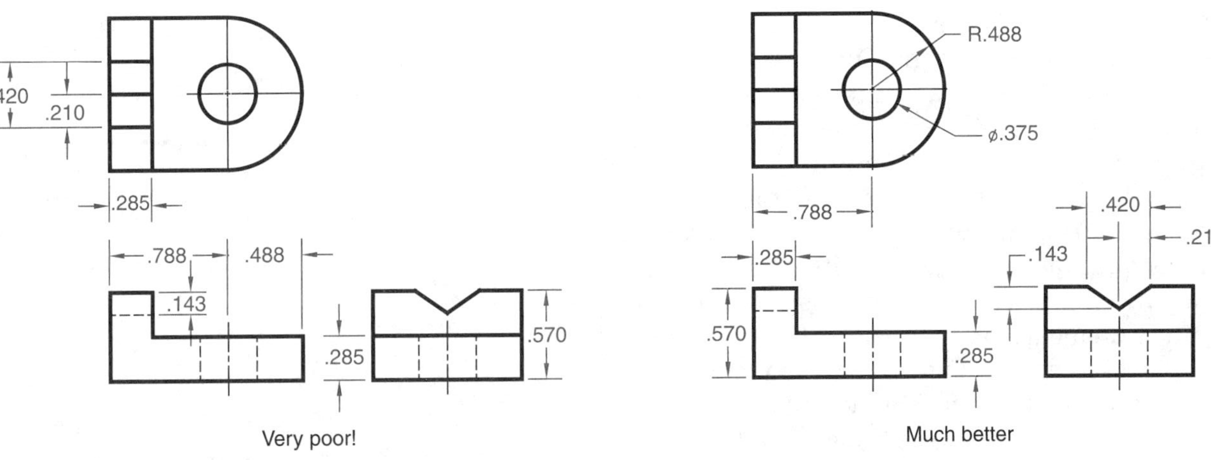


**Figure 9-7.**
Whenever possible, dimensions should be placed in the view that shows the shape of the feature being dimensioned. Also, try to avoid dimensioning to a hidden line. This is not always possible.

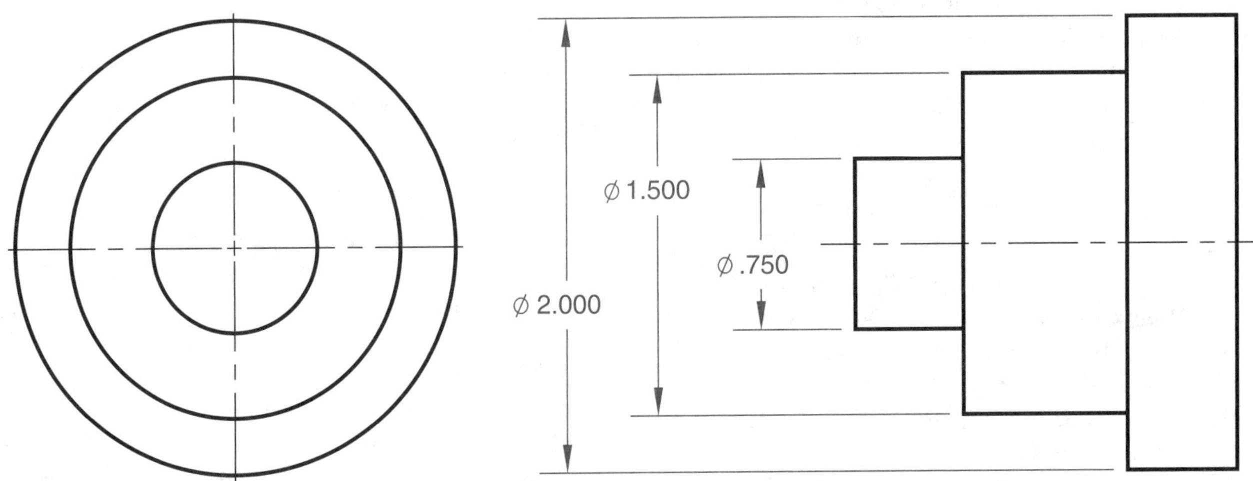


**Figure 9-8.**
External cylindrical features should be dimensioned in the rectangular view.

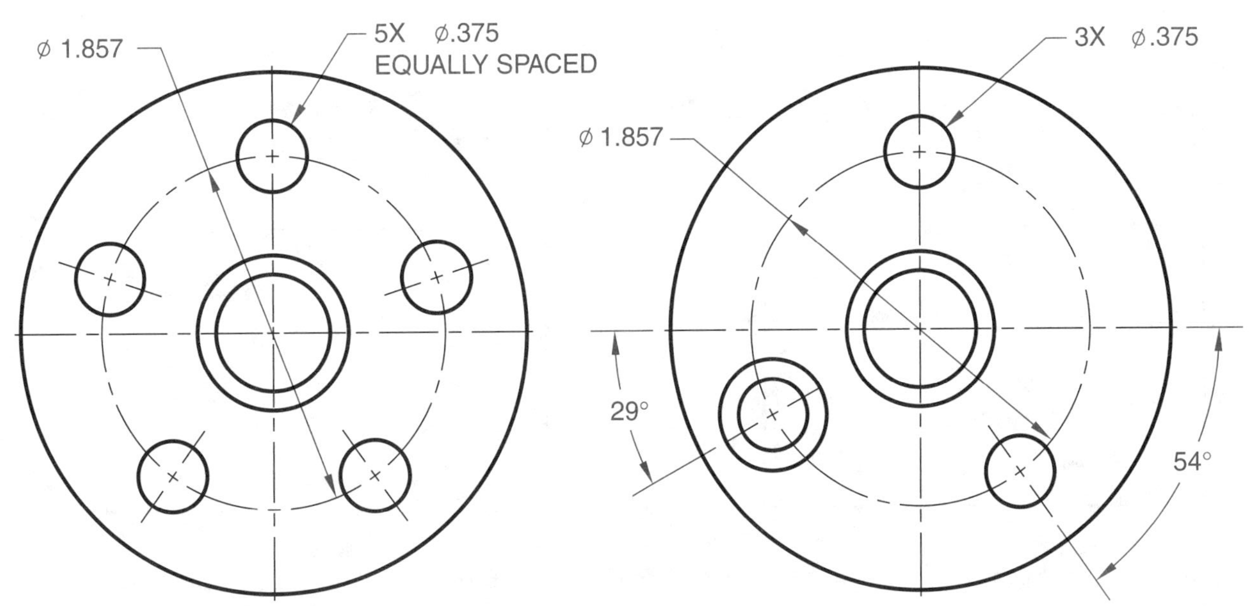


**Figure 9-9.**
Some common methods of locating holes on a bolt circle are shown here.

A drawing should never be "over dimensioned." There should be one way and one way only for a print reader to determine a feature's exact location or size. Sometimes a reference dimension is given to assist in analyzing the design or drawing. However, these dimensions are *not* to be used for manufacturing or quality control. Reference dimensions are discussed later in this unit.

An angled surface can be dimensioned by two offset measurements or by an angular value and an offset measurement, Figure 9-10. Sometimes an angular surface, by the nature of the angle, creates an edge that does not need a size dimension. If a print seems to be missing a dimension, check to see if this condition exists. Again, hopefully the drawing is dimensioned with the function and purpose of the angled surface in consideration. Manufacturing or inspection methods often dictate a particular choice of dimensions. In some cases, of course, the angled feature may not be critical. Unit 13, Geometric Dimensioning and Tolerancing, discusses various ways that geometric features, such as angled features, can be controlled with more precision and intent.

***Chain dimensioning,*** wherein dimensions are linked together end-to-end, can result in the accumulation of tolerances, if applied. To prevent this accumulation, features can be dimensioned from a common datum surface or feature, Figure 9-11. This is called ***baseline dimensioning*** or ***datum dimensioning.*** Tolerancing and datum references are covered in detail in Unit 10.

Features with rounded ends provide many choices for dimensioning. See Figure 9-12 for examples. The "correct" choice is often based on the accuracy required or the machining process used to create the feature. Parts with rounded ends are often not dimensioned overall. In this case, a radius is given and a center location is also required to determine overall measurements. The rounded ends of internal slots are often dimensioned as the distance across the slot, instead of the radius at the end of the slot. The reason for this could be that the slot will be machined with a cutting bit, therefore the diameter of the bit is the more valuable dimension.

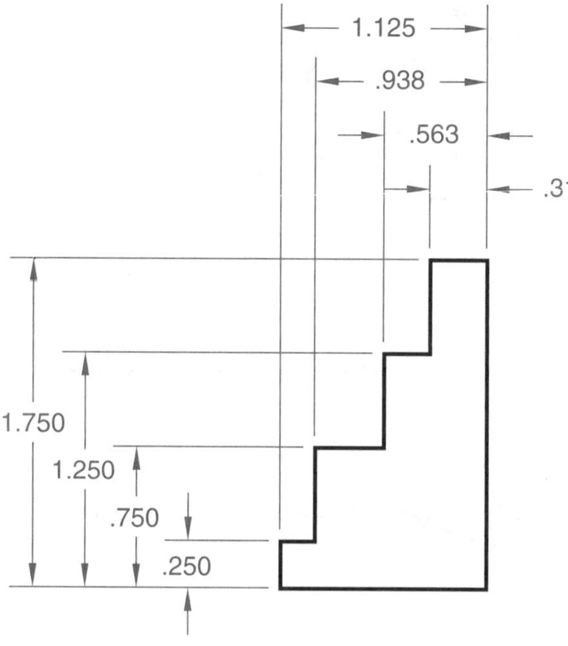


Figure 9-11.
Features are dimensioned from a common datum surface or feature to prevent the accumulation of tolerances.

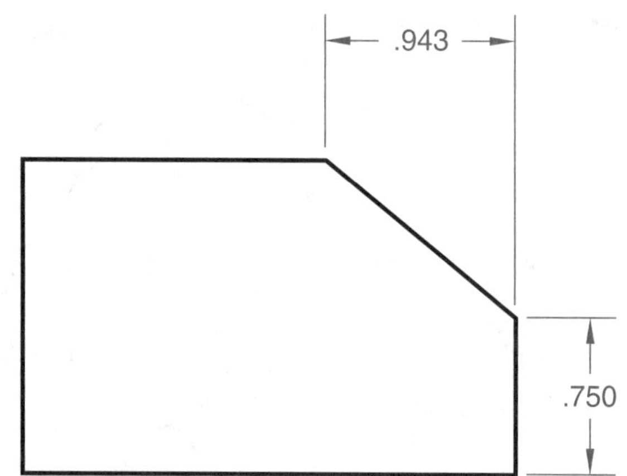


1.500
38°

Figure 9-10.
An angled surface can be dimensioned by two offset measurements (left) or by an angular value and an offset measurement (right).

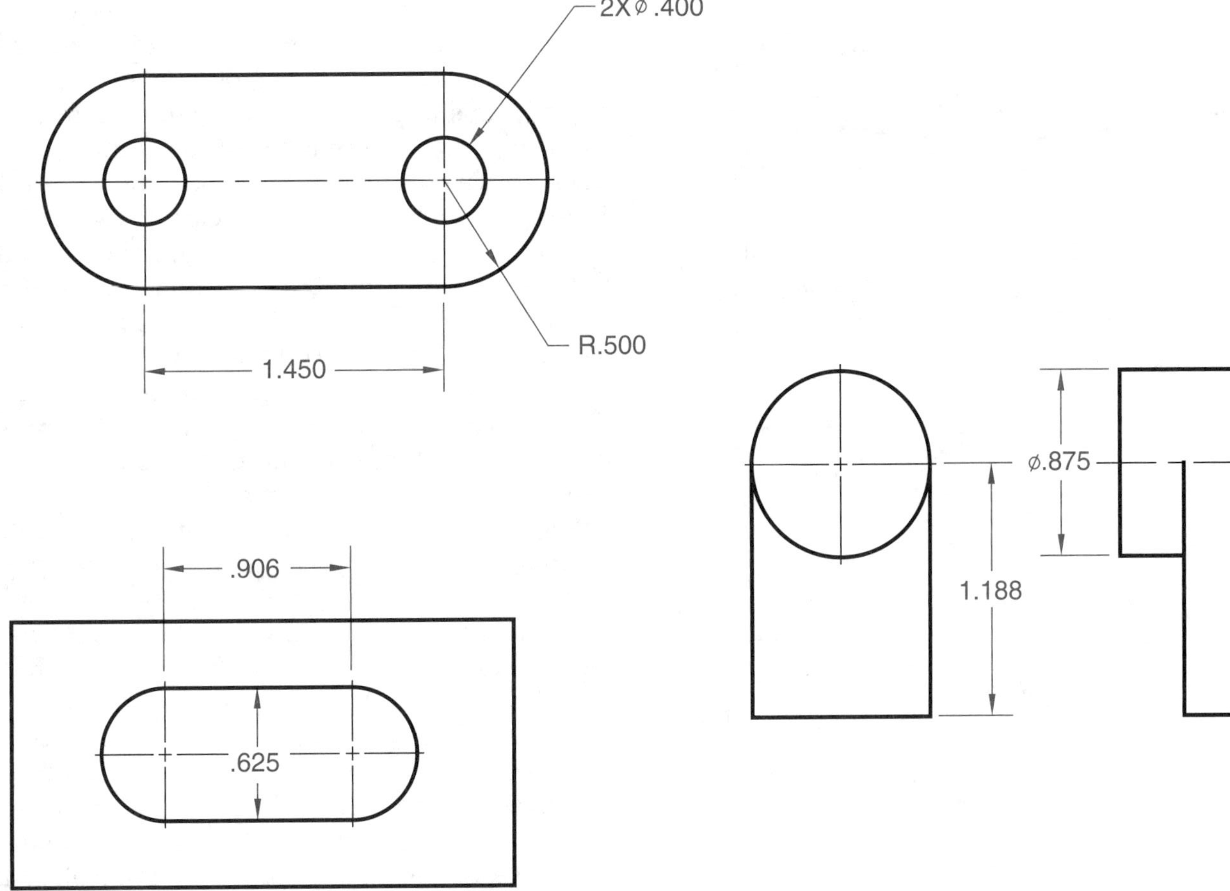


**Figure 9-12.**
Features with rounded ends provide many choices for dimensioning.

Another dimensioning rule is that center axes on corner arcs are usually self-locating. The location of the arc's center axis is based on the radius of the arc and the edges to which the curved surface is tangent. However, if a hole is located at the center of the arc, a location dimension must be given for the center axis of the hole, which coincides with the rounded corner axis. Even though this is the design intent, under close inspection the hole and curved surface axes may not perfectly coincide.

In summary, the guidelines in this section have been recommended throughout the years to help drafters create clear, concise size descriptions on drawings. There are too many special cases to cover within the scope of this textbook. Therefore, if a feature does not have a dimension or it appears a dimension is missing, do *not* make an assumption or measure the drawing for critical information. Simply put, as a print reader, you should *never* calculate or assume any dimension. Omissions should be reported to the department responsible for the prints.

## Additional Dimensioning Techniques

On a dimensioned drawing, there are usually both size and location dimensions. ***Size dimensions*** indicate the size of the part and the size of its various geometric features, such as holes, fillets, and slots. ***Location dimensions*** indicate the location of features such as holes, slots, and grooves.

As mentioned earlier, reference dimensions are occasionally given on drawings to assist in knowing certain general distance information, like the basic total size of a feature or sum of dimensions. ***Reference dimensions*** are not toleranced and are *not* to be used for layout, machining, or inspection operations. These dimensions are marked by parentheses or followed by REF, Figure 9-13.

***Tabular dimensions*** are placed on the drawing as reference letters within the dimension lines. A table on the drawing lists the corresponding dimensions, Figure 9-14. Tabular dimensions are useful when a company manufactures a series of sizes of an assembly or part. With this system, more than one part can be featured on one drawing.

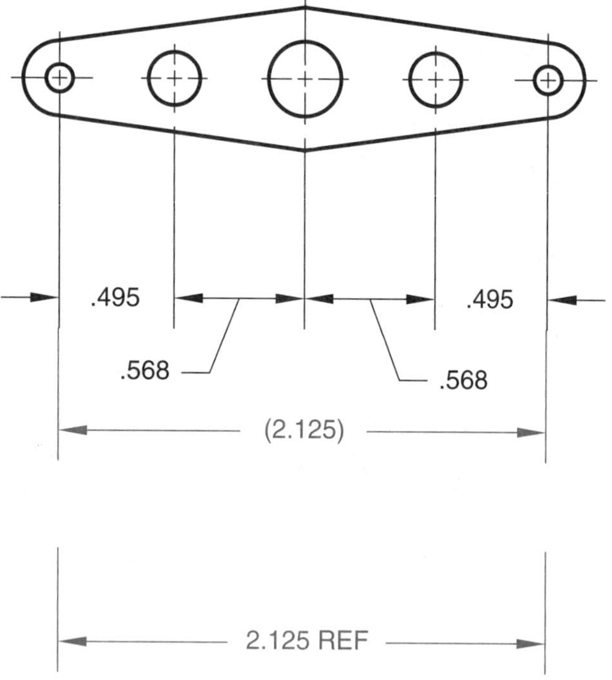


Figure 9-13.
Reference dimensions are marked by parentheses or followed by REF.

| PART NO. | A | B | C | D |
|---|---|---|---|---|
| 41–8706 | .750 | .500 | 1.312 | .875 |
| 41–8707 | 1.125 | .750 | 1.812 | 1.000 |
| 41–8708 | 1.500 | 1.000 | 2.062 | 1.125 |

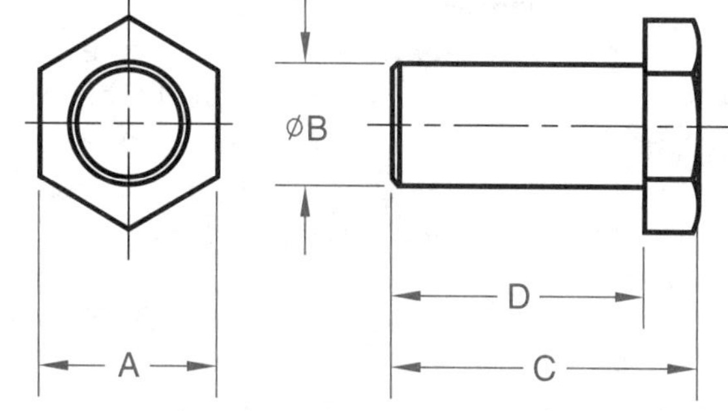


Figure 9-14.
In tabular dimensioning, a table on the drawing lists the corresponding dimensions for more than one part number.

Another use for tabular dimensions is dimensioning a part with a large number of repetitive features, such as holes, Figure 9-15. The chart in this application is sometimes referred to as a ***hole chart.*** Running extension and dimension lines to each hole would make the drawing difficult to read. To make the drawing clearer, each hole or feature can be assigned a letter, number, or letter with a subscript number. The dimensions of the feature and its location along the X and Y axes can then be given in a table on the drawing.

***Arrowless dimensioning*** is frequently used on drawings containing datum lines or planes, Figure 9-16. This practice eliminates numerous dimension and extension lines and improves the clarity of the drawing. Arrowless dimensioning is also called ***ordinate dimensioning, coordinate dimensioning,*** or ***datum dimensioning.*** This system of dimensioning is especially useful on drawings of parts that are to be machined on computer numerically controlled (CNC) equipment. Arrowless dimensioning requires "zero" reference planes be established from which all values are given in the X and Y directions.

***Dual dimensioning*** is used on a drawing to show both inch and millimeter values for some or all of the dimensions. Some companies place the millimeter above the dimension line and the inch value below the dimension line. Other companies use the bracket method and place the metric value in brackets following the inch value, as shown in Figure 9-17. In either case, the method used should be consistent throughout the drawing. A note should appear on the print to explain the method used.

## Notes and Special Features

Notes are almost always necessary to complement views and dimensions. Local notes use leader lines to point to special features, holes, and threads. Special features often include machining specifications, including counterbores, countersinks, tapers, knurls, keyways, and chamfers. General notes usually apply to the entire part. Examples of general notes include FILLETS AND ROUNDS R.125 UNLESS SPECIFIED, FINISH ALL OVER, and BREAK ALL SHARP CORNERS. Many of these special features, machining specifications, and other notes are examined in more detail in Unit 11, Machining Specifications and Drawing Notes.

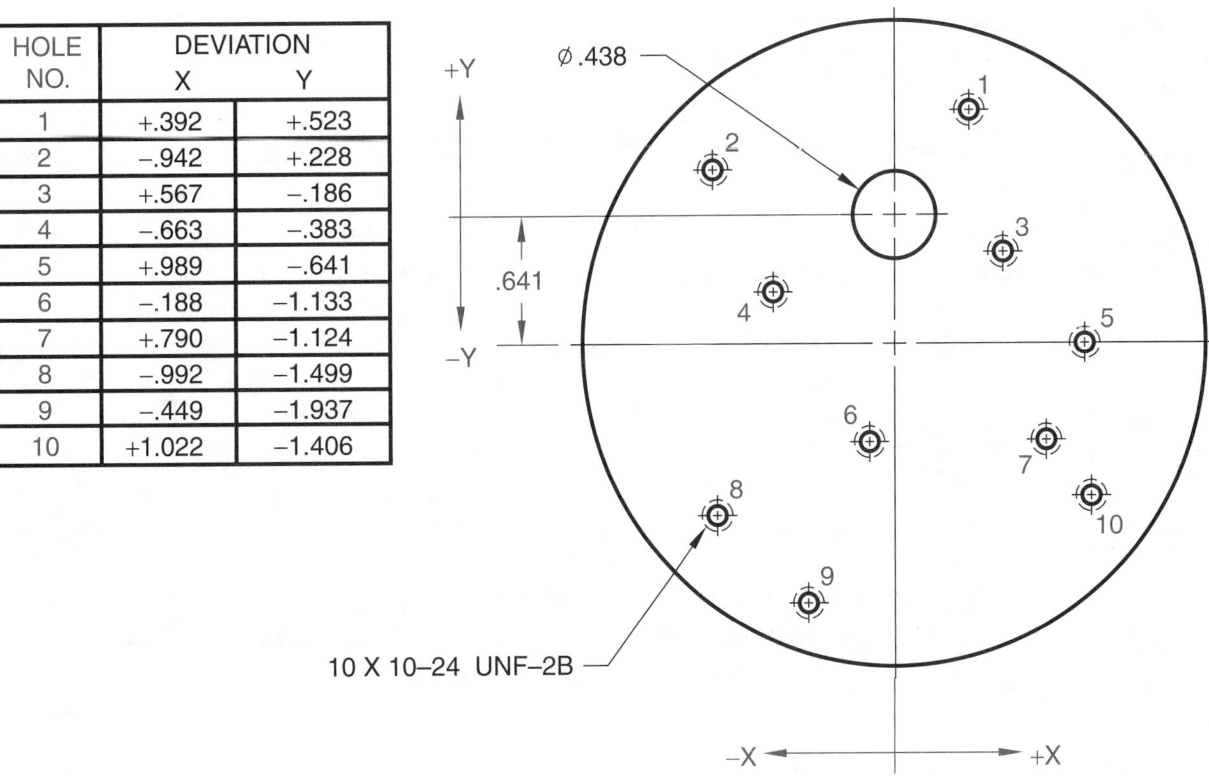

| HOLE NO. | DEVIATION X | Y |
|---|---|---|
| 1 | +.392 | +.523 |
| 2 | –.942 | +.228 |
| 3 | +.567 | –.186 |
| 4 | –.663 | –.383 |
| 5 | +.989 | –.641 |
| 6 | –.188 | –1.133 |
| 7 | +.790 | –1.124 |
| 8 | –.992 | –1.499 |
| 9 | –.449 | –1.937 |
| 10 | +1.022 | –1.406 |



**Figure 9-15.**
A hole chart is a type of tabular dimensioning for a part with a large number of repetitive features, such as holes.

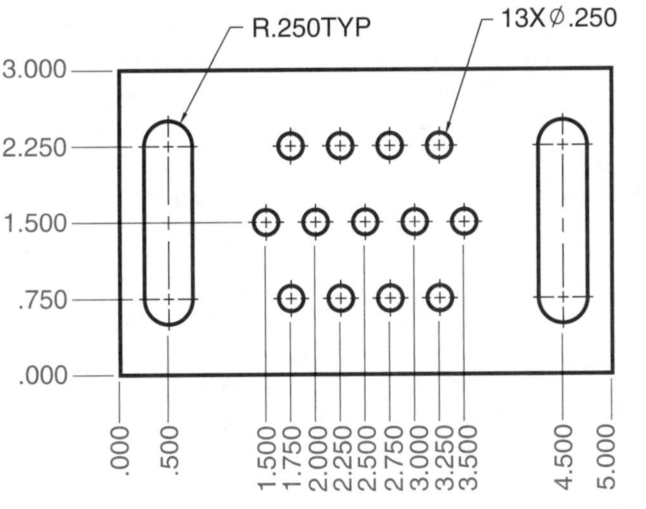



**Figure 9-16.**
Arrowless dimensioning eliminates numerous dimension and extension lines and improves the clarity of the drawing.

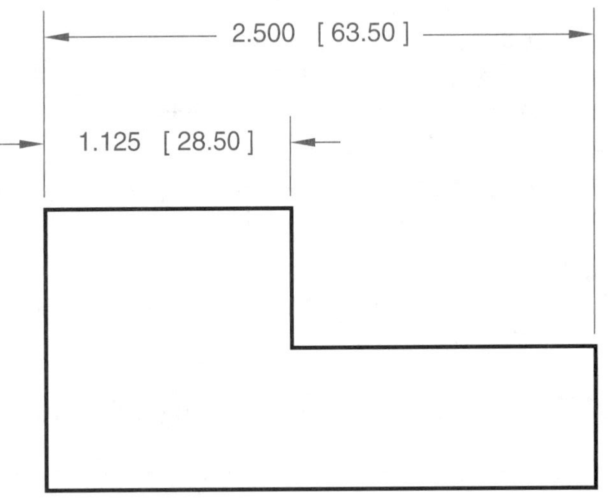



**Figure 9-17.**
Dual dimensioning is used on a drawing to show both inch and millimeter values.

## Review Questions

*Circle the answer of choice, fill in the blank, or write a short answer.*

1. Define dimensioning in two words. ______________________________

   ______________________________

2. *True or False?* Dimensioning is the easiest part of creating an industrial print.

3. Which of the following is the best definition for dimensioning mechanics?
   A. What dimensions look like.
   B. Where dimensions are placed.
   C. Which dimensions are used.
   D. All of the above.

4. For each of the following, list the recommended size.
   A. Amount of gap between extension lines and the object. ______________
   B. Amount of extension line beyond the arrowhead. ______________
   C. Height of the lettering. ______________
   D. Length of the arrowhead. ______________

5. The dimension line should be a minimum distance of ______________ from the object.

6. Leader lines have a(n) ______________ on one end and a(n) ______________ on the other.

7. Sketch the symbol for each of these words.
   A. Diameter ______________
   B. Radius ______________
   C. Square ______________
   D. Deep ______________

8. *True or False?* It is permissible for extension lines to cross each other but, if possible, dimension lines should not cross each other.

9. If the dimension line is an arc, that is an indication the dimension is a(n) ______________ dimension.

10. What term is applied to placing dimensions in the view where the shape is shown? ______________

    ______________________________

11. *True or False?* Cylinders are dimensioned by a leader line pointing to the circle view and a linear dimension that shows the length or height.

12. What is the term applied to aligning sub-dimensions arrowhead-to-arrowhead across the view?

    ______________________________

13. All dimensions can be grouped into two categories: ______________ and ______________.

14. *True or False?* Reference dimensions should be indicated by placing the numeric value in parentheses.

15. What type of note uses a leader line? ______________________________

______________________________

## Review Activity 9-1

*Sketch the dimensions required to completely describe the size of the block shown below without over dimensioning. If feasible, discuss the choice and placement with others to evaluate the available options. Your instructor will indicate whether or not you should use real number values and measurements or simply use "X" for the number values to focus on the mechanics and the choice and placement options.*

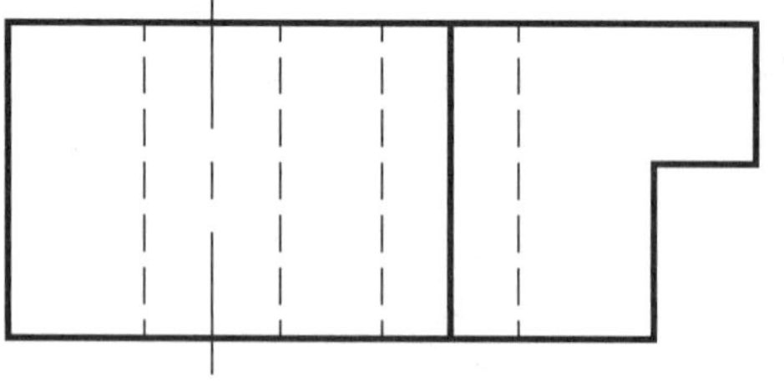

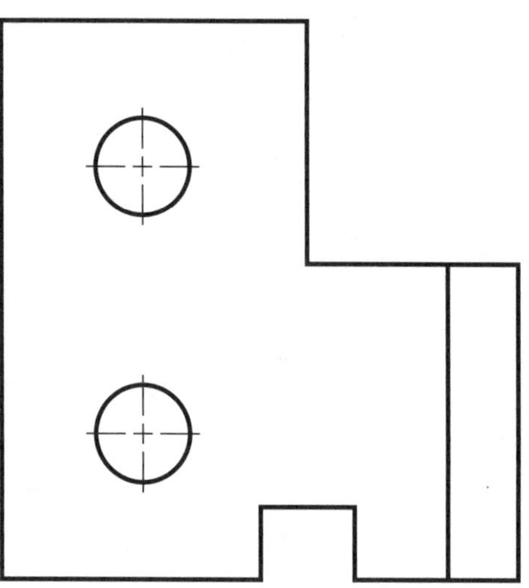

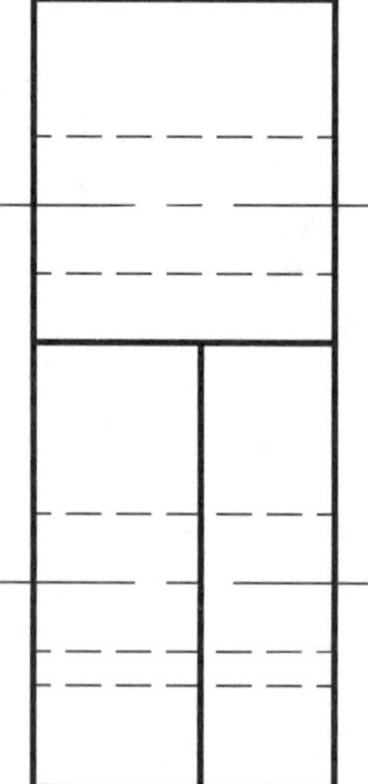

## Review Activity 9-2

*Sketch the dimensions required to completely describe the size of the block shown below without over dimensioning. If feasible, discuss the choice and placement with others to evaluate the available options. Your instructor will indicate whether or not you should use real number values and measurements or simply use "X" for the number values to focus on the mechanics and the choice and placement options.*

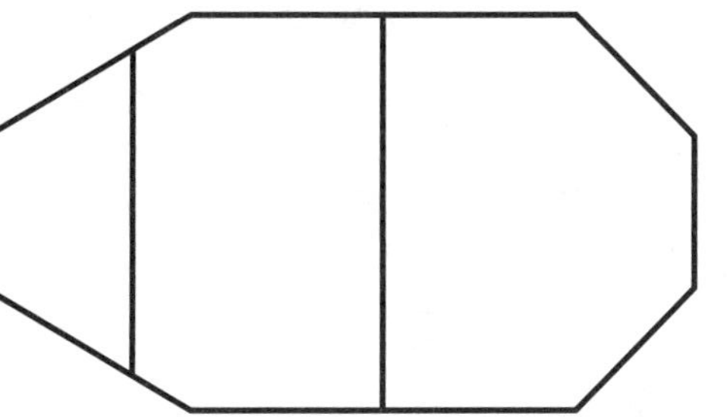

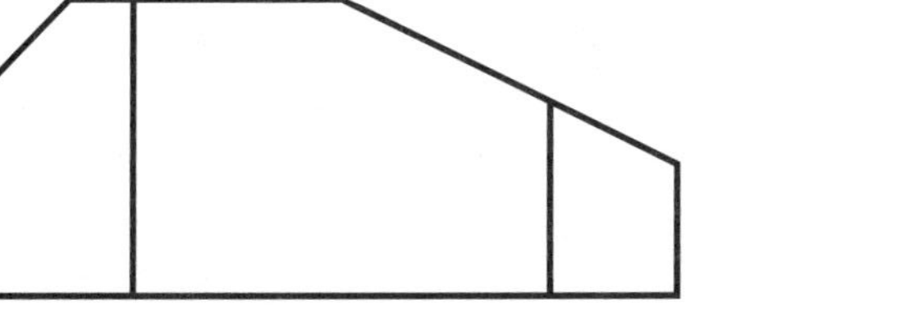

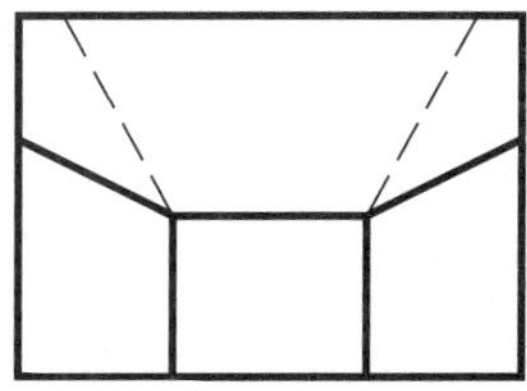

## Review Activity 9-3

*For the drawings on the next page, determine the minimum number of size and location dimensions required to completely dimension the given views.*

1. Minimum number of dimensions:
   A. 6
   B. 7
   C. 8
   D. 9
   E. 10

2. Minimum number of dimensions:
   A. 6
   B. 7
   C. 8
   D. 9
   E. 10

3. Minimum number of dimensions:
   A. 6
   B. 7
   C. 8
   D. 9
   E. 10

4. Minimum number of dimensions:
   A. 8
   B. 9
   C. 10
   D. 11
   E. 12

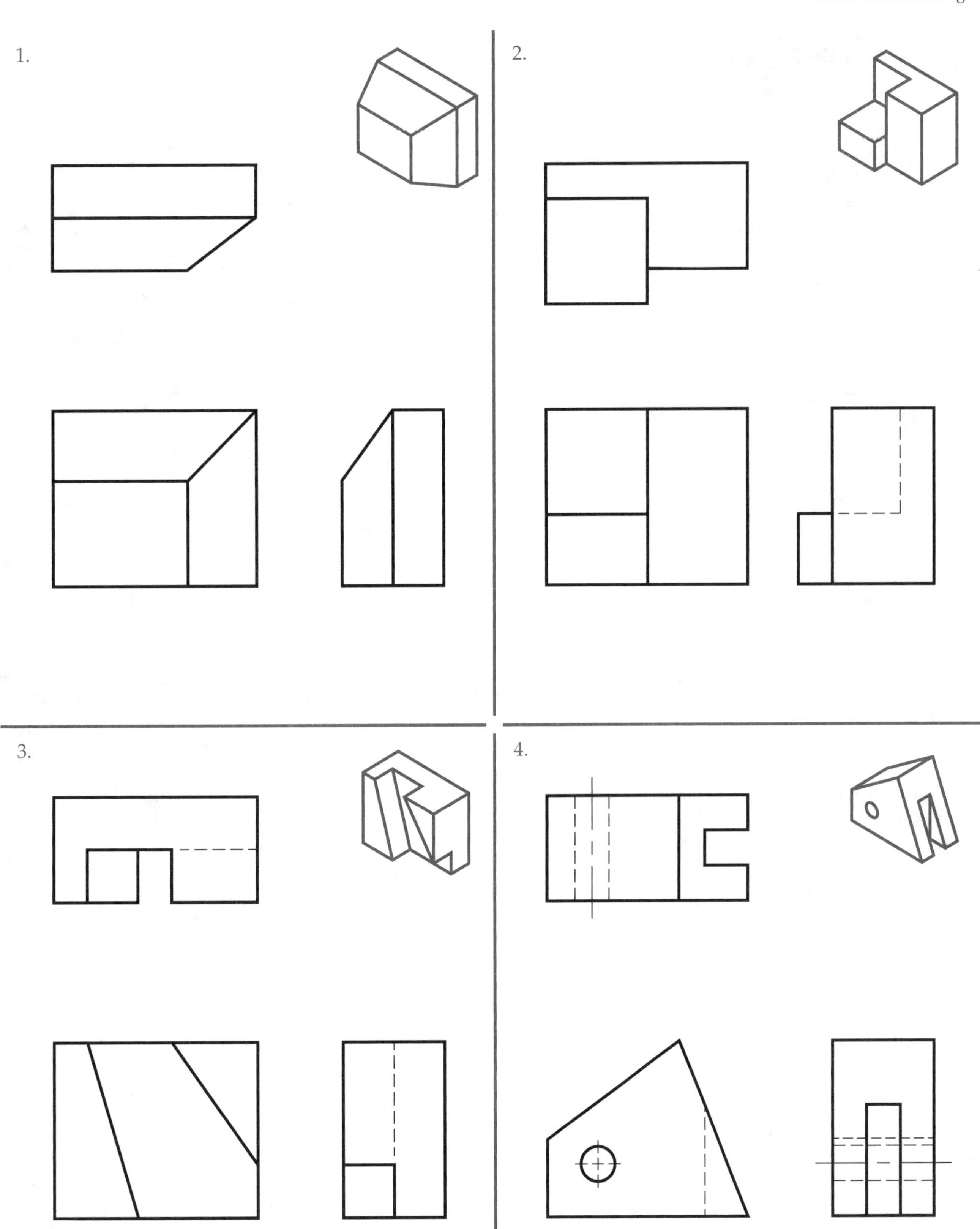
1.
2.
3.
4.

## Review Activity 9-4

*For the drawings on the next page, determine the minimum number of size and location dimensions required to completely dimension the given views.*

1. Minimum number of dimensions:
   A. 8
   B. 9
   C. 10
   D. 11
   E. 12

2. Minimum number of dimensions:
   A. 6
   B. 7
   C. 8
   D. 9
   E. 10

3. Minimum number of dimensions:
   A. 6
   B. 7
   C. 8
   D. 9
   E. 10

4. Minimum number of dimensions:
   A. 7
   B. 8
   C. 9
   D. 10
   E. 11

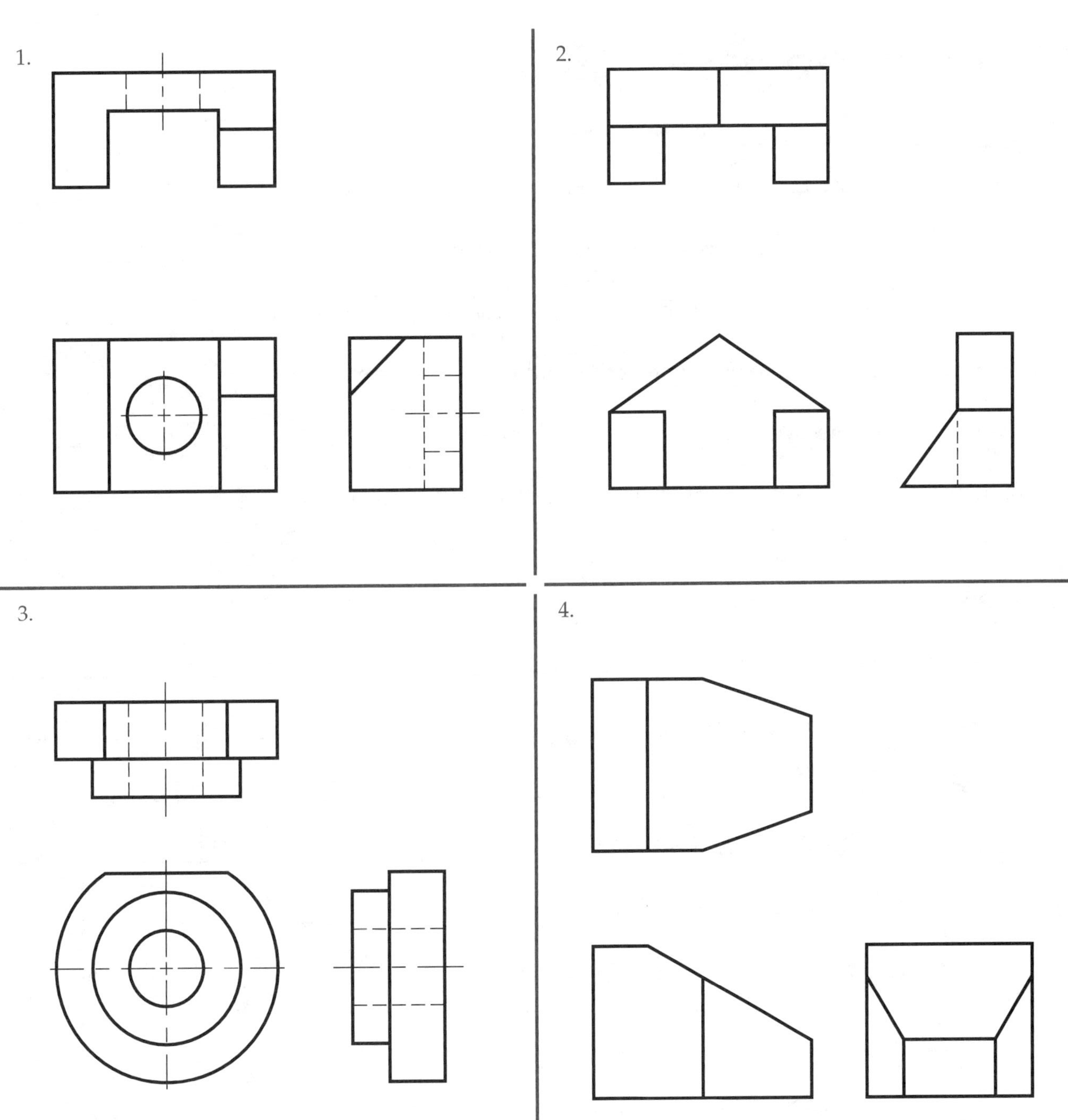
1.
2.
3.
4.

## Industry Print Exercise 9-1

*Refer to the print PR 9-1 and answer the questions below.*

1. How many angular dimensions are shown on this print? ______
2. How many reference dimensions are shown on this print? ______
3. Did the creator of this print follow the cylinder rule for the overall diameter? ______
4. Does this print use the aligned system or the unidirectional system? ______
5. How many local notes with leader lines are there on this print? ______
6. Counting reference dimensions, how many diameters are specified? ______

*Review questions based on previous units:*

7. What paper size is the original version of this print? ______
8. What is the material for this part? ______
9. What is the name of this part? ______
10. What is the part number of this print? ______
11. To what scale is the part drawn on the original drawing? ______
12. If the far-left circular view is called the left side view, what names would be given to the other three views? ______
13. How many "through holes" does this object have? ______
14. Counting all surfaces of the object, how many are "flat" planar surfaces? ______
15. Does this drawing feature any section views? ______

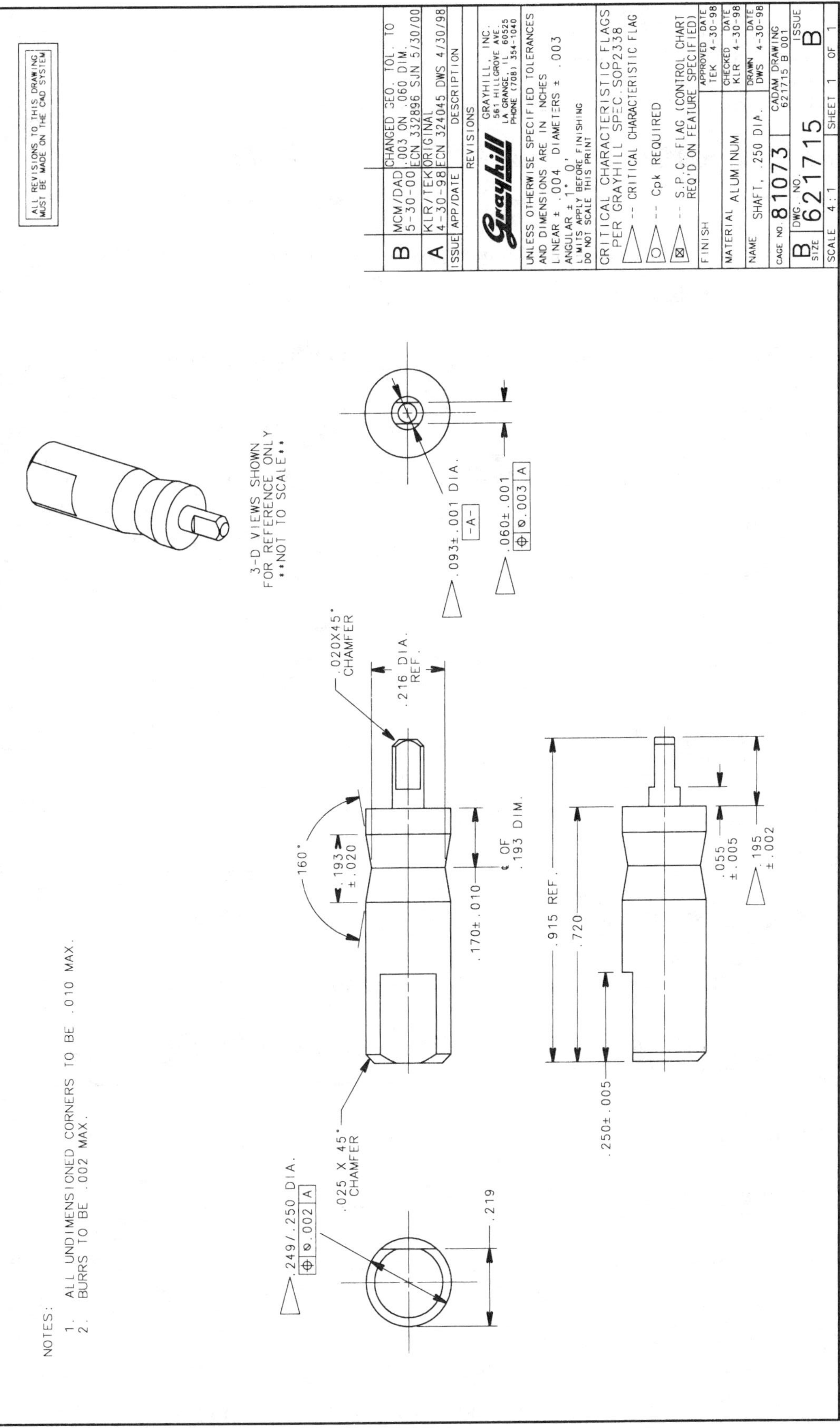


PR 9-1.
Print supplied by Grayhill, Inc.

## Industry Print Exercise 9-2

*Refer to the print PR 9-2 and answer the questions below.*

1. According to the front view, what is the center-to-center width between the two "posts" of this object? ______
2. On the enlarged Detail A, two dimensional values begin with 4X. What does that mean? ______
3. On the enlarged Detail A, what dimensional value is repeated as a reference measurement?
4. Of all of the diameters specified on this print, which is the largest? ______
5. Of all of the arcs that are specified by radius, which is the smallest? ______
6. The main stem of each post is tapered with a smaller diameter at the tip. What is the angle of the taper? ______

*Review questions based on previous units:*

7. Are there any section lines shown on this drawing? ______
8. What is the name of this part? ______
9. What is the number of this part? ______
10. What geometric 3D shape is created due to the 4° dimension shown in Section A-A? ______
11. Does this object have any features that could be described as cylindrical? ______
12. If the view that contains the cutting-plane line is the front view and the view above it the top view, what view direction is removed Section A-A? ______
13. What scale is used for Detail A? ______
14. Does this drawing feature any auxiliary views? ______
15. Does this part feature any threads? ______

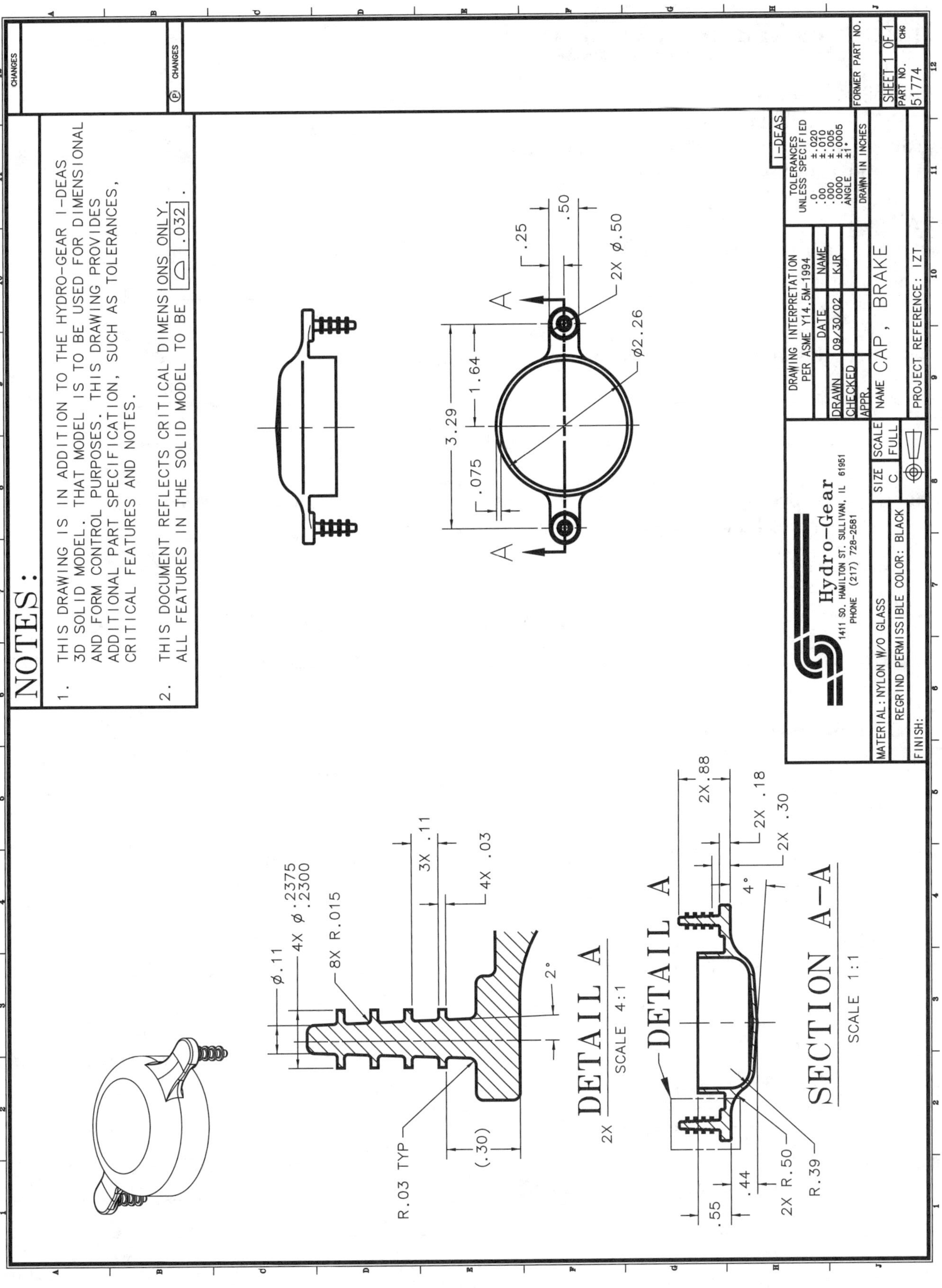


PR 9-2.
Print supplied by Hydro-Gear.

## Bonus Print Reading Exercises

The following questions are based on various bonus prints located in the folder at the back of this textbook. Refer to the print indicated, evaluate the print, and answer the question.

**Print AP-001:**

1. What is the diameter of the hole that is inclined 15°? _______________

**Print AP-002:**

2. Why does this drawing only have nine dimensions and those are all in parentheses? _______________

**Print AP-003:**

3. Many companies have a symbol, used much like an asterisk, that points out critical dimensions. How many critical dimensions are specified on this print? _______________
4. For the dimensions, does this print use the aligned method or unidirectional method? _______________

**Print AP-004:**

5. Why is the total width not given in the front view (central on drawing)? _______________
6. List the overall height, width, and depth of this part. Some calculation may be necessary. _______________

**Print AP-005:**

7. Did the person creating this print use the cylinder rule? _______________
8. Not counting leader line callouts, how many cylindrical features in Section G-G are specified as linear dimensions? _______________

**Print AP-007:**

9. How many reference dimensions are specified on this print? _______________
10. How many angular dimensions are specified on this print? _______________

**Print AP-008:**

11. How many diameter dimensions are given that are smaller than 17 mm? ________________

12. Looking at all of the left-to-right linear dimensions on this part, what is different about the way they are created from the ASME standard illustrations in this textbook? ________________

**Print AP-012:**

13. This print (although just 1 of 3 sheets) features a chart for two dimensions, A and B. What is dimension A locating? ________________

14. There are some dimensions on this print that are missing the dimension line and arrow on one side of the dimension. To what would the other arrow point to or touch if it were visible on the print?

**Print AP-013:**

15. Unless otherwise specified, how thick are the walls on this part? ________________

**Print AP-014:**

16. Leader lines are used for local notes, but also used to specify the size of a hole or set of holes. How many leader lines in this print are associated with hole size information? ________________

17. In the circular right side view, how many dimensions shown are location dimensions? ________________

**Print AP-015:**

18. How many radius dimensions in this print are specifying a radius value larger than 1″? ________________

19. One hole on this part is a blind hole drilled before threads are cut. What depth is specified for that hole?

**Print AP-018:**

20. One of the dimensions on this part has a double arrow and no extension line. What feature is the double arrow indicating, which is not shown? ________________

21. In the top view of this assembly, the hole surface specified for datum E also has an inch value in parentheses and a metric value in parentheses and brackets. What do the parentheses indicate?

22. This print features dual dimensioning. What is the significance of the comma within the bracketed values? ________________

**Print AP-019:**

23. What diameter is specified for the circle of centers (bolt circle)? ______

24. Does this print feature unidirectional or aligned dimensioning? ______

**Print AP-021:**

25. One leader line on this print does not feature an arrowhead. Why? ______

26. How many dimensions of this print are specifying an angular measurement? ______

**Print AP-023:**

27. How many holes on this print have a diameter of .125″? ______

28. Examine the holes marked A on this print. Left to right, are the three holes along the top edge lined up with the three holes along the bottom edge? ______

29. This one-view drawing does not show the depth in a side view. What is the depth? ______

**Print AP-024:**

30. One dimension indicates a distance of .300 TYP. What does TYP stand for? ______

31. For some holes, this print also implements a type of location dimensioning known as ______.

**Print AP-025:**

32. What letters are used to specify the datums on this print? ______

33. What method is used to specify the diameters of the holes? ______

**Print AP-026:**

34. Why are several features of this part not dimensioned on this print? ______

UNIT 10
# Tolerancing

*After completing this unit, you will be able to:*

**Define** terms related to tolerancing.

**Explain** how tolerances are expressed on a drawing.

**Identify** tolerance values for dimensions on the drawing, regardless of the tolerancing expression.

**Compare** customary inch tolerance classes with metric tolerance classes.

**Calculate** tolerances or limits for mating parts based on maximum material conditions and allowance.

Most companies do not manufacture all of the parts and subassemblies required in their products. Frequently, the parts are manufactured by "specialty industries" or subcontractors that work from specifications provided by the company. The parts and subassemblies are then assembled into the final product. The key to the successful operation of the final product is that duplicate parts must be interchangeable and still satisfactorily function. A common term for this is ***interchangeable manufacturing.*** To achieve this, all parts must be manufactured to within specified limits of size. The range of size is defined by the tolerance. The meaning of various terms, symbols, procedures, and techniques relating to tolerancing are presented in this unit.

## Tolerancing Terms

The terms presented in this section are frequently used in industry. In order to satisfactorily read and interpret prints, it is necessary to understand their true meaning and application. Other terms relating to conditions and applications of tolerancing can be found in the glossary of this text.

The ***tolerance*** is the total amount by which a dimension can vary. You can think of this as how much you can "tolerate" the dimension being less than 100% precise. Tolerance may be expressed as the desired size followed by a bilateral or unilateral tolerance expression or as a set of limits, Figure 10-1. Notice that all four of the examples given create the same range of size for the hole. Tolerances can be applied to all feature dimensions, including size dimensions and location dimensions, Figure 10-2.

A ***unilateral tolerance*** is a tolerance that permits variation from the dimensional value in one direction only. ***Bilateral tolerance*** is a tolerance that permits variation from the dimensional value in both directions. The permitted variation in a bilateral tolerance can be equal or unequal values. An example of a bilateral tolerance is ±.004″. A common error in reading a bilateral tolerance, such as this example, is to say that the tolerance is .004″. In reality the tolerance is .004″ + .004″, or .008″ total. The method of expressing tolerances with a basic size and a plus/minus value is called the ***plus-or-minus dimensioning*** method.

***Limits*** are the extreme maximum and minimum dimensions of a part allowed by the application of a tolerance. Two limit dimensions are always involved, which are a maximum size and a minimum size. For example, the design size of a part may be 1.375. If a tolerance of plus or minus two thousandths of an inch (±.002″) is applied, then the two limits are maximum limit 1.377 and minimum limit 1.373. This method of expressing a tolerance with limits is called the ***limit dimensioning*** method.

***Nominal size*** is a general term used to designate size for a commercial product, Figure 10-3. It may or may not express the true numerical size of the part or material. For example, a seamless wrought-steel pipe of 3/4″ (.75″) nominal size diameter has an actual inside diameter of .824″ and an actual outside diameter of 1.050″. Another example of nominal size is dimensional lumber used in construction.

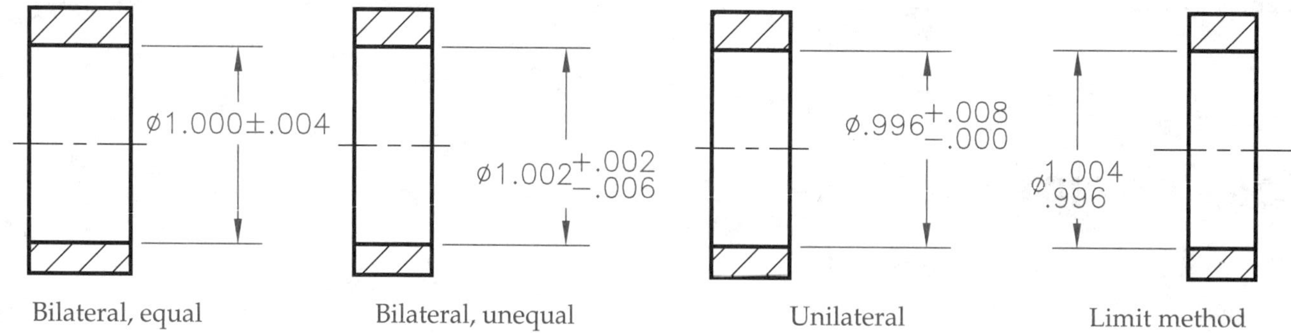


**Figure 10-1.**
Tolerance may be expressed as the desired size followed by a bilateral or unilateral tolerance or as a set of limits. In all four examples shown here, the tolerance is .008″. Therefore, the limits are the same.

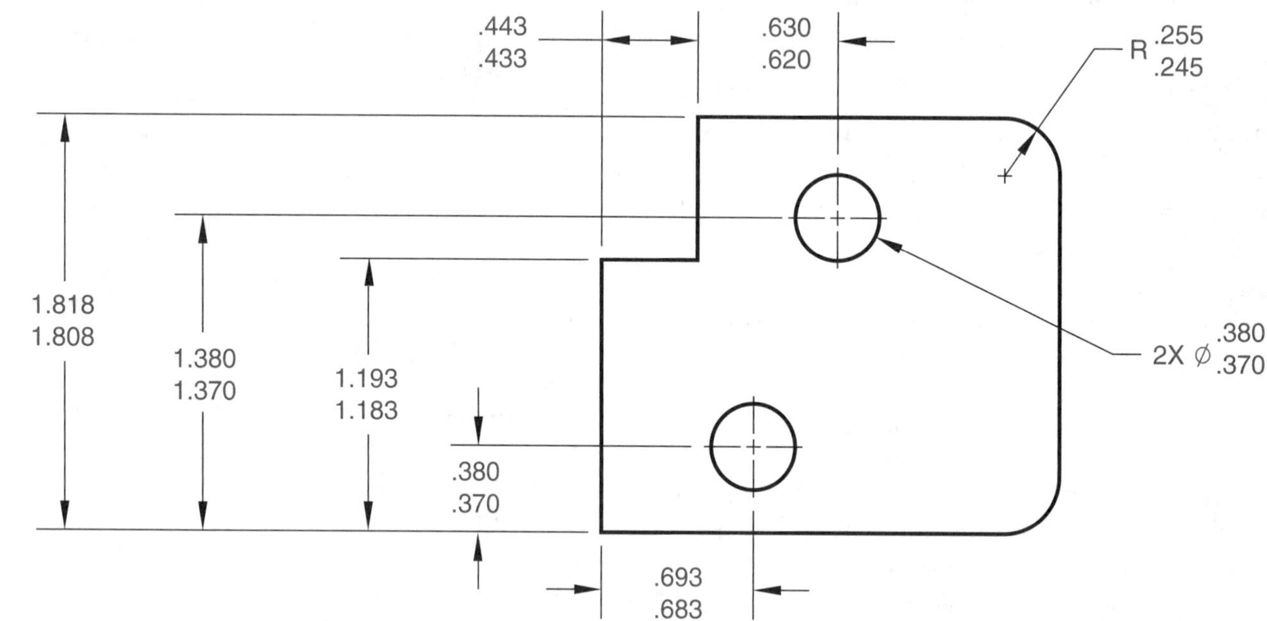


**Figure 10-2.**
Tolerances can be applied to all feature dimensions, including size dimensions and location dimensions.

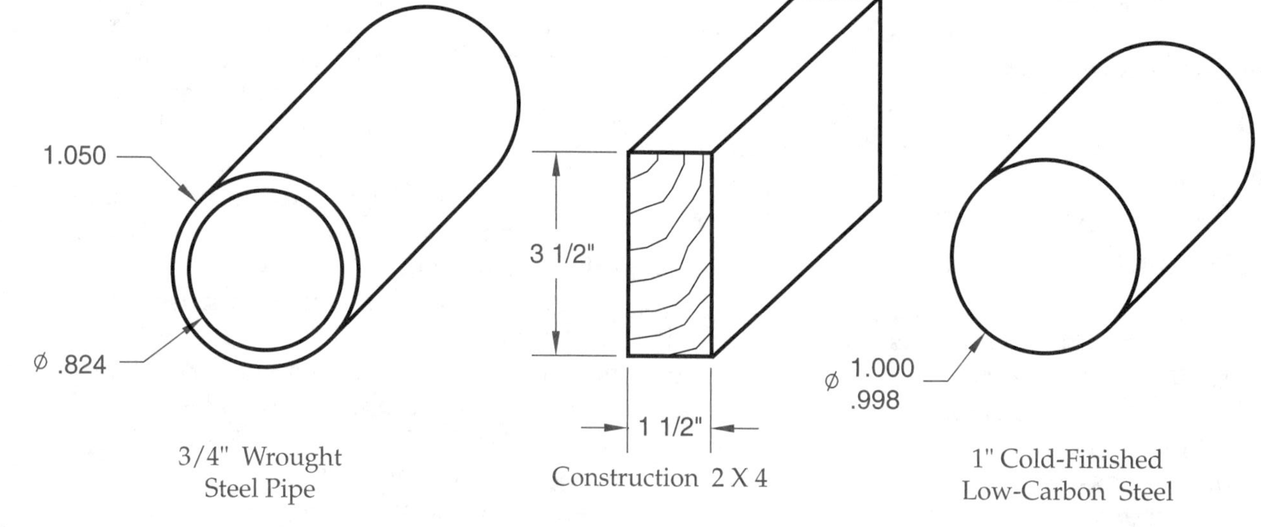


**Figure 10-3.**
Nominal size may or may not express the true, actual size of the part or material.

A 2×4 lumber size actually measures 1-1/2″×3-1/2″. In some cases, such as cold-finished, low-carbon steel rounds, the nominal 1″ size comes within two-thousandths (.002″) tolerance of actual size. Therefore, nominal size may or may not be an accurate numerical size of a material. Be aware that terms are not used the same in all regions of the country or from company to company. That is why standards are so important. Within some settings, the term "nominal size" may mean the same as basic size.

In general, the ***basic size*** is the size determined by engineering and design requirements for a part or set of mating parts. For a part that does show a dimensional value with a plus and minus tolerance, the basic size is the target value indicated. Basic size may also mean the size from which allowances and tolerances are applied.

For example, strength and stiffness may require a 1″ diameter shaft and hole for two parts to work well together. Using a system known as the ***basic hole system***, the basic 1″ size is applied to the hole as the smallest value. A slightly smaller size is applied to shaft's larger value. Tolerances are also applied to both parts to determine the looser fit. See Figure 10-4. In actuality, both parts are calculated based on the nominal size of 1″. Each part then can be dimensioned with a basic dimension with plus and minus tolerances applied or the dimensions can be expressed with the limit method.

Geometric dimensioning and tolerancing (GDT) is covered in Unit 13. Drawings that incorporate GDT apply location tolerances differently. In short, the location dimensions are expressed with a box around them and are called basic dimensions, but are defined as being theoretically exact. In these cases, the location tolerance is specified in a feature control frame, so the tolerance is applied in a different fashion. The GDT application of the term basic dimension is similar, but unrelated to the term basic dimension as discussed in this unit.

***Maximum material condition (MMC)*** is useful in determining how precisely parts fit together. The MMC of a feature is the size limit value representing the feature when the most material is present. For an external feature, such as a pin or post, MMC is the *largest* value. However, for an internal feature, such as a slot or hole, MMC is the *smallest* value. At first, this may seem like an unusual way to express sizes. Be careful to not talk about "maximum size", but rather "maximum material." Using the terms "upper limit" and "lower limit" to speak about

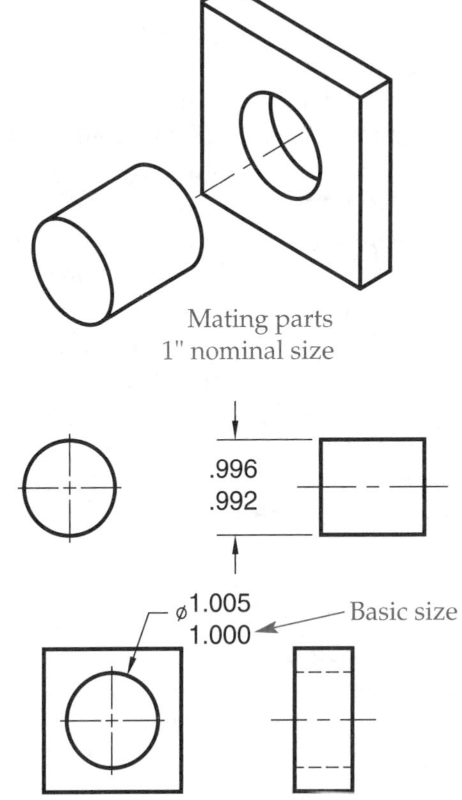


**Figure 10-4.**
With the basic hole system, the nominal or basic size is used as the MMC of the hole and tolerances are calculated for both the hole and the shaft based on the requirements of the fit.

the larger and smaller values allows the word "maximum" to be reserved for MMC. The MMC of an external feature is the upper limit, or larger value, because there is "more material" when the part is larger. The MMC of an internal feature is the lower limit, or smaller value, because there is "more material" when the hole is smaller. Making the hole larger requires removing material. One of the advantages of using the term MMC is the ability to determine the tightest fit between two parts simply by comparing the MMC values.

While not as critical to design calculations, the ***least material condition (LMC)*** of a feature is also an appropriate term. The least material condition of an internal feature is the larger size. The least material condition of an external feature is the smaller size. The loosest fit can be determined by comparing the least material conditions.

***Allowance*** is the intentional difference as expressed by the MMC of the dimensions of mating parts to provide for different classes of fits. There are different ways of describing allowance; "tightest fit" is perhaps the most appropriate. However, allowance can also be described as the MMC of the external feature subtracted from the MMC of the internal feature. Therefore, if allowance is a positive number, the fit is a clearance fit. If allowance is a negative number, the fit is an interference fit. In Figure 10-4, the MMC of the hole is 1.000" and the MMC of the pin is .996". Therefore, the allowance (tightest fit) is +.004". As the hole size departs from MMC (larger hole) or as the pin size departs from MMC (smaller pin), there is more clearance. If the parts pass inspection based on their size tolerances, the "loosest fit", or worst-case scenario, is a clearance of +.013".

A ***datum*** is the origin from which the location of features are established or the reference plane or axis to which a geometric control is referenced. A datum is established by real features or surfaces of a part. Datums are essential to many of the GDT controls in Unit 13. A datum is assumed to be exact for the purpose of reference. The datum identification symbol, per the ASME Y14.5 standard, is indicated by a letter in a small rectangle or box attached to the views in some manner by a stem that features a triangle, usually filled, Figure 10-5. On older drawings, the datum may be indicated in a different manner. The older standard for datum identification is still common in many CAD systems and remains in use in some companies. The older datum symbol is a letter preceded and followed by a dash enclosed in a small rectangle or box, but does not have a connecting stem.

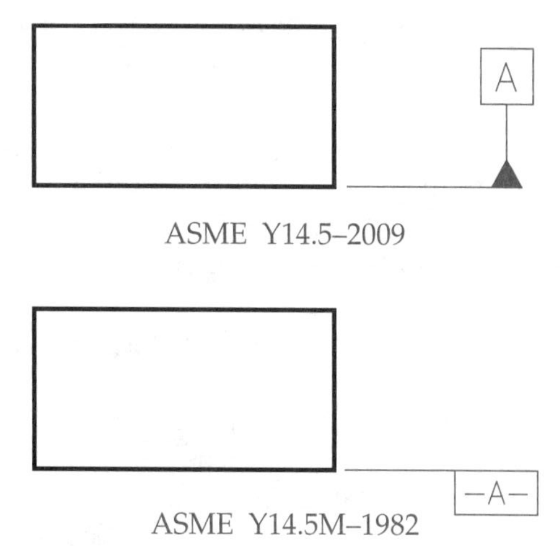


**Figure 10-5.**
A datum is identified with a symbol and the feature can then be used as the origin from which the location of other features are established or the reference plane or axis to which a geometric control is referenced.

Sometimes it is important to specify from which surface a dimension originates. By definition and unless otherwise specified, the surface points of either surface can be used to establish a measurement plane. Then, the surface points of the opposite surface are determined to be a certain distance away from the plane. With some features, such as illustrated in Figure 10-6, it may be that a shorter surface is the mounting surface and the longer surface must be within a tolerance range as measured from the shorter surface. The ***dimension origin*** symbol provides a concise and clear way of stating this on the print without establishing a datum surface and using GDT controls.

Another tolerancing technique that needs to be addressed is the MAX note. In some cases where the dimension value is small, such as rounding or "breaking" off corners, it may be convenient for the designer to indicate the maximum value for the radius or break, such as R.03 MAX. This is a unilateral expression that designates the largest value. The smallest amount is left up to the machinist. Usually in these cases, it is the intent of the designer that the radius or break *not* be zero, although the specification seems to indicate that zero is the lower limit with the value given as the upper limit.

## American Standard Tolerances

Engineers and designers are responsible for assigning tolerances to parts in an assembly so the assembly can function properly. Tolerances should always be as large as possible to reduce manufacturing costs, but should also ensure that any random selection of two parts does not have a situation that is too loose or too tight. Therefore, engineers and designers must know the appropriate level of accuracy for a given application and process. Fits between plain (non-threaded) cylindrical parts are recommended within the ASME B4.1 standard, which sets forth preferred limits and fits based on application. Within this standard, there are five types of fit. The fits and the letter symbols used to identify them are:

- Running or Sliding Clearance Fits (RC)
- Locational Clearance Fits (LC)
- Transitional Clearance or Interference Fits (LT)
- Locational Interference Fits (LN)
- Force or Shrink Fits (FN)

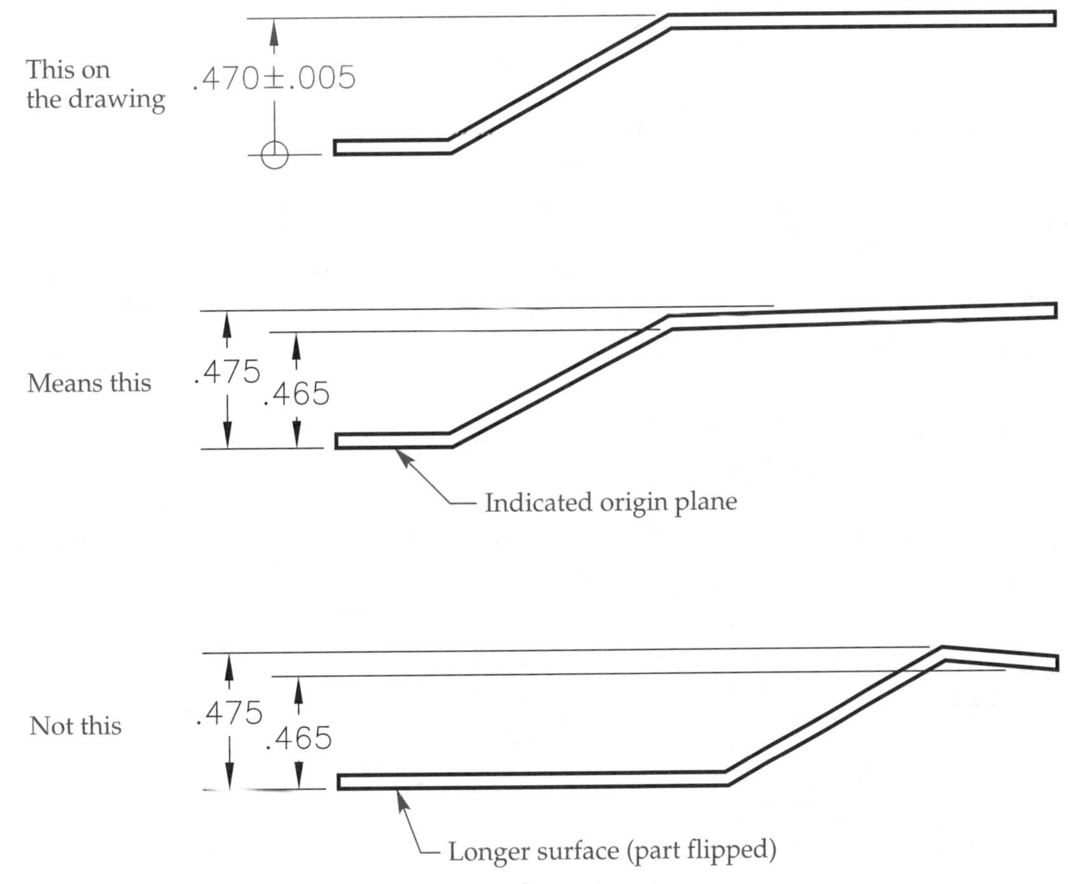


**Figure 10-6.**
A dimension origin symbol is sometimes used to specify which surface should be used for the measuring plane. Especially in cases without a datum reference, this could make a difference when inspecting the size measurement to see if it is within tolerance.

The abbreviations for each of these fits are not intended to be shown on industrial prints. Therefore, this section provides a general understanding of tolerances and how they may be derived, but does not explain what the print reader will see on the drawing.

Specifying one of the standard fits is a quick and easy way for an engineer or designer to communicate a desired amount of tolerance to a drafter or technician. For example, the engineer may simply specify on a sketch that a hole and its mating part should be a 3/4″ RC3 class fit. The technician then uses the tables to calculate the limits for each of the mating parts. See Figure 10-7. Finally, a drawing is created using the proper values and an appropriate tolerance method (limit or plus and minus). These dimensions are sometimes referred to as precision fits and may have a tolerance value carried to four decimal places. Less critical dimensions are typically to three-place decimal precision on drawings measured in inches.

To use the table, find the row that matches the size range of the nominal diameter. If the nominal diameter is 1″, the size range is 0.71–1.19. Notice that the values are given as thousandths of an inch, so move the decimal three places to the left, adding zeros as needed. If the class of fit specified by the designer or engineer is RC2, then use the Hole column to determine the tolerance for the hole. Using the basic hole system described earlier, the MMC of the hole should be 1.0000″, while the other limit is 1.0005″ (+0.5 means add .5 thousandths, which equals .0005″). The Shaft column indicates that 0.3 and 0.7 should be subtracted from the basic size to derive limits of .9997″ to .9993″ for the shaft. The MMC of the shaft is, therefore, .9997″ and the MMC of the hole is 1.0000″. This means the allowance for this RC2 class fit is .0003″. The loosest fit for a 1.0007″ hole and a .9993″ shaft would be .0014″. As you can see, this is a very precise fit. The ***limits of clearance*** is only 1.4 thousandth of an inch, which is a very tight range.

**Basic Hole System**
**Limits are in thousandths of an inch**
**Apply limits to the basic size**

| Nominal Size Range Inches Over To | RC1 | | | RC2 | | | RC3 | | | RC4 | | |
|---|---|---|---|---|---|---|---|---|---|---|---|---|
| | Limits of Clearance | Standard Limits | | Limits of Clearance | Standard Limits | | Limits of Clearance | Standard Limits | | Limits of Clearance | Standard Limits | |
| | | Hole | Shaft | | Hole | Shaft | | Hole | Shaft | | Hole | Shaft |
| 0.00 - 0.12 | 0.1<br>0.45 | +0.2<br>−0 | −0.1<br>−0.25 | 0.1<br>0.55 | +0.25<br>−0 | −0.1<br>−0.3 | 0.3<br>0.95 | +0.4<br>−0 | −0.3<br>−0.55 | 0.3<br>1.3 | +0.6<br>−0 | −0.3<br>−0.7 |
| 0.12 - 0.24 | 0.15<br>0.5 | +0.2<br>−0 | −0.15<br>−0.3 | 0.15<br>0.65 | +0.3<br>−0 | −0.15<br>−0.35 | 0.4<br>1.12 | +0.5<br>−0 | −0.4<br>−0.7 | 0.4<br>1.6 | +0.7<br>−0 | −0.4<br>−0.9 |
| 0.24 - 0.40 | 0.2<br>0.6 | +0.25<br>−0 | −0.2<br>−0.35 | 0.2<br>0.85 | +0.4<br>−0 | −0.2<br>−0.45 | 0.5<br>1.5 | +0.6<br>−0 | −0.5<br>−0.9 | 0.5<br>2.0 | +0.9<br>−0 | −0.5<br>−1.1 |
| 0.40 - 0.71 | 0.25<br>0.75 | +0.3<br>−0 | −0.25<br>−0.45 | 0.25<br>0.95 | +0.4<br>−0 | −0.25<br>−0.55 | 0.6<br>1.7 | +0.7<br>−0 | −0.6<br>−1.0 | 0.6<br>2.3 | +1.0<br>−0 | −0.6<br>−1.3 |
| 0.71 - 1.19 | 0.3<br>0.95 | +0.4<br>−0 | −0.3<br>−0.55 | 0.3<br>1.2 | +0.5<br>−0 | −0.3<br>−0.7 | 0.8<br>2.1 | +0.8<br>−0 | −0.8<br>−1.3 | 0.8<br>2.8 | +1.2<br>−0 | −0.8<br>−1.6 |
| 1.19 - 1.97 | 0.4<br>1.1 | +0.4<br>−0 | −0.4<br>−0.7 | 0.4<br>1.4 | +0.6<br>−0 | −0.4<br>−0.8 | 1.0<br>2.6 | +1.0<br>−0 | −1.0<br>−1.6 | 1.0<br>3.6 | +1.6<br>−0 | −1.0<br>−2.0 |
| 1.97 - 3.15 | 0.4<br>1.2 | +0.5<br>−0 | −0.4<br>−0.7 | 0.4<br>1.6 | +0.7<br>−0 | −0.4<br>−0.9 | 1.2<br>3.1 | +1.2<br>−0 | −1.2<br>−1.9 | 1.2<br>4.2 | +1.8<br>−0 | −1.2<br>−2.4 |
| 3.15 - 4.73 | 0.5<br>1.5 | +0.6<br>−0 | −0.5<br>−0.9 | 0.5<br>2.0 | +0.9<br>−0 | −0.5<br>−1.1 | 1.4<br>3.7 | +1.4<br>−0 | −1.4<br>−2.3 | 1.4<br>5.0 | +2.2<br>−0 | −1.4<br>−2.8 |
| 4.73 - 7.09 | 0.6<br>1.8 | +0.7<br>−0 | −0.6<br>−1.1 | 0.6<br>2.3 | +1.0<br>−0 | −0.6<br>−1.3 | 1.6<br>4.2 | +1.6<br>−0 | −1.6<br>−2.6 | 1.6<br>5.7 | +2.5<br>−0 | −1.6<br>−3.2 |
| 7.09 - 9.85 | 0.6<br>2.0 | +0.8<br>−0 | −0.6<br>−1.2 | 0.6<br>2.6 | +1.2<br>−0 | −0.6<br>−1.4 | 2.0<br>5.0 | +1.8<br>−0 | −2.0<br>−3.2 | 2.0<br>6.6 | +2.8<br>−0 | −2.0<br>−3.8 |
| 9.85 - 12.41 | 0.8<br>2.3 | +0.9<br>−0 | −0.8<br>−1.4 | 0.8<br>2.9 | +1.2<br>−0 | −0.8<br>−1.7 | 2.5<br>5.7 | +2.0<br>−0 | −2.5<br>−3.7 | 2.5<br>7.5 | +3.0<br>−0 | −2.5<br>−4.5 |
| 12.41 - 15.75 | 1.0<br>2.7 | +1.0<br>−0 | −1.0<br>−1.7 | 1.0<br>3.4 | +1.4<br>−0 | −1.0<br>−2.0 | 3.0<br>6.6 | +2.2<br>−0 | −3.0<br>−4.4 | 3.0<br>8.7 | +3.5<br>−0 | −3.0<br>−5.2 |

Figure 10-7.
A table similar to this is used to calculate tolerances for standard inch-based fits. These calculations are not usually done by the print reader, rather by the drafter or technician creating the drawing.

## Metric Tolerances and Fits

The international scope of many companies has created the need for international standards to help American companies be more productive with their operations. For metric dimensions, standards from the International Organization for Standardization (ISO) have been incorporated into ASME B4.2, entitled *Preferred Metric Limits and Fits.* This standard does not use the same notations of RC, LC, etc., nor are the tables quite the same, but the concepts are the same. In Figure 10-8, a partial table is shown. Since the precision of millimeters is usually two or three decimal places, the tables do not require decimal manipulation. As well, preferred basic sizes are given so designers can choose directly from the table values that are preset. For example, a close running fit for a 5 mm diameter could simply be specified on the print as 5.000–5.018 (or 5.000 +.018) for the hole and 4.990–4.978 (or 4.990 –.012) for the shaft. This means the allowance is .01 mm and the loosest fit is .04 mm. In metric applications it is common to specify the MMC and then a unilateral tolerance for mating features.

Some companies may prefer to use the symbolic representations of metric precision fits. Instead of calling it the basic hole system, the term used in this metric standard is ***hole basis system.*** Tolerances can be determined with a shaft basis system. As you can see in Figure 10-9, hole or shaft relationships fall into ten different levels of tightness and each level has an alphanumeric designation. A ***free running fit*** for a hole has an H9 designation, while

**Preferred Metric Hole Basis Clearance Fits – Partial Table**

| Basic Size | | Loose Running | | | Free Running | | | Close Running | | | Locational Clearance | | |
|---|---|---|---|---|---|---|---|---|---|---|---|---|---|
| | | Hole H11 | Shaft c11 | Fit | Hole H9 | Shaft d9 | Fit | Hole H8 | Shaft f7 | Fit | Hole H7 | Shaft h6 | Fit |
| 1 | Max | 1.060 | 0.940 | 0.180 | 1.025 | 0.980 | 0.070 | 1.014 | 0.994 | 0.030 | 1.010 | 1.000 | 0.016 |
| | Min | 1.060 | 0.880 | 0.060 | 1.000 | 0.955 | 0.020 | 1.000 | 0.984 | 0.006 | 1.000 | 0.994 | 0.000 |
| 1.2 | Max | 1.260 | 1.140 | 0.180 | 1.225 | 1.180 | 0.070 | 1.214 | 1.194 | 0.030 | 1.210 | 1.200 | 0.016 |
| | Min | 1.200 | 1.080 | 0.060 | 1.200 | 1.155 | 0.020 | 1.200 | 1.184 | 0.006 | 1.200 | 1.194 | 0.000 |
| 1.6 | Max | 1.660 | 1.540 | 0.180 | 1.625 | 1.580 | 0.070 | 1.614 | 1.594 | 0.030 | 1.610 | 1.600 | 0.016 |
| | Min | 1.600 | 1.480 | 0.060 | 1.600 | 1.555 | 0.020 | 1.600 | 1.584 | 0.006 | 1.600 | 1.594 | 0.000 |
| 2 | Max | 2.060 | 1.940 | 0.180 | 2.025 | 1.980 | 0.070 | 2.014 | 1.994 | 0.030 | 2.010 | 2.000 | 0.016 |
| | Min | 2.000 | 1.880 | 0.060 | 2.000 | 1.955 | 0.020 | 2.000 | 1.984 | 0.006 | 2.000 | 1.994 | 0.000 |
| 2.5 | Max | 2.560 | 2.440 | 0.180 | 2.525 | 2.480 | 0.070 | 2.514 | 2.494 | 0.030 | 2.510 | 2.500 | 0.016 |
| | Min | 2.500 | 2.380 | 0.060 | 2.500 | 2.455 | 0.020 | 2.500 | 2.484 | 0.006 | 2.500 | 2.494 | 0.000 |
| 3 | Max | 3.060 | 2.940 | 0.180 | 3.025 | 2.980 | 0.070 | 3.014 | 2.994 | 0.030 | 3.010 | 3.000 | 0.016 |
| | Min | 3.000 | 2.880 | 0.060 | 3.000 | 2.955 | 0.020 | 3.000 | 2.984 | 0.006 | 3.000 | 2.994 | 0.000 |
| 4 | Max | 4.075 | 3.930 | 0.220 | 4.030 | 3.970 | 0.090 | 4.018 | 3.990 | 0.040 | 4.012 | 4.000 | 0.020 |
| | Min | 4.000 | 3.855 | 0.070 | 4.000 | 3.940 | 0.030 | 4.000 | 3.978 | 0.010 | 4.000 | 3.992 | 0.000 |
| 5 | Max | 5.075 | 4.930 | 0.220 | 5.030 | 4.930 | 0.090 | 5.018 | 4.990 | 0.040 | 5.012 | 5.000 | 0.020 |
| | Min | 5.000 | 4.855 | 0.070 | 5.000 | 4.970 | 0.030 | 5.000 | 4.978 | 0.010 | 5.000 | 4.992 | 0.000 |
| 6 | Max | 6.075 | 5.930 | 0.220 | 6.030 | 5.970 | 0.090 | 6.018 | 5.990 | 0.040 | 6.012 | 6.000 | 0.020 |
| | Min | 6.000 | 5.885 | 0.070 | 6.000 | 5.940 | 0.030 | 6.000 | 5.978 | 0.010 | 6.000 | 5.992 | 0.000 |
| 8 | Max | 8.090 | 7.920 | 0.260 | 8.036 | 7.960 | 0.112 | 8.022 | 7.987 | 0.050 | 8.015 | 8.000 | 0.024 |
| | Min | 8.000 | 7.830 | 0.080 | 8.000 | 7.924 | 0.040 | 8.000 | 7.972 | 0.013 | 8.000 | 7.991 | 0.000 |
| 10 | Max | 0.090 | 9.920 | 0.260 | 10.036 | 9.960 | 0.112 | 10.022 | 9.987 | 0.050 | 10.015 | 10.000 | 0.024 |
| | Min | 0.000 | 9.830 | 0.080 | 10.000 | 9.924 | 0.040 | 10.000 | 9.972 | 0.013 | 10.000 | 9.991 | 0.000 |
| 12 | Max | 12.110 | 11.905 | 0.315 | 12.043 | 11.950 | 0.136 | 12.027 | 11.984 | 0.061 | 12.018 | 12.000 | 0.029 |
| | Min | 12.000 | 11.795 | 0.095 | 12.000 | 11.907 | 0.050 | 12.000 | 11.966 | 0.016 | 12.000 | 11.989 | 0.000 |

Figure 10-8.
A table similar to this is used to calculate basic clearance fits for metric applications.

| ISO Symbol | | |
|---|---|---|
| Hole Basis | Shaft Basis | Description |
| H11/c11 | C11/h11 | Loose–Running Fit Wide commercial tolerances or allowances on external members. |
| H9/d9 | D9/h9 | Free Running Fit Not for use where accuracy is essential, but good for large temperature variations, high running speeds, or heavy journal pressures. |
| H8/f7 | F8/h7 | Close Running Fit For running on accurate machines and for accurate location at moderate speed and journal pressures. |
| H7/g6 | G7/h6 | Sliding Fit Not intended to run freely, but to move and turn freely and locate accurately. |
| H7/h6 | H7/h6 | Locational Clearance Fit Provides snug fit for locating stationary parts; but can be freely assembled and disassembled. |
| H7/k6 | K7/h6 | Locational Transition Fit For accurate location, a compromise between clearance and interference. |
| H7/n6 | N7/h6 | Locational Transition Fit For more accurate location where greater interference is permissible. |
| H7/p6 | P7/h6 | Locational Interference Fit For parts requiring rigidity and alignment with prime accuracy of location but without special bore pressure requirements. |
| H7/s6 | S7/h6 | Medium Drive Fit For ordinary steel parts or shrink fits on light sections, the tightest fit usable with cast iron. |
| H7/u6 | U7/h6 | Force Fit Suitable for parts which can be highly stressed or for shrink fits where the heavy pressing forces required are impractical. |

Figure 10-9.
Symbols made up of letters and numerals that classify the tolerance of a metric feature can be used in the local note. These are based on ISO standards, but are incorporated into ASME standard B14.2.

the mating shaft has a d9 designation. One method of calling out the hole measurement is to specify the size of the hole, but then place the basic size and tolerance symbol in parentheses: 8.000–8.036 (8 H9). Some companies may choose to reverse those components and specify the hole as: 8 H9 (8.000–8.036). In the case of 8 H9 being the only callout, the print reader needs to use the standard charts or a machinist handbook to determine the limits for that feature.

## Tolerancing Methods

Each company determines the best practice for tolerancing. At one time, it was common to use the limit method for every dimension. To reduce clutter on the drawing, other companies may elect to use limit dimensions only for important and critical dimensions. In these cases, basic sizes are given for most dimensions. The print reader should examine title block notes or general notes to determine the tolerance amount. This method also uses decimal precision to indicate tolerance precision. For example, two-place decimals may have a tolerance of plus and minus .01″, while three-place decimals are plus and minus .005″. Many drawings combine tolerancing methods, depending on the processes for checking or manufacturing the parts or features. As you will see in Unit 15, geometric dimensioning and tolerancing also impacts the way in which tolerances and dimensions are expressed on a drawing.

## Review Questions

*Circle the answer of choice, fill in the blank, or write a short answer.*

1. What term describes the assurance that when you select one of part A from a storage bin containing hundreds of those parts and likewise a mating part B from a separate storage bin the two parts will fit together properly? ______________________________

______________________________

2. *True or False?* Tolerance is the total amount by which a dimension can vary.

3. Which term applies to the method of showing both the upper and lower value of a dimension?
   A. Plus-or-minus dimensioning.
   B. Bilateral tolerance.
   C. Unilateral tolerance.
   D. Limit dimensioning.

4. *True or False?* A dimension that specifies 1.500″ ±.005″ has a tolerance of .010″.

5. The ______________________ size of a feature or object may or may not be the actual size.

6. The ______________________ size of a feature or object is often defined as the target value for the size, with tolerances applied in one or both directions.

7. *True or False?* The maximum material condition of a feature or a part is another way of saying the "upper limit" or the "largest size."

8. The loosest fit of two parts can be determined by subtracting the ______________________ of the shaft from the ______________________ of the hole.

9. ______________________________ is another term for tightest fit.

10. What term applies to a surface that has been identified as a reference surface from which measurements are to be taken?
    A. Axis
    B. Datum
    C. Flat
    D. Reference

11. What symbol can replace an arrowhead to indicate which surface should be the reference plane for measurements? ______________________________

    ______________________________

12. *True or False?* To save time and space, standard fits such as RC, LC, and FN are often found in the local notes of a print. ______________________________

    ______________________________

13. The term for the total variation in two parts that fit together is:
    A. allowance.
    B. loosest fit.
    C. limits of clearance.
    D. basic hole system.

14. What standards organization impacts ASME standards for metric applications? ______________________________

    ______________________________

15. *True or False?* Metric fits have an alphanumeric code that express the tolerance classes and these may be indicated on a print.

## Review Activity 10-1

*For each nominal size given below, calculate the maximum material condition and least material condition for the hole and the shaft. Then, calculate the allowance.*

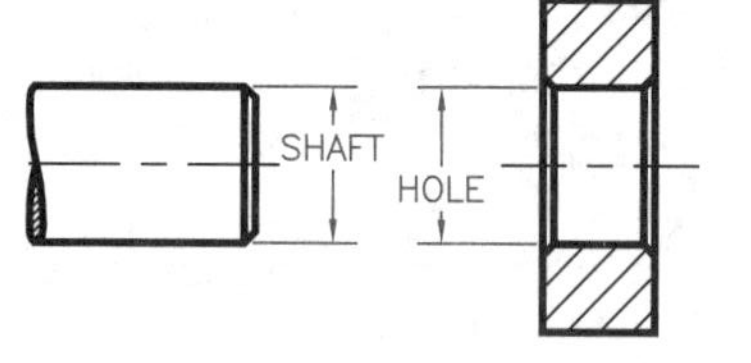


Using the basic hole system, calculate the dimensions as established for standard fits. Use the table in Figure 10–7 for values, and use four (.XXXX) decimal places of accuracy for all values.

| Nominal Fit "Basic" Size | | Hole Limits | | Shaft Limits | | Allowance |
|---|---|---|---|---|---|---|
| | | MMC | LMC | MMC | LMC | |
| ØA | 5/8″ RC3 | | | | | |
| ØB | 1 – 1/2″ RC2 | | | | | |
| ØC | 1/4″ RC4 | | | | | |
| ØD | 2 – 3/8″ RC1 | | | | | |
| ØE | 1″ RC2 | | | | | |

## Review Activity 10-2

*For each of the mating part pairs below, calculate the upper and lower limit for each part. For A and B, express the final dimension using the limit method. For C and D, express the final dimension using the plus and minus method (equal bilateral expression).*

A.

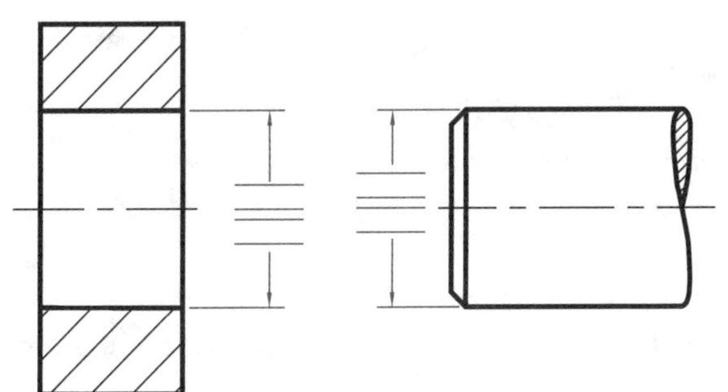

Basic hole size [MMC] is 5/8″
Allowance [A] = .005
Shaft tolerance = .011
Hole tolerance = .006

B.

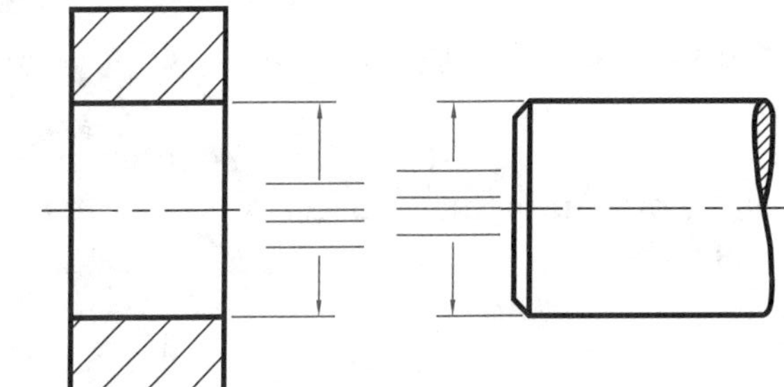

Basic hole size [MMC] is 1 1/2″
Allowance [A] = .010
Shaft tolerance = .010
Hole tolerance = .010

C.

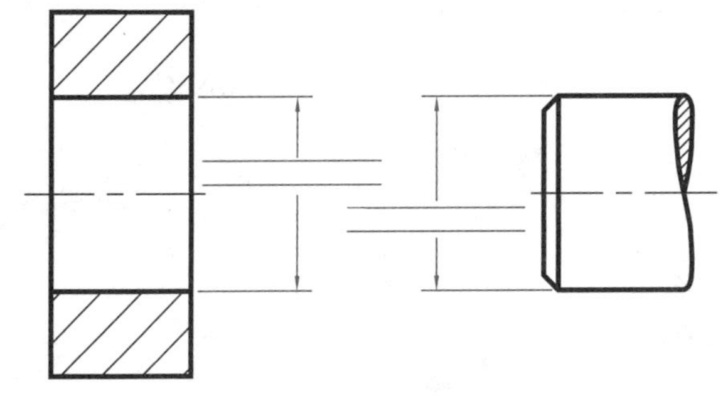

Basic hole size [MMC] is 3/4″
Allowance [A] = .008
Shaft tolerance = .008
Hole tolerance = .008

D.

Basic hole size [MMC] is 1″
Allowance [A] = −.006
Shaft tolerance = .002
Hole tolerance = .002

## Industry Print Exercise 10-1

*Refer to the print PR 10-1 and answer the questions below.*

1. Note 3 in the upper-left area of the drawing is referring to the crosshatched area on the front view. What is the maximum material condition of the hole in the middle of that area? ____________

2. Where on the drawing is the tolerance specified for the R.50 and the 2X R.12 dimensions and how much is this tolerance? ____________

3. How many dimensions on this print are expressed with the limit method? ____________

4. This drawing shows a fastener inserted into a hole, as indicated by note 5 in the top view. What is the maximum material condition of that pin? (Note: the front view indicates the size of the hole.) ____________

5. With reference to question 5, the hole in the part is to be .451–.453. What is the allowance for the pin and the hole? ____________

6. Two of the dimensions, 3.165, and 2.114, have boxes around them. As discussed in this unit, what are those dimensions called? ____________

7. Note 1 indicates 10 GAGE 1010-1020 sheet metal can be used to make the main body of this part. According to the left side view, how much tolerance is given to the thickness of the metal? ____________

8. The boxes that indicate an A or a B are called ____________ identification symbols.

9. The top view shows a bend radius of the thin metal. Would a bend radius of 1/8″ meet the requirements? ____________

*Review questions based on previous units:*

10. What paper size is the original version of this print? ____________

11. Are there any section views shown on this print? ____________

12. What is the name of this part? ____________

13. What is the part number of this print? ____________

14. How many chamfered edges are noted by leader lines on this drawing? ____________

15. Which system for placing dimension values is used, aligned or unidirectional? ____________

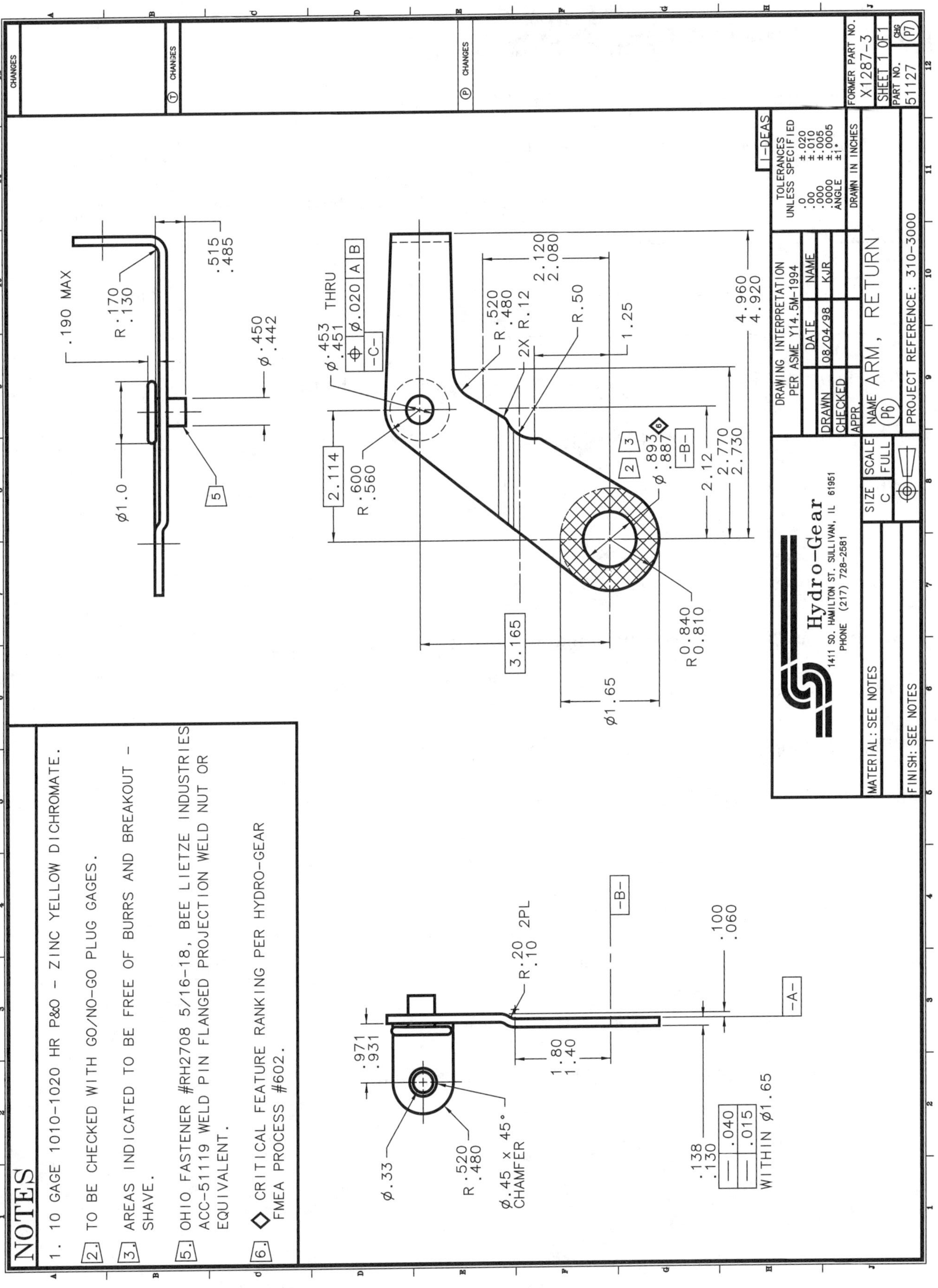


PR 10-1.
Print supplied by Hydro-Gear.

## Industry Print Exercise 10-2

*Refer to the print PR 10-2 and answer the questions below.*

1. The basic depth of this part is .715. What is the tolerance on that dimension? ____________
2. In the auxiliary section view, there is a chamfer note. What is the tolerance on the chamfer? ____________
3. The central, smallest hole has a target size of .1577″. What is the tolerance on that hole? ____________
4. What is the maximum material condition of the largest body diameter of this part? ____________
5. What is the angular tolerance for angles, unless otherwise specified? ____________
6. What is the tolerance for three-place decimal values, unless otherwise specified? ____________

*Review questions based on previous units:*

7. Are there any hidden lines shown on this drawing? ____________
8. What is the name of this part? ____________
9. What scale are the views on the original drawing? ____________
10. Is Section B-B shown in alignment with other views or as a removed view? ____________
11. Does this object have any features or surfaces that could be described as conical? ____________
12. How many cutting planes are indicated? ____________
13. Does this drawing feature any auxiliary views? ____________
14. Are there any reference dimensions on this print? ____________
15. As dimensioned, does this drawing follow the cylinder rule? ____________

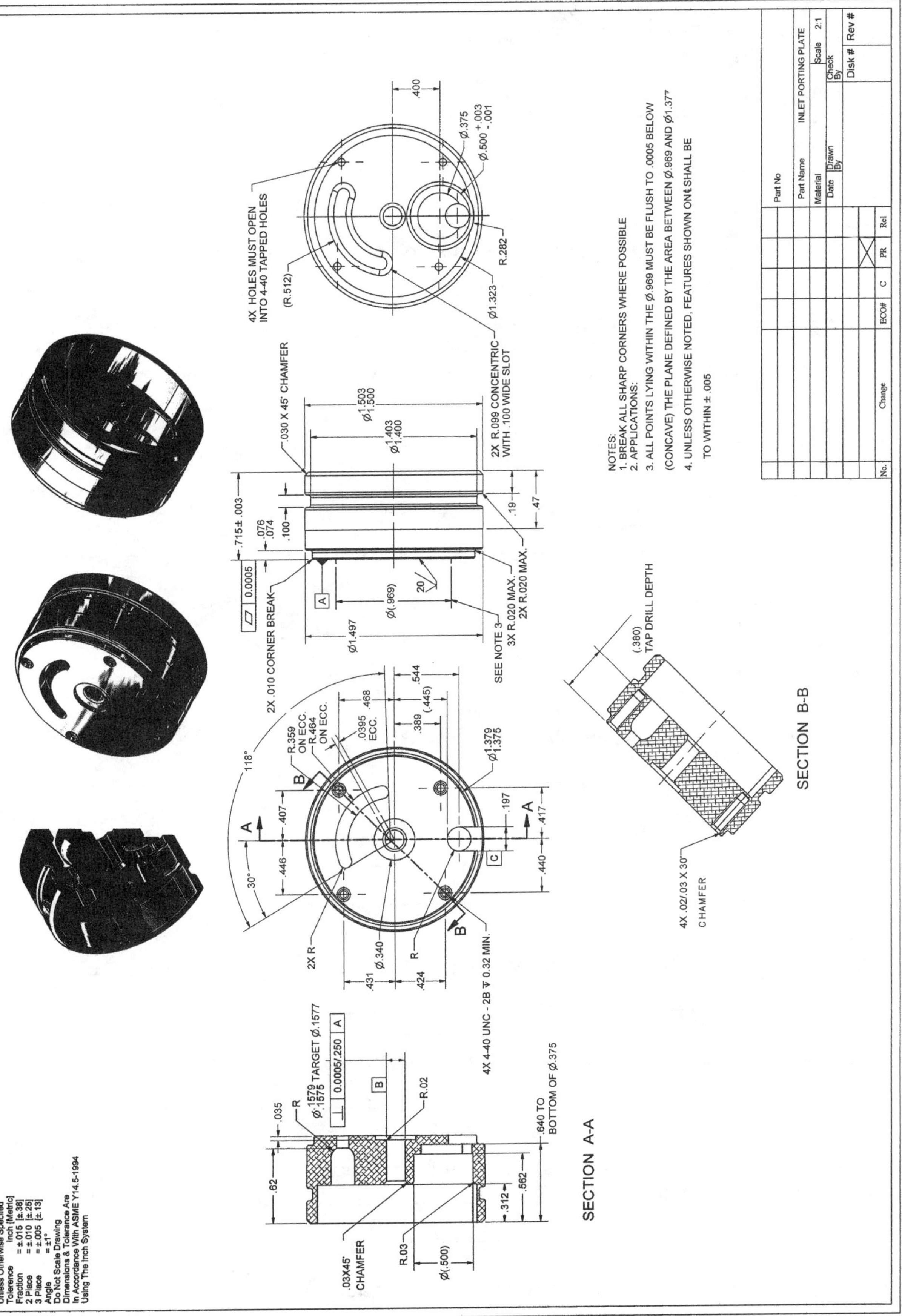


PR 10-2.
Print for use with print reading exercise 10-2.

## Bonus Print Reading Exercises

The following questions are based on various bonus prints located in the folder at the back of this textbook. Refer to the print indicated, evaluate the print, and answer the question.

**Print AP-001:**

1. For this print, there are special triangles floating next to a few dimensions. What tolerance is applied to those dimensions? ______

2. How many dimensions on this print express the tolerance using the limit method? (Note: exclude thread notes.) ______

**Print AP-003:**

3. For the overall diameter dimension, what name is given to the tolerancing method used? ______

4. In detail A, the angular measurements are given in degrees and minutes. How much tolerance is specified for the angles? ______

**Print AP-004:**

5. For the two holes identified as datum D, what is the total size tolerance specified for those holes? ______

6. The two holes identified as datum D have a counterbore on one end. What size tolerance is specified for the counterbore diameter? ______

**Print AP-005:**

7. For those dimensions that have a tolerance specified, what name is given to the method used? ______

8. What tolerance is specified for the depth of the groove featured in the enlarged detail? ______

**Print AP-006:**

9. Some hole size tolerances are specified as D7, N6, and H8. Where does the print reader find the values for those specifications? ______

10. Do the dimensions and tolerances apply before the finish is applied or after? ______

11. This print has a series of size ranges that impact the tolerance amount to be applied unless otherwise specified. What tolerance applies to the 65 mm diameter? ____________

**Print AP-009:**

12. For the two holes with a .875″ diameter counterbore, is the counterbore diameter tolerance specified as unilateral, equilateral, or bilateral? ____________

13. For the two holes with a .875″ diameter counterbore, what tolerance is specified for the counterbore depth? ____________

**Print AP-011:**

14. When checking the diameter over balls, what total tolerance is specified? ____________

15. Calculate the total width of the sectional view. Specify your answer as a range. ____________

**Print AP-014:**

16. How much tolerance is specified for the ring groove that was modified in Revision B? ____________

**Print AP-019:**

17. Is the tolerance on the 1.656″ length specified as unilateral, equilateral, or bilateral? ____________

18. What is the maximum material condition of the .3120″ diameter feature? ____________

**Print AP-020:**

19. What tolerance is specified for the total depth of this object, as featured left-to-right in the section view? ____________

20. What tolerance is specified for the cylindrical surface used as datum A? ____________

**Print AP-022:**

21. What tolerance is specified for the overall width of .860″? ____________

**Print AP-025:**

22. *True or False?* The size of the holes marked A have a tighter tolerance than holes marked B.

23. The holes marked B are located 9.8 mm away from datum Y. How much tolerance is allowed for that value? ____________

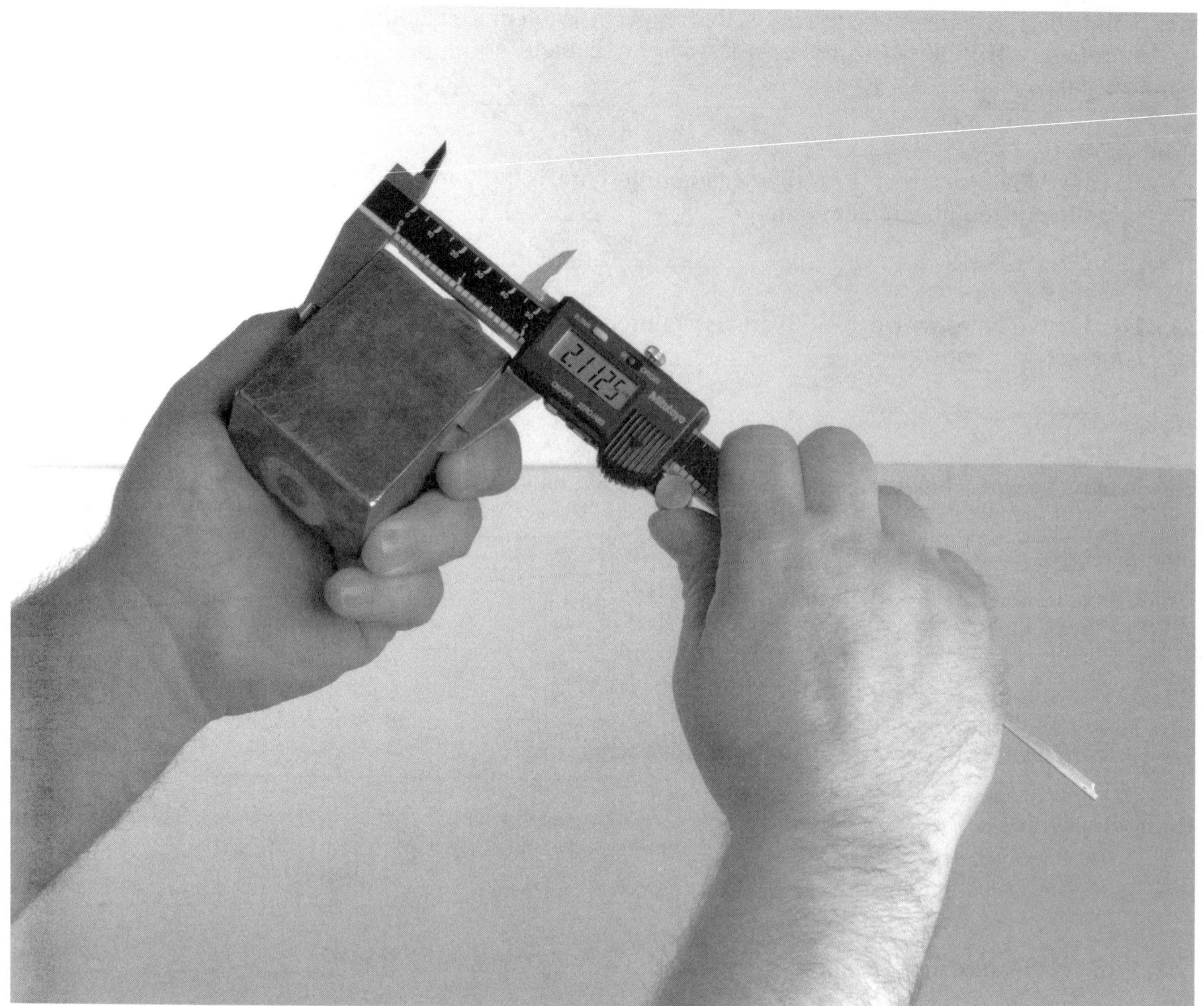


A machinist must be able to read and interpret prints in order to be successful.

UNIT 11

# Machining Specifications and Drawing Notes

*After completing this unit, you will be able to:*

**Identify** and **interpret** general notes on a print.

**Identify** and **interpret** local notes on a print.

**Read** and **interpret** specifications for holes and additional processes such as counterbores and countersinks.

**Read** and **interpret** callouts for common machining processes such as necks, keyways, and knurls.

Additional information must often appear on a drawing to provide information and instructions beyond the title block information, list of materials, graphic shape description, and basic dimensioning. These additional annotations are usually classified as *notes, specifications,* or *callouts.* Notes can be used to eliminate repetitive information or to give more information about the size of holes, fastener types, or other special specifications for removal of machining burrs.

Sometimes notes can contain so much information that placing the note on the drawing makes the drawing unreadable. This is often the case for architectural and structural drawing specifications. For these notes, the information is typed or printed on separate sheets. The sheets are then included with the set of drawings. This is from where the term "drawings and specifications" is derived.

Many large industries have internal process-specification manuals. These manuals may specify how to annotate a drawing so the machinist has information on how to perform machining processes—the machine, tools, and cutters to be used, as well as the tolerances. Most current drafting standards discourage the practice of putting all of these processes on the drawing, allowing any vendor or supplier to choose their own processes, as long as the part matches size and location dimensions.

As is common, older drawings will use different standards and must be read by the print reader. You should become familiar with the standards used by your company and the processes involved in the work you are required to perform. This unit will help provide you the knowledge needed for this.

## Notes

Basically, notes are classified as either general or local, but both types of notes can contain similar information. The type of note is determined by the application of the note and how it is placed on the drawing. ***General notes*** apply to the entire drawing. They are usually placed in a horizontal position above or to the left of the title block. General notes are not referenced in the list of materials or from specific areas of the drawing. Some examples of general notes are given in Figure 11-1.

Sometimes there are exceptions to general notes indicated in local notes. In this case, the general note is followed by the phrase EXCEPT AS SHOWN or UNLESS OTHERWISE SPECIFIED. The general note then applies to the entire part or to the entire drawing, except where a difference is noted by a local note.

1. THIS PART SHALL BE PURCHASED ONLY FROM SOURCES APPROVED BY THE ENGINEERING DEPARTMENT.
2. BREAK SHARP EDGES R.030 MAX UNLESS OTHERWISE SPECIFIED.
3. REMOVE BURRS.
4. FINISH ALL OVER.
5. METALLURGICAL INSPECTION REQUIRED BEFORE MACHINING.

**Figure 11-1.**
General notes apply to the entire drawing.

***Local notes,*** also referred to as *specific notes* or *callouts,* apply only to certain features or areas. They are positioned near the feature or area to which the note apply. A leader is used to indicate the feature or area. See Figure 11-2.

Sometimes local notes use numbers enclosed in an equilateral triangle. This is called a ***flag.*** See Figure 11-3. The actual note text appears in a central location with other "flagged" local note text. This technique can help keep the area near the views more free from clutter.

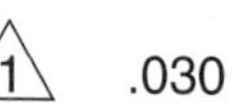 .030 X Chamfer.

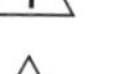 This surface to be coplanar with surface marked "Y".

 Rubber stamp part number here.

 Mount in chuck using this surface.

**Figure 11-3.**
A flagged local note is a number enclosed in an equilateral triangle. The actual note text appears in a central location with other local note text, as shown here.

## Holes

Often, a hole has specifications associated with it. For example, a hole may be counterbored or countersunk. Additionally, especially on older drawings, the type of operation may be specified. These specifications are usually identified in a note.

### Drilled and Reamed Holes

***Drilled holes*** are created with a twist drill, not a mill cutter. As mentioned, ASME standards discourage specifying whether or not the hole must be manufactured by drilling. If specified, drilled holes are usually specified by the diameter of the drill bit.

The number of holes and depth may also be specified. Figure 11-4 shows how symbols are used in local callouts for holes. Standard practice recommends specifying the holes as 2X Ø.250 with the leader line only pointing to one hole. The depth symbol is used in front of the depth value. If the depth is specified as THRU (for through), the hole passes entirely through the feature. Also illustrated in Figure 11-4 are local notes for counterbore and countersink, which are enlargements to the hole opening.

Review Figure 9-3 in the dimensioning unit for a thorough review of symbols used in local notes for holes. The symbols in current standards explain

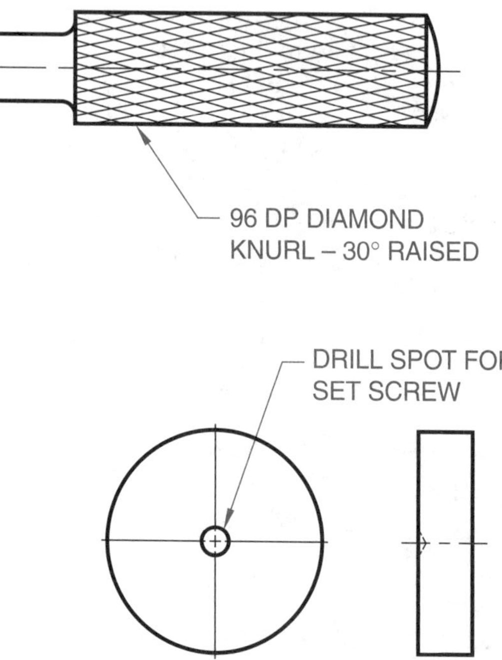


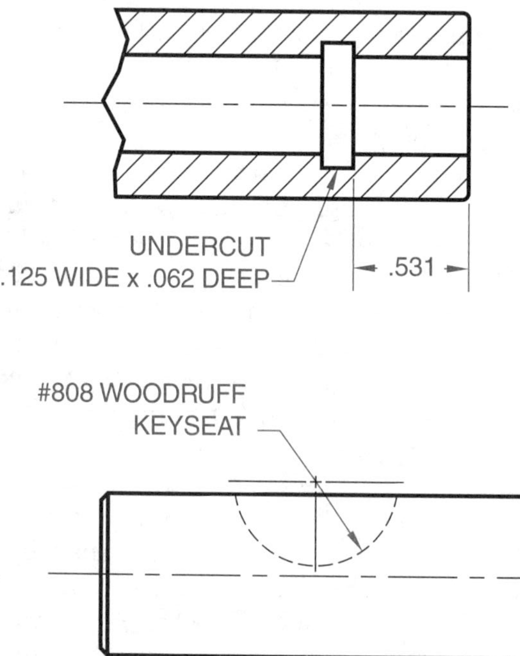

**Figure 11-2.**
Local notes should use symbols that have been standardized for several years. In former practice, words were in common use, as shown here.

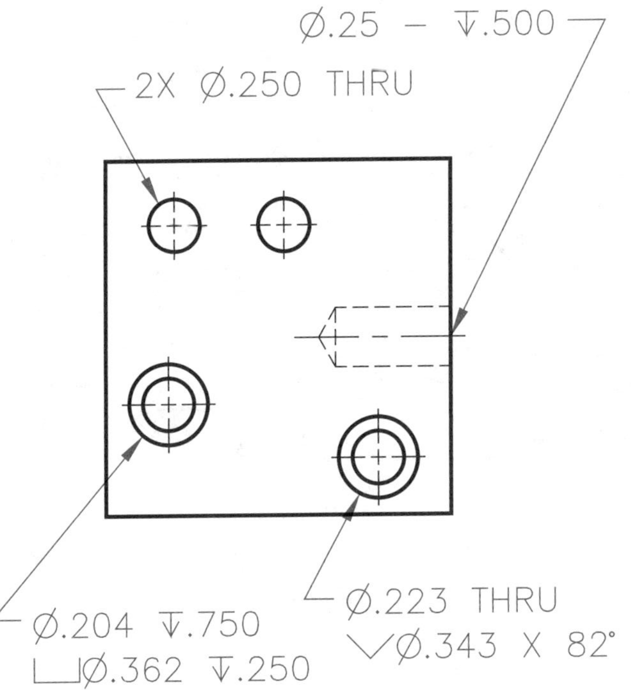


**Figure 11-4.**
These callouts use standard symbols for number of places, diameter, depth, counterbore, and countersink.

common terms symbolically or with alphabetic characters and should be used on drawings in lieu of words when feasible. In former practice, the local note for two holes may have specified .250 DRILL, TWO PLACES and the word DEEP, if used, would be after the value instead of the deep symbol in front of the value.

Some hole notes contain the term TYP. This stands for typical. When this term appears in the specification, the note applies to all similar features on the print, unless otherwise noted.

Slotted holes are dimensioned various ways, as discussed in the unit on dimensioning. Usually, a slotted hole is dimensioned by diameter, with a center-to-center linear dimension, but the radius is indicated as FULL R or just simply R, Figure 11-5.

***Reamed holes*** are created with a machine tool called a ream or reamer. This tool primarily applies to metal manufacturing and creates a very true, smooth, and accurately sized hole. To create a reamed hole, the hole is initially created as a drill hole slightly smaller (.010″ to .025″) than the finished size. Then, the hole is reamed to the specified finished size. The specification for a reamed hole, if required, is shown in Figure 11-6. The drilled hole size may be omitted, but if the process to create the hole is critical, it may also be shown in the specification.

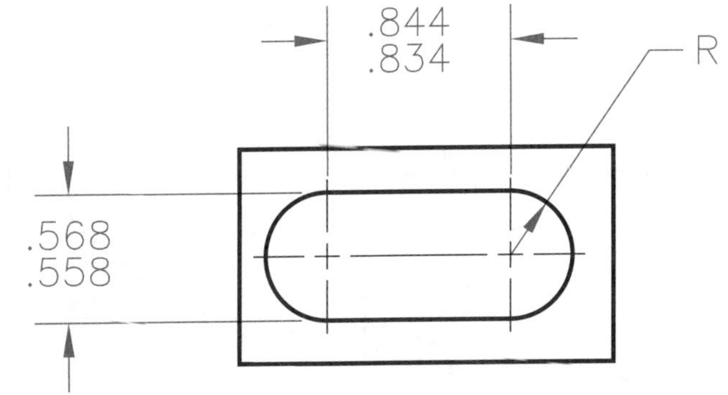


**Figure 11-5.**
For a slotted hole, the diameter is usually given along with the center-to-center linear distance. The radius is simply indicated as R or FULL R.

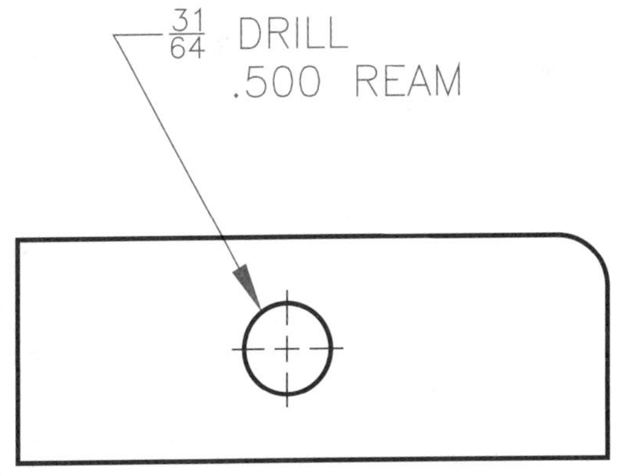


**Figure 11-6.**
If the specification of the manufacturing processes is deemed necessary, a drill and ream note may be used. In former practice, drawings often specified the manufacturing process.

## Counterbores and Countersinks

***Counterbored holes*** have been cylindrically enlarged on the one or both ends to form a recessed flat shoulder. A counterbore is often used to recess a bolt or machine screw head below the surface of the part. A counterbore specification is shown in Figure 11-7. Current standards provide a symbol for counterbore and depth. The counterbore diameter, the through hole, and the depth of the counterbore are often given in one note that points to the outermost circle representing a plan view of the hole. However, separate notes can be used. Although discouraged, C'BORE may be used as the abbreviation in the callout instead of the symbol.

***Spotfaced holes*** are simply very shallow counterbores. The spotface provides a flat surface on rough stock for the purpose of a bearing or seating surface (for a bolt head, nut, etc.). See Figure 11-8. The

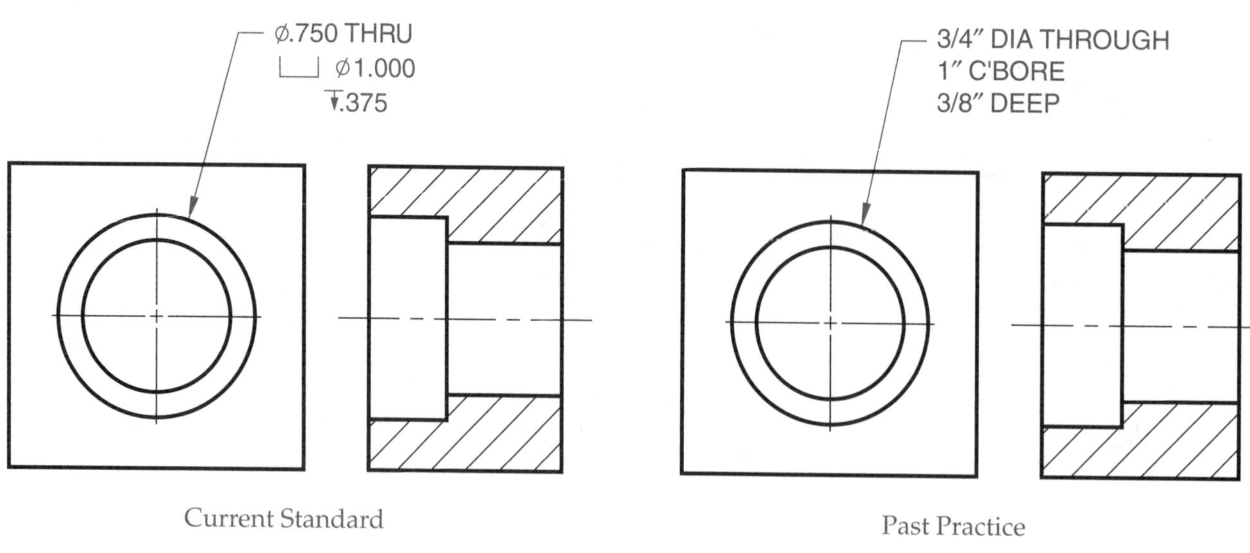


**Figure 11-7.**
The figure on the left shows the current practice for specifying a counterbore hole. The figure on the right shows the older practice.

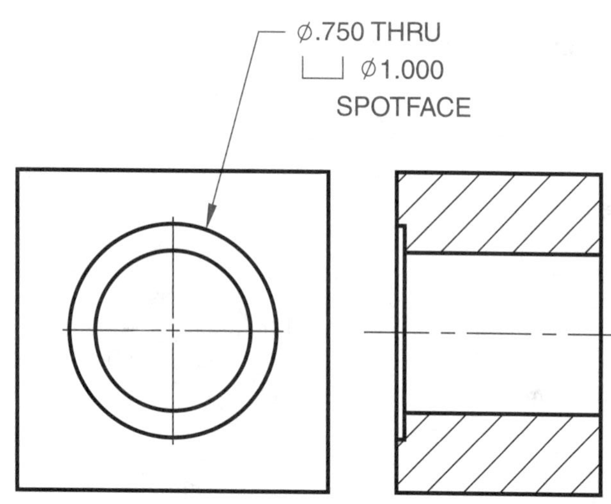


**Figure 11-8.**
The specification for a spotface is shown here.

depth is often omitted. The machinist understands to simply "break off" the rough surface at a depth of roughly .062″ or less. A new ASME standard symbol introduced in 2009 features an "SF" inside the counterbore bracket symbol as a way of specifying the counterbore is a spotface. As with most new symbols, it may not be present on older prints.

***Countersunk holes*** have a cone-shaped enlargement on the end. These holes are usually used to provide a seat for conical screw heads and rivets. Current standards provide a symbol for countersinking holes. See Figure 11-9. Older practices specified countersink with C'SINK and this may still be found on some drawings. A countersunk hole is often drawn with an included angle of 90°, even if the angle is 82°. There are different ways to specify the size of a countersink. Usually the included angle is called out in the local note, which is determined by the bit. The machinist also then needs a specification for the larger diameter of the rim or the depth of the countersink.

When a larger hole has a conical transition to a smaller hole, this is called a ***counterdrill.*** The transition is usually shown at the same angle as a drill bit tip (120°). Dimensioning the angle is considered optional. The depth of the counterdrill is considered to be the portion of the larger hole that is full diameter. See Figure 11-10.

Holes are sometimes drilled and countersunk for the purpose of holding the part between lathe centers or in a machining fixture. These are called ***machining centers*** and have no other function than providing a place to hold the part. Machining centers can be indicated by local note or callout. Machining center surfaces can also be identified as datum surfaces to establish a datum axis.

## Common Machining Operations

There are various machining operations that can be performed on a part. In the past, these operations were performed on manual machining equipment. Today, in many cases the machining process has been automated with CNC equipment. This section covers various machining processes. The processes are the same whether performed by CNC equipment or on manual machining equipment.

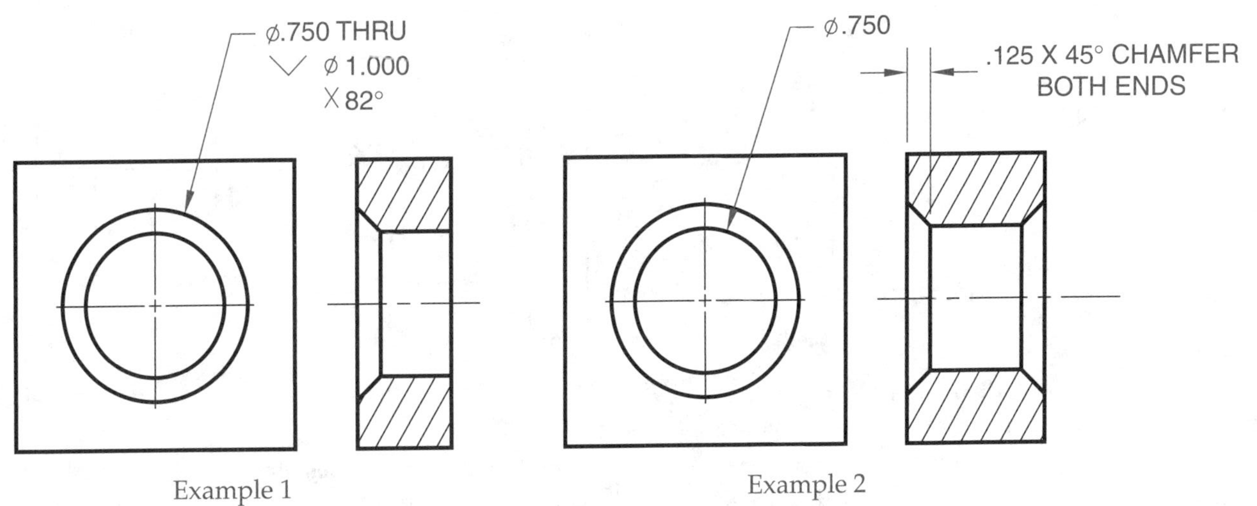


**Figure 11-9.**
Countersunk holes are holes with a cone-shaped enlargement and can be dimensioned by note. Chamfer depth can also be dimensioned as shown on the right in example 2.

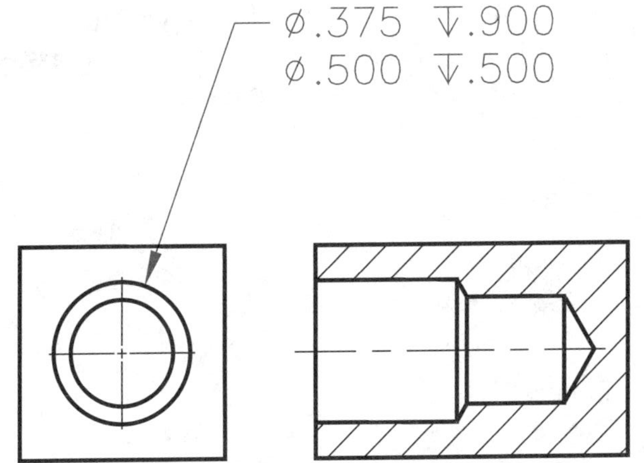


**Figure 11-10.**
A counterdrilled hole has a conical transition that does not require an angular dimension. The depth measurements are for the full-diameter distances.

## *Chamfered Edges*

Beveled edges on parts are common for a variety of reasons and are often called ***chamfers.*** One reason for chamfering an edge is to remove burrs. In essence, a chamfered edge around a hole is the same as a countersink, but the term countersink is more commonly used with smaller holes. A countersink bit can perform the operation on smaller holes. The term chamfer is more often applied for larger holes. It also describes a straight edge that has been beveled. The end of a cylinder is also chamfered in many cases to allow for easier assembly.

Figure 11-11 illustrates different methods for specifying a chamfer on the print, either by using a local note or by using linear and angular dimensions. If a chamfer is 45°, the *leg* distance is the same on both sides. For angles other than 45°, a *leg* distance must be specified. Figure 11-12 shows recommended standard practice for chamfered edges where surfaces are not 90° to each other.

## *Necks and Undercuts*

Sometimes a groove or ***neck*** is cut into a cylindrical surface for a retaining ring or to provide a good transition between two features. The groove is dimensioned by a local note, such as .06 WIDE X .03 DEEP. Linear measurements that specify the resulting diameter within the groove

**Figure 11-11.**
Chamfers on the end of a cylinder can be dimensioned by note (example 1) or by dimensions (example 2).

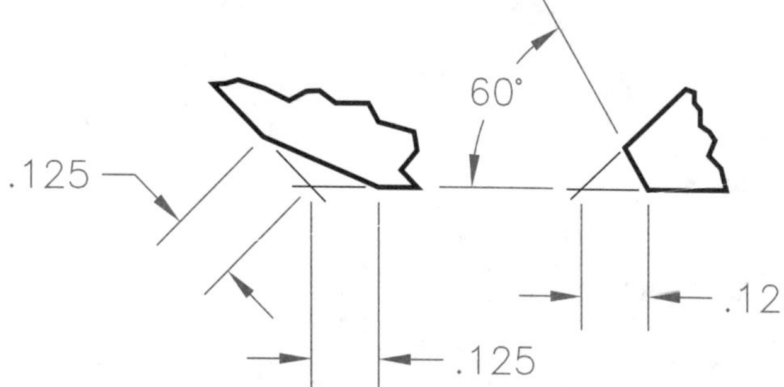


**Figure 11-12.**
Chamfers on non-perpendicular surfaces should *not* be dimensioned by note, but rather with dimensions.

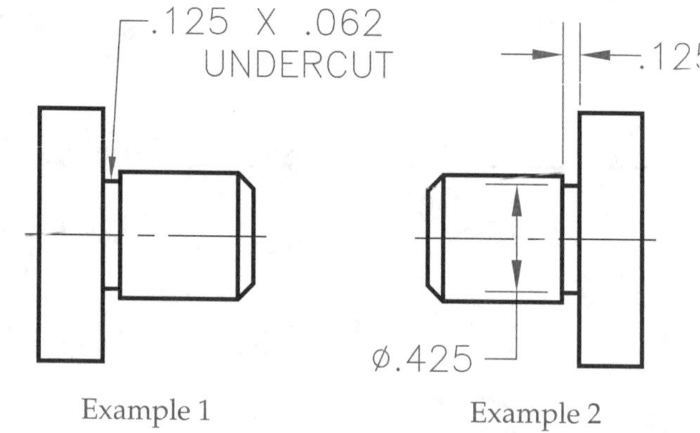


**Figure 11-14.**
An undercut is sometimes used to help prevent corner interference where a larger cylinder intersects a smaller cylinder.

provide the inspector with an easier checking distance, Figure 11-13. In similar fashion, sometimes an undercut groove is used at the head of a cylindrical part to eliminate the chance for a fillet being created due to tool tip wear or a rounded cutting tool. While local notes can be used to specify the undercut, linear measurements that are easily checked are preferred. See Figure 11-14.

## Keyways

A **key** is a fastener used to prevent rotation of gears, pulleys, and rocker arms on rotating shafts. The key is a piece of metal that fits into a ***keyseat*** in the shaft, which holds the key, or a ***keyway*** in the hub, which allows the hub to slide over the key. There are a variety of keys used in industry for different situations. Examples of keys are shown in Figure 11-15.

While a common practice at one time was to indicate a keyway with a note, newer standards promote the use of linear dimensions to specify exact distances that are logical inspection distances. See Figure 11-16. This technique eliminates confusion regarding the size of the keyway. The older method allowed the depth to be interpreted as the height above the center of the curve, but it was also sometimes interpreted to be the keyway sidewall.

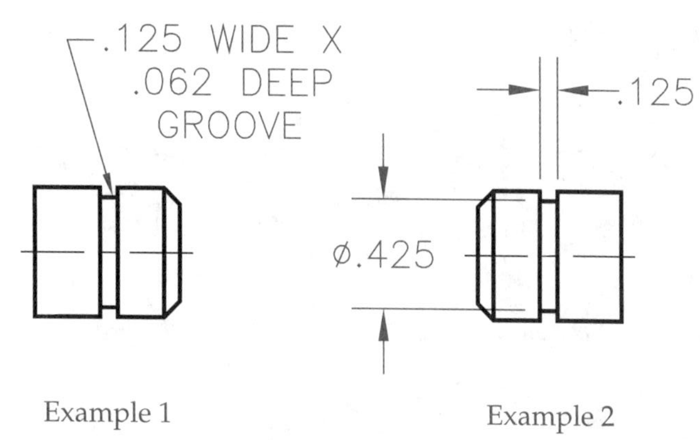


**Figure 11-13.**
A ring groove or neck may be dimensioned by note (example 1) or by dimensions (example 2).

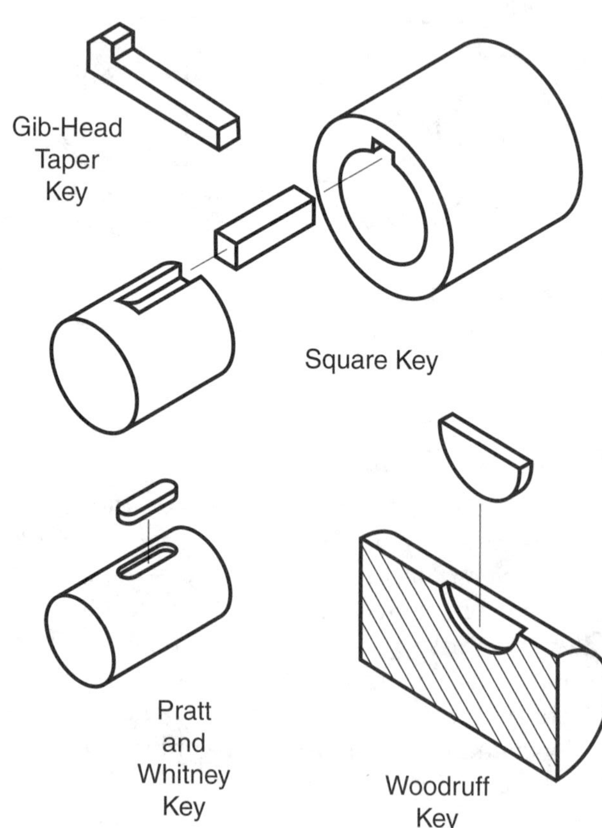


**Figure 11-15.**
There are a variety of keys used in industry for different situations.

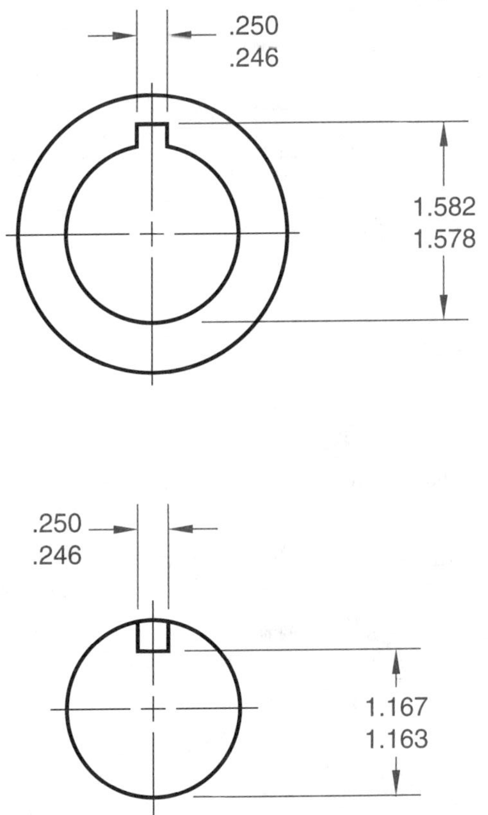


Figure 11-16.
Linear dimensions are used to specify exact distances for keyways.

To eliminate confusion, keyways should not be dimensioned by a callout, but should be dimensioned as illustrated in Figure 11-16.

## Knurls

***Knurls*** are patterns of ridges and grooves cut into a cylindrical surface for the purpose of providing a grip or increasing the diameter of a part. The pattern is often crisscrossed, leaving a diamond-like series of ridges. A metal part may feature a knurl so that it can be force fit into a plastic part. Knurling standards are covered in ASME B94.6. A knurl should be specified with respect to pitch, type, and possibly the diameters before and after the knurl is made. See Figure 11-17.

## Broaching

***Broaching*** is done on a special machine using a machine cutter called a broach. The broach progressively "punches" a shape with a series of cutting teeth. The teeth are set in such a way that each tooth is a few thousandths of an inch higher than the preceding one. Broaching can produce holes of circular, square or irregular outline; keyways; internal gear teeth; splines; or flat external contours. Broaching is fast, accurate, and produces a good-quality finish. As discussed several times in this unit, the practice of specifying the machine tool to make a particular feature is discouraged, but if a broaching operation is specified, it may appear as shown in Figure 11-18.

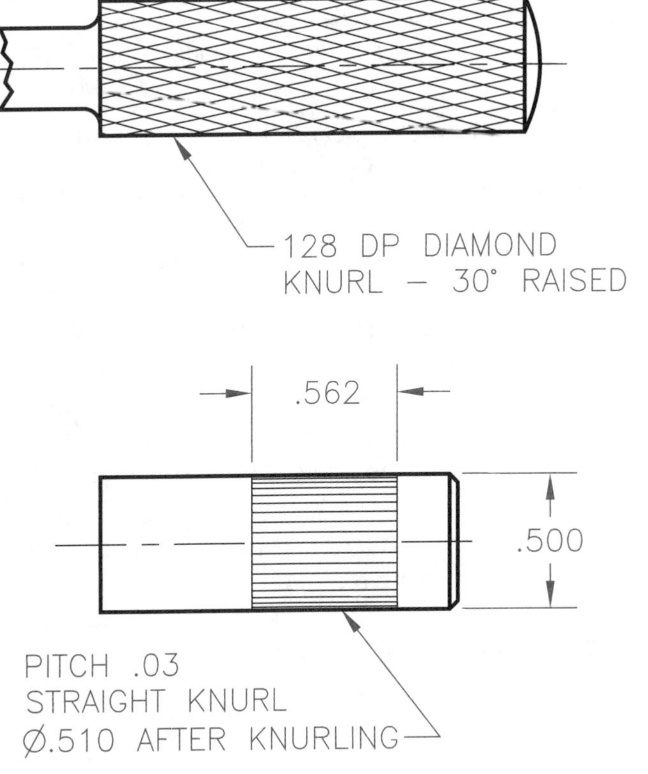


Figure 11-17.
A callout for a knurl may specify several items, including pitch, type, and before and after diameters.

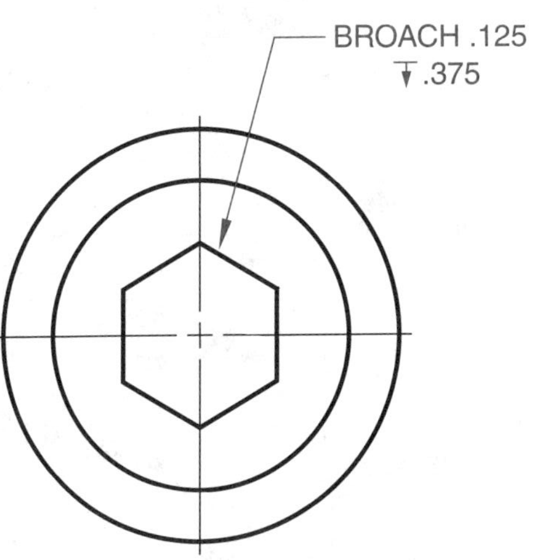


Figure 11-18.
A callout for a typical broaching operation is shown here.

## Review Questions

*Circle the answer of choice, fill in the blank, or write a short answer.*

1. What are the two classifications for notes? ____________________

____________________

2. *True or False?* Local notes can also be referred to as callouts.

3. Which of the following is the best method to create a hole?
   A. Twist drill.
   B. Mill cutter bit.
   C. Laser.
   D. None of the above can be specified as the best method without additional information.

4. *True or False?* If a hole is to be reamed, the word REAM must appear on the drawing preceding the diameter of the reamer.

5. A cylindrical enlargement on one end of a hole is called a(n) ____________________.

6. *True or False?* A spotface is simply a counterbore with a locked-in depth of .125″.

7. A conical enlargement on one end of a hole is called a(n) ____________________.

8. *True or False?* If a counterbore transitions to a smaller hole with a 120° tapered conical surface, it might be referred to as a counterdrill.

9. If a cylindrical object has a countersunk hole on each end that is not very deep and not dimensioned in size, but rather simply noted and identified as datum surfaces, they are ____________________

____________________.

10. A beveled edge is called a:
    A. fillet.
    B. chamfer.
    C. machining center.
    D. datum.

11. What purpose might there be for a cylindrical part to feature a neck, groove, or undercut?

____________________

12. What is the difference between a keyseat and a keyway? ____________________

____________________

____________________

13. *True or False?* The preferred method for dimensioning a keyway is with a local note.

14. A pattern of ridges cut or formed into a metal cylindrical handle to provide grip or increase a diameter is a:
    A. broach.
    B. thread.
    C. knurl.
    D. spline.

15. *True or False?* Broaching is a process that cuts metal away by force.

## Review Activity 11-1

*For each local note or callout given on the next page, use lettering to put the notes into words that you can read. An example has been done for you. Also refer to Figure 9-3 as needed.*

1. 1/2″ diameter hole, 3/8″ deep, two places
2. ______________________________
3. ______________________________
4. ______________________________
5. ______________________________
6. ______________________________
7. ______________________________
8. ______________________________
9. ______________________________
10. ______________________________

1. 

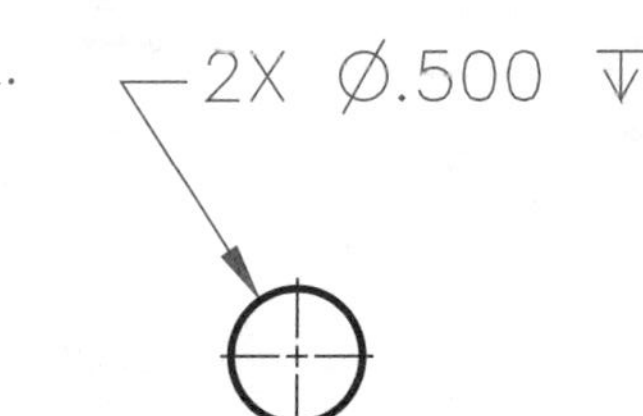


2. 

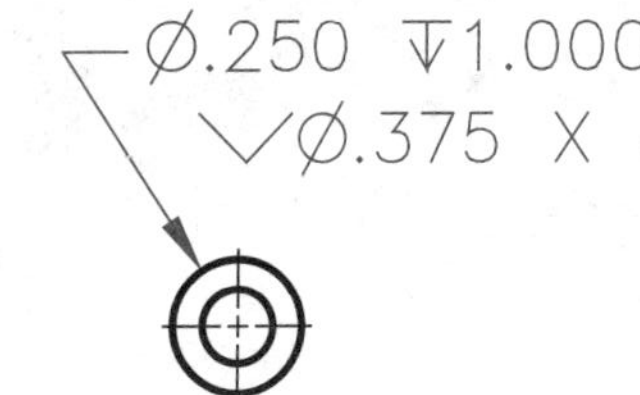


3. ⌀.750 THRU
⌴⌀1.250 ↧.375

4. SR.562

5. 2X ⌀.750 ↧1.500
⌴⌀1.000 ↧.250
⌴⌀1.250 ↧.125

6. .250 X 45°

7. 

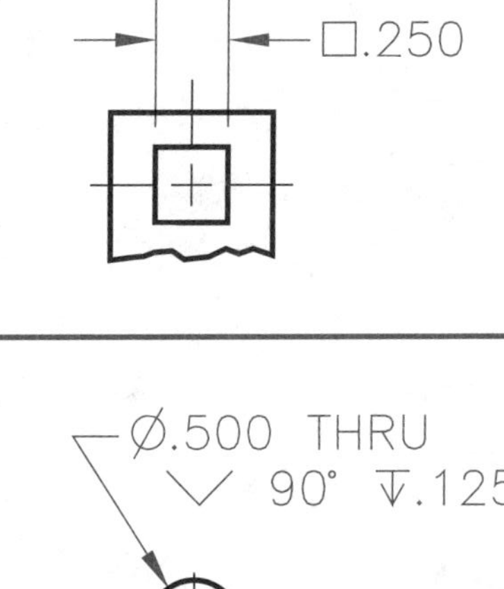


8. 4X ⌀.562 THRU
⌴⌀.625

9. ⌀.500 THRU
⌵ 90° ↧.125

10. 5X ⌀.500, EQUALLY
SPACED

## Review Activity 11-2

*For each of the written descriptions on the next page, sketch ASME standard symbols, if there are any, along with the numeric values next to the leader line shoulder. An example has been done for you. Also refer to Figure 9-3 as needed.*

1. Diameter 1/4″ through hole, three places.
2. Diameter 3/8″ hole, 1/2″deep, with a countersink diameter of 5/8″ and an included angle of 90 degrees.
3. Diameter 1″ through hole with a 1 1/2″ diameter counterbore 1/2″ deep, two places.
4. Spherical diameter of 3/4″.
5. Diameter 5/8″ hole, 1″ deep, with a 7/8″ diameter counterbore 1/2″ deep and a 1″ counterbore 1/4″ deep.
6. Chamfer the edge 45 degrees with a 1/8″ leg distance (1/8″ x 1/8″).
7. Square, 3/4″ x 3/4″.
8. Five 1 9/16″ diameter through holes each with a 2″ diameter spotface.
9. Diameter 3/4″ through hole with a 90 degrees countersink 1/8″ deep.
10. Six 5/8″ diameter holes equally spaced (through hole understood or obvious).

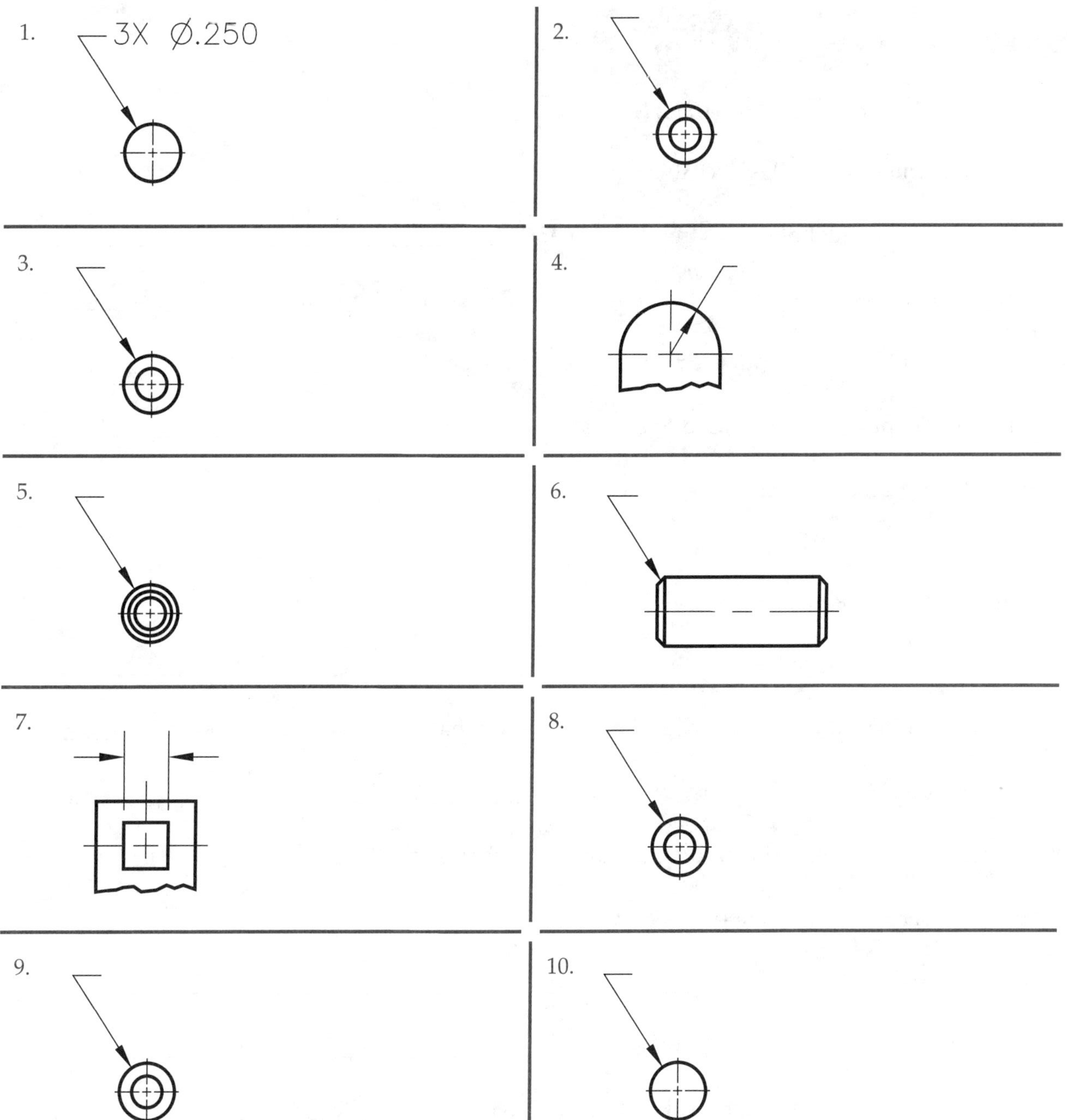
1.
3X Ø.250
2.
3.
4.
5.
6.
7.
8.
9.
10.

## Industry Print Exercise 11-1

*Refer to the print PR 11-1 and answer the questions below.*

1. How many counterbored holes does this part have? ______
2. How many countersunk holes does this part have? ______
3. Does this drawing feature symbols dictated by current standards for local note callouts? ______
4. What process will be used to put the part number on the part? ______
5. What machining process will be used to eliminate sharp corners? ______
6. What machining process is specified for the slot? ______

*Review questions based on previous units:*

7. How many leader lines are there on this print? ______
8. If the view with the cutting-plane line is the front view, what is the width of the object? ______
9. How many threaded holes does this part have? ______
10. Of the threads that feature a major diameter of 5/8″, what is the pitch? ______
11. What type of section view is the right side view? ______
12. What is the name of this part? ______
13. What size paper was the original drawing for this print? ______
14. Is the left side of this part symmetrical with the right side? ______
15. What scale was used for the views on the original drawing for this print? ______

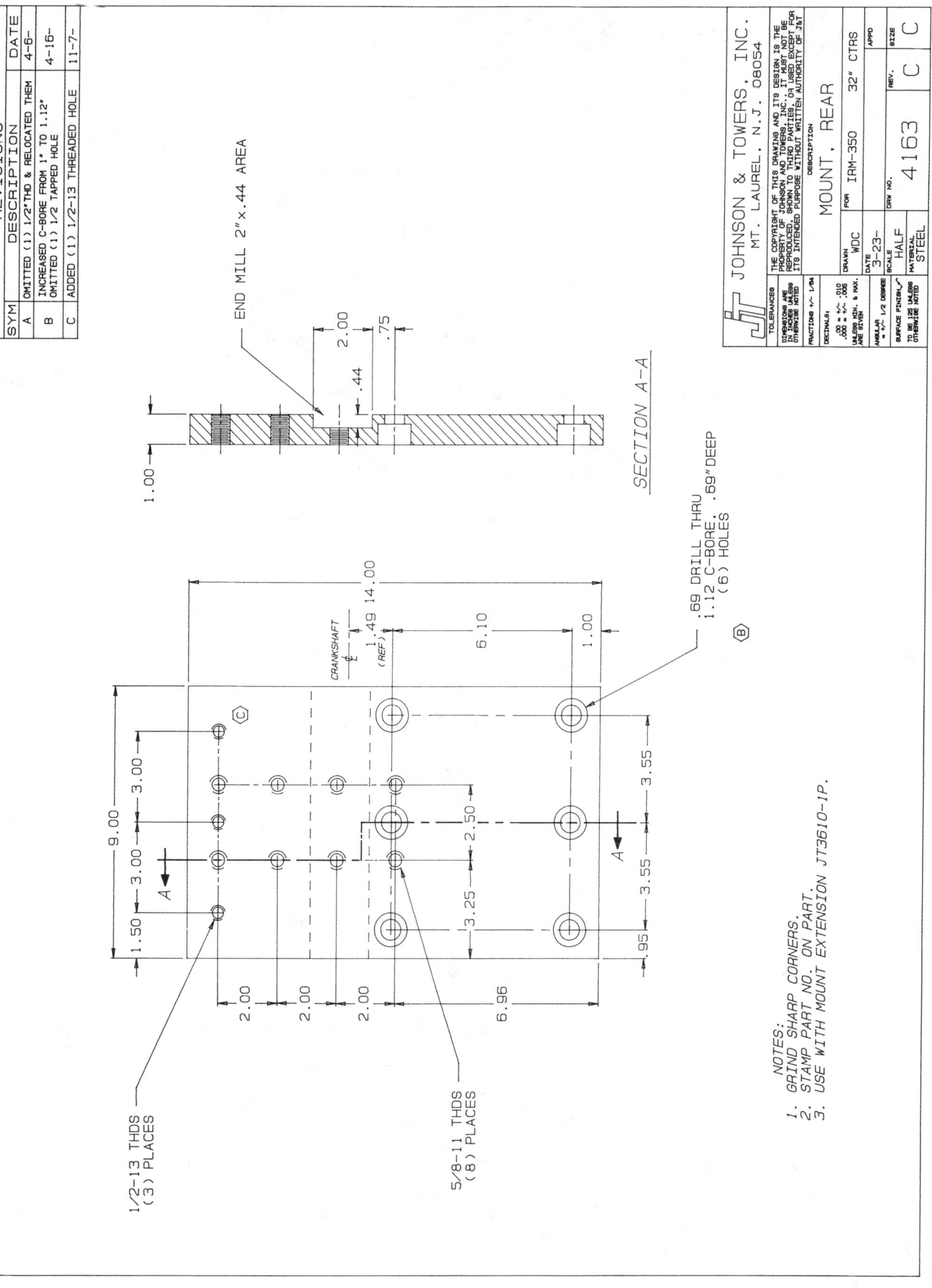


PR 11-1.
Print supplied by Johnson and Towers, Inc.

## Industry Print Exercise 11-2

*Refer to the print PR 11-2 and answer the study questions below.*

1. What is the primary manufacturing process for this part? ______
2. What is the height specification for the "Made in USA" letters? ______
3. How many local notes does this drawing have? ______
4. How many general notes does this drawing have, not counting those in the title block? ______
5. Is any of the lettering for this part recessed into the part? ______
6. One of the specifications for this part is to "wheelabrate," a process that is similar to sand blasting. What other finishing processes are covered in the same note? ______

*Review questions based on previous units:*

7. Who was the engineer for this drawing? ______
8. How many auxiliary views were used in this drawing? ______
9. How many cutting-plane lines are shown in this drawing? ______
10. If the section view is called the top view, what are the other views called? ______
11. Does this object have any features or surfaces that could be described as conical? ______
12. On this drawing, are visible lines shown thicker than center lines? ______
13. Are the dimension values placed on the drawing with the aligned system or the unidirectional system ? ______
14. During what month was this drawing issued? ______
15. What is the smallest radius value specified on this drawing for a fillet or round? ______

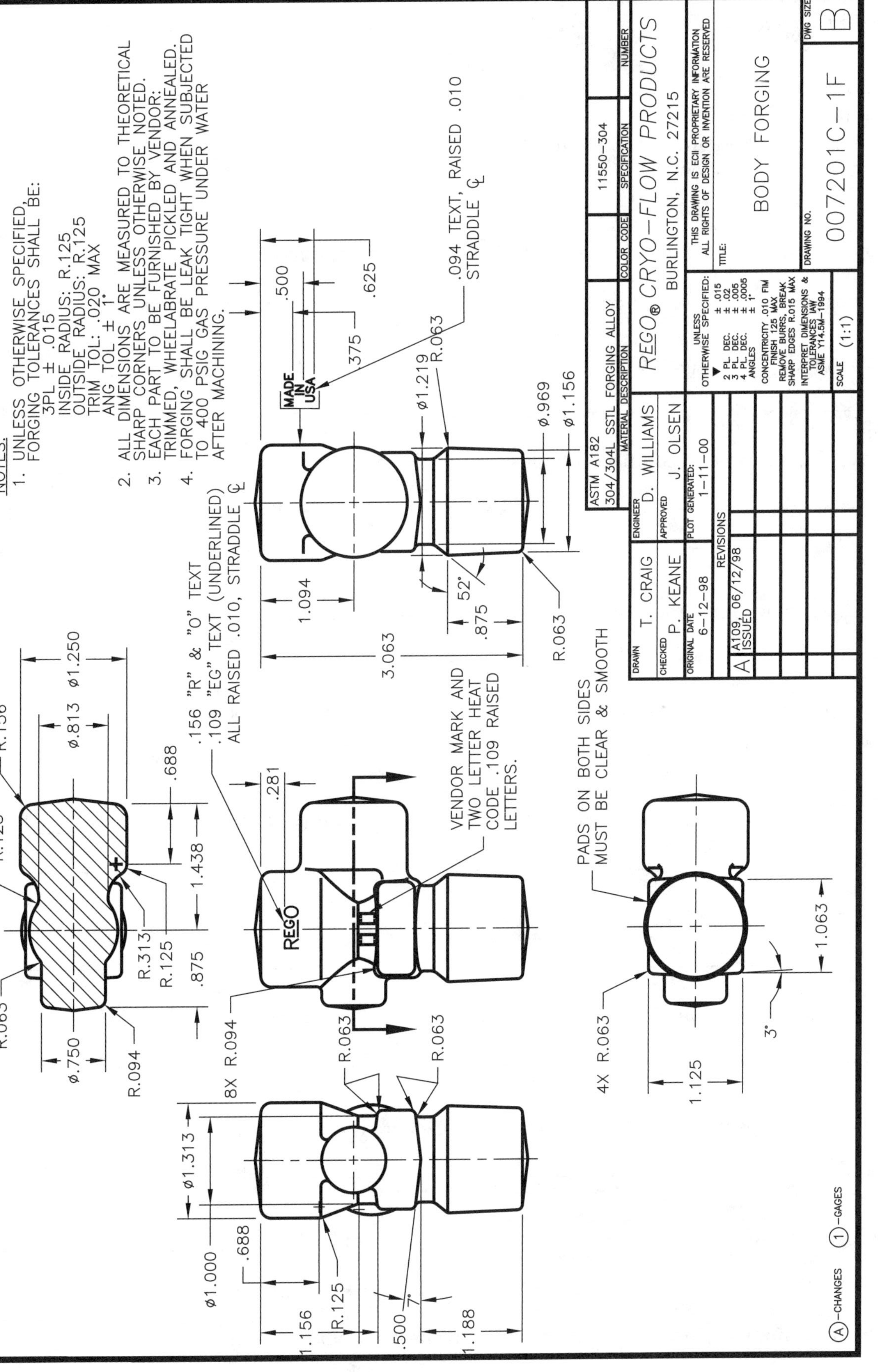


PR 11-2.
Print supplied by RegO Cryo-Flow Products.

## Bonus Print Reading Exercises

The following questions are based on various bonus prints located in the folder at the back of this textbook. Refer to the print indicated, evaluate the print, and answer the question.

**Print AP-004:**

1. According to the notes, what must be performed per military standard A-8625? ____________

**Print AP-007:**

2. What will be placed in the groove that must be free of parting line flash? ____________
3. List three conditions considered unacceptable for the valve seat surface. ____________

**Print AP-008:**

4. To help facilitate machining and inspection, what is permitted on both ends of this object? ____________
5. What type of line is used to indicate the cylindrical surfaces that are to be heat treated? ____________

**Print AP-009:**

6. How many counterbore operations are called out in View F-F for this part? ____________
7. How many spot drill operations are called out in View F-F for this part? ____________
8. Many of the holes specified in View F-F are located on a basic diameter B.C. What does B.C. stand for? ____________

**Print AP-010:**

9. What substance will be used on the set screws when they are installed? ____________

**Print AP-011:**

10. Is the part number stamp covered by local note, general note, or both? ____________
11. What is the maximum amount of break to be used on sharp corners and edges? ____________

**Print AP-012:**

12. How many general notes are listed on this drawing? ______

13. According to the notes, is this drawing inches or metric? ______

14. According to the notes, in what state must datum A be when checking flatness of the datum surface? ______

**Print AP-014:**

15. Some of the corners have leaders indicating an R value, and some have a C value. What manufacturing term does the C stand for? ______

**Print AP-016:**

16. How much torque is to be applied when installing the flange screws? ______

17. What process is specified for part #15 before assembly? ______

**Print AP-017:**

18. Which note was added with revision B? ______

19. This assembly drawing is for a cutting tool. According to the notes, what is the maximum speed for it? ______

**Print AP-018:**

20. What finish is specified for this part? ______

21. What is the approximate weight of this assembly? ______

22. According to note 4, how many minutes away from vertical is the maximum allowable angle for the edge of the insert? ______

**Print AP-019:**

23. There are three optional holes on this part. If used, what are they for? ______

**Print AP-020:**

24. Which one of the following processes does not apply to this part?
    A. Forging.
    B. Carburizing.
    C. Grinding.
    D. Welding.

25. What term is applied to breaking off the sharp edges of the teeth? ____________

**Print AP-021:**

26. Which of the general notes specifies a geometric control and, thus, could be replaced with a feature control frame? ____________

**Print AP-022:**

27. What is the basic radius for fillets and rounds, unless otherwise specified? ____________

28. How many leader lines are used in this print for notes? ____________

**Print AP-023:**

29. What finish is specified for this part? ____________

30. How are burrs to be removed? ____________

**Print AP-025:**

31. Unless specified, what imperfection along the edges is acceptable? ____________

**Print AP-026:**

32. What do the circular hidden lines represent in the top view? ____________

# UNIT 12
# Surface Texture Symbols

*After completing this unit, you will be able to:*

**Explain** common terms related to surface quality and surface texture symbols.

**Identify** and **interpret** the components of surface texture symbols.

**Identify** lay symbols used in surface texture symbols.

**Describe** the units used by various values within surface texture symbols.

**Explain** standard practices for applying surface texture symbols on a print.

**Identify** special designations related to surface quality that may appear on a print.

Products with metal parts may have specific requirements for how smooth the metal surfaces must be. For any particular feature of the part, the degree of roughness (or degree of smoothness) may need to be specified to ensure a proper function or finish. By nature, objects cast in sand molds have some rough surfaces. Some features of the part will be "finished" by machining and holes may need to be smoothed and made functional by drilling or boring. ASME B46.1 sets forth the terms and definitions for surface texture. ASME Y14.36 sets forth standards for how the surface texture symbols are to be specified on the print, which is the primary focus of this unit. There are other symbols and systems for controlling the general flatness of surfaces covered in the next unit, Geometric Dimensioning and Tolerancing.

## Surface Texture Terms

Prior to the development of current surface texture symbols, drafters indicated a finished surface with an italic "f" or a simple sans serif "V" symbol, Figure 12-1. In those cases, notes or company guidelines were available to set forth the precise standard of work. In today's manufacturing enterprise, standards have been developed that provide a thorough system for not only noting finished surfaces, but for specifying the roughness, waviness, and lay of a surface.

V mark
Traditional F mark
Finished surface

**Figure 12-1.**
This figure shows finish marks that were in common use before surface texture symbols were standardized.

A fundamental understanding of the standard surface symbols requires an understanding of a few terms. This unit does not attempt to explain the manufacturing or inspection processes required to measure surface roughness, but rather prepares the print reader to interpret the symbols on the print. Use Figure 12-2 to help you understand the following terms.

***Surface texture,*** or ***finish,*** refers to the overall roughness, waviness, lay, and/or flaws of a surface. The surface roughness, sometimes stated as "surface finish," is the specified smoothness required on the finished surface of a part. It is usually obtained by machining, grinding, or lapping. In this context, lapping is polishing a surface to achieve smoothness. Honing is another term that indicates a high-precision finishing process.

***Surface roughness*** is the fine irregularities in the surface texture. It is a result of the production process used. Included in surface roughness measurements are irregularities that result from the machine

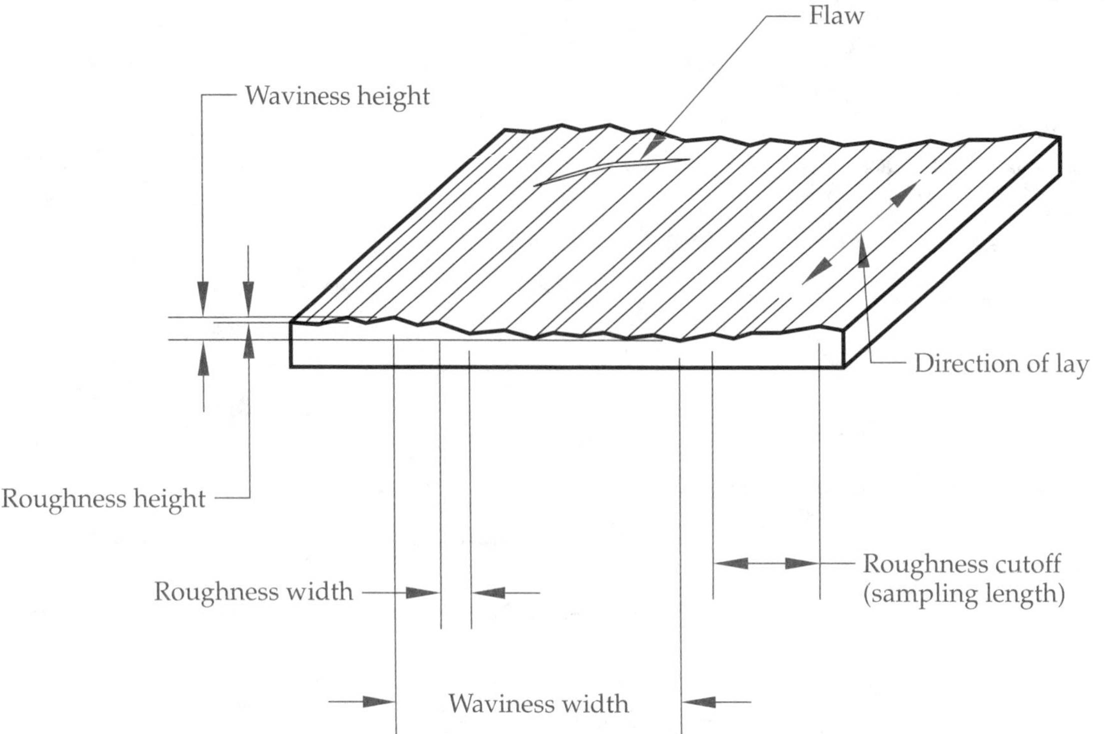


**Figure 12-2.**
Roughness, waviness, and lay are the three most important terms for understanding surface texture symbols. Flaws are also a concern if precise surface quality is required.

production process, such as traverse feed marks. Roughness height is measured by a profilometer in microinches (millionths of an inch) or micrometers (millionths of a meter). To put these terms into decimal values, 32 microinches is equal to .000032″ and .8 micrometers is equal to .0008 millimeters. So, it is evident surface texture measurements are working with microscopic distances. Most profilometers measure surface roughness height from 1 to 1000 microinches.

***Surface waviness*** is the widely-spaced component of surface texture due to such factors as machine chatter, vibrations, work deflections, warpage, and heat treatment. Waviness is rated in inches. If a spline curve were drawn through the average roughness peaks and valleys, surface waviness would be the variation in the total curve.

***Lay*** is the term used to describe the direction of the predominant surface pattern. There are seven lay symbols that are used to indicate the direction of the lay, if that is considered essential to a particular surface finish.

***Flaw*** is a term used to describe an unintentional interruption in the surface, such as a crack, pit, or dent. In general, surface texture symbols do not deal with flaws, rather just the roughness, waviness, and lay of a surface.

A surface texture specification of 8 microinches requires a roughness average less than .000008″. In order to create this finish, some very fine honing or lapping of a surface must be done. The table in Figure 12-3 indicates some recommended roughness height values (in both micrometers and microinches) with a description of the surface and the process or processes by which the surface may be produced.

## Surface Texture Symbol

The surface texture symbol resembles a check mark, as shown in Figure 12-4. There are five basic forms the symbol can take. In Figure 12-4A, the basic surface texture symbol indicates the surface may be produced by any method. Often, especially with former practice, only the surface roughness average specification is used with this symbol. Other values are assumed to be a default value specified by standards. In Figure 12-4B, the horizontal bar indicates material removal is required by machining. In Figure 12-4C, a value is placed directly to the left of the short leg of the check mark. The value indicates a minimum amount of material that must be removed, but this is specified in inches or millimeters. In Figure 12-4D, the circle indicates

| Roughness Height Rating | | Surface Description | Process |
|---|---|---|---|
| Micrometers | Microinches | | |
| 25.2 √ | 1000 √ | Very rough | Saw and torch cutting, forging, or sand casting. |
| 12.5 √ | 500 √ | Rough machining | Heavy cuts and coarse feeds in turning, milling, and boring. |
| 6.3 √ | 250 √ | Coarse | Very coarse surface grind, rapid feeds in turning, planning, milling, boring, and filing. |
| 3.2 √ | 125 √ | Medium | Machining operations with sharp tools, high speeds, fine feeds, and light cuts. |
| 1.6 √ | 63 √ | Good machine finish | Sharp tools, high speeds, extra-fine feeds and cuts. |
| 0.8 √ | 32 √ | High-grade machine finish | Extremely fine feeds and cuts on lathe, mill, and shapers required. Easily produced by centerless cylindrical and surface grinding. |
| 0.4 √ | 16 √ | High-quality machine finish | Very smooth reaming or fine cylindrical or surface grinding or coarse hone or lapping of surface. |
| 0.2 √ | 8 √ | Very-fine machine finish | Fine honing and lapping of surface. |
| 0.05 0.1 √ | 2–4 √ | Extremely smooth machine finish | Extra-fine honing and lapping of surface. |

**Figure 12-3.**
This table indicates recommended roughness height values in microinches.

that material removal is prohibited. In this case, processes that do not remove material, such as powder metallurgy, injection molding, cold finishing, casting, or forging, are to be used to manufacture the part with the surface texture as specified. In Figure 12-4E, a horizontal bar is added to the long leg of the check mark to help locate parameters that define the surface texture to the print reader. Above the horizontal bar, the production method will be indicated. Below the bar, the roughness cutoff, otherwise known as sampling length, can be specified.

A key to the location of numeric values, parameters, and notes that can be added to the surface texture symbol is illustrated in Figure 12-5. In this diagram, "a" indicates the average roughness value and can be given in microinches or micrometers. The arithmetical mean roughness is referred to as Ra. Ry is the parameter for maximum peak and Rz is the parameter for a ten-point mean roughness. If it is important to indicate the production method, such as MILL, GRIND, or LAP, this is placed in position "b". Under the bar, "c" indicates the roughness cutoff,

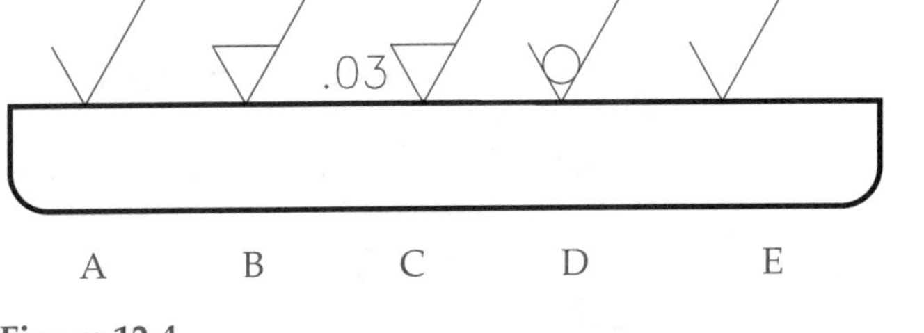


**Figure 12-4.**
These are the five basic forms of the surface texture symbol, minus numeric or symbolic annotations, as explained in the text.

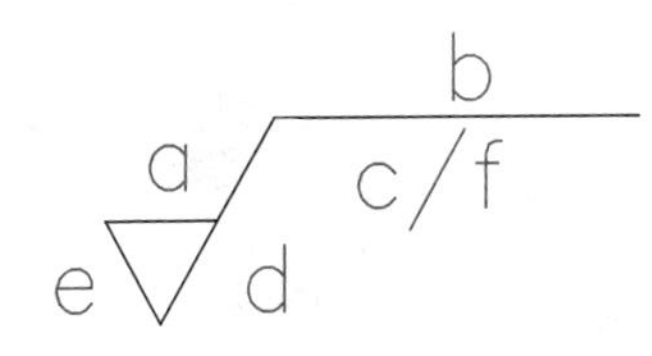


**Figure 12-5.**
The values that are applied to a surface texture symbol must be appropriately located with respect to the linework. The positions are explained in the text.

which means the sampling length. Prior to ASME B46.1-1995, the default rating was 0.8 mm if no rating was stated. Location "d" is for a lay symbol. Lay symbols are illustrated in Figure 12-6 and specify the acceptable direction of the lay. Location "e" is for the minimum material removal requirement amount, specified in inches or millimeters. Location "f" is for roughness parameters other than the arithmetic mean average (Ra). For example, waviness height can be specified with $W_1$ and maximum roughness spacing can be indicated with $S_m$.

There are various techniques for locating the surface texture symbol on the print. The V-point of the symbol is placed on a line that is considered the "edge view" of the surface being controlled. To improve readability of the drawing, the symbol can also be attached to an extension line leading out from the surface or to the shoulder of a leader line that points to the surface. As shown in Figure 12-7, notes can also help alleviate drawing clutter by alphabetically flagging the surface texture symbol near the views and specifying the numeric specifications in a general note.

| Lay Symbol | Meaning | Example Showing Tool Marks |
|---|---|---|
| = | Lay is somewhat parallel to the "edge view" line to which the symbol is applied. | = |
| ⊥ | Lay is perpendicular to the "edge view" line to which the symbol is applied. | ⊥ |
| X | Lay is angular in both directions to the "edge view" line to which the symbol is applied. | X |
| M | Lay is multidirectional or random. | M |
| C | Lay is approximately circular to the center of the surface to which the symbol is applied. | C |
| R | Lay is approximately radial to the center of the surface to which the symbol is applied. | R |
| P | Lay is particulate, non-directional, or protuberant (bulging). | P |

Figure 12-6.
One of the seven lay symbols illustrated here may be indicated within the surface texture symbol. The lay symbol is based on the direction or pattern of the lay.

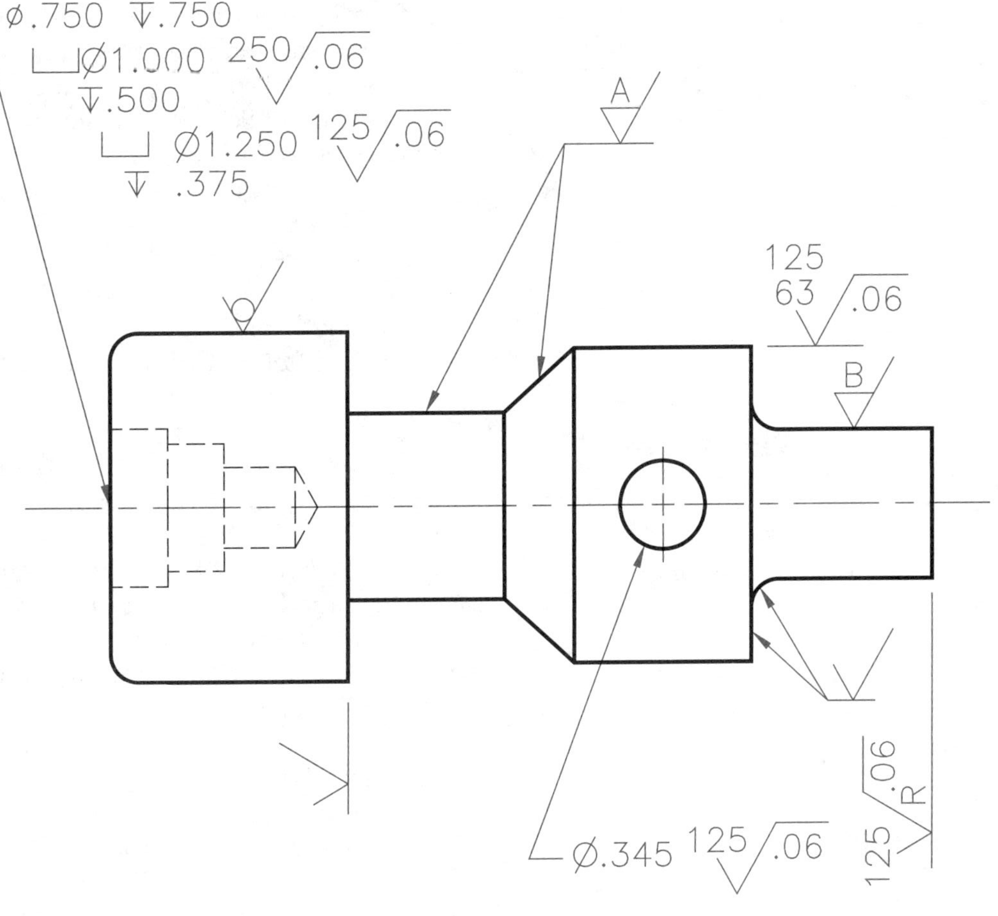


Figure 12-7.
The surface texture symbol should be placed on the "edge view" of the surface being specified, but extension lines, leader lines, and notes can be used to help keep the shape description from being too cluttered.

## Application Examples for Surface Texture Symbols

Since many of the standards for application of surface texture symbols are relatively new, you may not see a full use of all available designations. As a print reader, your value within the company setting may be increased by helping spread the word of new and more in-depth standards that help prints specify additional information to others within the manufacturing enterprise. Examples of surface texture symbol applications are given in Figure 12-8. The first three are shown in the U.S. customary system and the next five shown in the metric system.

In Figure 12-8A, the symbol specifies a roughness average rating of 63 microinches with a sampling length of .05 inches. Since there is only one roughness value specified, the number indicates the

U.S. Customary System

Microinches

Inches

A 63 / .05

B 63 32 / .10

C 63 / .05

Metric System

Micrometers (microns)

Millimeters

D 0.2 / 0.8

E 0.2 / 0.8 2.5/$R_Z$ 0.8

F 0.8 / 0.8

G 0.8 / 0.8/$S_M$0.5

H 1.6 6 / 0.8 NOTE X

**Figure 12-8.**
Specific applications of the surface texture symbol are shown here to help you appreciate the multitude of options for applying the symbol.

maximum amount of roughness (63 microinches). In Figure 12-8B, the range of roughness average is specified as 32–63 microinches with the sampling length of .10 inches. In Figure 12-8C, the roughness average is specified as in Figure 12-8A, but this specification indicates no removal of material is allowed.

The symbols are the same for metric values, but the values are different. It is common practice in the metric system to apply a leading zero in front of the decimal. The values specified in Figure 12-8 are for print reading practice only. Recommended values for various scenarios may be indicated in ASME standards or perhaps company standards.

In Figure 12-8D, the 0.8 below the bar indicates a roughness sampling length of .8 millimeters and the roughness average (Ra) specified as .2 micrometers. In Figure 12-8E, the additional 2.5/Rz0.8 specification indicates a ten-point mean roughness value should also be held at no more than .8 micrometers sampled over a length of 2.5 millimeters. This inspection method is different than the usual arithmetic mean roughness average. Also shown in Figure 12-8E is the lay symbol for the parallel direction.

In Figure 12-8F, the lay designation is the perpendicular symbol. This means the lay runs perpendicular to the line to which the symbol is applied. In Figure 12-8G, the Sm parameter is placed after the sampling length. The average roughness is still held at .8 micrometers per sampling length of .8 millimeters, although the maximum roughness is .5 millimeters. In Figure 12-8H, the symbol indicates material removal is required to produce the surface. The minimum amount of stock to be removed is 6 millimeters, the roughness average is 1.6 micrometers, and the sampling length is .8 millimeters. The NOTE X directs the print reader to a general note. In place of this note, a default process such as MILL or GRIND could be specified.

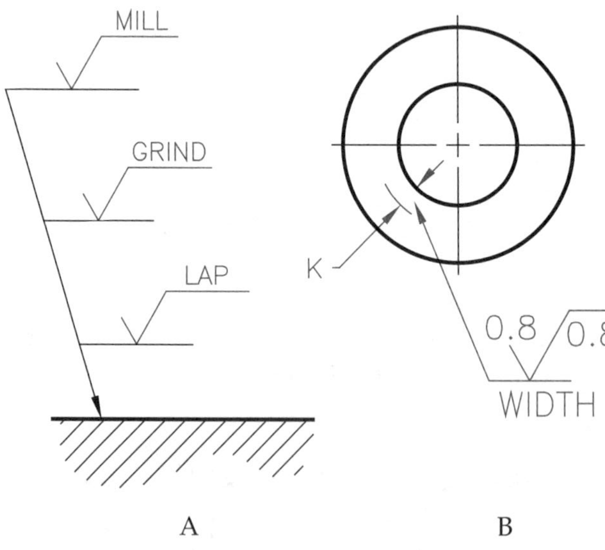


**Figure 12-9.**
Special applications of the surface texture symbol may be applied as at A where surface quality is specified for each process step and as at B where a designated area is specified on the surface.

## Special Surface Finish Designations

As with other print reading symbols and notes, many special designations may be needed to cover special situations for surface finish. Figure 12-9A shows how specific processes can each be applied, with increased surface quality as each process is completed. Figure 12-9B shows how an area can be designated on the surface, in this case the rim of a hole that may have a special relationship with a mating part.

Another special case that sometimes is noted on a drawing is surface mismatch. ***Surface mismatch*** may result when multiple passes of the cutter create an uneven "blend" on a machined surface. Unless otherwise indicated, surface mismatch is not allowed. As shown in Figure 12-10, a local note or dimension may be used at a specific location to indicate allowable surface mismatch.

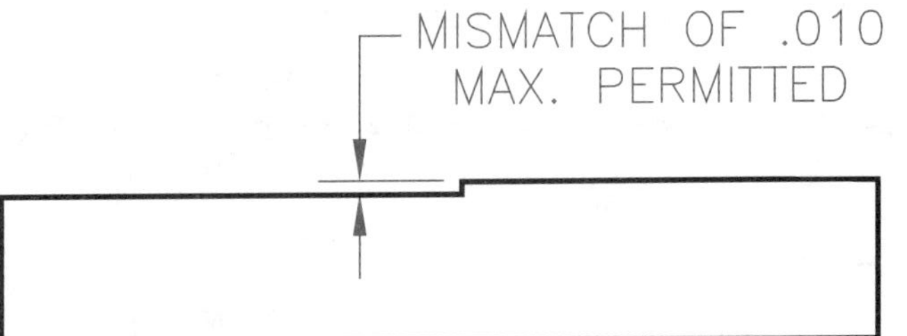


**Figure 12-10.**
Mismatch results when multiple cutter passes leave an uneven surface. The specification shown here is a common method of specifying allowable mismatch.

## Review Questions

*Circle the answer of choice, fill in the blank, or write a short answer.*

1. What is the main difference between ASME B46.1 and ASME Y14.36? ____________________
____________________
____________________

2. *True or False?* Lapping and honing are processes that create rougher surfaces than grinding or milling.

3. A device for measuring surface texture in microinches is a:
   A. profilometer.
   B. micrometer.
   C. calipers.
   D. coordinate measuring machine.

4. *True or False?* Waviness can be described as the overall high and low of the average roughness charted over a specified distance.

5. ____________________ is the term used to describe the direction of the predominant surface pattern.

6. A(n) ____________________ is some unintentional interruption in the surface and is not usually covered by a surface texture symbol.

7. The current standard surface texture symbol most resembles a(n):
   A. V.
   B. check mark.
   C. f.
   D. encircled S.

8. *True or False?* If a circle is included in the surface texture symbol, that means material removal is prohibited.

9. Another name for roughness cutoff is ____________________.

10. How many lay symbols are there?
    A. Five.
    B. Six.
    C. Seven.
    D. Eight.

11. *True or False?* Surface texture symbols can be attached to extension lines and leader lines.

12. Which of the following is placed near the short stem of the surface texture symbol if a roughness average range is desired?
    A. The upper number, such as 63.
    B. The lower number, such as 32.
    C. Both numbers, such as 63 over 32.
    D. An average number with a tolerance, such as 47±15.

13. What is the primary difference between surface texture symbols for the metric system versus the U.S. customary system? ____________________

____________________

____________________

____________________

14. *True or False?* One drawback to surface texture symbols is that current standards do not allow for a portion of a surface to be specified. The symbol must be applied to an entire surface.

15. An uneven transition on a surface, where multiple machining passes did not blend well, may be permitted if the surface is dimensioned and labeled as having permissible ____________________.

## Review Activity 12-1

*For each surface texture symbols given below, fill in the maximum roughness height average, the sampling length, the lay symbol, and the minimum amount of material that must be removed. If the value is not specified, then fill in the blank with N/A. The first five are in the U.S. customary system and the second five are in the metric system.*

| | Maximum Roughness Average | Sampling Length | Lay Symbol | Minimum Material Removal |
|---|---|---|---|---|
| 1. 125 / .05, X | ________ | ________ | ________ | ________ |
| 2. 63, 32 / .10 | ________ | ________ | ________ | ________ |
| 3. 16, lay symbol circle | ________ | ________ | ________ | ________ |
| 4. 250, .06, C | ________ | ________ | ________ | ________ |
| 5. 8 / .05 | ________ | ________ | ________ | ________ |
| 6. 1.6 / 0.8 | ________ | ________ | ________ | ________ |
| 7. 0.2 / 0.8, 2.5/$R_Z$ 0.8 | ________ | ________ | ________ | ________ |
| 8. 0.8 / 2.5, ⊥ | ________ | ________ | ________ | ________ |
| 9. 3.2 / 0.8/$S_M$ 0.5, ⊥ | ________ | ________ | ________ | ________ |
| 10. NOTE X, 6.3 / 0.4, 6 | ________ | ________ | ________ | ________ |

## Review Activity 12-2

*For each of the written descriptions below, sketch a surface texture symbol with the proper notation(s) required to accomplish that description. The first five are in U.S. customary units and the second five are metric units.*

1. Maximum roughness average is 250 microinches with a sampling length of .05″.

2. Roughness average is 32 to 63 microinches with a sampling length of .05″.

3. No material can be removed and the maximum roughness average is 32 microinches with a sampling length of .10″.

4. Maximum roughness average is 250 microinches, the lay is multidirectional, and no sampling length is specified.

5. Minimum material removal is .06″, maximum Ra is 8 microinches, sampling length is .05″, and the lay is circular.

6. Maximum roughness average is 0.8 micrometers with a sampling length of 0.8 mm.

7. Instead of roughness average, use Rz at 0.8 with a sampling length of 2.5 mm.

8. Maximum roughness average is 0.8 microns, sampling length is 2.5 mm, and the lay is parallel to the indicated line.

9. Maximum roughness average is 3.2 microns, sampling length is 0.8 mm, and the lay is radial.

10. Maximum roughness average is 6.3 microns, sampling length is 0.4 mm, and 6 mm or more must be removed by milling.

## Industry Print Exercise 12-1

*Refer to the print PR 12-1 and answer the questions below.*

1. How many surface texture symbols are there on the views of this drawing? ____________

2. The two face surfaces perpendicular to the axis of this part that have surface texture specification should have a surface texture of ____________ microinches or smoother.

3. For the surface texture symbols that have a horizontal bar, what does the .05 indicate? ____________

   ____________

4. According to the lay symbols shown in the round view, should the lay on the two flat surfaces be in the same direction as the axis of the cylinder or perpendicular to the axis? ____________

   ____________

5. What surface texture value should be applied to all surfaces not specified on the drawing views? ____________

*Review questions based on previous units:*

6. What scale are the views on the original drawing for this print? ____________

7. For the fractional dimensions, what tolerances should be applied? ____________

8. What is the part name? ____________

9. Are there any section lines on this drawing? ____________

10. What type of section view is used in this drawing? ____________

11. What is the pitch of the external thread? ____________

12. How are the threads represented? ____________

13. Why is the countersink hole on the right end not dimensioned? ____________

    ____________

14. There is a limit-method dimension labeled P.D. What does P.D. stand for? ____________

15. What is the MMC of the counterbore diameter? ____________

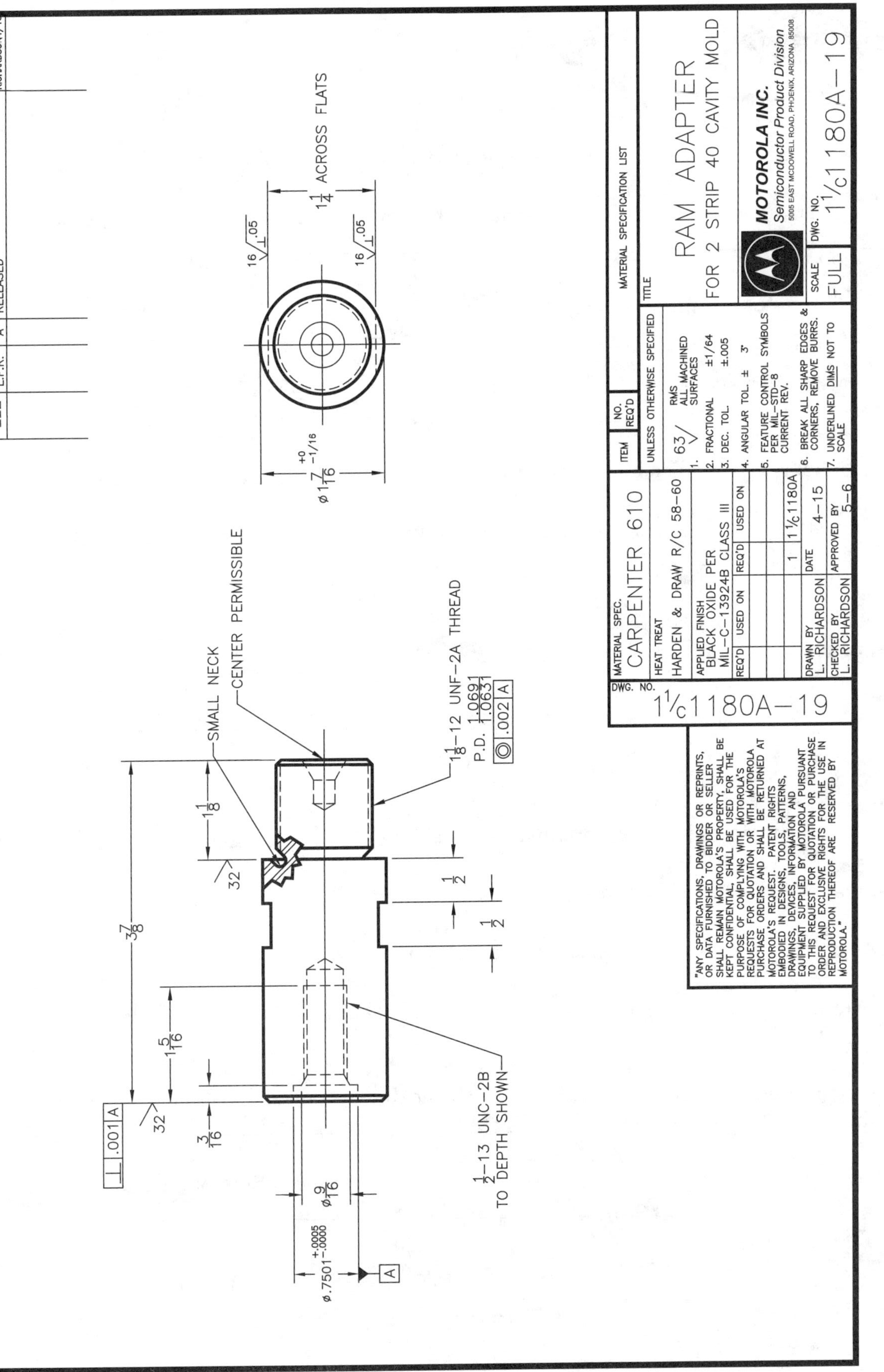


PR 12-1.
Print supplied by Motorola, Inc.

## Industry Print Exercise 12-2

*Refer to the print PR 12-2 and answer the questions below.*

1. How many surface texture symbols have been specified on the views of this drawing? __________

2. Which surface is to be smoother, the inside of the .500″ diameter hole or the 15° tapered conical surface? __________

3. Which machining process, if any, is specified for the surface text symbol applied to the 1.5″ diameter face surface? __________

4. For the surface texture symbols indicated on this drawing, are the values in microinches or micrometers? __________

5. What lay symbol is indicated on one of the surface texture symbols? __________

6. Do any of the surface texture symbols indicate the surface texture should be obtained without removing material? __________

*Review questions on previous units:*

7. Are there any cutting-plane lines shown on this drawing? __________

8. What material is specified for this part? __________

9. What scale are the main views on the original drawing? __________

10. How many threaded features does this part have? __________

11. The details are shown in two circles made with a line type known as: __________

12. Looking over the entire drawing, how many dimensions are expressed using the limit method? __________

13. There is a solid, equilateral triangle floating next to some dimension values. What tolerance applies to those values? __________

14. What is the maximum diameter of the drill spot that may exist at the bottom of the main hole in this part? __________

15. In the view that shows the hexagon shape, why is the dimensional value in parentheses?

__________

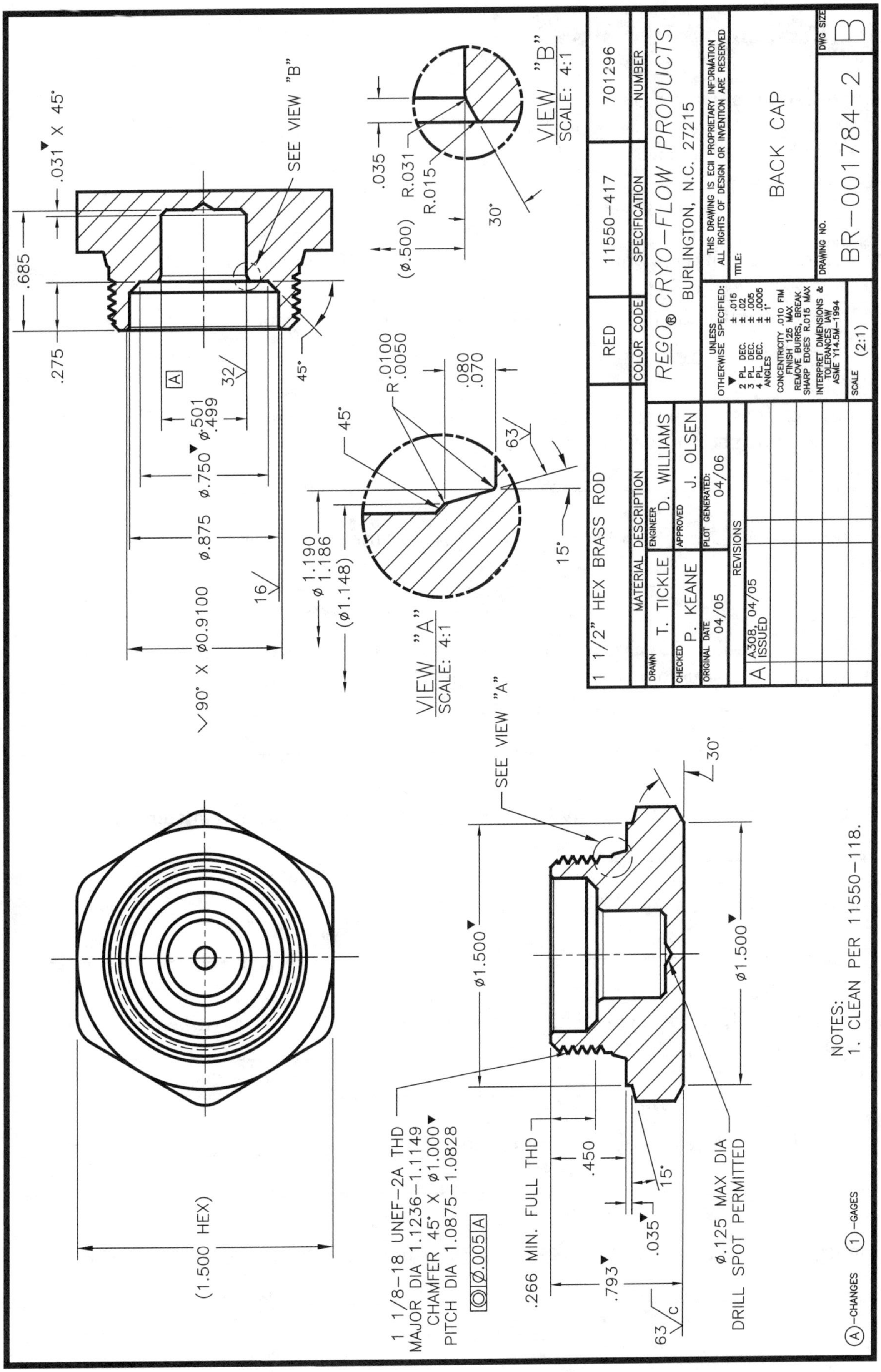

NOTES:
1. CLEAN PER 11550-118.

Ⓐ-CHANGES ①-GAGES

1 1/8-18 UNEF-2A THD
MAJOR DIA 1.1236-1.1149
CHAMFER 45° X ⌀1.000
PITCH DIA 1.0875-1.0828

⌀.125 MAX DIA
DRILL SPOT PERMITTED

.266 MIN. FULL THD

SEE VIEW "A"

SEE VIEW "B"

⌀.875 ⌀.750 ⌀.501/.499

⌀1.148 ⌀1.190/1.186

R.0100/.0050 .080/.070

R.031 R.015 (⌀.500) .035

.685 .275 .031 X 45°

(1.500 HEX) ⌀1.500 .793 .450 .035

90° X ⌀0.9100

VIEW "A"
SCALE: 4:1

VIEW "B"
SCALE: 4:1

| MATERIAL DESCRIPTION | COLOR CODE | SPECIFICATION | NUMBER |
|---|---|---|---|
| 1 1/2" HEX BRASS ROD | RED | 11550-417 | 701296 |

REGO® CRYO-FLOW PRODUCTS
BURLINGTON, N.C. 27215

| | | |
|---|---|---|
| DRAWN: T. TICKLE | ENGINEER: D. WILLIAMS | |
| CHECKED: P. KEANE | APPROVED: J. OLSEN | |
| ORIGINAL DATE: 04/05 | PLOT GENERATED: 04/06 | |

REVISIONS
A A308, 04/05 ISSUED

UNLESS OTHERWISE SPECIFIED:
▼ ± .015
2 PL. DEC. ± .02
3 PL. DEC. ± .005
4 PL. DEC. ± .0005
ANGLES ± 1°
CONCENTRICITY .010 FIM
FINISH 125 MAX
REMOVE BURRS, BREAK SHARP EDGES R.015 MAX
INTERPRET DIMENSIONS & TOLERANCES IAW ASME Y14.5M-1994

SCALE (2:1)



TITLE: BACK CAP

DRAWING NO. BR-001784-2

DWG SIZE B

PR 12-2.
Print supplied by RegO Cryo-Flow Products.

## Bonus Print Reading Exercises

The following questions are based on various bonus prints located in the folder at the back of this textbook. Refer to the print indicated, evaluate the print, and answer the question.

**Print AP-005:**

1. The surface texture specification (in microinches) for the groove surfaces (see detail) is ____________.

2. The surface texture specification (in microinches) for surface texture symbols that do not feature a number is ____________.

**Print AP-006:**

3. The groove surfaces are specified to have a surface texture of .8 micrometers. Are any surfaces on the object specified to have a better quality than that? ____________

4. What maximum surface texture (in micrometers) is specified for all surfaces? ____________

**Print AP-008:**

5. This print features international N values on the surface text symbols. N5 is equivalent to 0.4 micrometers and N6 is equivalent to 0.8 micrometers. What value (in micrometers) is the next logical equivalent for N7? ____________

**Print AP-011:**

6. What is the surface texture specification for the cylindrical surface that has a diameter of 31.7–31.8?

____________

7. What is the smoothest surface texture specified on this print? ____________

**Print AP-013:**

8. In very general terms, what is the symbol next to the .09 value, as found on the top and bottom edges of the front view? ____________

9. What is the smoothest surface texture specified on this print? ____________

**Print AP-015:**

10. According to the notes, what surface texture specification applies to the symbols in this print?

____________

**Print AP-019:**

11. What is the maximum roughness specification for machined surfaces? ______

**Print AP-020:**

12. The surface texture symbols on this print have a horizontal bar rather than an open check mark appearance. What does that specify? ______

13. What is the surface texture specification for the cylindrical surface to be used as datum A?

A computer hard drive is manufactured with very close tolerance. Geometric dimensioning and tolerancing can be used to specify tolerances for critical parts.

UNIT 13

# Geometric Dimensioning and Tolerancing

*After completing this unit, you will be able to:*

**Describe** the purpose and objectives of geometric dimensioning and tolerancing (GD&T).

**Identify** current and former ASME Y14.5 symbols used in GD&T.

**Define** terms related to GD&T.

**Explain** the purpose and function of datums.

**Identify** proper datum identification techniques on a print.

**Explain** basic dimensions as featured in drawings that use GD&T methods.

**Read** and **interpret** the use of modifiers as they apply to basic GD&T applications.

**Read** and **interpret** basic applications of feature control frames for each of the GD&T control symbols.

**Explain** how composite tolerances are applied in a basic GD&T application.

Modern manufacturing requires more preciseness or exactness in the design and production of parts than formerly required. Simply using traditional dimensioning and tolerancing methods set forth in earlier units will not meet the preciseness required for many applications. For the last few decades, industry has been using an advanced system of print annotation known as ***geometric dimensioning and tolerancing (GD&T)*** to control the quality of mass-produced parts with a system of geometric control symbols, feature control frames, basic dimensions, and modifiers.

In Unit 10, you studied about tolerances of size and location. However, these methods do not address the preciseness of the geometry. In other words, how flat is flat? How round is round? The design team can implement GD&T methods to address these issues. With GD&T, the print can express the geometric form required for the part to function effectively.

The ASME standard for GD&T is ASME Y14.5, entitled *Dimensioning and Tolerancing*. This standard is by far the most extensive of the ASME standards, including over 200 pages of text and illustrations for standard dimensioning and tolerancing practices as well as the preferred practices for GD&T. While the material presented in this unit is based on the current standard, examples of former practice are also covered to help you read older prints. In some cases, CAD systems have not yet been adapted to the more recent standards for various aspects of GD&T. This makes it difficult for a company to do a complete switch to the newest symbology.

GD&T is an in-depth, intermediate-level field of knowledge that often requires intensive study for a complete understanding. It is recommended you check into some of the complete texts available solely for GD&T. For those who wish to be certified in GD&T, certification is available through ASME. With this preface, the goal of this unit is to provide an overview of the basics of GD&T to get you started on the road to reading and interpreting prints containing GD&T symbology.

As GD&T has developed, many changes have made it a complicated topic for the print reader. Rules of interpretation have changed on more than one occasion, especially with respect to modifier use and datum identification. You should study GD&T beyond the material presented in this textbook, but pay particular attention to older standards as well as the current standards. This will help you interpret the symbols and specifications used by designers and drafters on industrial prints.

In summary, geometric dimensioning and tolerancing is an extensive, and sometimes complex, system for describing the quality of geometry and geometric locations on industrial prints. As standards continue to evolve for this field, you need to keep abreast of the latest standards to help your company maintain clear and concise dimensioning and tolerancing practices, resulting in high-quality manufactured parts that function well.

## Geometric Dimensioning and Tolerancing Symbols

Simply glancing at a drawing will quickly indicate if it contains GD&T specifications. A drawing with GD&T will have square or rectangular "boxes" associated with local notes and around some dimensions. Often, these symbols will be throughout the drawing, although they may be isolated to just one or two features. See Figure 13-1. The "boxes" are either datum identification symbols, basic dimensions, or feature control frames. Each of these components is addressed in this unit.

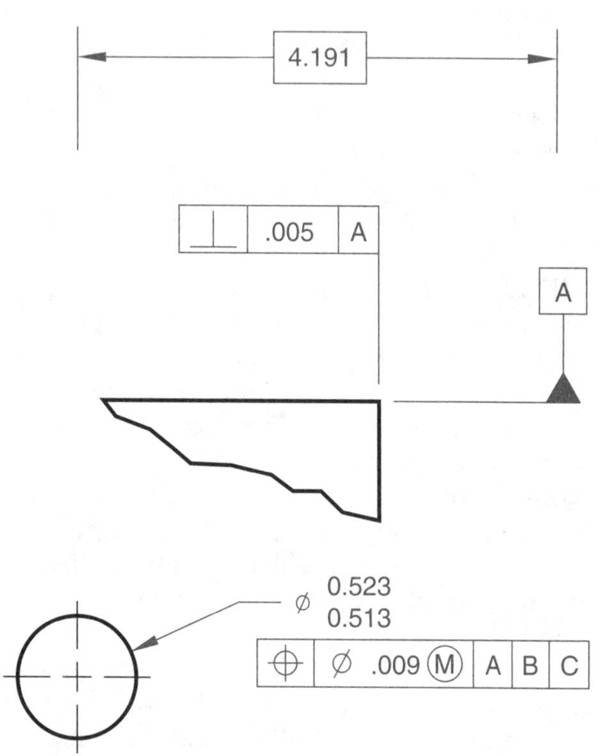


**Figure 13-1.**
The "boxes" in local notes and around some dimensions quickly identify a drawing as containing GD&T specifications.

Feature control frames are long rectangular boxes attached to extension lines, leader lines, or floating near certain dimensions. Inside of the feature control frames are geometric characteristic symbols that specify "control" of the geometry for a particular feature. Figure 13-2 shows the 14 characteristics introduced in this unit.

## GD&T Terminology

A ***datum,*** or datum target, is a theoretically exact point, axis, or plane used as the origin from which the location or geometric characteristics of a feature are established. The datum is identified on the drawing by a box on a leader "stem" terminating with a triangle. See Figure 13-3. This is called the ***datum identification symbol.*** For datum axes and datum center planes, the datum identification symbol must be *aligned* with the dimension line of the feature used to determine the axis or center plane. The datum identification symbol *cannot* be simply attached to a centerline without reference to any measurements or surfaces. If you are reading a print incorrectly drawn in this manner, you must seek clarification of which feature (or set of features) is to used to establish the datum reference.

In former practice, the datum identification symbol was drawn as a rectangular box and the datum identification letter was preceded and followed by a dash. See Figure 13-4. The box had no stem, but was attached to the view with an extension line or leader line. For a datum axis or center plane identification, the box was to be "floating" next to a measurement of a size feature, such as the diameter of a hole or width of a slot. In these cases, the dimension indicated the real features of the object that were to be used to establish the hypothetical datum axis or datum center plane.

A ***basic dimension*** is a numeric value used to describe the theoretically exact size, profile shape, orientation angle, or location distance for a feature or datum target. Tolerances are *not* directly applied to basic dimensions. In the case of hole or feature locations, the tolerance is given in the feature control frame. In other cases, such as for datum target location, gage makers and inspection personnel use their own tolerance guidelines at a higher level of precision. The symbol for a basic dimension is a rectangle enclosing the dimension, as shown in Figure 13-5. Again, just because there is no tolerance shown with the dimension, it does not mean there is no size or locational tolerance. The tolerance is just applied differently through standard GD&T practices.

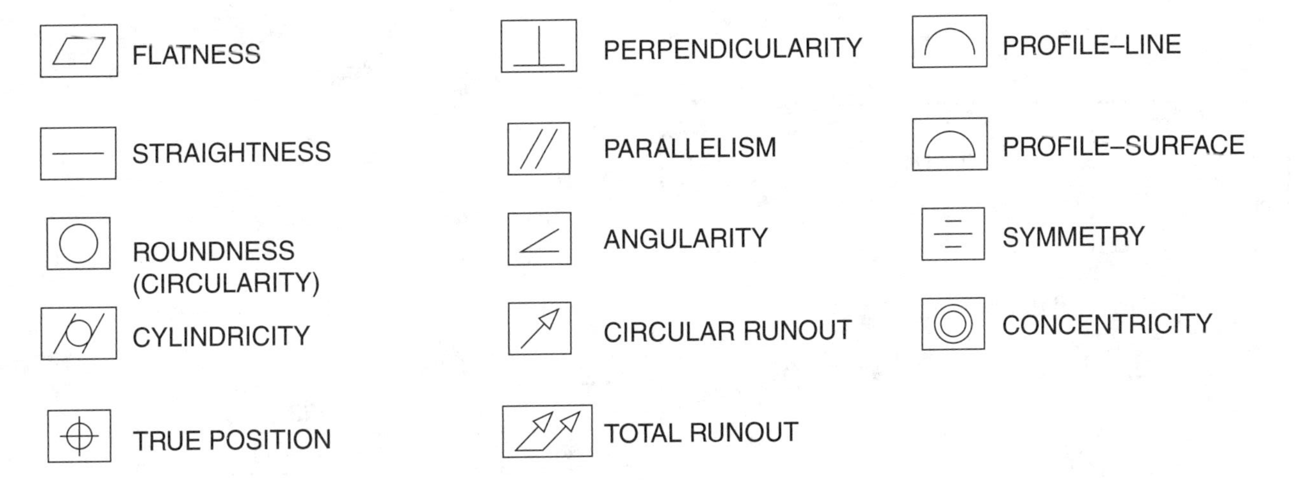


**Figure 13-2.**
These are the 14 geometric control characteristics used in GD&T.

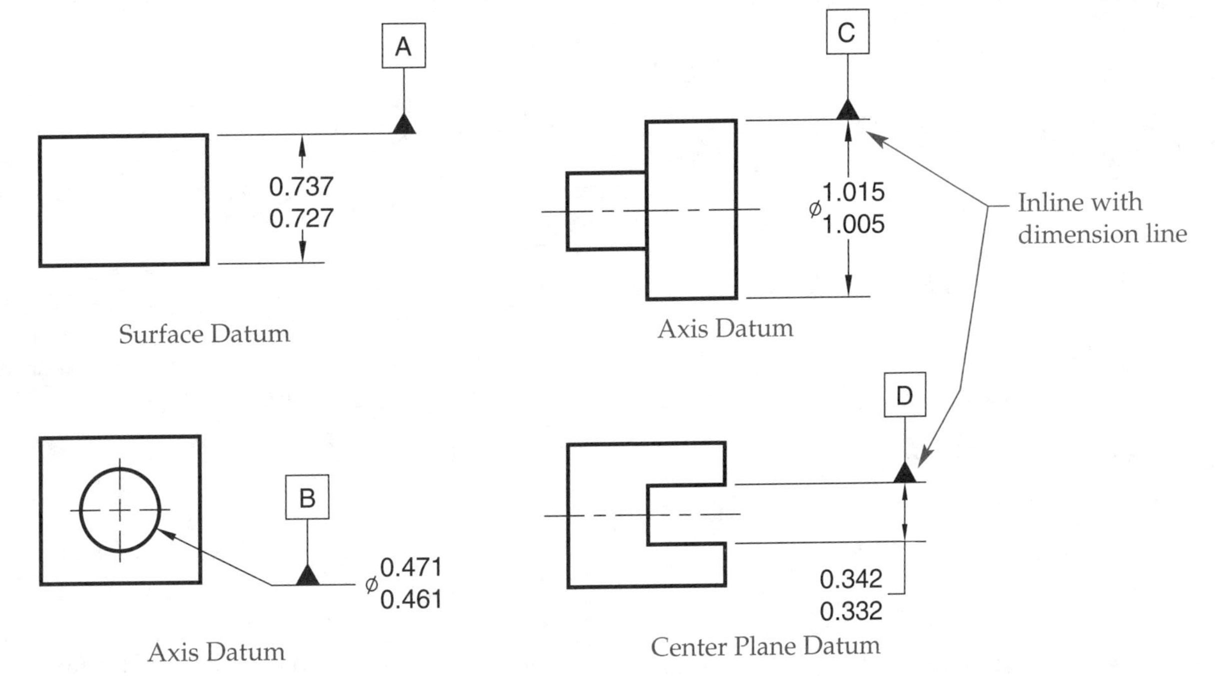


**Figure 13-3.**
The datum identification symbol is a box on a leader "stem" terminating with a triangle.

The term ***feature*** refers to a physical portion of a part. For example, a surface is a feature. Slots, tabs, posts, holes, keyways, and threaded parts of an object are also called features. Any object has many features. An important concept in GD&T is also the term ***size feature***, which is simply a feature that has a center plane or center axis. These "features of size" are things such as holes, slots, tabs, and posts. There are certain benefits in GD&T that can be applied to a size feature.

A ***feature control frame*** is a rectangular box with compartments enclosing the geometric characteristic symbol, specified geometric tolerance, and datum reference, Figure 13-6. The feature control frame can be very simple for some controls, but very complex for others. For a positional tolerance, three datums are often required to establish a "locked-in" three-dimensional position. If referenced, each datum is in a separate compartment in the feature control frame. On occasion, two features can form a datum. In this case, the datum reference letters are placed

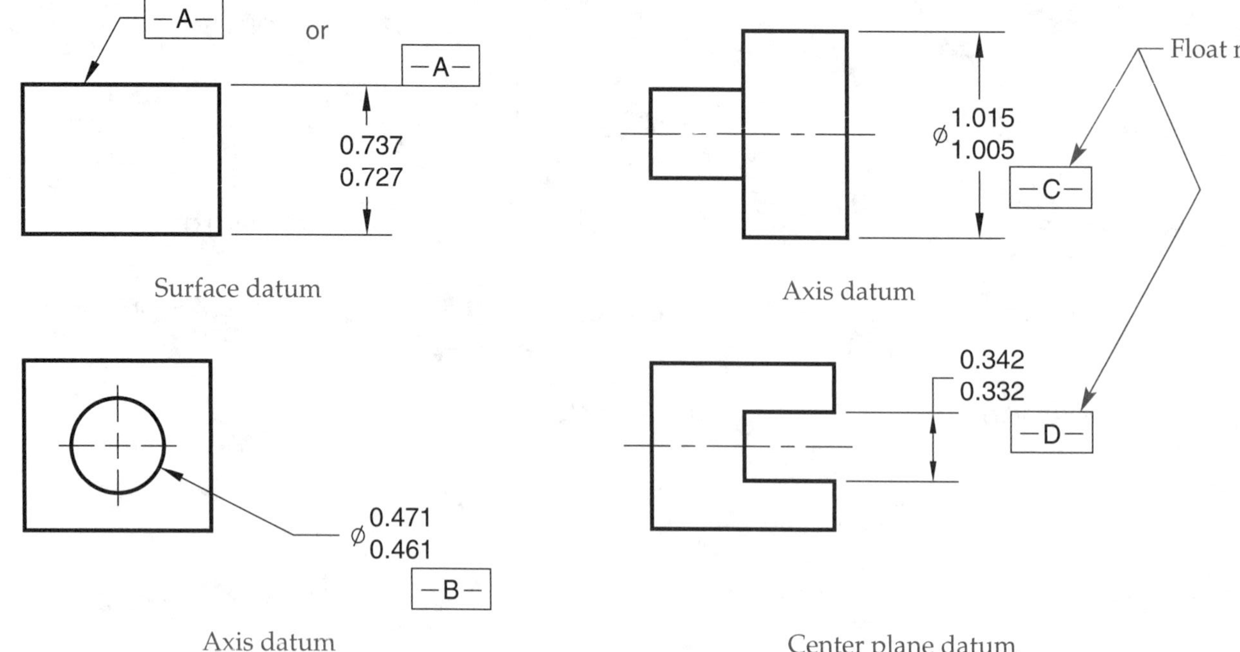


**Figure 13-4.**
In older standards, the datum identification symbol was a rectangular box with the datum letter preceded and followed by a dash.

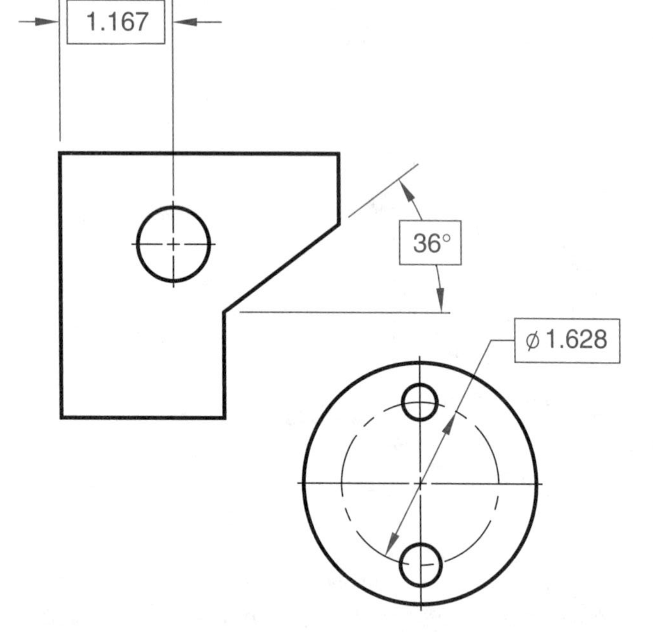


**Figure 13-5.**
A basic dimension features a rectangle enclosing the dimension value.

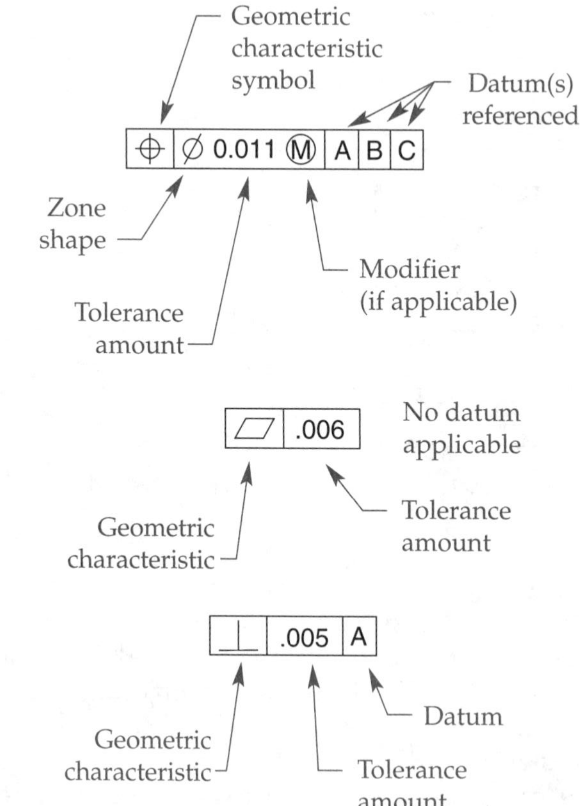


**Figure 13-6.**
A feature control frame is a rectangular "box" with compartments enclosing the geometric characteristic symbol, the specified geometric tolerance, a possible modifier, and one or more possible datum references.

in a single compartment separated by dashes. Some of the early GD&T standards call for the datum reference to precede the amount of tolerance, but current standards place the datum reference at the end.

***Tolerance*** is the range within which a specified dimensional value is permitted to vary. As discussed in Unit 10, the tolerance is the difference between the larger and smaller limits of a size or location value. For GD&T, tolerance should be considered as two-dimensional or three-dimensional space within which the geometry of a feature can vary. Each feature of the object can be specified to fall within an envelope of space determined by geometric tolerance specifications. For example, a straightness tolerance zone is the area between two line segments (2D zone), while a flatness tolerance zone is the area between two flat planes (3D zone). A locational tolerance zone for an axis is most often a cylindrical location zone. The tolerance zone size is indicated in a feature control frame along with the geometric characteristic.

***True position*** is the theoretically exact location of the axis or median plane of a feature. True position is established by basic dimensions, which locate a three-dimensional locational zone. Due to the extensive and sometimes complex nature of true position control, an entire chapter of the standards is entirely dedicated to tolerances of location.

***Material condition modifiers*** are used to indicate a bonus tolerance is to be applied, with the bonus based on the actual size of the feature as it compares with either the maximum material condition or the least material condition. The modifiers can also be applied to datum references, giving the designer or drafter the ability to specify a more flexible inspection setup. The encircled M for maximum material condition or the encircled L for least material condition, are the symbols used in feature control frames. In former practice, an encircled S indicated "regardless of feature size." However, that condition is now understood for any parts not modified with the least material condition or maximum material condition symbol. Figure 13-7 shows the material condition modifiers as well as other modifying symbols that can be applied to help describe the geometric control or datum setup. Some of these modifiers are explained in more detail later in the unit.

| GD&T Symbols | |
|---|---|
| Ⓜ | Maximum material condition modifier |
| Ⓛ | Least material condition modifier |
| Ⓢ | Regardless of feature size (old standard; obsolete) |
| Ⓟ | Projected tolerance zone |
| Ⓣ | Tangent plane |
| Ⓕ | Free state |
| ⟨ST⟩ | Statistical tolerance |
| ◄► | Between |

**Figure 13-7.**
Modifying symbols are used in a variety of ways, including material condition modifiers, projected tolerance zones, and free state specification.

## Datums

With respect to datums, the terminology should be carefully considered. For example, there is a fine distinction to be made between a datum *feature* and a datum *plane* or *axis.* One is real and one is hypothetical. A surface is a real datum feature used to establish a theoretically exact datum plane. The surface of a hole is a real datum feature used to establish a theoretically exact datum axis. In summary, ***datum features*** are actual geometric features of the part, such as flat surfaces or cylindrical surfaces, used as a reference from which to establish the measurements and readings. Datum features used to set up the part for inspection are identified on the drawing. The hypothetical datum planes and axes are represented by items such as precision inspection equipment, gauges, gauge pins, the axes of a coordinate measuring machine, or the surface of a marble inspection block. For example, the real points of an object rest on the marble block surface and the marble surface then becomes the hypothetically perfect datum plane from which flatness or parallelism can be checked.

A ***datum reference framework*** consists of three mutually perpendicular planes that represent the features important to the design or function of the part. See Figure 13-8. A datum reference framework is necessary to lock the object in space so certain geometric characteristics or hole locations can be checked. These planes are simulated by positioning the part on appropriate datum features and restricting the motion of the part. See Figure 13-9.

**Figure 13-8.**
A datum reference framework consists of three mutually perpendicular planes representing the features of the part that are the most important in relation to the geometric controls being applied.

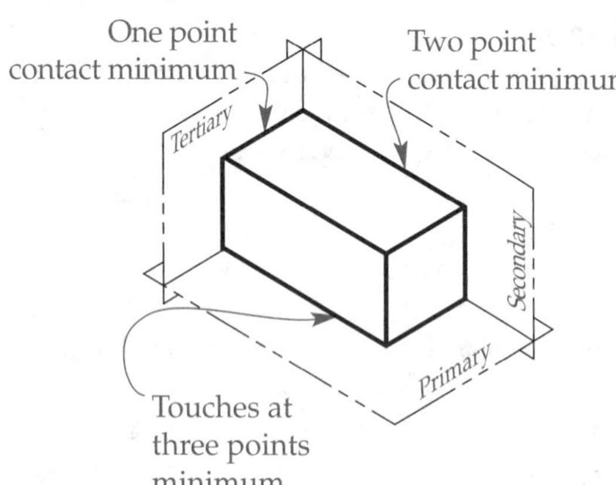


**Figure 13-9.**
A datum reference framework is necessary for checking certain geometric characteristics.

Not all geometric controls require three datums to be established. Also, cylindrical datum features have two theoretical planes intersecting at right angles on the datum axis, and one additional surface for perpendicularity. So, a cylindrical object may be set into a datum reference framework with two datum features. On the other hand, some parts require more than three datums to be specified. Perhaps three datums are used to establish the three-dimensional framework, but additional datums are specified for other geometric or relational controls.

***Datum precedence*** is the order in which an object is placed into the datum reference framework. The precedence is indicated by the order of the datum reference letters, from left-to-right, in the feature control frame. From the left, these datums are designated as primary, secondary, and tertiary datum features. They are selected in order of functional design importance, if applicable. The drafter usually attempts to use the letters A, B, and C in the primary-secondary-tertiary order, but that is not required, nor of great importance.

In theory, the primary datum feature relates the part to the datum reference framework with three points on the surface in contact with the first datum plane. The secondary datum feature relates with two points in contact with the second datum plane. The tertiary datum feature relates with one point into contact with the third datum plane.

***Datum targets*** are specific points, lines, or areas of contact on a part that can be specified to establish the datum reference framework. Datum targets can be used to correct for the irregularities of some features, such as nonplanar, thin, or uneven surfaces, especially when it is not economically feasible to machine or stabilize them. Datum targets are identified by an X on the drawing or a target area of the desired shape is specified. See Figure 13-10. The datum targets are basically the fixture pins that hold a part in place during manufacturing or

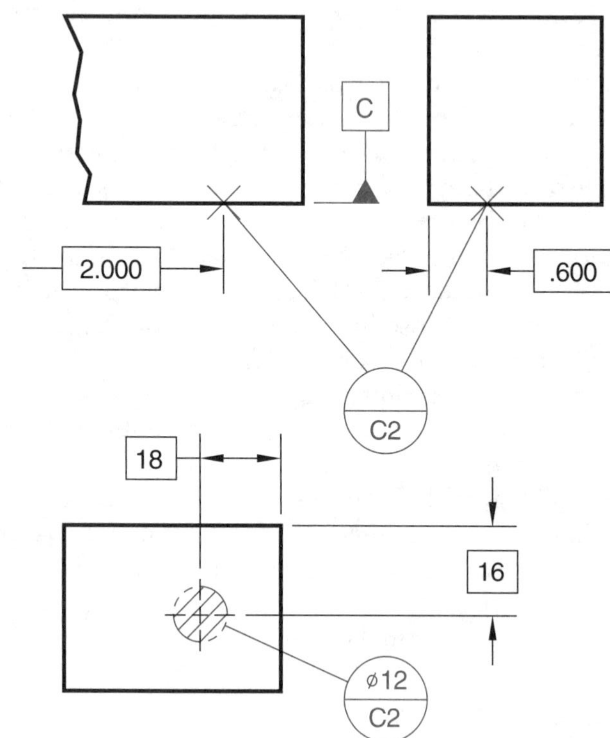


**Figure 13-10.**
Datum targets are identified by an "X" on the drawing or a target area of the desired shape is specified.

inspection. Sometimes, checking gauges are built to hold the part during inspection. In these cases, the datum targets are used by the gauge maker to establish how the part is to be mounted. The basic dimensions on the print are used by the gauge maker, who in turn applies tolerances according to gauge maker standards or guidelines.

## Standardized Meaning of Size

Before applying geometric controls to an object, it is critical to standardize the meaning of "size." Unless otherwise specified, the limits of size control the geometry and GD&T symbols are not needed. This standardized definition of size, also referred to as ***Rule 1,*** can be paraphrased to read, "where only a tolerance of size is specified, the limits of size prescribe the extent to which variations in size are allowed." In other words, a part cannot geometrically deviate beyond the perfect form of its maximum material condition.

Therefore, by definition, the size of a cylinder controls the shape of that cylinder. If a cylinder has specified limits of 1.490″ to 1.500″ and if its actual size is at the maximum material condition of 1.500″, then it has to be perfectly straight. If it is .005″ undersize, some of the elements technically could deviate up to .005″. The problem arises when a size tolerance amount is so large that geometric variance is possible, so then GD&T controls can be used to control the geometric variance.

The exceptions to Rule 1 are beyond the scope of this textbook. Where it is desirable to permit a surface or surfaces of a feature to exceed the boundary of perfect form, a note such as PERFECT FORM AT MMC NOT REQUIRED can be specified on the drawing.

## Material Condition Applicability

The principles of geometric dimensioning and tolerancing can also provide a flexible bonus tolerance for certain features. As defined earlier, a size feature is a feature with a center axis or a center plane. For example, a hole is a size feature. Slots and tabs are also size features. When controlling the geometry of a size feature or when using a size feature as a datum, additional consideration can be given based on the material condition of the feature. There are two material conditions under which geometric control or datum reference setup can be applied or modified: 1) maximum material condition and 2) least material condition.

The ***maximum material condition (MMC)*** occurs when the object contains the most material allowed by a size dimension. That is, a hole's MMC is the smallest diameter, while a shaft's MMC is the largest diameter. The symbol for MMC is an M contained within a circle. The symbol is used in the feature control frame to modify an individual tolerance, a datum reference, or both. See Figure 13-11.

Maximum material condition is commonly applied to the positional tolerance of a hole. In this case, the designer is specifying that as the hole gets bigger (away from MMC), it can be more out of position and still be functional. In the case of a pin, as the pin gets smaller (away from MMC), it should be logical that the pin can be more out of position and still clear a mating hole. The mathematical calculations are easy. For each .001″ of diameter departure from MMC, the diameter of the positional zone can increase by .001″. Before GD&T, there was not a way to specify this bonus tolerance.

The ***least material condition (LMC)*** occurs when the object contains the least amount of material allowed by a size dimension. That is, a hole's LMC is the largest diameter, while a shaft's LMC is the smallest diameter. The symbol for LMC is an L contained within a circle. The symbol is used in the feature control frame to modify an individual tolerance, a datum reference, or both. See Figure 13-12.

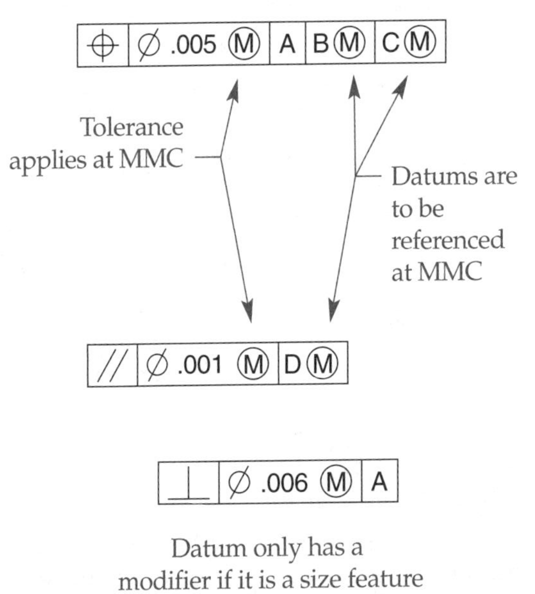


**Figure 13-11.**
The symbol for MMC is an M contained within a circle. It indicates bonus tolerance is given as the feature departs from MMC. It can also be applied to allow a datum feature to be established in a more flexible manner.

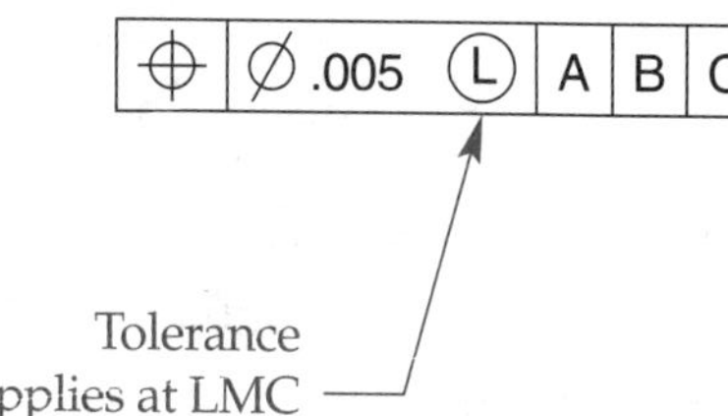


**Figure 13-12.**
The symbol for LMC is an L contained within a circle. It indicates bonus tolerance is given as the feature departs from LMC. It can also be applied to allow a datum feature to be established in a more flexible manner.

Using LMC as a modifier is not nearly as common as using MMC. It is often applied in situations where a designer wants to guarantee a minimum edge distance between two features. In the case of a clearance hole, for example, as the hole gets smaller (away from LMC), it can be more out of position without endangering the minimum thickness of the material around the hole.

For years, another modifier known as ***regardless of feature size (RFS)*** was used to specify those situations wherein an MMC or LMC modifier could be applied, but the desire was not to keep the tolerance the same regardless of the feature size. This was represented by an S contained in a circle. See Figure 13-13. Under current standards, if an MMC or LMC modifier is not present, RFS is automatically assumed. The RFS modifier is not to be used.

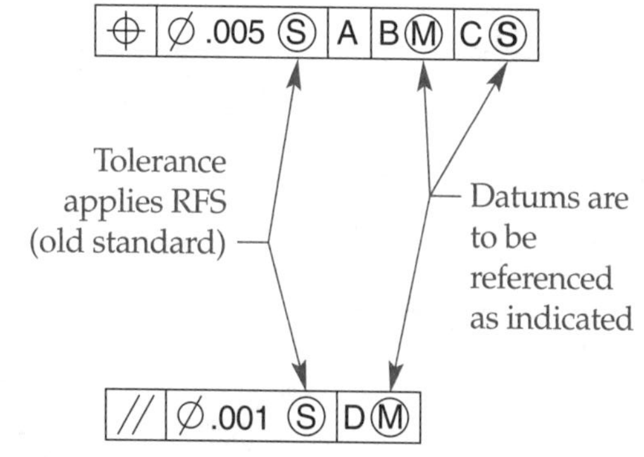


**Figure 13-13.**
In former practice, the symbol for "regardless of feature size" was an S contained in a circle. It used to be a required specification for positional tolerance and for a datum that was a size feature, unless MMC or LMC was desired. Now, RFS is assumed and the symbol is no longer needed.

## Historical Summary

A summary of GD&T symbol history may be helpful to those who deal with older prints. At one point in time, MMC was so common for true position tolerancing the "rule" was that if no modifier symbol appeared in a positional tolerance specification, MMC was *assumed.* This apparently caused some confusion, so for another period in time, the standard "rule" was that a symbol for MMC, LMC, or RFS was *always* required for the positional tolerance specification, but not for the other characteristics. This was formerly known as Rule #2. In current standards, Rule #2 states that RFS is to be applied, with respect to individual tolerances, datum references, or both, if the controlled feature or datum is a size feature that is not modified.

In summary, under current standards, MMC or LMC must be specified and never assumed. RFS should not be specified unless a company feels it is important to do so. As always, consult with the proper personnel if an interpretation is unclear. The concept of material condition modification takes time to understand. You need to carefully study the geometric characteristics to see how the modifiers have an impact on reading a print.

## Form Tolerances

***Form tolerances*** are used to control straightness, flatness, roundness (circularity), and cylindricity. They are perhaps the easiest GD&T controls to understand. Form tolerances are *not* related to datums, but are applicable to single (individual) features or elements of single features. Form tolerances are not required on the drawing if the size tolerance is sufficient to control the geometry, as discussed earlier. For example, if a basic block has a size tolerance of .010", then the individual surfaces will be, by definition, flat within .010".

***Straightness*** is a measure of an element of a surface or an axis existing in a straight line. Straightness tolerance specifies a tolerance zone within which the considered element or axis must lie. See Figure 13-14. Straightness is one control that has two possible applications for cylindrical features: element straightness and axis straightness. For axis straightness, Rule 1 (perfect form at MMC) does not apply. The straightness of all elements as a composite is summarized as the straightness of the axis. Straightness can also be applied to non-cylindrical features.

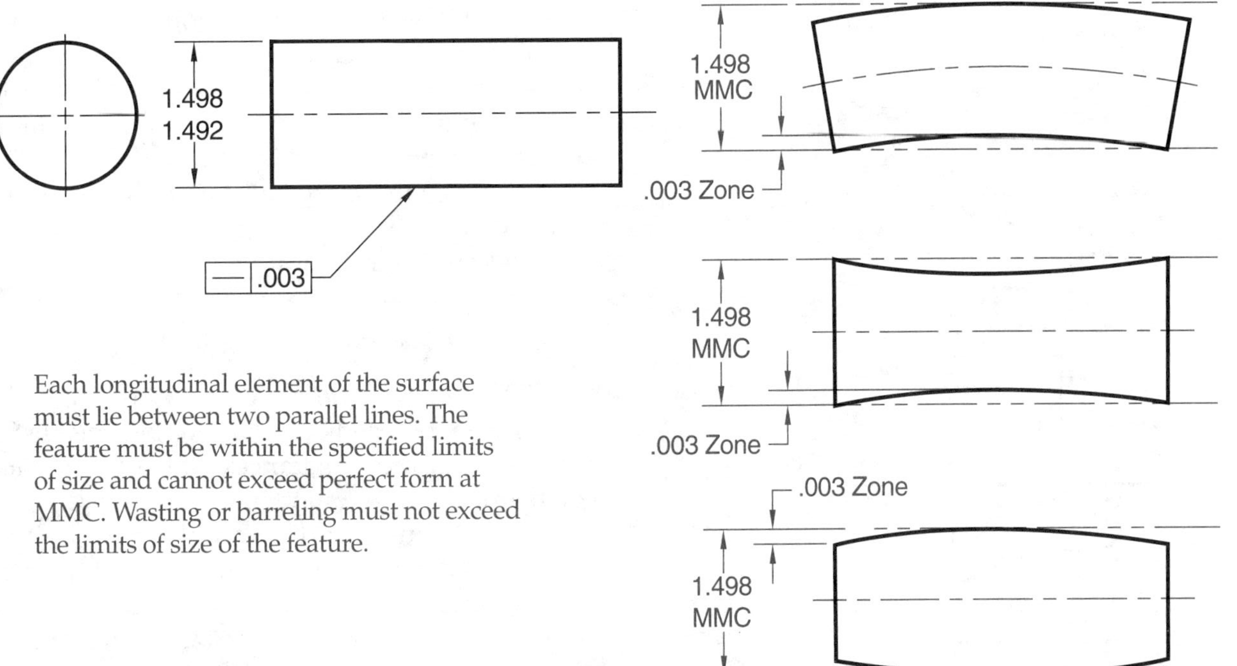


**Figure 13-14.**
Straightness is a measure of an element of a surface or an axis existing in a straight line.

***Flatness*** is the condition of a surface having all elements in one plane. Flatness tolerance specifies a tolerance zone defined by two parallel planes within which the surface must lie. See Figure 13-15. Flatness can be thought of as a three-dimensional version of straightness. Flatness is often used to "qualify" a datum surface. In these cases, the datum identification symbol is often attached to the feature control frame, but the flatness control is not referencing the datum. See Figure 13-16. Flatness is to be determined first and, if the feature passes inspection, then the surface can be used as a datum feature.

***Roundness*** or ***circularity*** for a cylinder or cone is a condition where all points of the surface intersected by any plane perpendicular to a common axis are equidistant from that axis. See Figure 13-17. For a sphere, all points of the surface intersected by any plane passing through a common center are equidistant from that center. Simply put, roundness asks the question, "how round is any given cross section of a round feature, regardless of a datum axis?" Circularity tolerance specifies a radius value for a tolerance zone bounded by two concentric circles within which each circular element of the surface must lie.

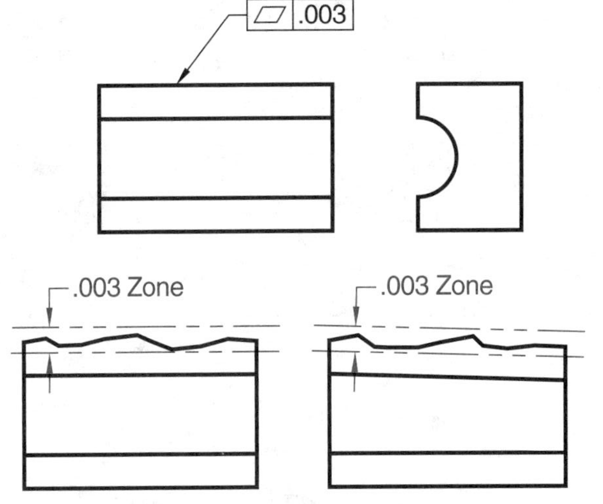


**Figure 13-15.**
Flatness is the condition of a surface having all elements in one plane.

(Old standard)

**Figure 13-16.**
A datum identification symbol attached to a surface or feature control frame indicates the feature controlled by the geometric tolerance is to be used as a datum, assuming it passes inspection.

***Cylindricity*** is a condition of a surface of revolution in which all points on the surface are equidistant from a common axis. See Figure 13-18. Cylindricity tolerance specifies a radius value for a tolerance zone bounded by two concentric cylinders within which the controlled surface must lie. Simply put, cylindricity asks the question, "how close to a true cylinder is the feature?"

## Orientation Tolerances

***Orientation tolerances*** control angularity, parallelism, perpendicularity, and, in some instances, profile. Since these tolerances control the orientation of features to one another, they always refer the controlled feature to at least one datum. As with form tolerances, orientation tolerances are only specified when size tolerances do not adequately control orientation. For example, a size tolerance of .020″ provides a parallelism of .020″. If this is acceptable, a feature control frame specifying parallelism is not needed.

***Parallelism*** is the condition of a surface or axis being equidistant at all points from a datum plane or datum axis. A parallelism tolerance zone for a surface feature is defined by two planes or lines parallel to a datum plane or axis within which the line elements of the feature must lie. See Figure 13-19.

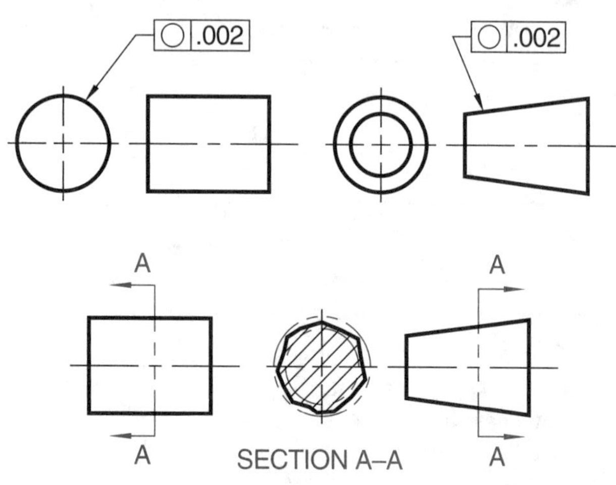


**Figure 13-17.**
Roundness or circularity for a cylinder or cone is a condition where all points of the surface intersected by any plane perpendicular to a common axis are equidistant from that axis.

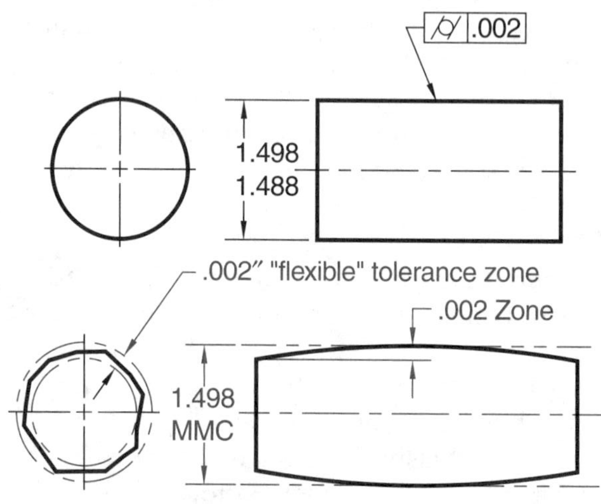


**Figure 13-18.**
Cylindricity is a condition of a surface of revolution in which all points on the surface are equidistant from a common axis.

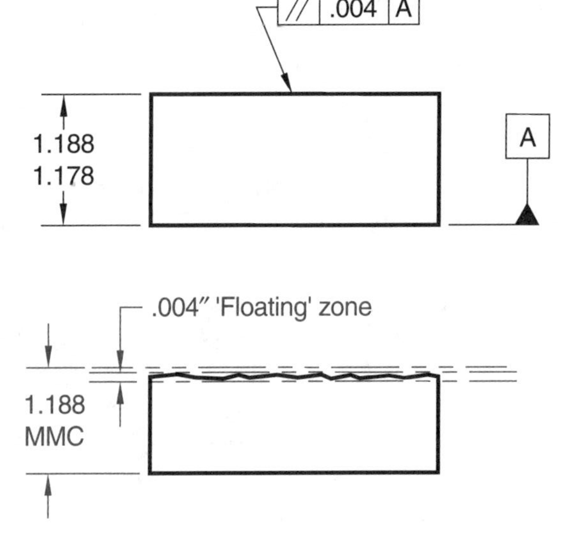


**Figure 13-19.**
Parallelism is the condition of a surface or axis being equidistant at all points from a datum plane.

A parallelism tolerance zone for a cylindrical feature is defined by a cylindrical zone whose axis is parallel to a plane or datum axis. The axis of the feature must lie within this zone, Figure 13-20. A feature control frame, in this case, will probably be located on the bottom of a positional tolerance feature control frame to refine the positional tolerance, as shown in the figure.

***Perpendicularity*** is the condition of a surface, median plane, or axis existing at a right angle to a datum plane or axis. A perpendicularity tolerance is defined by two parallel planes perpendicular to a datum plane or axis between which the surface or axis of the feature must lie. See Figure 13-21. A perpendicularity tolerance for a cylindrical feature is a cylindrical zone perpendicular to a datum plane within which the axis of the feature must lie. See Figure 13-22. A diameter symbol appears in the feature control frame. Like parallelism of an axis, perpendicularity control of an axis is often a refinement of a positional tolerance.

***Angularity*** is the condition of a surface or axis existing at a specified angle to a datum plane or axis, wherein the angle is not 90°. An angularity tolerance zone is defined by two parallel planes at the specified basic angle from a datum plane or axis between which the surface or axis of the feature must lie. See Figure 13-23. The GD&T angularity control creates a uniform tolerance zone, whereas plus and minus degrees always creates a "fan-shaped" zone.

## Profile Tolerance

***Profile tolerance*** is a control that most often describes the amount of tolerance to be maintained for an irregular outline of an object or feature. The elements of a profile can be straight lines, arcs, and other curved lines. The profile tolerance specifies a uniform boundary along the true profile within which the elements of the surface must lie. See Figure 13-24. The profile can be referenced to a

In addition to the .010″ position tolerance, a parallelism tolerance of .004″ is applied at MMC to further control the parallel relationship of the holes. The datum reference is also to be applied at MMC.

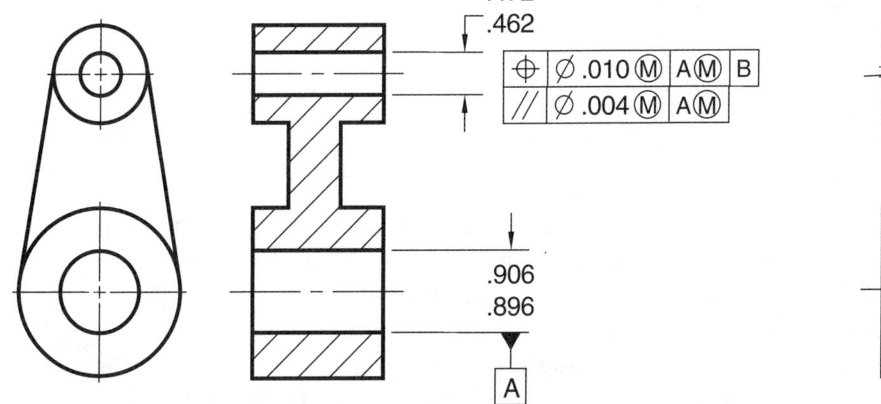


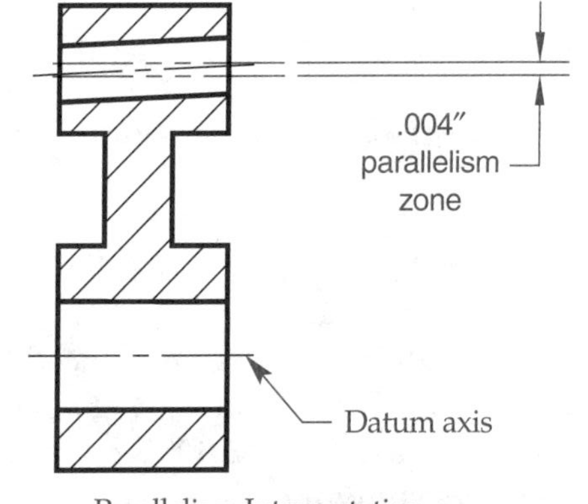


**Figure 13-20.**
A parallelism tolerance zone for a cylindrical feature is defined by a cylindrical tolerance zone whose axis is parallel to a datum plane or datum axis.

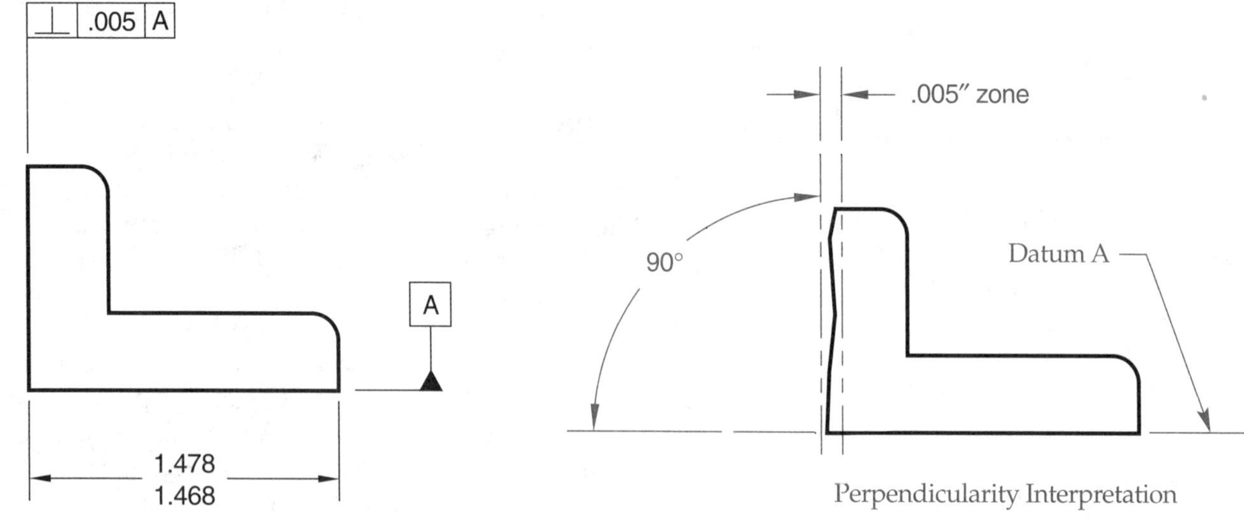


**Figure 13-21.**
Perpendicularity is the condition of a surface, median plane, or axis existing at a right angle to a datum plane or axis.

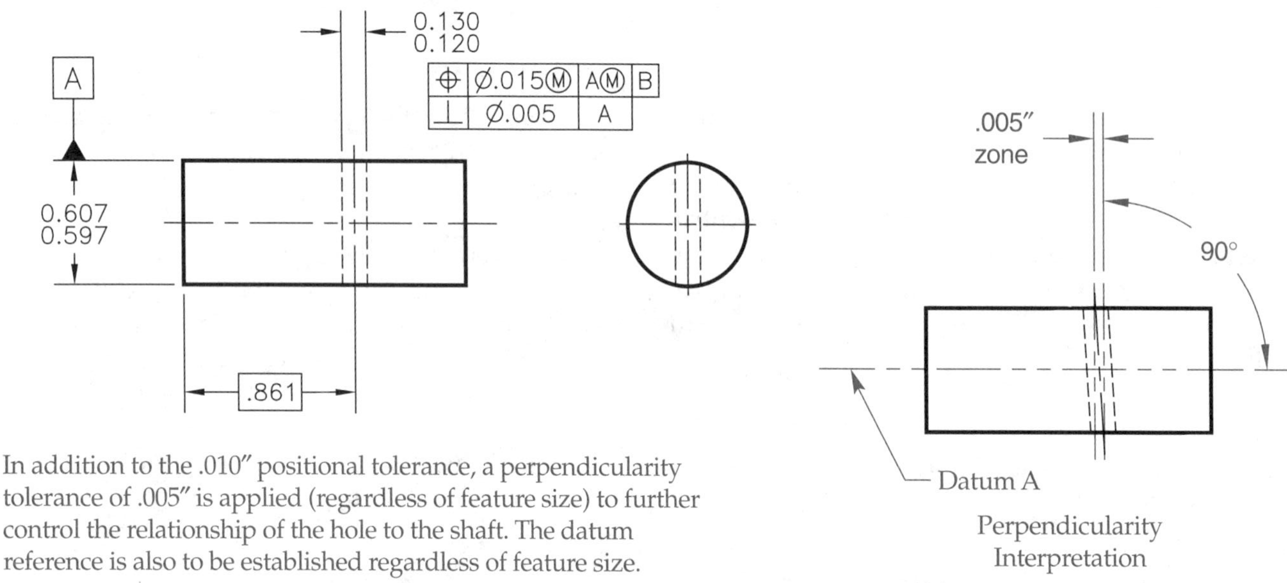


**Figure 13-22.**
A perpendicularity tolerance for a cylindrical feature is a cylindrical tolerance zone perpendicular to a datum plane within which the axis of the feature must lie.

datum similar to the way parallelism is controlled. The profile can also "float" in a way similar to flatness and does not have to be referenced to a datum. In some applications, profile is between two specified points, such as BETWEEN X & Y. A newer symbol for between is shown in Figure 13-24.

Profile is divided into two categories—profile of a line and profile of a surface. See Figure 13-25. ***Profile of a line*** is a two-dimensional control of the *elements* of an irregular surface. It can be thought of in the same way as straightness or roundness, but for shapes defined by basic dimensions. ***Profile of a surface*** is a three-dimensional control that extends throughout the irregular surface, so at least one datum needs to be established that helps locate the tolerance zone profile.

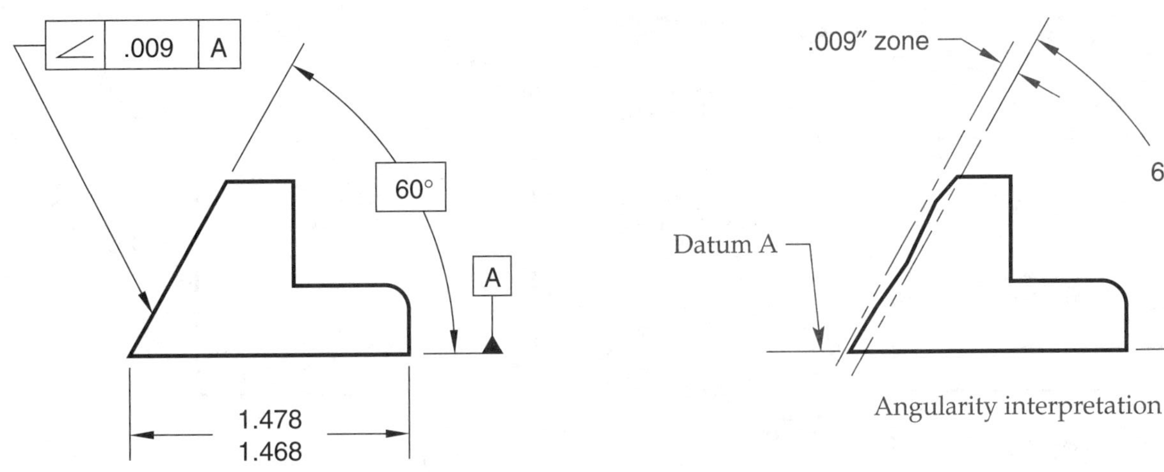


**Figure 13-23.**
Angularity is the condition of a surface or axis existing at a specified angle to a datum plane or axis.

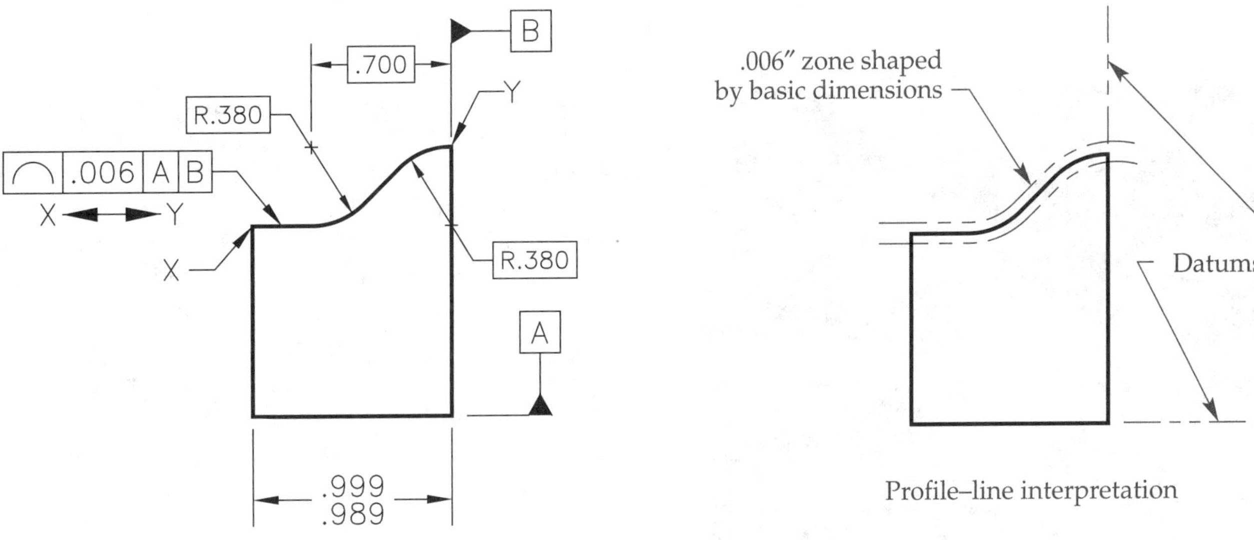


**Figure 13-24.**
A profile of a line tolerance specifies a uniform boundary along the true profile within which the elements of the surface must lie, but each element is individually controlled.

## Runout Tolerance

***Runout tolerance*** is a composite tolerance used to control the functional relationship of one or more features of a part to a datum axis. This includes those surfaces constructed around a datum axis and those constructed at right angles to a datum axis. The two types of runout control are circular runout and total runout. Runout is always measured with a dial indicator and is measured regardless of feature size. A dial indicator provides a needle reading as the part is rotated. See Figure 13-26. ***Full indicator movement (FIM)*** is the total movement of the dial needle in measuring the variance of a surface. FIM has the same meaning as the older terms full indicator reading (FIR) and total indicator reading (TIR).

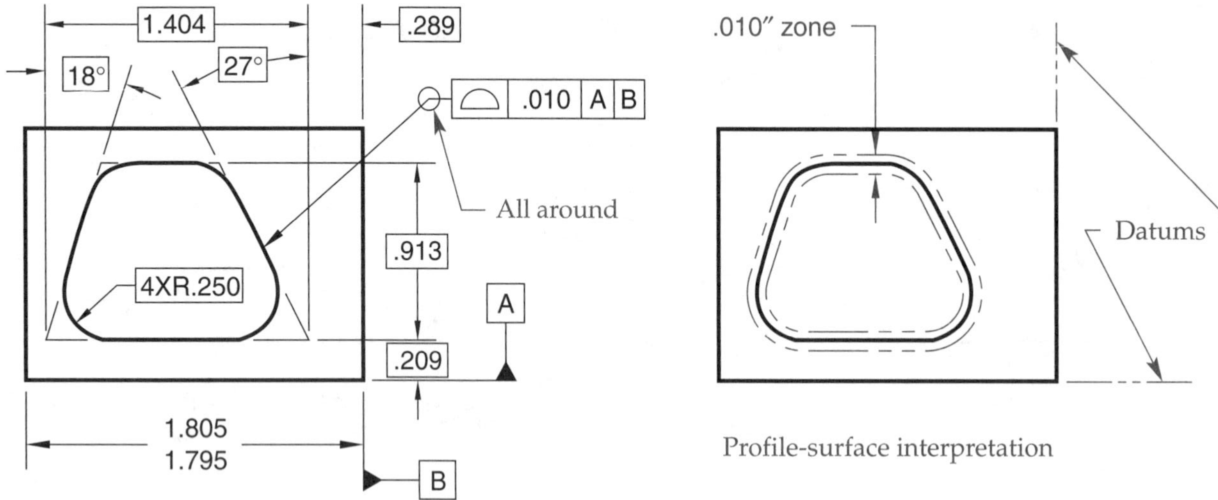


**Figure 13-25.**
A profile of a surface tolerance zone is often defined by basic dimensions and considers all elements of the surface as a whole.

**Figure 13-26.**
Full indicator movement is the total movement of a dial indicator when measuring the variance of a surface as it rotates about a datum axis.

***Circular runout*** controls circular elements of a surface. For surfaces constructed around a datum axis, circular runout controls circularity and concentricity *with* respect to the datum. See Figure 13-27. Bonus tolerance can never be specified for runout, as the dial indicator always makes contact with the surface as it rotates about the datum axis.

***Total runout*** provides composite control of all surfaces. See Figure 13-28. For surfaces constructed around a datum axis, total runout is used to control cumulative variations of circularity, straightness, coaxiality, angularity, taper, and profile of a surface. It is important to note that the features themselves could be perfectly cylindrical, but if they are not *coaxial,* then runout can occur. It is also important to note that the features could be perfectly coaxial, but if surface irregularities exist, then runout can occur. Also, where a runout tolerance applies to a specific portion of a surface, a thick chain line is drawn adjacent to the surface profile on one side of the datum axis for the desired length, as specified with a basic dimension.

The engineer or designer must understand the various ways GD&T can be used to control coaxiality. In the early years of GD&T, controls such as concentricity were sometimes applied to parts and features that were simply to be inspected for runout.

## Tolerances of Location

***Tolerances of location*** are used to control position, symmetry, and concentricity. Since locational tolerances are often applied to features of size, be sure to understand how material condition modifiers are applied. Some of the examples in this section have material condition modifiers to provide bonus tolerance as a feature varies from its maximum or least material condition.

The elements of the surface must be within the specified tolerance as determined by FIM reading perpendicular to the measured surface and with the feature revolving around the applicable datum.

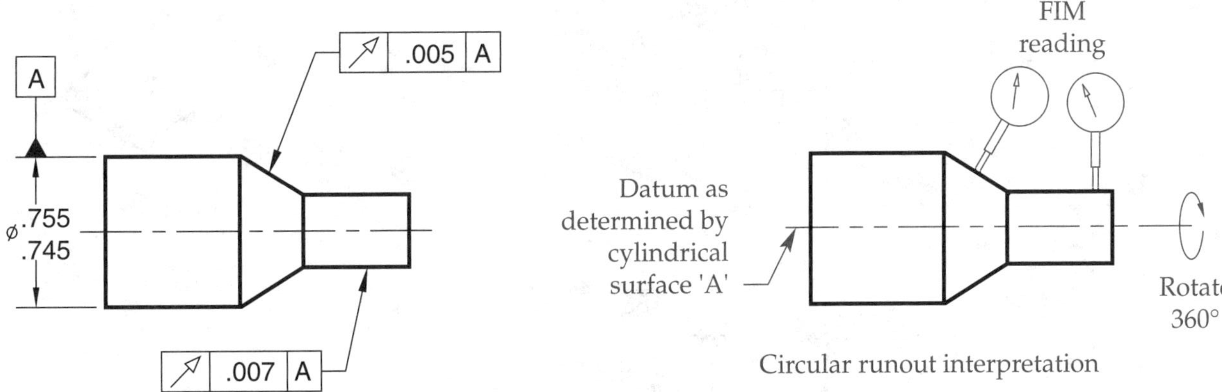


**Figure 13-27.**
Circular runout controls individual circular elements of a surface as the part is rotated about a datum axis. The dial indicator is positioned perpendicular to the surface elements as the surface is rotated.

The elements of the surface must be within the specified tolerance as determined by FIM reading perpendicular to the measured surface. The dial indicator moves parallel to the axis and the feature revolves around the applicable datum.

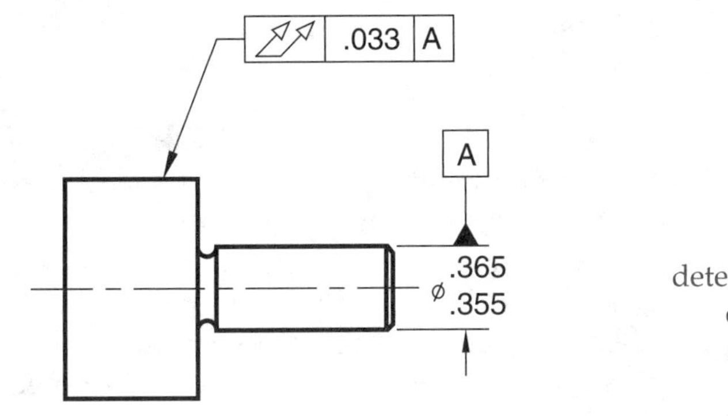


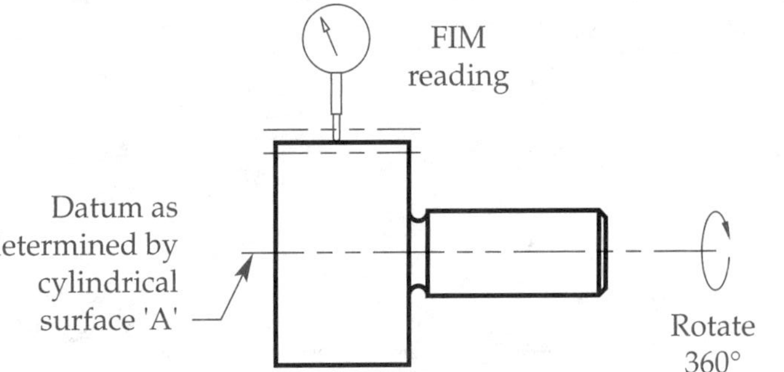


**Figure 13-28.**
Total runout provides composite control of all elements of a surface as the part is rotated about a datum axis. Total runout can be applied to an "end" surface that is perpendicular to the axis.

A ***positional tolerance,*** sometimes called ***true position,*** defines a zone within which the center, axis, or center plane of a size feature is permitted to vary from the theoretically exact position. See Figure 13-29. The true position is established by basic dimensions from specified datum features and between interrelated features. However, the true position is also oriented to a third plane of the datum reference framework. A positional tolerance is specified by the position symbol, a tolerance, a modifier, and then the appropriate datum reference, all in a feature control frame. The tolerance is three dimensional in nature. The tolerance zone is cylindrical for a hole or the distance between two planes for a slot or tab. This one GD&T control alone is one of the most beneficial controls available. Several pages of application examples and explanations are given in ASME Y14.5.

***Symmetry*** is a form of positional tolerance where a feature is controlled identically about a center plane of a datum feature, as shown in Figure 13-30. In the early years of GD&T, symmetry was used for

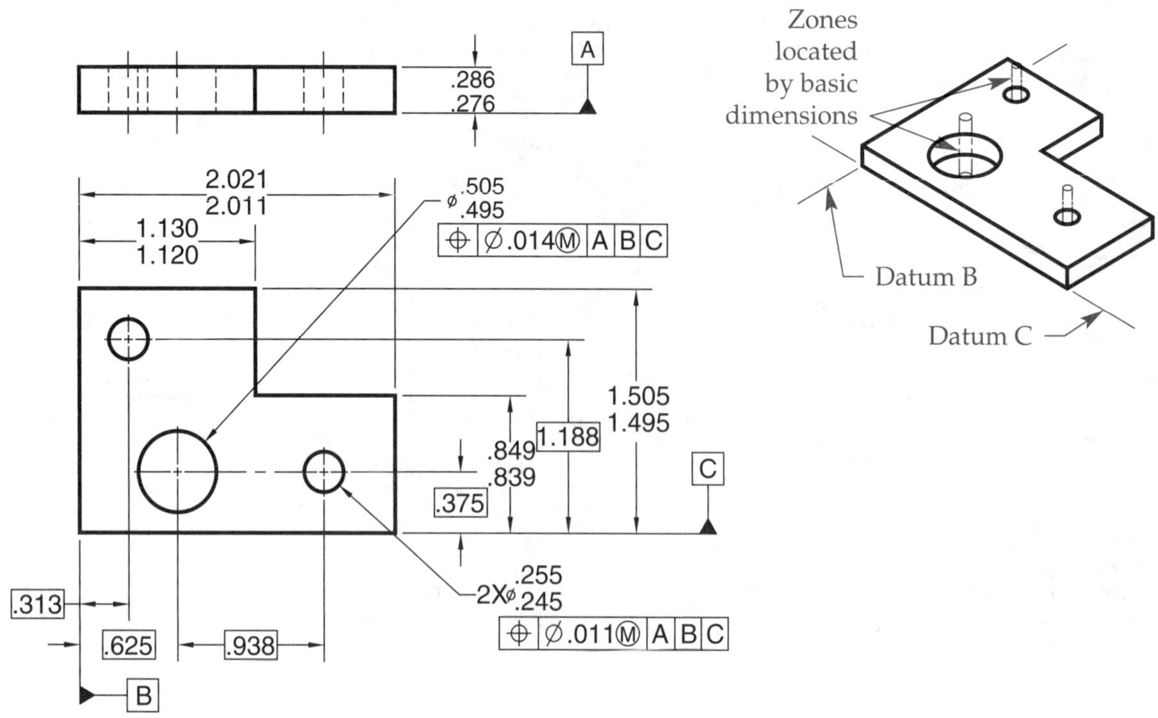

The elements of the hole surfaces will be used to determine if hypothetical axes are within the specified cylindrical tolerance zones. The object must be set within the datum reference framework.

**Figure 13-29.**
A positional tolerance defines a zone within which the center axis or center plane of a size feature is permitted to vary from the theoretically exact position. In most cases, bonus tolerance can be applied.

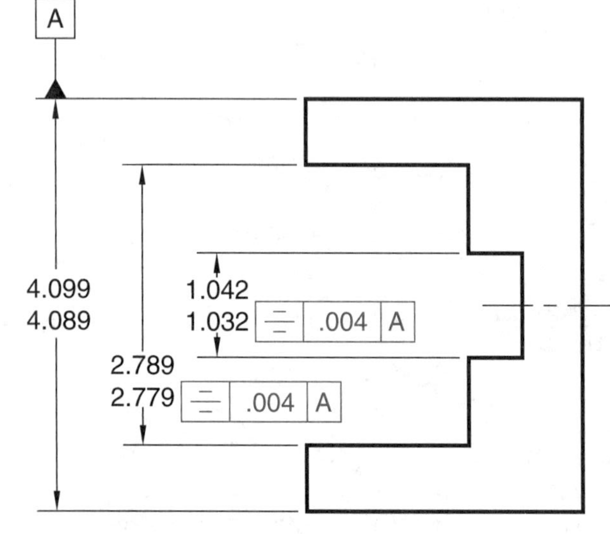


Symmetry is to be applied where true balance on each side of the center plane is desired, regardless of feature size. If the part simply needs center-based spacing, then use true position with MMC modifiers.

**Figure 13-30.**
Symmetry is a form of locational tolerance where a feature is controlled identically on each side of a center plane established by a datum feature, regardless of feature size.

most "center plane" applications. However, true position was later adopted for all center plane applications, especially in the case of bonus tolerance applications at MMC. For a period of time, American standards eliminated the symmetry symbol and the positional tolerance at RFS was used to specify true symmetry. However, to move back toward international standards, the symmetry symbol has been reinstated and is in use again under certain guidelines. Now, symmetry should be used if the tolerance is applied regardless of feature size and true balance is the desired result, regardless of the feature's shape on each side of the center plane. For more detailed information on interpreting symmetry, refer to the current ASME standards.

The changes to the standards that have been made in recent decades can make it challenging to read older prints. This also points out the importance of studying the standards under which the print was developed, as well as the importance of continuing education in an industrial setting.

***Concentricity*** describes the condition where the axes of all cross-sectional elements of a surface of revolution are common to the axis of a datum feature. See Figure 13-31. A concentricity tolerance is indicated by a cylindrical tolerance zone whose axis coincides with a datum axis. All cross-sectional axes of the feature must lie within this zone. Like symmetry, this tolerance should only be applied regardless of feature size. Apply concentricity only when needed to control the true balance in regard to the center axis, regardless of the shape of the part. To specify two parts as generally being coaxial, a positional tolerance or runout may be more appropriate specifications.

## Composite Tolerancing

Sometimes, a feature control frame is a double-layer frame with a single entry of the geometric characteristic in the first section, but two individual layers that specify different tolerance amounts or datum references. See Figure 13-32. When this happens for a positional tolerance and a pattern of holes, the top layer indicates the positional tolerance for each individual hole within the pattern, with respect to the feature's datum reference framework. The bottom layer applies to each individual feature within the pattern, in reference to other members of the pattern. For example, if the top layer specifies true position within a diameter of .010″ and the bottom layer specifies true position within a diameter of .004″, the .010″ tolerance zone is located by a datum reference framework that includes datums A, B, and C. For the pattern itself, the .004″ diameter zones "float" within the other zones, which prevents each feature within the pattern from drifting too far apart. To check this, a pattern reference framework is also established that checks the location of the holes to each other, while still remaining true to primary datum A.

Composite tolerancing may also be used with the straightness characteristic to specify total straightness on one layer, but straightness per unit on the other layer. Profile of a surface may also be specified in a composite manner. For example, the tolerance zone for the irregular shape of a vent opening on a dashboard may have a very generous tolerance with respect to the datum reference framework, but the surface profile of the shape must be maintained more tightly. The composite feature control frame allows for both specifications.

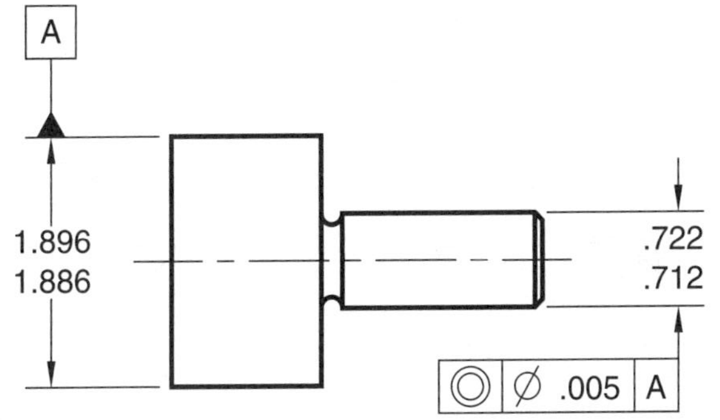


Concentricity is to be applied where true balance of the part about the datum axis is desired, regardless of feature shape or size. If the part simply needs coaxial spacing, then use true position with MMC modifiers.

**Figure 13-31.**
Concentricity describes the condition wherein the axis established by cross-sectional elements of a surface is coaxial with the axis of a datum feature. This control should only be used regardless of feature size. It examines how well the feature is "balanced" along the datum axis, regardless of the surface shape.

## Other Specifications

There are numerous additional GD&T examples and applications that are beyond the scope of this textbook. You should consult additional resources and references to have a full understanding of GD&T as used in your company setting. In Figure 13-33, a few additional GD&T specifications are shown.

In Figure 13-33A, a projected tolerance zone is shown. This specification is especially useful for threaded features. It "removes" the inspection from the threaded part to an area above the hole, which is usually the area that coincides with a mating part, such as the hole of a cover that must match the fastener that will be screwed into the threaded hole.

In Figure 13-33B, profile of a surface is being used to indicate a coplanar relationship between two flat surfaces. In Figure 13-33C, the free state modifier is being used with roundness. This indicates the check should be made with the part in a "free," or unconstrained, state. In Figure 13-33D, the tangent modifier indicates the parallelism control is to be applied with respect to a plane that is tangent with all the high points of the surface. This modification specifies a different checking procedure.

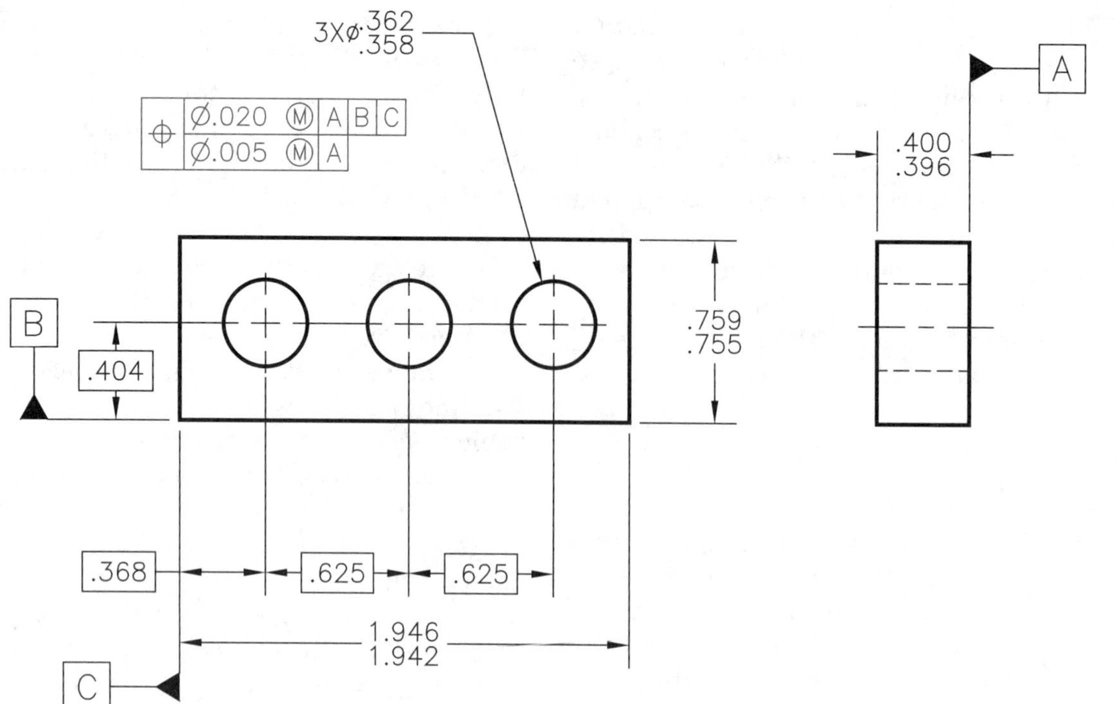


**Figure 13-32.**
The upper row of a composite tolerance feature control frame applies to the individual features with respect to the datum reference framework, while the lower row applies to the individual features with respect to each other within the pattern.

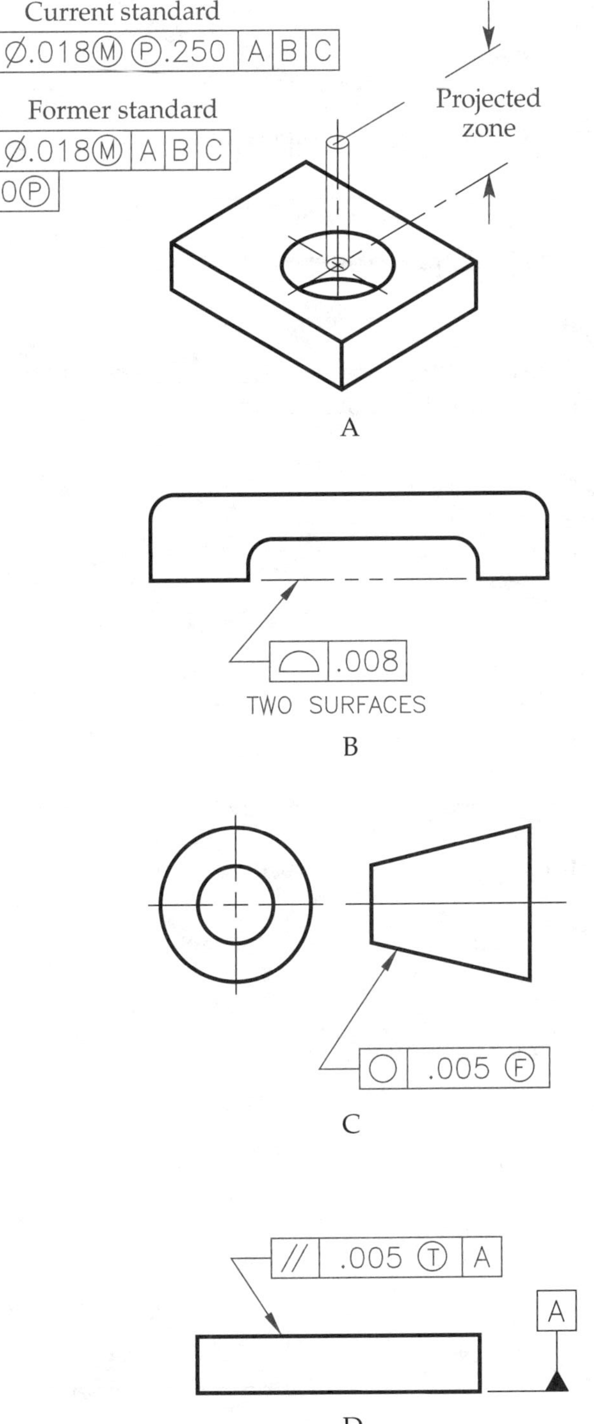


**Figure 13-33.**
Additional GD&T specifications include (A) a projected tolerance zone, (B) profile of a surface for coplanar relationships, (C) free state modifier, and (D) the tangent plane modifier.

## Review Questions

*Circle the answer of choice, fill in the blank, or write a short answer.*

1. What ASME standard is the resource for geometric dimensioning and tolerancing? ______________

______________________________________________

2. *True or False?* Geometric dimensioning and tolerancing is simply a different way to specify tolerances, it does not provide for more precise control.

3. One way to know that a drawing has GD&T is the presence of ______________ associated with local notes, around dimension values, or floating next to size dimensions.

4. Which of the following does *not* have a "boxed" notation:
   A. Datum identification symbol.
   B. Feature control frame.
   C. Maximum material condition modifier.
   D. Basic dimension.

5. Define datum. ______________________________________________

______________________________________________

______________________________________________

6. *True or False?* A basic dimension means there is no tolerance associated with that dimension.

7. *True or False?* A feature control frame always has three sections, one for the characteristic, one for the value, and one for the datum reference.

8. A geometric tolerance zone can be ______________, such as between two lines, or ______________, such as between two planes.

9. Which of the following is considered to be a modifier that specifies bonus tolerance?
   A. Maximum material condition.
   B. Least material condition.
   C. Regardless of feature size.
   D. Both A and B.

10. What name is given to the framework that "locks" the object in space? ______________

______________________________________________

11. *True or False?* By definition, the size of an object also controls the geometric form.

12. Which of the following geometric controls is *not* in the category of individual form tolerances that do not require reference to a datum?
    A. Parallelism
    B. Flatness
    C. Roundness
    D. Straightness

13. List the three orientation tolerances. ______________________________

______________________________

14. *True or False?* A .010 angularity tolerance zone for a 30° angle surface is about the same as ±1°.

15. A geometric tolerance control known as ______________ is used to control the shape of irregular surfaces, perhaps with a datum reference.

16. Runout requires an FIM reading. What does FIM stand for and what instrument is used to obtain the FIM reading? ______________________________

______________________________

17. Which of the following geometric controls is *not* in the category of locational tolerances?
    A. True position.
    B. Runout.
    C. Symmetry.
    D. Concentricity.

18. *True or False?* If a hole is located with a positional tolerance, the location dimensions will all be basic dimensions, since the tolerance is a cylindrical zone specified in the feature control frame.

19. A(n) ______________ tolerance is specified in a double-layer feature control frame.

20. The letter P with a circle around it indicates a(n) ______________ tolerance zone.

## Review Activity 13-1

*For each feature control frame given below, write out the words that would be used to read the feature control frame.*

1. | // | .003 | A |

2. | ∠ | .012 | B |

3. | ⌒ | .010 | A |

   X ◄—► Y

4. | ⌖ | Ø.009Ⓜ | A | BⓂ |

5. | ⌭ | .008 |

6. | ▱ | .0003 |

7. | ⌰ | .004 | A |

8. | ⊥ | Ø.005Ⓜ | A |

9. | — | .003 |

10. | ⌖ | Ø.020Ⓜ | A | B | C |

    | // | Ø.003 | B |

## Review Activity 13-2

*For each of the written descriptions below, sketch a feature control frame that matches the description given.*

1. This feature must be flat within .003″.
2. The cross-sectional shape of this feature must be round within a radial tolerance zone of .002″.
3. FIM reading of this surface should be within .005″, measured at any location along the axis, with respect to datum A.
4. This hole should be located within a cylindrical zone of .005 at MMC with respect to datums A, B, and C.
5. The axis of this feature should be coaxial within .006″ with the datum axis, as measured RFS with the datum RFS.
6. The profile of this surface should be within a .010″ zone with respect to datums A, B, and C between X and Y.
7. The surface elements of this cylinder must be straight within .002″.
8. This surface must be perpendicular to datum C within .005″.
9. The runout of this cylindrical surface should be .003″ or less, measured totally along the surface, to datum A.
10. The axis of this post should be located within a cylindrical zone of .015″ at MMC with respect to datums A, B, and C, but the perpendicularity to datum A must be within .002″ at MMC.

## Industry Print Exercise 13-1

*Refer to the print PR 13-1 and answer the questions below.*

1. Are the datums identified following the current standard or an older standard? ____________
2. Before datum A can be used as a datum, what qualification must it meet? ____________
3. What feature is used to establish datum B? ____________
4. Why do the coordinate location dimensions have boxes around them? ____________
5. Do any of the feature control frames specify a bonus tolerance? ____________
6. The depth (front-to-back) of the part has a tolerance specified by the title block as .02″ (±.01″). Nevertheless, the front surface must still be parallel to the back by what amount? ____________
7. What is the MMC of datum feature B, which has a title block tolerance specification of .010″ (±.005″)? ____________

*Review questions based on previous units:*

8. What is the part number of this print? ____________
9. How many threaded holes are there on this part? ____________
10. What is the radius of the four corners of this part? ____________
11. What type of metal should be used for this part? ____________
12. What is the size tolerance on the four largest counterbore diameters? ____________
13. What is the overall height, width, and depth of this part? H________ W________ D________
14. What type of section view is the right side view? ____________
15. What type of representation is used to show the threads in the section view? ____________

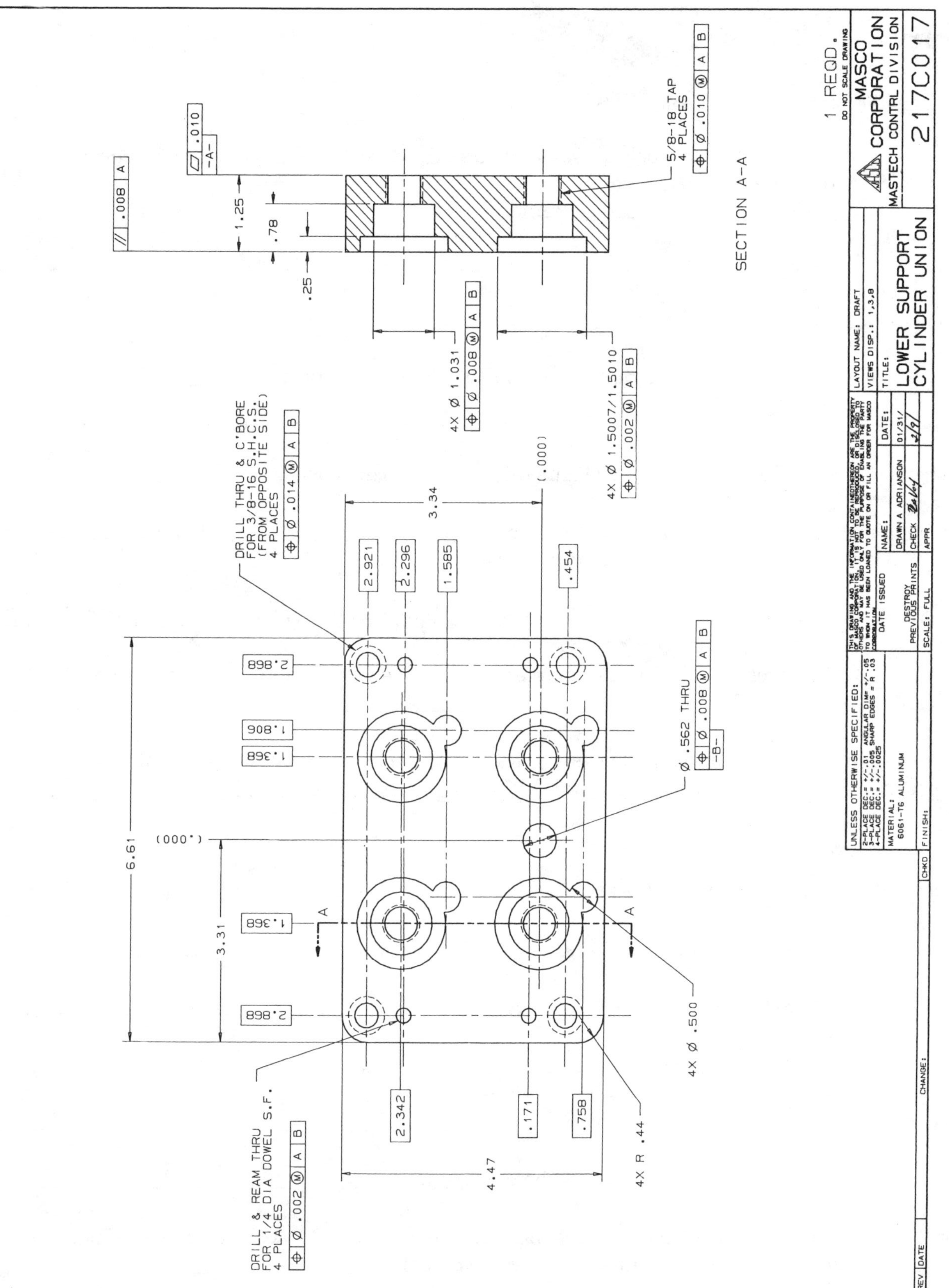


**PR 13-1.**
Print supplied by MASCO Corporation.

## Industry Print Exercise 13-2

*Refer to the print PR 13-2 and answer the questions below.*

1. For each of the datum references below, check whether the datum feature is used to establish a datum axis, datum plane, or datum center plane:
   Datum A ____axis or ____plane or ____center plane
   Datum B ____axis or ____plane or ____center plane
   Datum C ____axis or ____plane or ____center plane
   Datum D ____axis or ____plane or ____center plane
   Datum E ____axis or ____plane or ____center plane

2. Besides true position, what geometric controls are specified on this print? ____________

3. Are the basic dimensions of this print size dimensions, location dimensions, or both? ____________

4. For the four threaded holes not marked "A", a(n) ____________ tolerance is applied, as indicated by the double-layer feature control frame.

5. What is the perpendicularity tolerance used to qualify datum feature A? ____________

6. Are there any instances wherein the datum reference is a size feature, allowing it to be referenced with an MMC modifier? ____________

7. There is a small circle on the leader line elbow for one of the feature control frames that has a profile of surface control. What does that circle indicate? ____________

*Review questions based on previous units:*

8. What is the scale of the original print? ____________

9. What is the name of this part? ____________

10. Are there any auxiliary views shown? ____________

11. How many threaded holes are there total? ____________

12. What is the major diameter of the threaded holes? ____________

13. For the two slotted holes shown best in the top view, there are two MMC measurements, one in the depth direction and one in the width direction. What are they? ____________

14. In the upper-right corner, outside of the border line near the revision block, what does the number T146479 indicate? ____________

15. What is the overall height, width, and depth of this part? H________ W________ D________

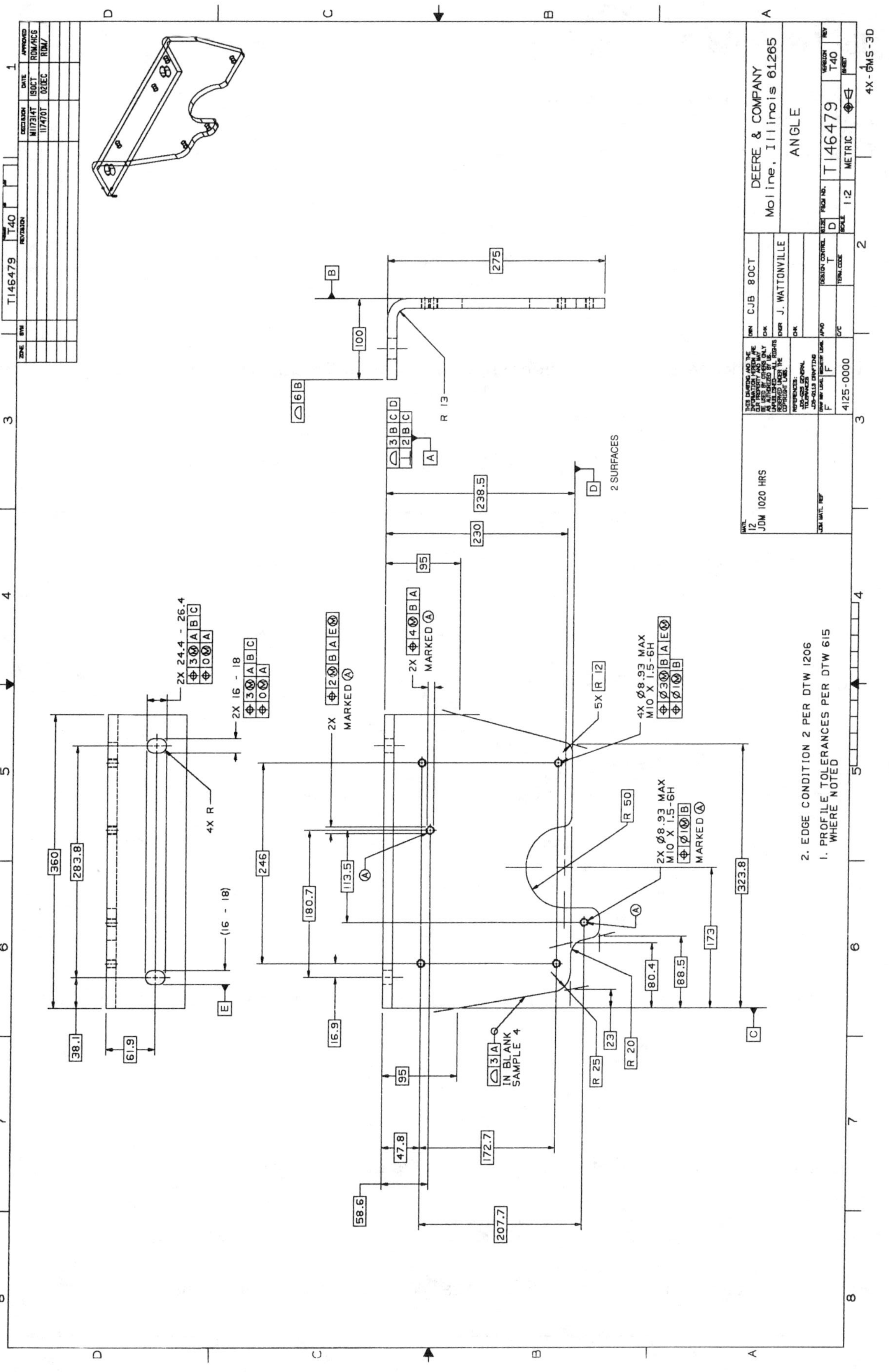


PR 13-2.
Print supplied by Deere and Company.

## Bonus Print Reading Exercises

The following questions are based on various bonus prints located in the folder at the back of this textbook. Refer to the print indicated, evaluate the print, and answer the question.

**Print AP-004:**

1. Are the datums drawn according to the current standard or an older standard? ______

**Print AP-005:**

2. Are the datums drawn according to the current standard or an older standard? ______
3. What is the maximum material condition of the hole that is to be used as datum D? ______
4. Before datum B can be established, the feature that establishes it must be perpendicular within .002″ to which other datum? ______
5. How many different geometric controls are used in this print? ______

**Print AP-006:**

6. For this part, is datum A established by a flat surface or by a cylindrical surface? ______
7. What is the basic size of the hole that is to be concentric within 0.5 mm to datum C? ______

**Print AP-008:**

8. What geometric control is also specified for the surfaces with an N5 surface texture specification? ______
9. For datum A to be acceptable, it must be cylindrical within ______.

**Print AP-009:**

10. For the true position controls specified in this print, what is the primary datum and is it an axis or a plane? ______
11. The location dimensions for the holes are inside boxes. What are these dimensions called? ______
12. The four threaded holes do not have an MMC modifier on the true position control. That means they must be located ______.

**Print AP-010:**

13. What geometric control is specified for this assembly? ______________________

______________________________________________

**Print AP-011:**

14. List the four geometric tolerancing controls specified on this print within feature control frames.

______________________________________________

15. Is datum D established as a plane or an axis? ______________________

______________________________________________

**Print AP-013:**

16. How flat must datum A be to pass inspection? ______________________

______________________________________________

17. What feature(s) of the object are used to establish datum B? ______________________

______________________________________________

18. This print features a feature control frame that is two layers tall. This is called a(n) ____________ tolerance.

**Print AP-014:**

19. What diameter is specified for the cylindrical feature to be used to establish datum C? ____________

______________________________________________

20. What diameter is specified for the cylindrical feature to be used to establish datum D? ____________

______________________________________________

**Print AP-018:**

21. The two feature control frames indicate datums C and E are to be used. Is this a primary datum-secondary datum specification or a simultaneous datum specification? ______________________

______________________________________________

22. There are two feature control frames on this print. One contains a single-arrow symbol and the other a double-arrow symbol. Identify these geometric controls. ______________________

______________________________________________

**Print AP-020:**

23. What geometric control is applied on this drawing? ______________________

______________________________________________

**Print AP-026:**

24. What modifier is applied to the true position tolerance of the two largest holes? ____________

______________________________________________

25. What geometric control is applied to indicate the two surfaces of the two T150620 parts must be coplanar? ______________________

______________________________________________

As products go through revisions, the associated drawings need to be updated. A drawing revision system is required to keep track of the various changes to a product.

UNIT 14

# Drawing Revision Systems

*After completing this unit, you will be able to:*

**Describe** drawing practices related to drawing revisions.

**Identify** revision information on an industrial print.

**Explain** the information contained within a revision history block.

**Explain** the information found in a "status of sheets" block for multisheet drawings.

As products and parts are designed, the activities that surround the process are documented in a variety of ways. As design ideas are developed, prototypes made, and drawings created, the approval process for the design activity is usually structured in such a way that all parties in the manufacturing enterprise "sign off" at various stages. In the case of drawings and prints, those with authority to do so add signatures to the prints. These signatures often include initials or names for DRAWN BY, CHECKED BY, and APPROVED BY. Within the CAD system, signatures may be issued in electronic form. This unit addresses the system for approving changes to the original design idea of a part or assembly. Standards for revision systems are covered by ASME Y14.35M entitled *Revision of Engineering Drawings and Associated Documents.*

The term ***drawing revision*** refers to any change made to a drawing after final approval has been given to the drawing. This term does not apply to changes made *prior* to final approval of the drawing. Revisions may occur for a variety of reasons. Changes may relate to an improvement in the product's function or reliability, a new method of manufacture, cost-reducing implementations, alterations to quality control standards, or correcting errors in the original drawing. In some cases, customers may request a change to incorporate a new or improved design into the product. In addition to changing the drawing, a ***revision authorization document*** is usually generated as part of the revision system within the company. Whatever the reason for a revision, an understanding of a company's drawing revision system is required in order to correctly interpret revised prints.

This unit explains some of the drawing practices that are impacted by revisions. Remember, considerable variation exists among industries in how print revisions are processed and recorded. The information presented in this unit will help you develop an understanding of the revision process in general. You should take additional steps to become thoroughly familiar with the details of the revision system used by your company.

## Revision Process

Revisions made to a drawing vary in the "degree of change." Revisions are usually permanent changes in the design or manufacturing process. However, a revision may be a temporary change to accommodate a special situation, with another revision required to "reverse" the temporary change. With electronic documents and computer systems, the process can be quite different than in a company where originals and prints are still output to paper and manually filed.

Revision authorization documents may take on a variety of terms, including ***engineering change order (ECO),*** which will be used in this unit for the sake of clarity. Other names for revision authorization documents include *alternation notice (AN), change in design (CID), engineering notice (EN), engineering change notice (ECN)* and *advance drawing change notice (ADCN).* Figure 14-1 shows an ECO form used by one company some years ago. Notice in the upper-left corner the ADCN box is checked to indicate this is an advance drawing change notice.

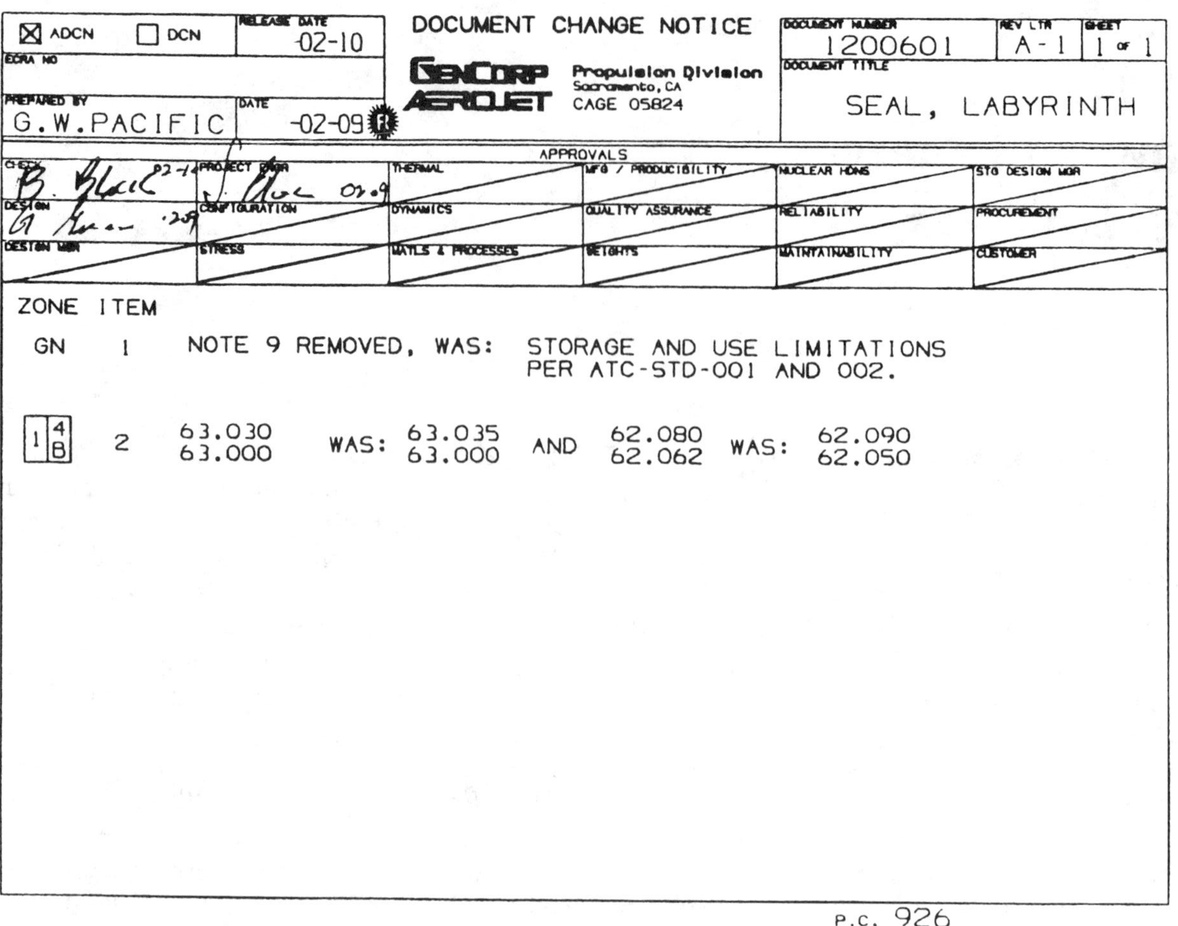
☒ ADCN ☐ DCN
RELEASE DATE -02-10
DOCUMENT CHANGE NOTICE
GENCORP AEROJET
Propulsion Division
Sacramento, CA
CAGE 05824
DOCUMENT NUMBER 1200601
REV LTR A-1
SHEET 1 of 1
ECRA NO
PREPARED BY G.W.PACIFIC
DATE -02-09
DOCUMENT TITLE SEAL, LABYRINTH

APPROVALS

| CHECK | PROJECT ENGR | THERMAL | MFG / PRODUCIBILITY | NUCLEAR HDNS | STD DESIGN MGR |
|---|---|---|---|---|---|
| DESIGN | CONFIGURATION | DYNAMICS | QUALITY ASSURANCE | RELIABILITY | PROCUREMENT |
| DESIGN MGR | STRESS | MATLS & PROCESSES | WEIGHTS | MAINTAINABILITY | CUSTOMER |

ZONE ITEM

GN 1 NOTE 9 REMOVED, WAS: STORAGE AND USE LIMITATIONS PER ATC-STD-001 AND 002.

1 4B 2 63.030 / 63.000 WAS: 63.035 / 63.000 AND 62.080 / 62.062 WAS: 62.090 / 62.050

P.C. 926

**Figure 14-1.**
This revision authorization document is identified as an advance drawing change notice (ADCN) by the check box in the upper-left corner. (GenCorp Aerojet)

It is very important to understand that each company has its own system for initiating and approving revisions to drawings. In addition to making parts in-house, drawings are also documents representing a binding contract between a supplier and the company. Therefore, one person seldom has the right to change the drawing once it has been adopted and approved. Most companies periodically have engineering-team meetings and proposals for changes can be brought up for discussion at these meetings. Smaller companies may have individuals who have the authority to initiate a change without a committee meeting.

In some companies, initiating and approving a change to a part, therefore a change to the print, requires the ECO document to be generated first. An ECO document might be a form filled out and registered by a clerk, secretary, or administrative assistant. A number is usually assigned to the ECO and then a drafter or technician is assigned to make the drawing or computer model changes indicated by the designer or engineer. In the case of CAD files, the revision letter is built into the file name. A file should exist for each "edition" of the part, with the existing CAD files being used until the ECO is approved.

In the case of prints, if a new plot or drawing is created, copies of the new "original" are usually circulated to several departments, whether affected by the change or not. Purchasing may need to know if new materials or different quantities of materials must be ordered. Sales may need to be notified if the appearance of a part changes so literature can be updated. Manufacturing and process control may need to be notified if the processes used to make the parts are affected. The product design department needs to be notified in case the current parts are used in new products in development.

Either by paper or electronic mail, the ECO revision document circulates through all appropriate departments. Approval signatures, either paper or electronic, are gathered as needed. While the revision approval is in process, something such as ECO PENDING might be stamped on current prints issued to personnel from the engineering or print-control group, as shown in Figure 14-2. In a situation where the documents are electronic, the company will have a system for flagging parts or prints that are progressing through the revision process.

The paperwork or electronic files being circulated may contain both the old and the new conditions. The old condition is called the ***was condition.*** The new condition is called the ***now condition.*** Since the design revision may include more than one part, therefore more than one print, all prints may be circulated with the ECO document. A sequence number can also be assigned to each element of the revision and the changes can be described on the drawing in addition to the ECO document.

## Noting the Revision

Once the ECO has been prepared, the drawing is ready to be changed, with the changes recorded in the revision block on the drawing. Items of the drawing can be changed by deleting, crossing out, or changing dimensional values and views. Removing information, such as a detail or dimension, should make room for the new items. However, it may be desirable to leave the information on the drawing so that it can still be read, but crossed through to allow the print reader to know that information has been replaced or deleted. Standards recommend the replaced information, such as a note, be crossed through with one or two lines or with a hatching pattern.

After changes have been made, revision letters are applied next to the specific areas of the drawing wherein the change occurred. As recommended by standards, uppercase lettering should be used for revisions beginning with the letter A and progressing through the alphabet. However, the letters I, O, Q, S, X, and Z are not used. These letters may be confused for numbers or drawing elements (i.e., the letter "I" looks like a one and the letter "O" looks like a zero). If the revision of a part progresses beyond the letter Y, dual letters AA, AB, AC, etc., are next in sequence.

The current standards recommend the revision letter be enclosed in a circle. Some companies may elect to use other geometric shapes. Leaders may be used to indicate a specific position or a multitude of positions, Figure 14-3. This figure also illustrates the use of revision item numbers. If a single revision contains several changes, the changes can be numerically itemized. The item number can then be combined with the revision letter, for example, B1 and B2.

Sometimes, the revision letter may simply "float" in an empty space where a detail once existed, but is now gone. In cases where a drawing has been redrawn without change, the revision letters can be omitted from the revised drawing, although the revision letter of the new drawing does not need to be changed. If revision letters might be confused with other symbols on the drawing, they may also be omitted. In cases where a drawing has been redrawn with change, the revision letters for the current change should be implemented. In both of these cases, former revisions can still be shown and listed. Be aware that each print may have its own unique traits and many factors are taken into consideration by the person who approved the print and revisions.

UNLESS OTHERWISE SPECIFIED ALL DIMENSIONS ARE IN INCHES
TOLERANCES:
1 PLACE DIMS: +/- .1
2 PLACE DIMS: +/- .03
3 PLACE DIMS: +/- .005
ANGULAR: +/- 1°
FRACTIONAL: +/- 1/32
MATERIAL CRS
FINISH NONE
PROJECT NO. PR - 119
APPROVALS DATE
DRAWN RKB 09 / 25
CHECKED
APPROVED
ISSUED
BROWN ENGINEERING
PLATE
SIZE B
FSCM NO.
DWG NO. 1114-55
REV A
SCALE 1 = 1
WEIGHT .234
SHEET 1 OF 1

**Figure 14-2.**
When ECO PENDING is stamped near the title block, an ECO is in process for the drawing.

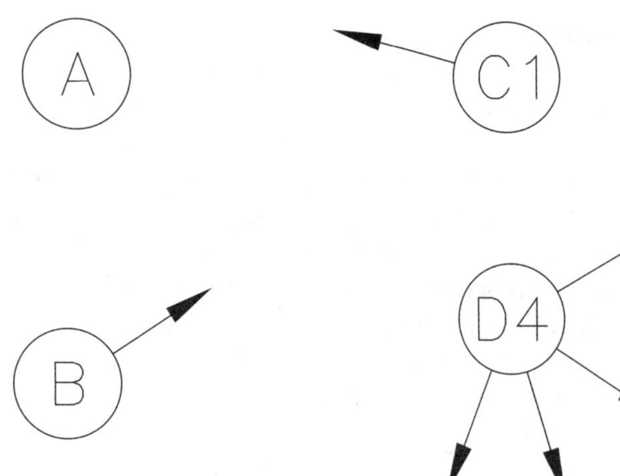


**Figure 14-3.**
Revision letters are usually enclosed within a geometric shape. Leaders can be used for clarity. Multiple changes in one revision can be itemized with numeric values.

Some companies prefer to apply the letter A to the initial release of a drawing, so the print begins with revision A as a new part. It is also permissible to begin the release of a new part or print with a dash as the revision indication. If a drawing contains a dash in the revision box, this indicates the part has never been revised.

Digital data have provided a new set of rules to the revision system. According to ASME standards, revisions of digital data files and copies of digital data files are to be considered redraws, except wherein the document is first converted from a manual drawing to a digital drawing file. It is also acceptable practice to indicate the revision letter within the electronic part file number. For example, the file names 236-2345-B or 236-2345-REV-B could be used to help track the revision letter in the computerized archive system.

## Revision History Block

As studied in Unit 3, the ***revision history block*** should be located in the upper-right corner for most sheet layouts. This is shown in Figure 3-10 in that unit. The revision history block often includes information such as the zone, or area, in the drawing in which the change is located, the alphanumeric character applied to the change, an itemized description of the change or changes, a reference to the ECO number, the date, and the approval initials as set forth by the company standards. Additional columns can also be added as necessary. For multisheet drawings, there may be a revision *status of sheets* block attached near the revision history block.

Figure 14-4 shows a typical revision history block. A horizontal dividing line should appear between revisions. The columns are used in this fashion:

- The first column (A) identifies the zone on the drawing where the change is found. If the revision covers several zones, the zone locations can be listed in the description.
- The second column (B) identifies the revision letter. Nothing else should be entered into this column. The revision many have several changes with several zones or itemized descriptions and perhaps even multiple sheets, but the revision letter is listed once in the revision history block.
- A description column (C) is used to provide a concise explanation of the change, if possible. For example, when a note is added to the drawing, the type or number of the note can be referred to in the description block, such as ADDED NOTE 1. When a dimension is changed, the old value can be given, such as WAS .999–1.003. If the description is too lengthy, the ECO number can be listed here. Some companies prefer to always list the ECO number and may even have a column for that.
- A date column (D) is for the date the change was made to the drawing.
- A column for initials of the approving engineer (E) is also recommended.

Figure 14-5 illustrates how the first line of the revision history block may indicate that the drawing was redrawn without change, as at A, or redrawn with change, as at B. The revision history block may also include information when a drawing supersedes or is superseded by another drawing. The drawing that supersedes, or replaces, another drawing may have REPLACES WITH CHANGE DRAWING 236-2345-REV B in the description area. An additional description of how this new drawing is different may also be included. If the description is too lengthy or cumbersome to place on the drawing, a reference may only be made the ECO number. The ECO then describes the differences.

With CAD systems, most drawings have associative dimensions that are automatically maintained by the software. These dimensions are always real size within the CAD database and, if a drawing is properly plotted, the views will always be to scale. In the case of manually created drawings, it is also desirable that the views of a drawing are always created to scale, with the scale specified on

| REVISION HISTORY | | | | |
|---|---|---|---|---|
| ZONE | REV | DESCRIPTION | DATE | APPROVED |
| | A | REDRAWN WITHOUT CHANGE | 05/09 | DCW |
| B3 | B | SEE ECO 8746 | 06/10 | DCW |
| C4 | C | REMOVED HOLE AND KEYWAY | 02/11 | RKB |
| Ⓐ | Ⓑ | Ⓒ | Ⓓ | Ⓔ |

**Figure 14-4.**
The revision history block provides a guide to each revision that has been made to the original drawing.

| REVISION HISTORY | | | | |
|---|---|---|---|---|
| ZONE | REV | DESCRIPTION | DATE | APPROVED |
| | C | REDRAWN WITHOUT CHANGE (ECO 24359) | 05/09 | DCW |

A

| REVISION HISTORY | | | | |
|---|---|---|---|---|
| ZONE | REV | DESCRIPTION | DATE | APPROVED |
| F4 | A | ADDED VENDOR SPECIFICATION – NOTE 4 | 01/08 | RKB |
| D3 | B | REDRAWN WITH CHANGE (ECO 24360) | 05/09 | DCW |

B

**Figure 14-5.**
A redraw, with or without change, should be indicated in the revision history block.

the drawing. However, there are occasions when a change to the views of the drawing is just too extensive and the drawing or views can be revised by simply changing the dimension value. In these cases, if the views are still clear, the ***not-to-scale method*** is used to indicate that the dimension values as shown are *not* to the drawing scale stated. In this method, a heavy line is drawn below the dimension value of the revised feature. This indicates the dimension is changed, but the feature is not redrawn to scale. See Figure 14-6.

In summary, the revision authorization document should contain a complete record of the changes that have been made to the original drawing. Throughout the life of the drawing and as drawings are redrawn, some of the revision history is lost from the original. The ECO files should be maintained throughout the life of the part to assist in tracking down older dimensions or other design features that may be associated with old inventory.

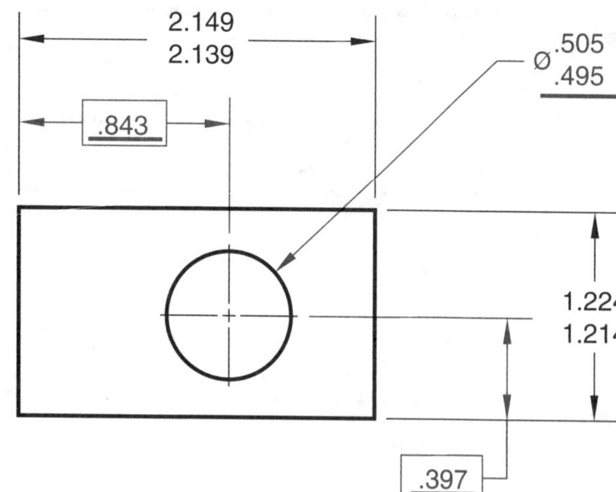


Figure 14-6.
In the not-to-scale method, a heavy line drawn below a dimension value indicates the value has been changed, but the feature was not redrawn to scale.

## Multisheet Revision History

If a part has multiple sheets, the revision history can become more complex. Standards set forth guidelines for three approaches to multisheet revision history: by drawing, by sheet, or all sheets the same. In the ***by drawing method,*** the revision history block is on sheet one and applies to the whole drawing set. The revision history block on sheet one can reference all sheets as illustrated in Figure 14-7, either separated with subheadings, as shown at A, or with zone markings that indicate page, as shown at B.

In the ***by sheet method,*** each sheet is treated independently. A ***status of sheets*** block is added to the left of the revision history block. Each sheet also has its own revision history block. For example, sheet 1 may be identified as revision B, sheet 2 may be identified as revision A, and sheet 3 may be identified as a dash to indicate no revisions have ever been made to that sheet. If the only sheet that changes is sheet 1 and a sheet 4 is added, sheet 1 becomes revision C, sheet 2 remains revision A, sheet 3 remains a dash, and sheet 4 begins with a dash in the revision block. Figure 14-8 illustrates the STATUS OF SHEETS table for a WAS and NOW of a multisheet drawing.

A slightly different scenario is a method referred to as the ***all sheets same*** approach. When this approach is used, a revision will cause all sheets that have any change to all be changed to the same letter. Figure 14-9 illustrates a WAS and NOW for a multisheet drawing that had four sheets, but a revision added a fifth sheet and also had changes on sheets 1, 2, and 4. As a result of the all sheets same approach, sheets 1, 2, 4, and 5 all became revision C and sheet 3, which was unchanged, remained at revision A status.

## Other Revision System Features

Although not expressly covered by ASME standards, there are many variations that have been used for revisions throughout the years. As a member of the manufacturing enterprise, you need to be aware of company standards that are unique to the company or perhaps used by suppliers and subcontractors. The next sections cover a few examples of additional revision system features.

### *Revision History Block Columns*

Other columns may be entered in a revision history block to clarify the revision history of a drawing. Sometimes this is to further document the changes made in the original drawing. Some other listings that may appear include:

- An authority or ECO column that can be used to record the engineering change order number.
- A microfilm column that can be used to indicate the date the revised drawing was placed on microfilm.
- An "effective on" column that can be used to give the serial number or ship number of the machine, assembly, or part on which the change becomes effective. Sometimes this is a separate block. The change may also be indicated as effective on a certain date.

### *Drawing Deviation*

Another possible term used by some companies is the ***drawing deviation (DD).*** This is basically a formal description of a temporary revision that authorizes a deviation from the print. The deviation is allowed only as described on a DD sheet or form. These can be issued for revisions such as:

- Temporary change in the standard part.
- Temporary substitution in the method of manufacture.
- A design change that requires rework of existing units.

A DD is prepared on the same type of sheet as an ECO, but a colored sheet may be used to call attention to the fact the revision is temporary. A sequence number is assigned the DD and may be recorded in the revision history. The DD remains in effect until altered or voided by another DD or a revision in the drawing.

| REVISION HISTORY | | | | |
|---|---|---|---|---|
| ZONE | REV | DESCRIPTION | DATE | APPROVED |
| | A | SHEET 1 | 01/08 | RKB |
| A7 | | CHANGED .500 TO .510 | | |
| D3 | | ADDED DATUM D | | |
| | | SHEET 2 | | |
| C4 | | CHANGED .495 TO .490 | | |
| C4 | | ADDED WELDING NOTE | | |
| | | SHEET 3 | | |
| | | NO CHANGE | | |
| | B | | 04/09 | RKB |

A

| REVISION HISTORY | | | | |
|---|---|---|---|---|
| ZONE | REV | DESCRIPTION | DATE | APPROVED |
| | A | | 01/08 | RKB |
| A7–1 | | CHANGED .500 TO .510 | | |
| D3–1 | | ADDED DATUM D | | |
| C4–2 | | CHANGED .495 TO .490 | | |
| C4–2 | | ADDED WELDING NOTE | | |
| | B | | 04/09 | RKB |

B

**Figure 14-7.**
For a multisheet drawing, the revision history block on sheet 1 can be used to note the revision status for the whole drawing set.

WAS

| REV STATUS | |
|---|---|
| SH | REV |
| 1 | B |
| 2 | A |
| 3 | – |

NOW

| REV STATUS | |
|---|---|
| SH | REV |
| 1 | C |
| 2 | A |
| 3 | – |
| 4 | – |

STATUS OF SHEETS

**Figure 14-8.**
For a multisheet drawing, a status of sheets block is added to the left of the revision history block. This status of sheets block treats all sheets independently. Each sheet progresses to the next letter if something on that sheet is changed.

SHEET 1 WAS

| REV STATUS | | REVISION HISTORY | | | | |
|---|---|---|---|---|---|---|
| SH | REV | ZONE | REV | DESCRIPTION | DATE | APPROVED |
| 1 | B | D3 | B | ADDED FLATNESS TOLERANCE (ECO 29037) | 11/10 | DCW |
| 2 | A | | | | | |
| 3 | A | | | | | |
| 4 | – | | | | | |

SHEET 1 NOW

| REV STATUS | | REVISION HISTORY | | | | |
|---|---|---|---|---|---|---|
| SH | REV | ZONE | REV | DESCRIPTION | DATE | APPROVED |
| 1 | C | D3 | B | ADDED FLATNESS TOLERANCE (ECO 29037) | 11/10 | DCW |
| 2 | C | E4 | C | ADDED SURFACE TEXTURE (ECO 29124) | 02/12 | DCW |
| 3 | A | | | | | |
| 4 | C | | | | | |
| 5 | C | | | | | |

**Figure 14-9.**
This status of sheets block uses the all sheets same method. Each sheet is independently revised, but when a revision affects a sheet, it uses the same letter as all other sheets affected by the same revision.

## Review Questions

*Circle the answer of choice, fill in the blank, or write a short answer.*

1. What ASME standard is the resource for drawing revision systems? ______________________

______________________________________________

2. *True or False?* All companies refer to their drawing revisions as engineering change orders (ECO).

3. A drawing revision may include three different ways of changing information; list them.

______________________________________________

4. Which of the following letters should be skipped in an on-going progression of revisions? Select more than one if necessary.
   A. I
   B. O
   C. S
   D. X

5. *True or False?* The first release of a drawing should always be identified as revision A.

6. List three items likely to be found in the revision history block. ______________________

______________________________________________

7. To indicate a dimensional value was changed on an original drawing, even though the actual geometry of the view was not changed, a(n) ______________________ should be placed under the dimension value to indicate the view is out of scale.

8. Which of the following is *not* an accepted method for maintaining revisions of multisheet drawings?
   A. Revisions are not shown on any sheets, but are kept in separate files.
   B. One revision letter for all sheets; one revision history block on sheet one.
   C. Each sheet maintains its own revision letter; each sheet is totally independent.
   D. Each sheet maintains its own revision letter; all sheets affected by a revision obtain the same revision letter as a result of that revision.

9. *True or False?* A company may elect to include other columns to a revision history block, such as additional authority columns or additional date columns for when the change became effective.

10. What name might be given to a documented and approved temporary change to a standard part?

______________________________________________

## Industry Print Exercise 14-1

*Refer to the print PR 14-1 and answer the questions below.*

1. How many columns are featured in the revision history block of this drawing? ________
2. What is the current revision letter assigned to this part? ________
3. What revision specified the surface texture of the tapered portion? ________
4. What are the initials of the person who implemented or approved ECO 68-557? ________
5. Which revision specified the vendor part number of the bushing? ________
6. Revision F changed the maximum axial deflection to .500 at 2000 pounds. What had it been? ________
7. Instead of revision D, E, and F being three letters, what other method of marking could have been implemented for these three changes? ________

*Review questions based on previous units:*

8. What is the thread form specified for the internal threaded holes? ________
9. What method was used to show external threads on this part? ________
10. Is this a drawing of multiple parts assembled? ________
11. Does this drawing feature any cutting-plane lines? ________
12. What is the radius of the undercut indicated by a local note? ________
13. What is the maximum material condition diameter of the largest cylindrical feature of this part? ________
14. This drawing has an italic F symbol on some surfaces in the left side view. What are these symbols? ________
15. What does the 125 near the check mark at revision A on the tapered surface mean? ________

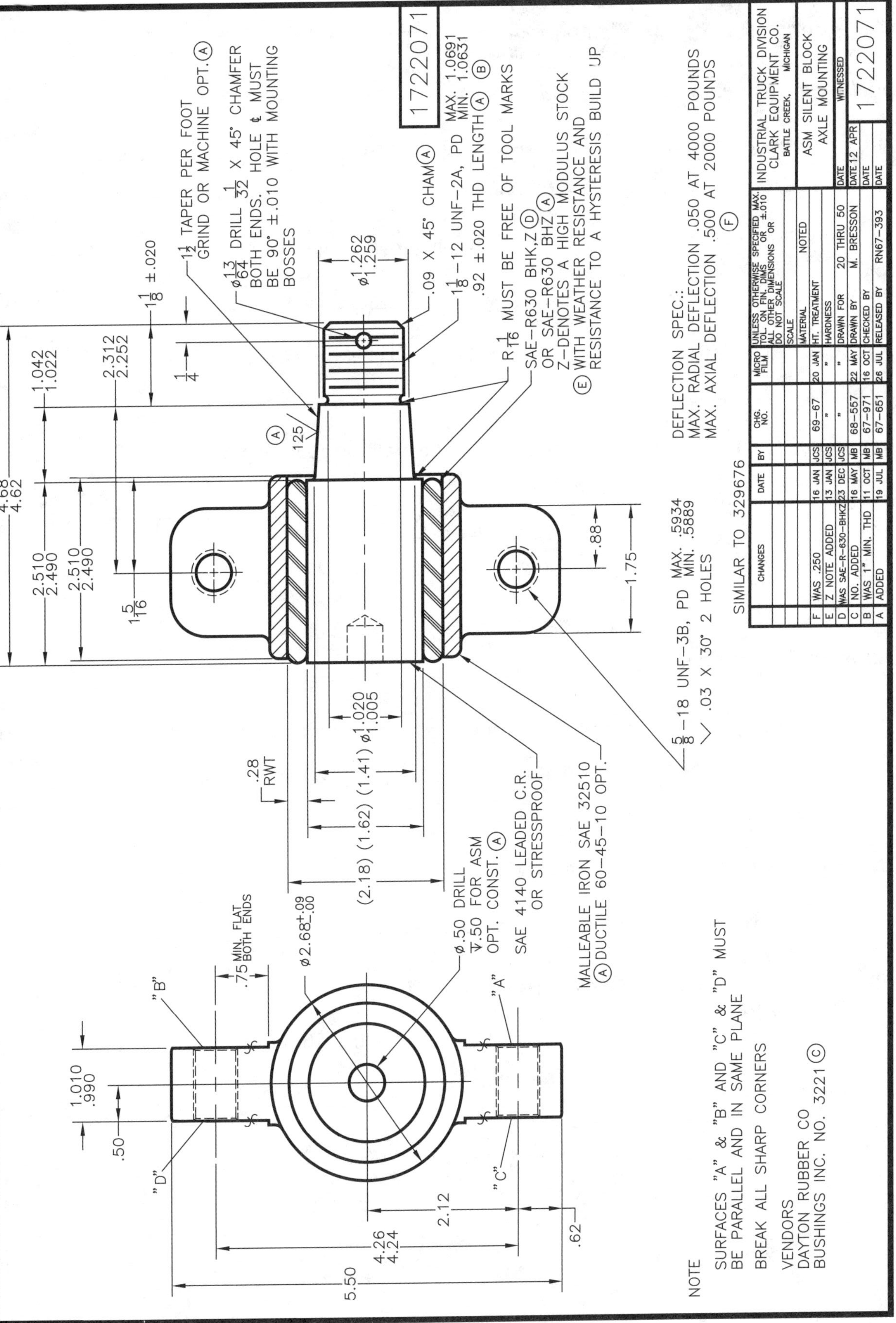


PR 14-1.
Print supplied by Clark Equipment Company.

## Industry Print Exercise 14-2

*Refer to the print PR 14-2 and answer the questions below.*

1. What are the three columns of the revision history block of this drawing? ______________________
2. What revision letter/item number is assigned to the addition of datum W? ______________________
3. Revision A9 is floating next to an empty space where ______________________ used to be.
4. Revision A6 added a triangle and a feature control frame. What does the triangle mean? ______________________
5. What revision letter and item number is assigned to the material specification? ______________________
6. In what zone is revision A2 located? ______________________
7. What type of pin is referred to in revision A1? ______________________

*Review questions based on previous units:*

8. What is the name of this part? ______________________
9. At the time this drawing was created, what standard was specified for dimensioning? ______________________
10. How many general notes are shown on this print? ______________________
11. If not specified, what is the tolerance for a three-place decimal value? ______________________
12. Are there any surface texture symbols shown? ______________________
13. What is the maximum material condition diameter of the main central hole? ______________________
14. At what scale is the enlarged view? ______________________
15. What four geometric tolerances are invoked on this drawing? ______________________

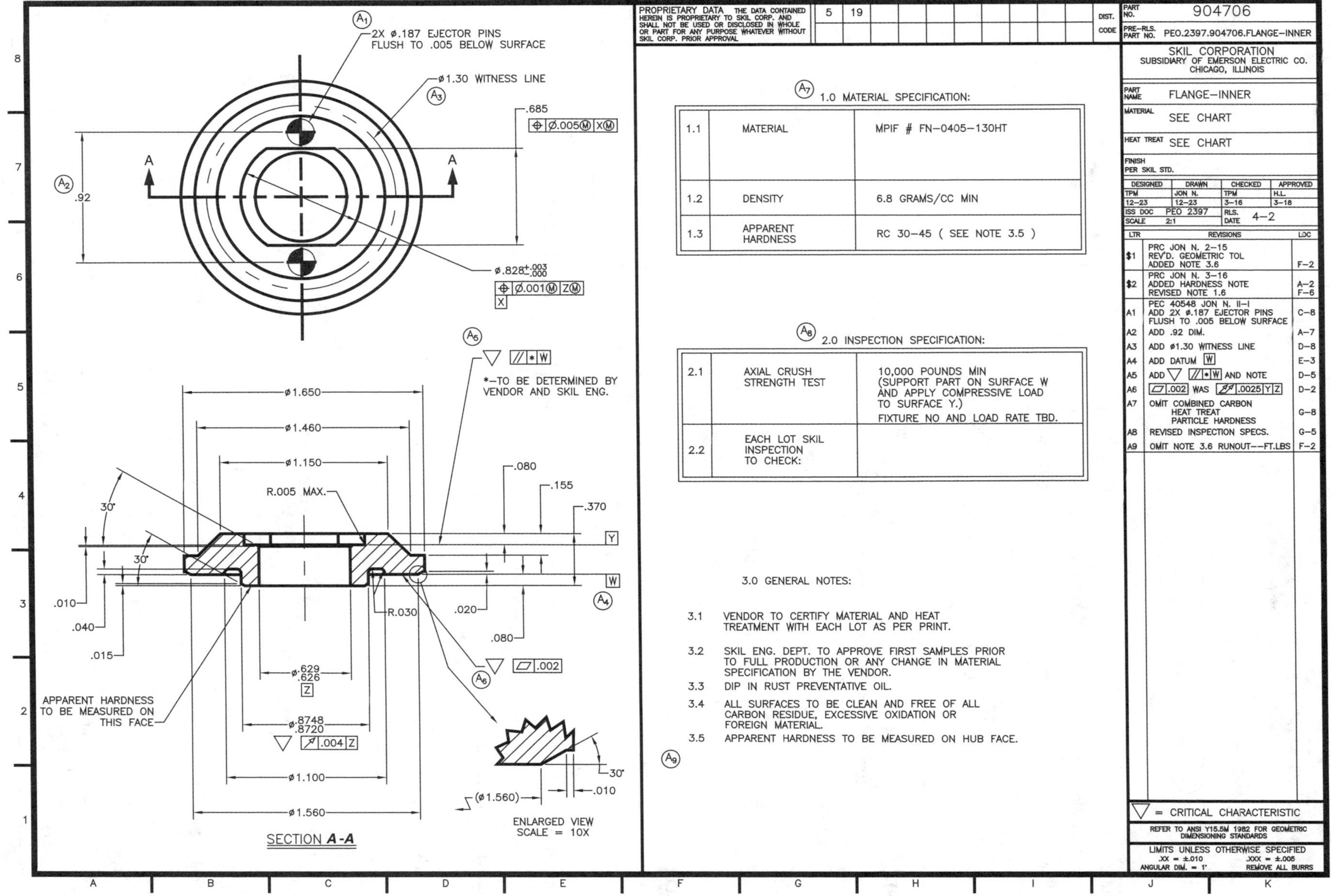


PR 14-2.
Print supplied by Skil Corporation.

## Bonus Print Reading Exercises

The following questions are based on various bonus prints located in the folder at the back of this textbook. Refer to the print indicated, evaluate the print, and answer the question.

**Print AP-002:**

1. How many revisions are indicated on this print? ______

**Print AP-003:**

2. In which view does revision C appear, upper left, upper right, Section A-A, or Section B? ______

**Print AP-004:**

3. What letter represents the revision status of this print? ______

**Print AP-005:**

4. What was the main focus of the last revision of this print? ______

**Print AP-006:**

5. In the title block, lower-right corner is a B. Is that the sheet size or the revision status? ______
6. What ECN number was assigned to the last revision of this print? What does ECN most likely stand for? ______

**Print AP-009:**

7. What is the current revision status of this print? ______

**Print AP-010:**

8. During what month was this drawing released for production as revision A? ______

**Print AP-011:**

9. How many individual changes are shown in the revision history for change number 510? ______

**Print AP-012:**

10. In addition to a revision history block in the upper, right-hand corner of the print, this print also features a block entitled ____, since it is a multisheet print. ______

**Print AP-013:**

11. What does the third column from the left of the revision history block indicate? ____________

12. How many different changes occurred on March 29th? ____________

**Print AP-014:**

13. What geometric shape is used to help identify revisions on the drawing? ____________

14. What new item is added as a result of revision E? ____________

**Print AP-017:**

15. What geometric shape is used to help identify revisions on the drawing? ____________

16. List the three things indicated by the initials in the revision block. ____________

17. How many changes occurred during revision 05? ____________

**Print AP-020:**

18. How many changes occurred during revision C? ____________

19. In general terms, what was changed as a result of revision D? ____________

**Print AP-021:**

20. During what month was the .015 × 45 chamfer added? ____________

21. In the revision history block, what does E.C.O. most likely stand for? ____________

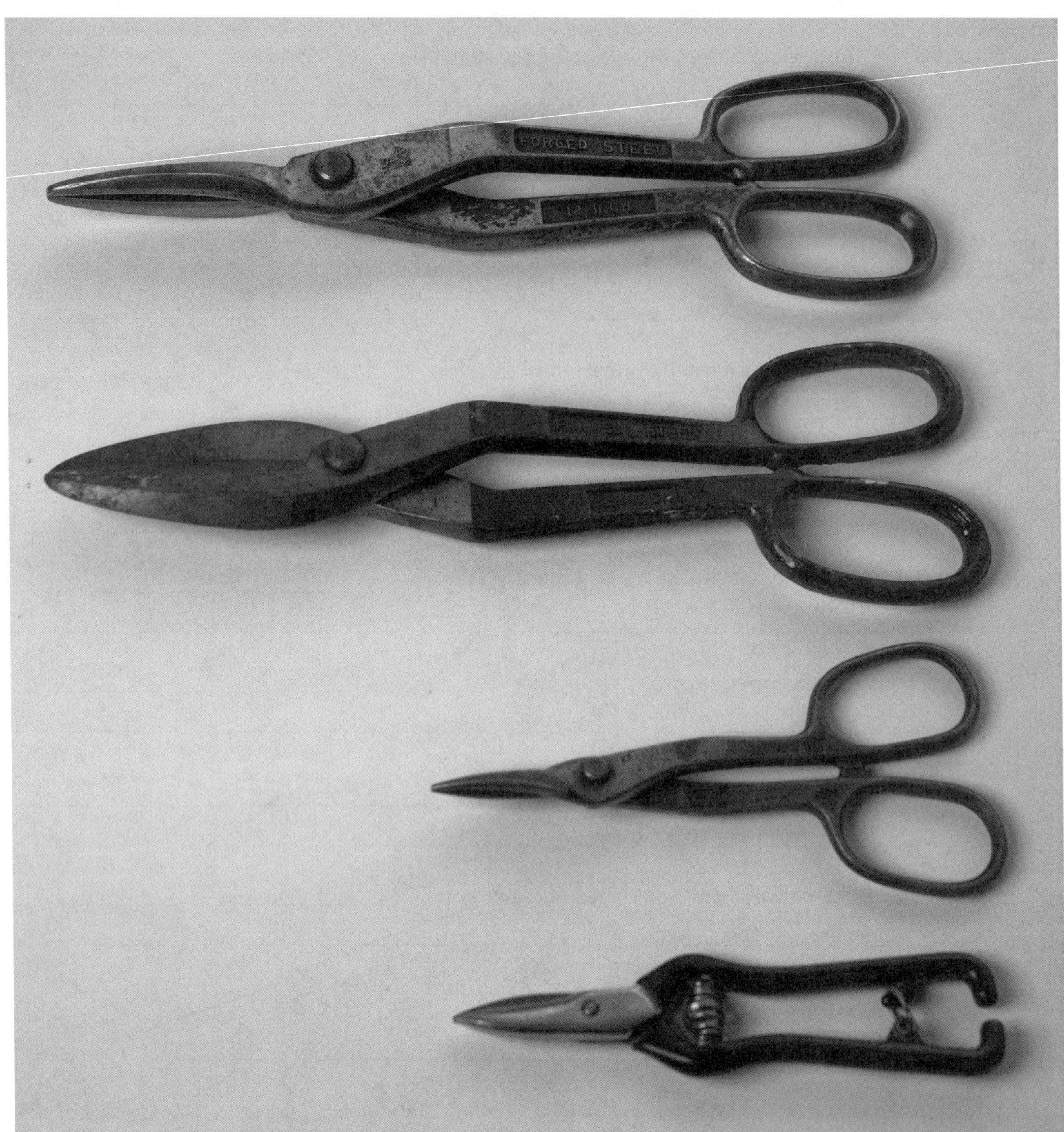


Every part and product requires a detail drawing, even products as simple as tin snips.

# UNIT 15
# Detail Drawings

*After completing this unit, you will be able to:*

**Explain** terms and standards related to various types of drawings used in industry.

**Describe** how detail drawings are defined and categorized in industry.

**Identify** the unique characteristics of casting and forging drawings.

**Identify** the unique characteristics of purchased part control drawings.

**Identify** the unique characteristics of modifying drawings.

**Describe** the unique characteristics of pattern development and welding drawings.

**List** and **describe** other specialized types of drawings used in industry.

To produce a part, assembly, or structure, a set of drawings is necessary to provide details for the production of each component and correct assembly of all parts. The terminology for drawing types is covered in ASME Y14.24, entitled *Types and Applications of Engineering Drawings.* This standard covers a wide array of drawing types, each classified according to its purpose or unique role in manufacturing. In drafting texts, the types of drawings are often simply referred to as ***working drawings*** and consist of two basic types—detail drawings and assembly drawings. While detail drawings provide information on how to produce the parts, assembly drawings provide information on how each unit or subunit is put together. In this unit, you will examine the nature and purpose of detail drawings. Assembly drawings are discussed in Unit 16.

## General Terminology

***Detail drawings*** provide all of the information necessary to produce a single part. Every part of every product has a detail drawing that is, in essence, the "contract" for how the part is to be made, including tolerance ranges and other specifications. Detail drawings supply a worker with:

- Name of the part.
- Shape description of the part.
- Dimensional size of the part and the part's features.
- Notes detailing such things as material, special machining, surface texture, and heat-treatment.

Detail drawings can be very complex or very simple. Usually only one part is placed on a detail drawing, as shown in Figure 15-1. The standard term for this is a ***monodetail drawing,*** but in practice it is often simply called a detail drawing. Very complex products may require several hundred detail drawings to fully explain their manufacture or construction. In some cases, one detail drawing may require more than one sheet.

Although a detail drawing is considered to be one drawing of one part, a series of parts can be shown on one sheet if the only difference between parts is a few dimensions. For example, a tabulated drawing would be useful for a series of threaded rods that differ only in length. In this case, several parts in a family of products can be included on one sheet with a table plan, as shown in Figure 15-2.

A ***multidetail drawing*** shows more than one multiview drawing on one sheet. While the standards caution against using multidetail drawings, there may be situations were two or more unique parts are integral to each other and it is deemed important to show both parts on one sheet. If this is done, one revision letter should apply to both parts on the sheet and records of both parts should be updated anytime

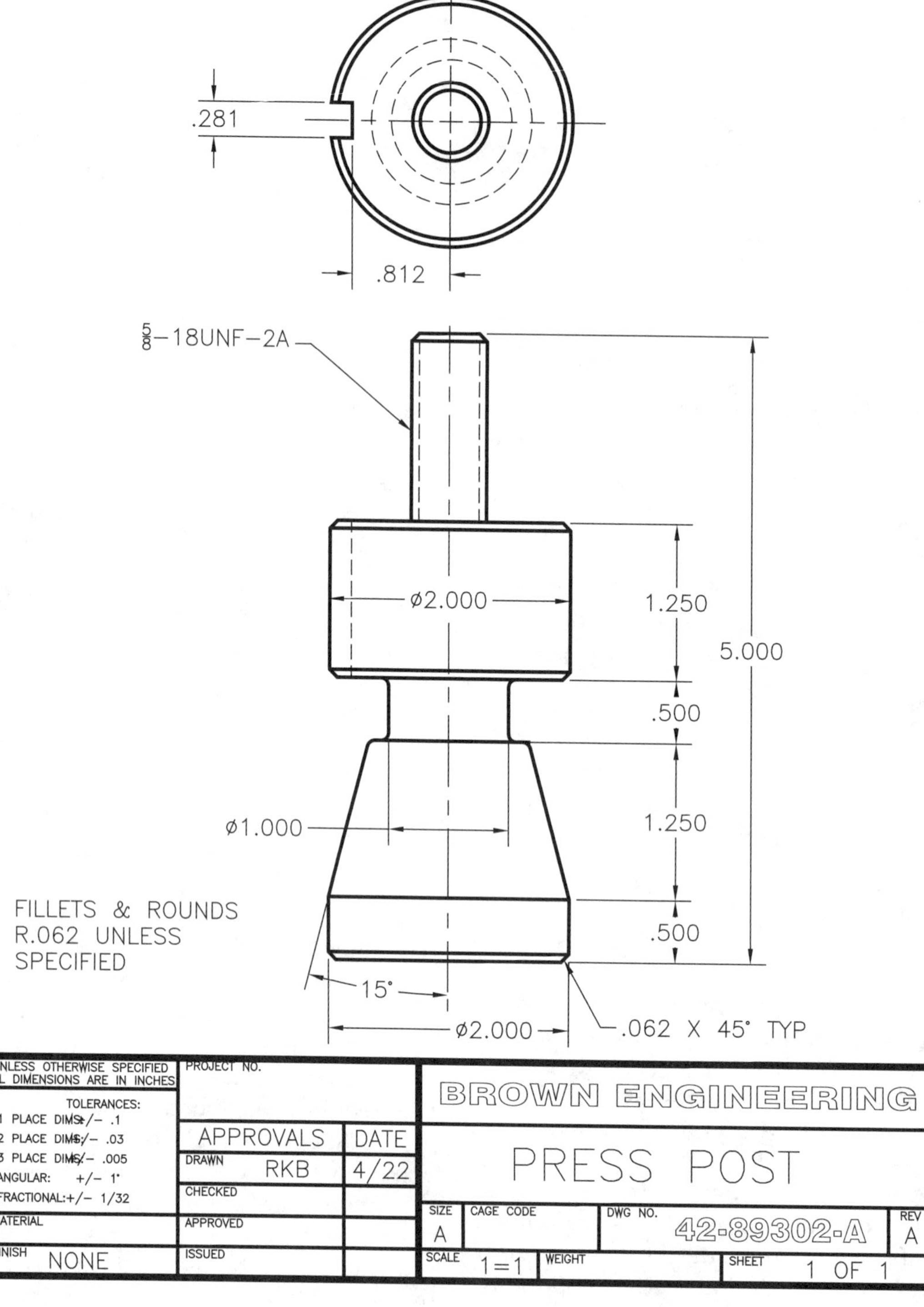


**Figure 15-1.**
A detail drawing can be a very simple drawing that shows the shape description, dimensions, specifications, material, and other contractual descriptions needed to manufacture the part.

the drawing is updated. In educational settings, drafting exercises and problems often incorporate many details on one large sheet to streamline the learning process, but, in industry, combining more than one part within one detail drawing can be problematic.

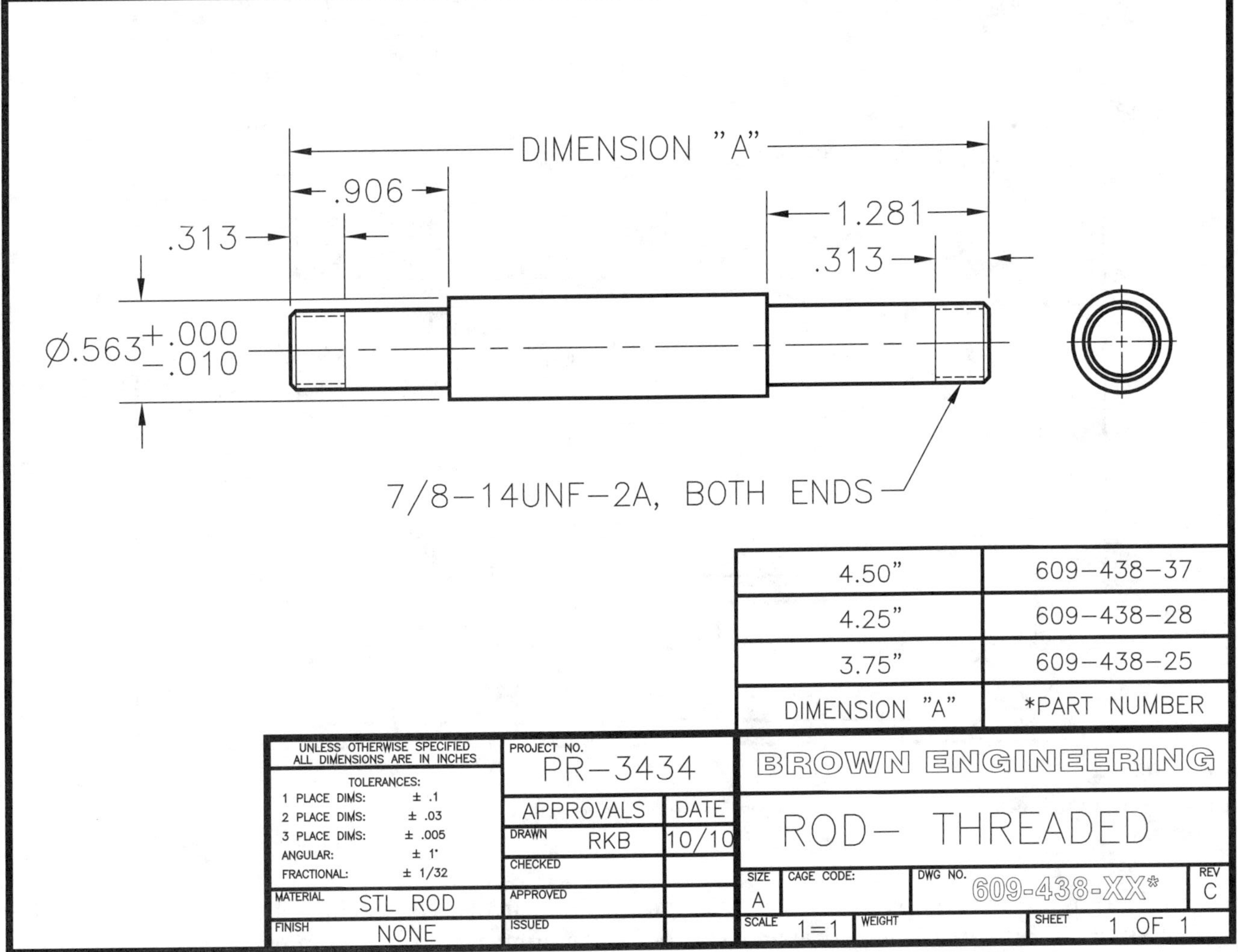


**Figure 15-2.**
The monodetail drawing allows more than one part to be covered with one flexible detail drawing featuring a tabulated dimension.

## Casting and Forging Drawings

Parts that are forged or cast create a unique scenario for the designer and drafter. Cast or forged components often have two sets of requirements: the dimensional size of the mold and the dimensional size of the finished product after machining. These parts often use a type of drawing called a ***casting drawing*** to show the dimensions necessary to create the mold. See Figure 15-3. Additional information about the mold parting line and other mold-related information is often provided. The material to be cast is called out in the title block or as a general note on the drawing.

***Machining drawings*** are the detail drawings that show the dimensions needed to convert a cast part into a finished part. A machining drawing describes the location and sizes of machined surfaces and features, such as threaded holes or datum reference surfaces. The drawing in Figure 15-4 is an example of a machining drawing for the cast part shown in Figure 15-3. For any one casting drawing, there may be several different machining drawings. Each different drawing may show a different way of machining the same casting to create a family of parts. In this example, the part is made by machining away a large section of the center post, which may be needed for a different part created from the casting. The material for the casting drawing, Figure 15-3, is specified as cast iron, but the material specified for the machining drawing, Figure 15-4, is casting FF-332.

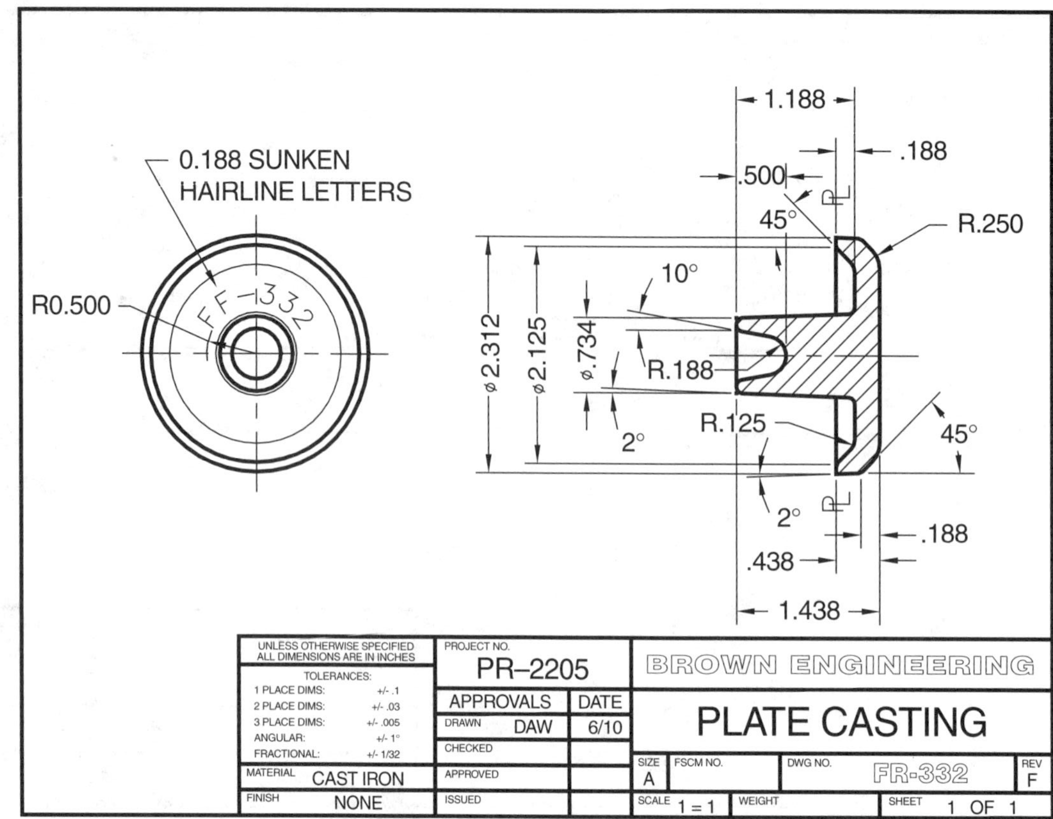


**Figure 15-3.**
A casting drawing shows the dimensions necessary to create a mold.

Depending on the complexity of the part, the casting drawing and machining drawing can be combined on one print. In this case, the finished size is represented by the visible lines and the outline of the casting shape is represented by phantom lines. See Figure 15-5. With this scenario, it is crucial that both the mold maker and the machinist be able to work from the one combination print. Also in this scenario, the cast object in its unfinished form does not have a separate part number, but may be assigned a process control number for inventory. Once the part is machined to the finished dimensions, it can then be stored by its part number.

Standards for documenting casting and forging drawings are set forth in ASME Y14.8. Additional information about parting lines, draft, mismatch, and datum referencing are all covered in more detail in that standard, but are beyond the scope of this textbook.

## Purchased Part Control Drawings

Components and parts that are purchased, like a switch or a timer, often do not require detail drawings. However, ***control drawings*** are often created for these items to document various engineering requirements to assure interchangeability of items each time they are purchased. This may include the appearance; the function, size envelope, mounting and mating dimensions; and such things as interface or performance characteristics. More importantly, a purchased part control drawing serves as a document from which bids can be obtained from suppliers. Figure 15-6 shows an example of a purchased part control drawing.

Standard industry items, such as nuts, bolts, and fasteners, usually do not require control drawings. These items are often well-documented in standard manuals and handbooks with regard to size, shape, strength, tolerances, etc. Libraries of CAD drawings and models also make these parts readily available to the design team.

## Modifying Drawings

By definition, some drawings are referred to as ***modifying drawings.*** These drawings describe objects that are not made from raw or bulk materials, but rather from other parts. Types of modifying drawings include altered-item drawings and modification drawings.

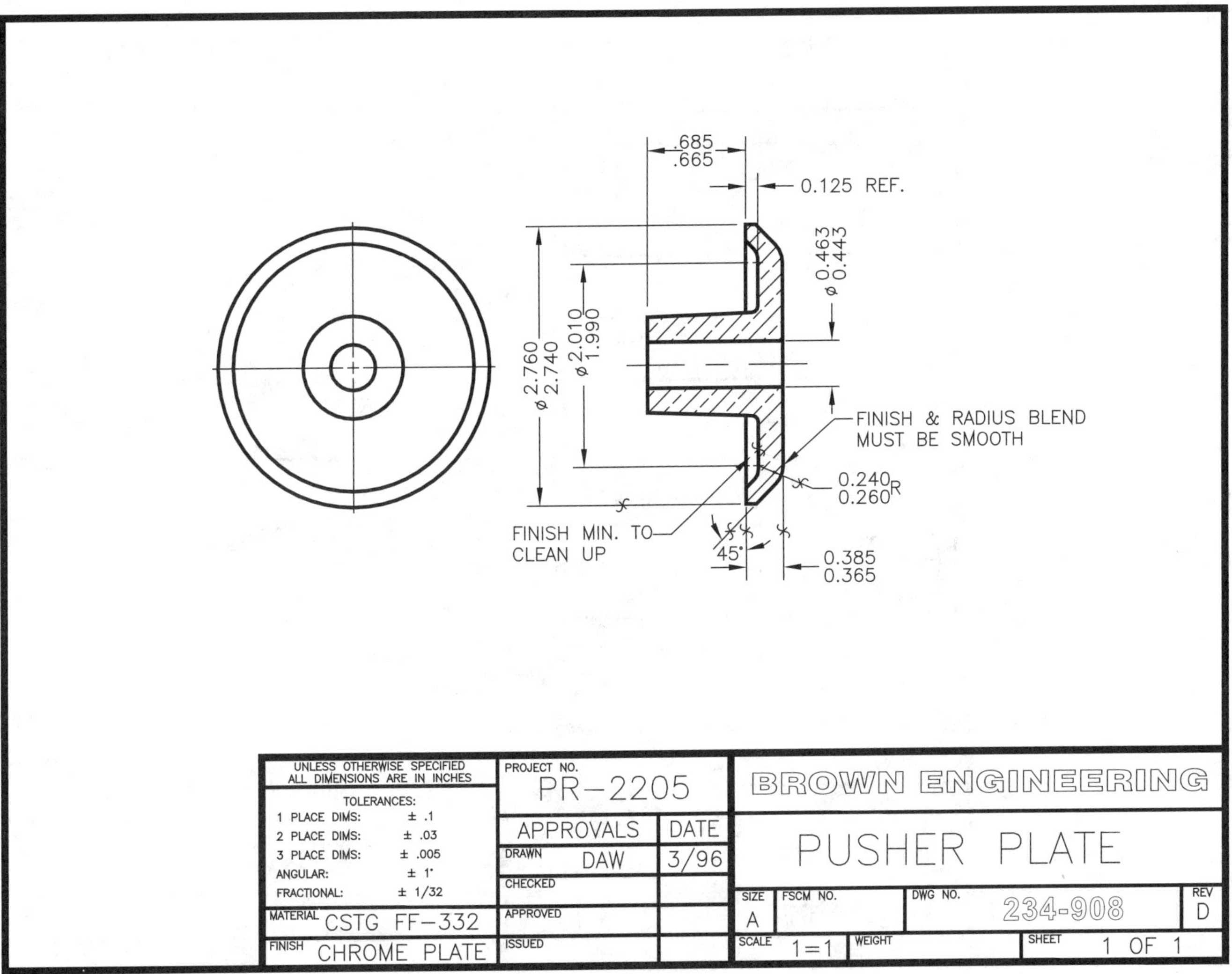


**Figure 15-4.**
A machining drawing shows the dimensions needed to convert a cast part into a finished part.

***Altered-item drawings*** are drawings that describe an alteration procedure that transforms one part into another part. This alteration can be performed by the original manufacturer or a third party and establishes a new part number for the altered part. Only the dimensions necessary in making the transformation are required. Some of the dimensions of the original part may be given as reference dimensions. Altered-item drawings are similar in nature to a machining drawing that describes only the changes to be made to a casting. The material is listed as the part being altered, not the raw material of the original part. ASME standards recommend the notation ALTERED ITEM DRAWING be given adjacent to the title block. The drawing in Figure 15-7 is an altered-item drawing that serves as the drawing for making a new part from the part shown in Figure 15-4.

***Modification drawings*** are created to describe changes to items after they have been delivered. By standard definition, modification drawings are often prepared to satisfy a sudden change in requirements and may involve adding to, removing from, or reworking items to satisfy a given situation. In this scenario, it is desirable to revise the original detail drawing specifications to accommodate future parts. ASME standards recommend the notation MODIFICATION DRAWING be given adjacent to the title block.

## Pattern-Development Drawings

***Pattern-development drawings*** apply dimensions to the shape of the part while it is a flat piece of metal or other stock. The detail drawing may show the final shape description of the part and include phantom lines to show the size of the object before bending.

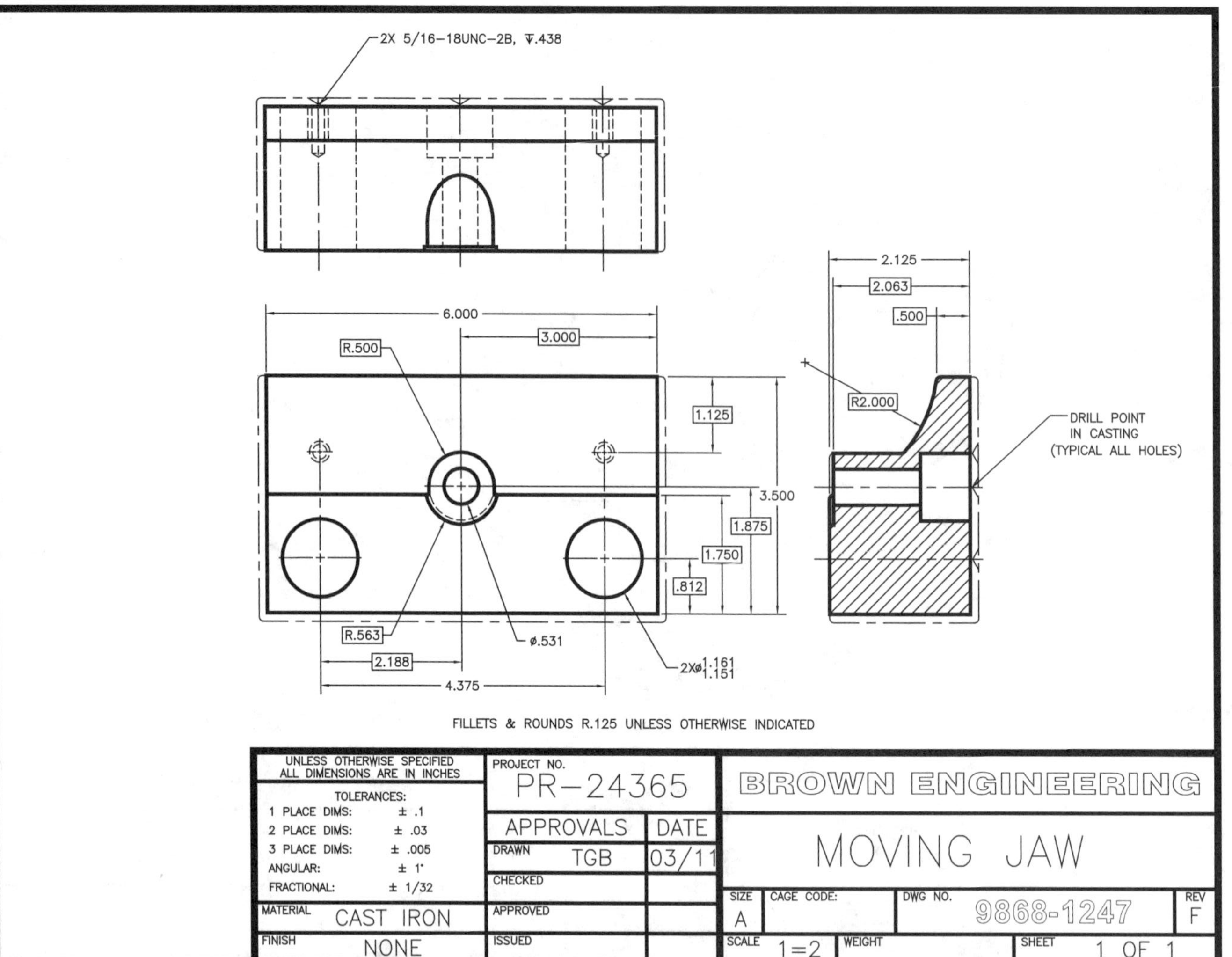


**Figure 15-5.**
A machining drawing may contain lines, notes, or dimensions that provide the mold maker with required information; therefore, a separate casting drawing is not necessary.

Sometimes the part is shown flat, or "unfolded," and a pictorial view is shown to indicate the final form. See Figure 15-8. This type of drawing is discussed in more detail in Unit 21, Precision Sheet Metal Parts.

## Welding Drawings

***Welding drawings*** are detail drawings that show all components of a welded part assembled and include welding specifications. The welding drawing takes on the appearance of an assembly drawing due to the multiple parts that are assembled in position to be welded. The drawing in Figure 15-9 is a welding drawing. Welding drawings and welding symbols are discussed in detail in Unit 22, Welding Prints. Other types of inseparable-assembly drawings are examined in Unit 16, Assembly Drawings.

## Other Types of Drawings

There are many other types of drawings classified by ASME Y14.24. Most of these drawings incorporate the same print reading skills covered in this textbook, so a thorough coverage of the drawing types is unnecessary. Here are a few other types of drawings that may help you understand the purpose and scope of other industrial prints:

- An ***installation drawing*** provides information about how to position or install parts in relation to another part or assembly.
- A ***selected-item drawing*** is a modifying drawing that specifies criteria for selecting a part from inventory that may have a further restriction or specification beyond the other parts. Perhaps the selected items survive a test or tolerance requirement. ASME standards recommend the notation SELECTED ITEM DRAWING be given adjacent to the title block.

- A ***tube-bending drawing*** is a drawing that may use pictorial representation or tables to describe the bends necessary to make a final product.
- A ***matched-set drawing*** describes parts that are uniquely required to be produced to function within a matching relationship. A matched-set drawing specifies the matching requirements by dimensions or other specifications. The dimensions of the individual parts not related to the matching features can be supplied on the matched-set drawing or on other drawings. ASME standards recommend the notation FURNISH ONLY AS A MATCHED SET be given adjacent to the title block.
- A ***contour-definition drawing*** contains the mathematical, numeric, or graphic definition required to define or build a contoured or sculptured surface. This type of drawing is needed for parts such as car fenders and airplane bodies. The definition of the contoured surface may require a series of graphic sections, a table of coordinates, or mathematical equations. In computer-integrated manufacturing, CAD models are usually the best way to control the data associated with contour definitions.

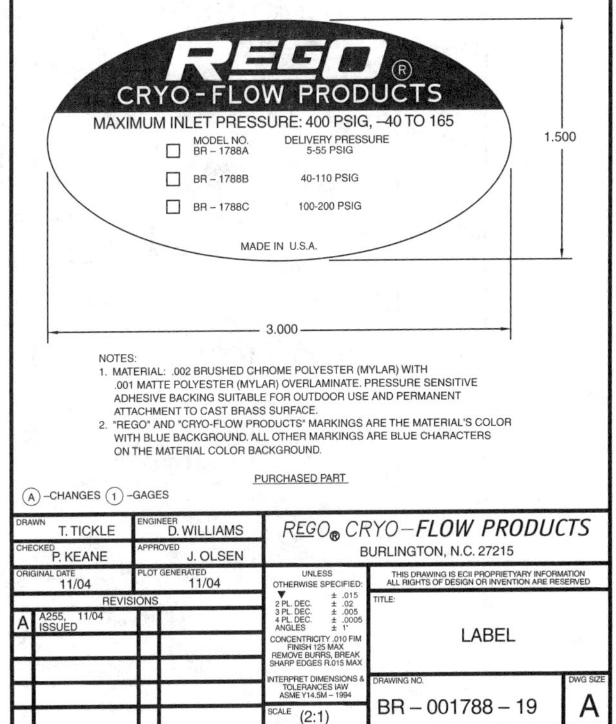


**Figure 15-6.**
Purchased part control drawings can be used to record the measurements of a purchased part or product.

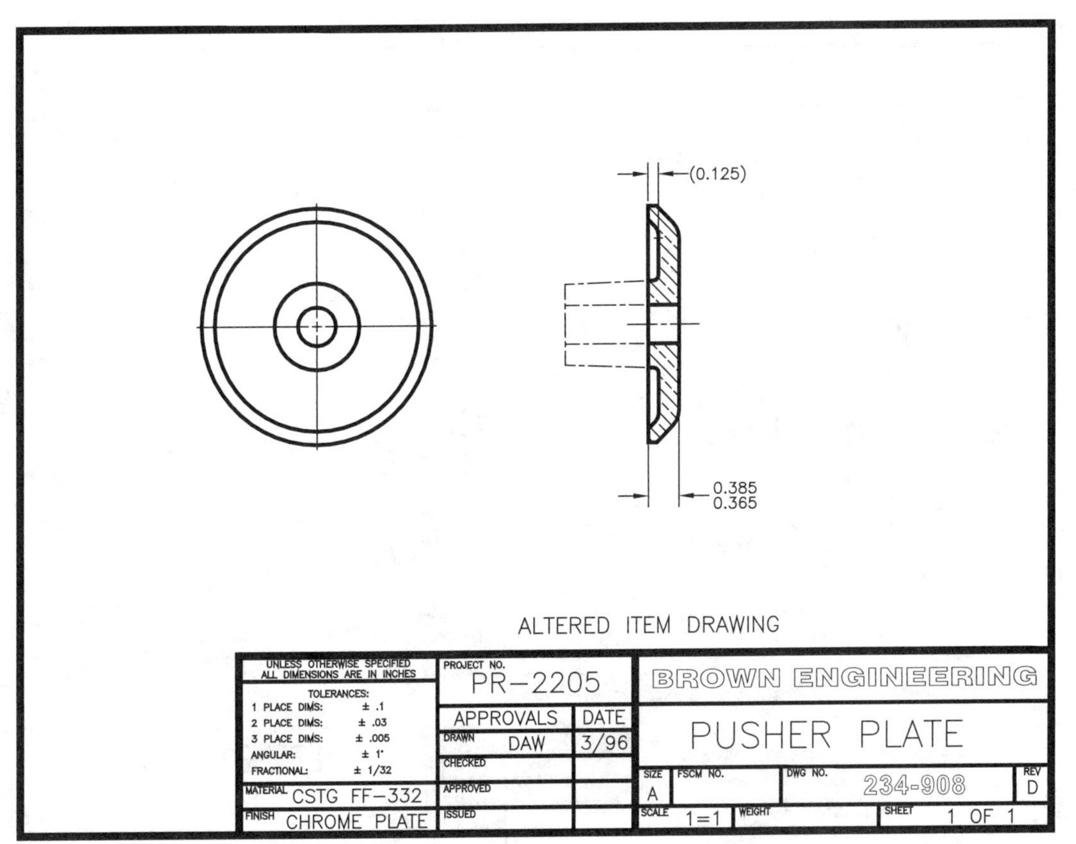


**Figure 15-7.**
Altered-item drawings describe the procedure used to transform a part into another part.

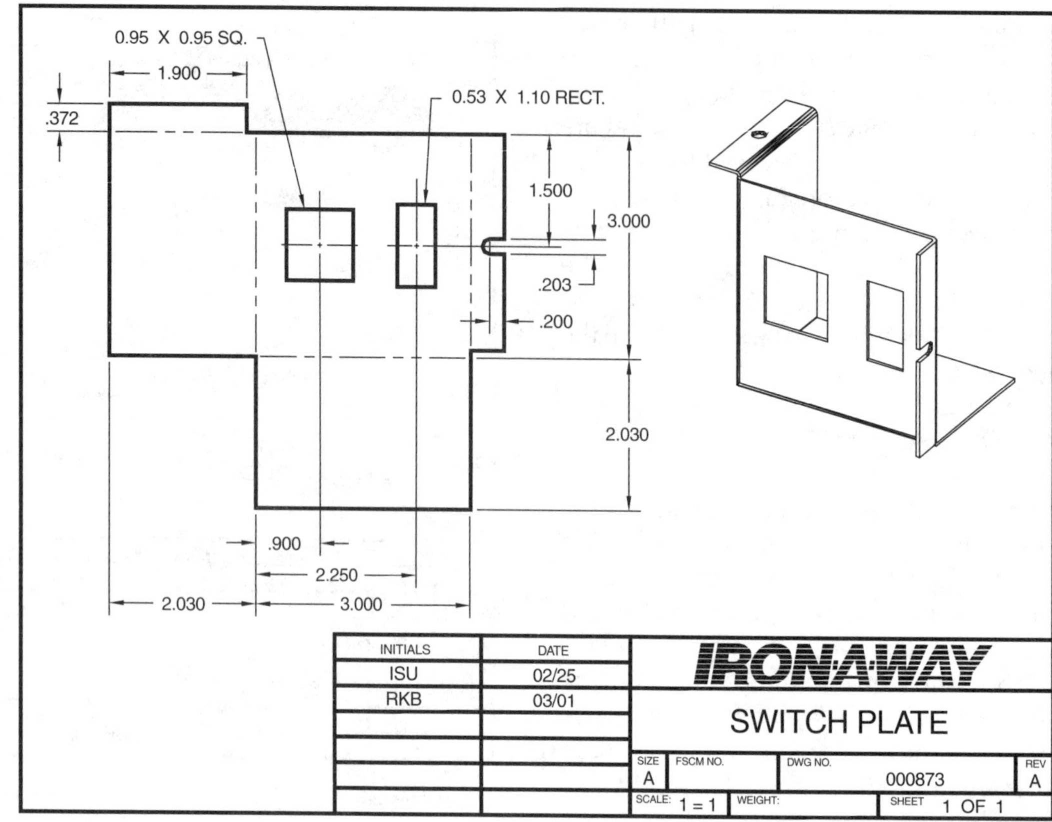


Figure 15-8.
Pattern-development drawings apply dimensions to the shape of the part while it is a flat piece. (Iron-A-Way)

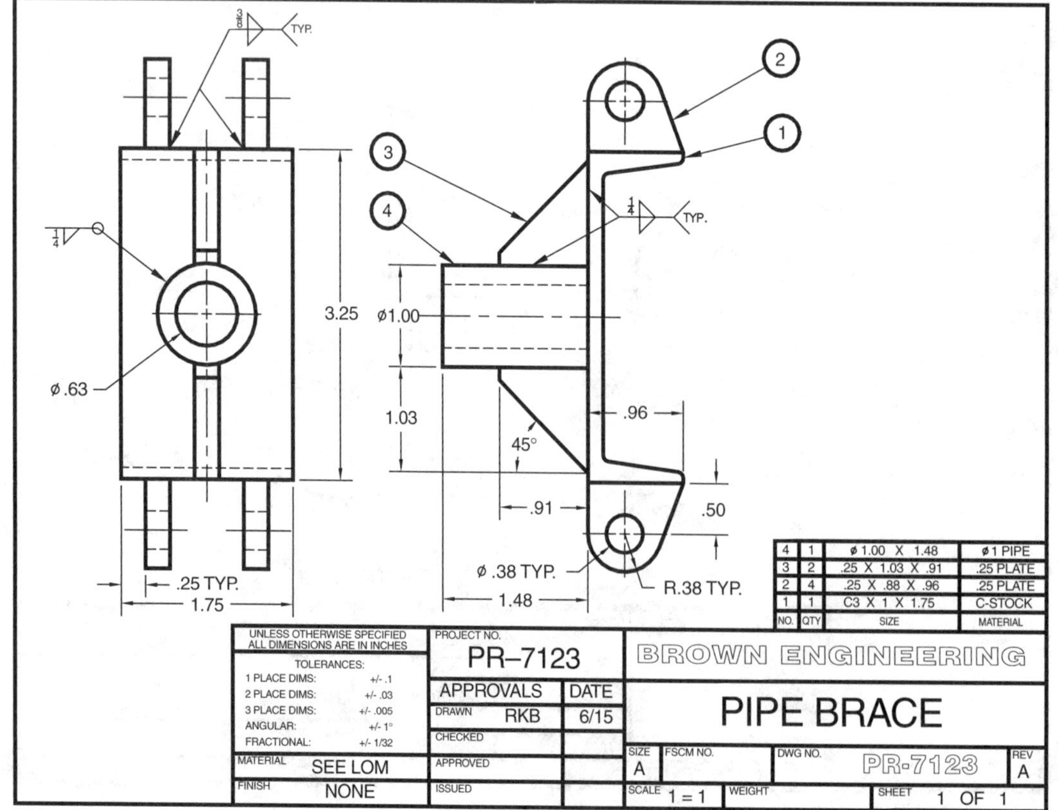


Figure 15-9.
Welding drawings show all components of a welded part assembled with welding specifications.

## Review Questions

*Circle the answer of choice, fill in the blank, or write a short answer.*

1. What are the two basic categories of working drawings? ____________________

____________________

2. *True or False?* Most detail drawings are monodetail—one multiview drawing of one part on one sheet.

3. Are there any conditions under which a multidetail drawing could be created? ____________________

____________________

4. Which of the following drawings is usually of an unfinished product?
   A. Purchased part control drawing.
   B. Machining drawing.
   C. Casting drawing.
   D. Altered-item drawing.

5. *True or False?* A machining drawing and a casting drawing can be combined in one.

6. List three examples of purchased parts that do not need detail drawings, but might still be illustrated in an assembly drawing. ____________________

____________________

7. A(n) ____________________ drawing transforms one part into another part, establishing a new part number, while a(n) ____________________ drawing is created to show how to modify a part that has already come in the door.

8. *True or False?* Another name for a flat pattern drawing is a development.

9. Which of the following drawing types is technically an assembly drawing?
   A. Metal
   B. Development
   C. Welding
   D. Installation

10. Refer to the ASME drawing types listed in the unit. Which one of the following is *not* an ASME drawing type?
    A. Tube bending.
    B. Contour definition.
    C. Matched set.
    D. Skeleton.

## Industry Print Exercise 15-1

*Refer to the print PR 15-1 and answer the questions below.*

1. Is this a detail drawing of just one part? ______
2. What material is specified for this part? ______
3. Is this drawing a casting drawing? ______
4. What are the overall dimensions of this part (diameter × length)? ______
5. Several of the dimensions must be checked with gauges. How does the inspector know which gauge to use for which dimension? ______

*Review questions based on previous units:*

6. What process is to be used to create the hexagon shaped cavity? ______
7. What is the name of this part? ______
8. What surface texture (in microinches) is specified for the cylindrical surface with a maximum material condition of 1.250″? ______
9. How many threaded holes are there on this part? ______
10. Why is the 1.50″ diameter a reference dimension? ______
11. How many dimensions are expressed using the limit method? ______
12. How many times is the countersink symbol used in this drawing? ______
13. Is the sectional view a full section or a half section? ______
14. What is the primary description of revision A? ______
15. What tolerance applies to the 20° angle specified in the detail view? ______

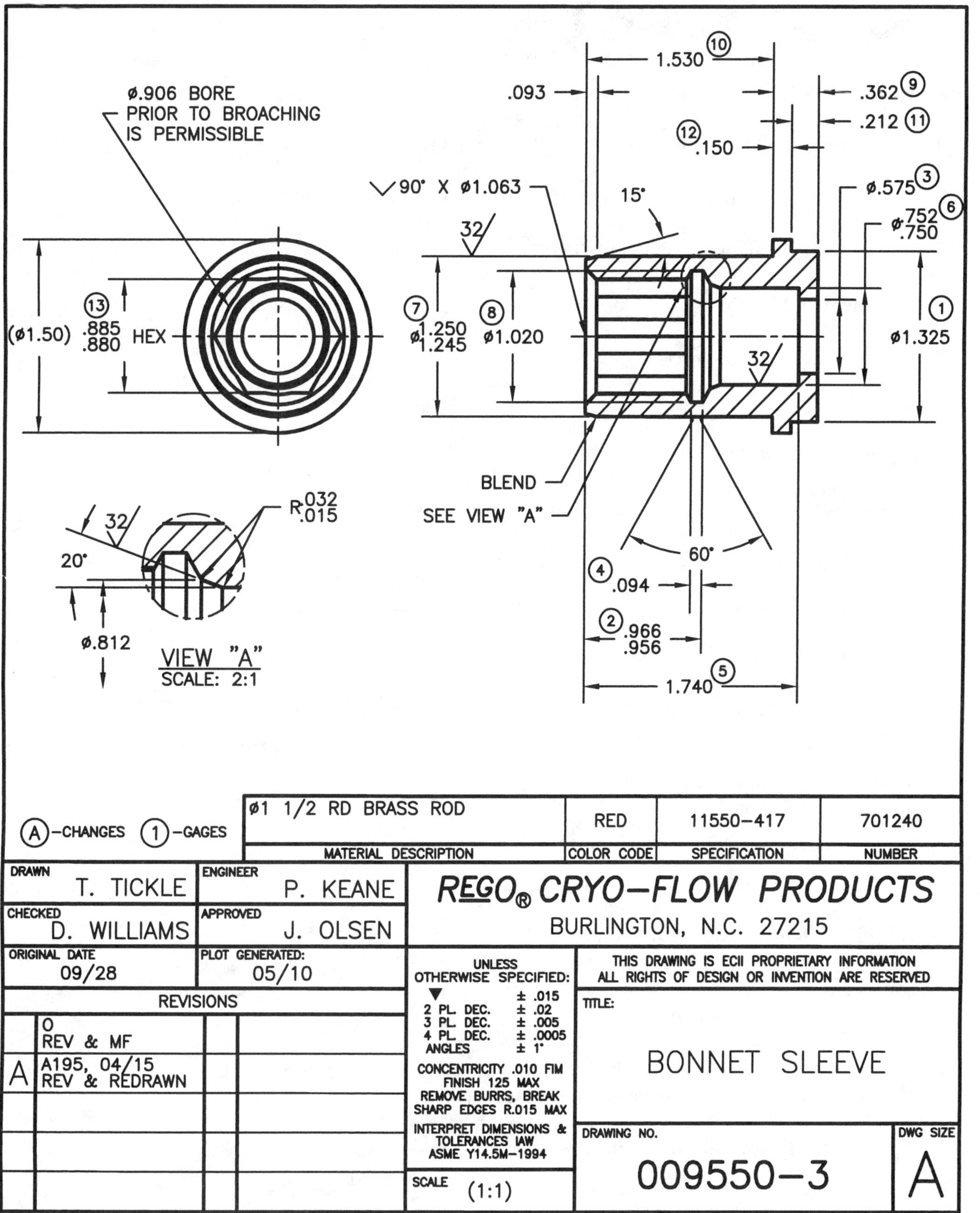


**PR 15-1.**
Print supplied by RegO Cryo-Flow Products.

## Industry Print Exercise 15-2

*Refer to the print PR 15-2 and answer the questions below.*

1. What company specification number is given to the material for this part? ______
2. What revision letter (issue) is the current revision status for this print? ______
3. Is this a monodetail or a multidetail drawing? ______
4. How many critical characteristics are specified in the dimensions and notes? ______
5. Is this a machining drawing? ______

*Review questions based on previous units:*

6. At what scale are the regular views on the original print? ______
7. What GD&T controls are specified for this part? ______
8. What type of section view is Section A-A? ______
9. What method of tolerance expression is used for the overall depth of this object (as shown in rectangular view)? ______
10. Are there any auxiliary views shown? ______
11. In the revision history block, what does ECN stand for? ______
12. What is the maximum material condition diameter of datum A? ______
13. Some companies use a special symbol to point out critical dimensions. Describe the basic shape used by this company on this print. ______
14. This part is a 16-position rotor. The contour of the outer surface determines the positions. How many degrees are there between each rotor position? ______
15. Approximately how tall should the recessed lettering be that marks the cavity I.D. on this part? ______

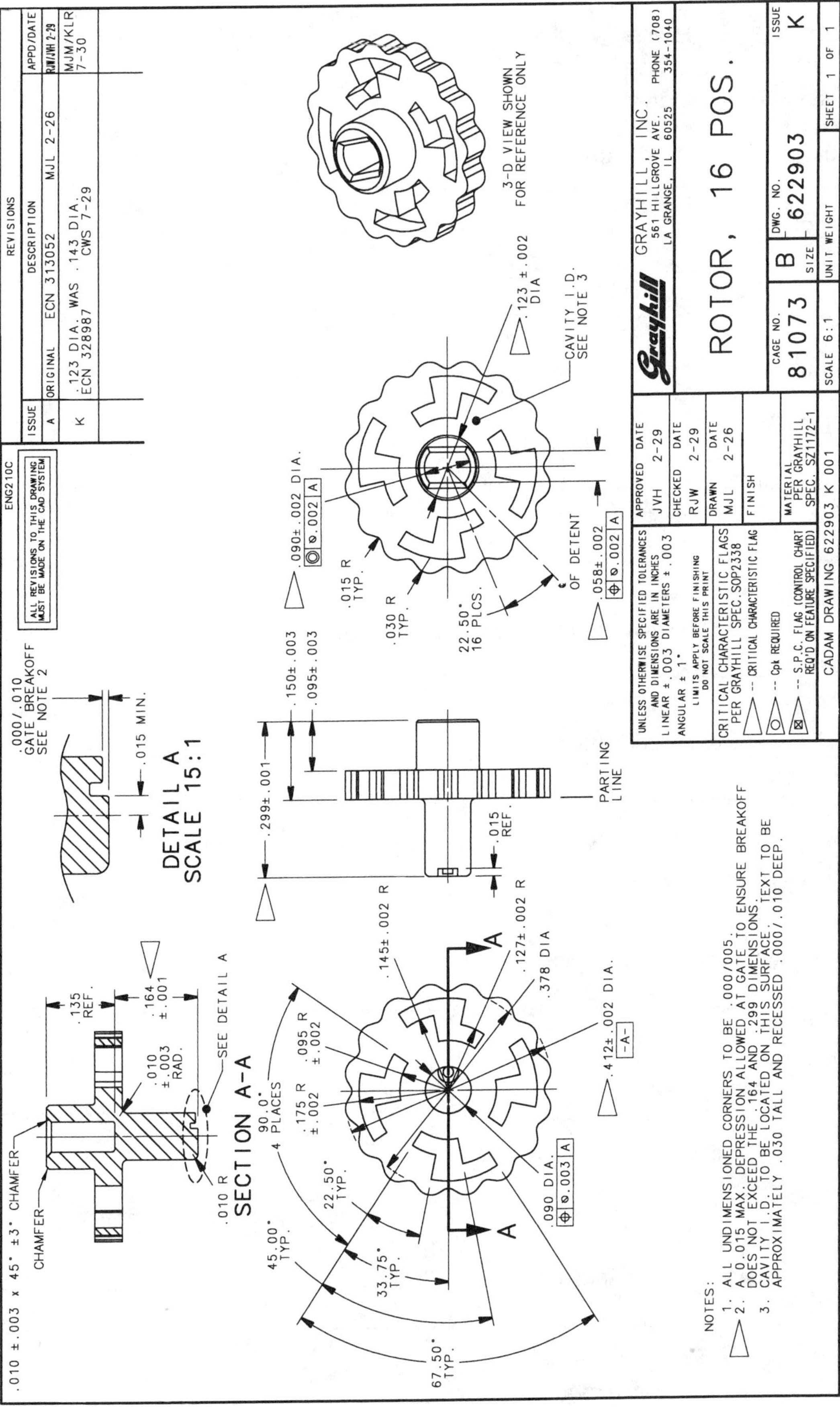


**PR 15-2.**
Print supplied by Grayhill, Inc.

## Bonus Print Reading Exercises

The following questions are based on various bonus prints located in the folder at the back of this textbook. Refer to the print indicated, evaluate the print, and answer the question.

**Print AP-001:**

1. What type of detail drawing is this? ______

2. What is the name and number of the specified material for this drawing? ______

**Print AP-003:**

3. The general notes indicate this is a cast part. Is this detail drawing showing all dimensions for the finished part or the casting dimensions for an unfinished part? ______

**Print AP-005:**

4. Is this print a casting drawing or a machining drawing? ______

**Print AP-007:**

5. Is this print a detail drawing or an assembly drawing? ______

6. This print features a pictorial drawing at a scale of 2:1. Is that a requirement for a detail drawing?

______

**Print AP-011:**

7. What is the name and number of the specified material for this detail drawing? ______

8. Is datum A specified as a machined or cast surface? ______

9. In Section A-A, what do the hidden lines represent? ______

**Print AP-013:**

10. Is this detail drawing just a machining drawing? ______

**Print AP-014:**

11. Is this a detail drawing or an assembly drawing? ______

**Print AP-015:**

12. If this part is still in the rough, unfinished state, its part number is ____________________.

**Print AP-023:**

13. What type of detail drawing is this print? ____________________

____________________

**Print AP-025:**

14. What type of detail drawing is this print? ____________________

____________________

**Print AP-026:**

15. What type of detail drawing is this print? ____________________

____________________

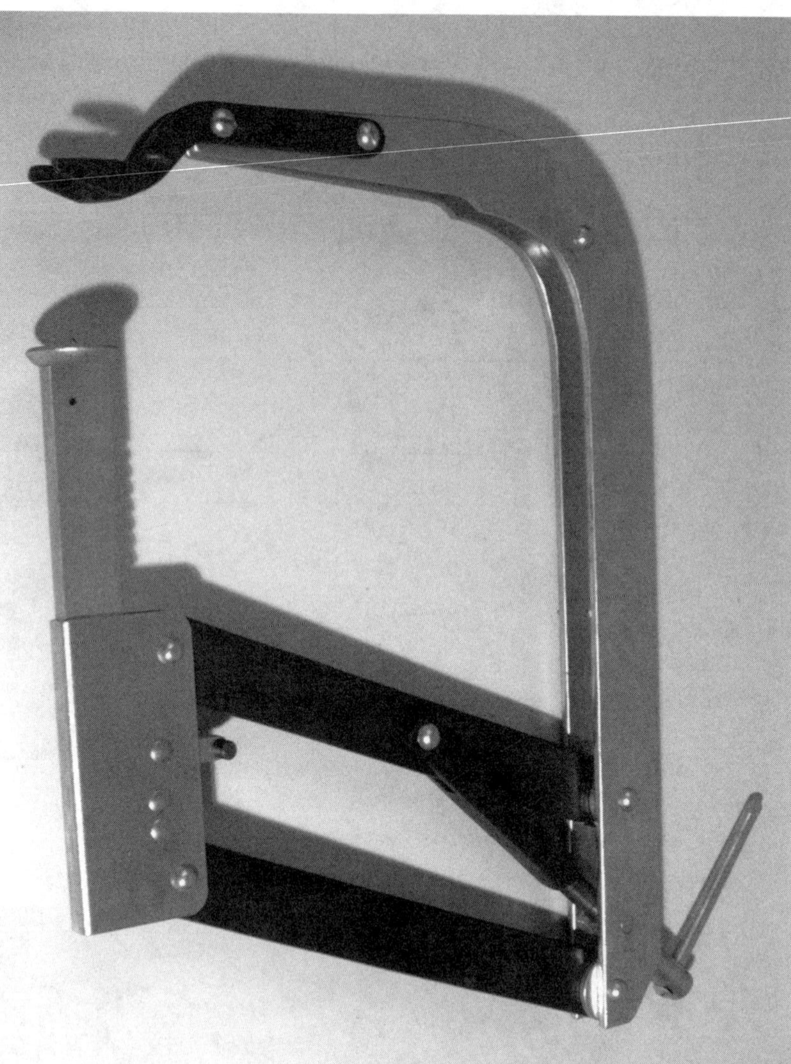

Assembly drawings depict how the parts of an assembly are put together. Every assembly, such as this valve spring compressor, requires an assembly drawing.

# UNIT 16
# Assembly Drawings

*After completing this unit, you will be able to:*

**Explain** the nature and role of assembly drawings in the industrial setting.

**Explain** different ways of creating pictorial and multiview assembly drawings used in industry.

**Discuss** the role of the subassembly drawing.

**Identify** and read information about the components within an assembly drawing.

**Identify** sectioning techniques used to delineate component parts in an assembly drawing.

**Identify** and read parts list information about assembly drawings that are drawn for multiple variations.

Unit 15 discussed the nature of detail drawings in industry. These drawings are necessary to manufacture every piece or part of a product. This unit examines assembly drawings. These drawings are helpful by showing how each unit or subunit is put together. It is important to note that the terminology and definitions related to assembly drawings can vary greatly depending on the industry. Terms are not clearly defined and textbooks do not all agree, although standards such as ASME Y14.24, *Types and Applications of Engineering Drawings*, are working toward that end.

Also, drawing types can be combined. For example, an assembly drawing may describe precise manufacturing processes that need to be applied after the assembly of two or more parts, but also contain detailed information about a part. In these cases, some drawings may function both as detail drawings and assembly drawings. It is common to express several specifications for the entire assembly using general notes.

In ASME Y14.24, the ***inseparable assembly drawing*** is defined as a drawing composed of two or more parts that, once assembled, are permanently joined and thus become one unit. Examples include welded or brazed parts, riveted parts, and parts with pressed or molded inserts. Within the context of this textbook, inseparable assembly drawings serve in the same capacity as detail drawings of single parts. Specific examples, such as welding drawings, are covered in other units.

## Assembly Drawings

In general, ***assembly drawings*** are those drawings that show the working relationship of the various parts of a machine, structure, or product as they fit and function together. Usually, each part in the assembly has its own unique part number. Within the assembly drawing, the part number is indexed and listed in a list of materials on the drawing, as described in Unit 3. The assembly drawing usually provides:

- Name of the subassembly or assembly mechanism, as well as a part number that can be used in higher assemblies and inventory.
- Visual relationship of one part to another in order to correctly assemble components.
- List of parts or bill of material.
- Overall size and location dimensions, when necessary to provide critical information.
- Information that cannot be determined from the separate part drawings.

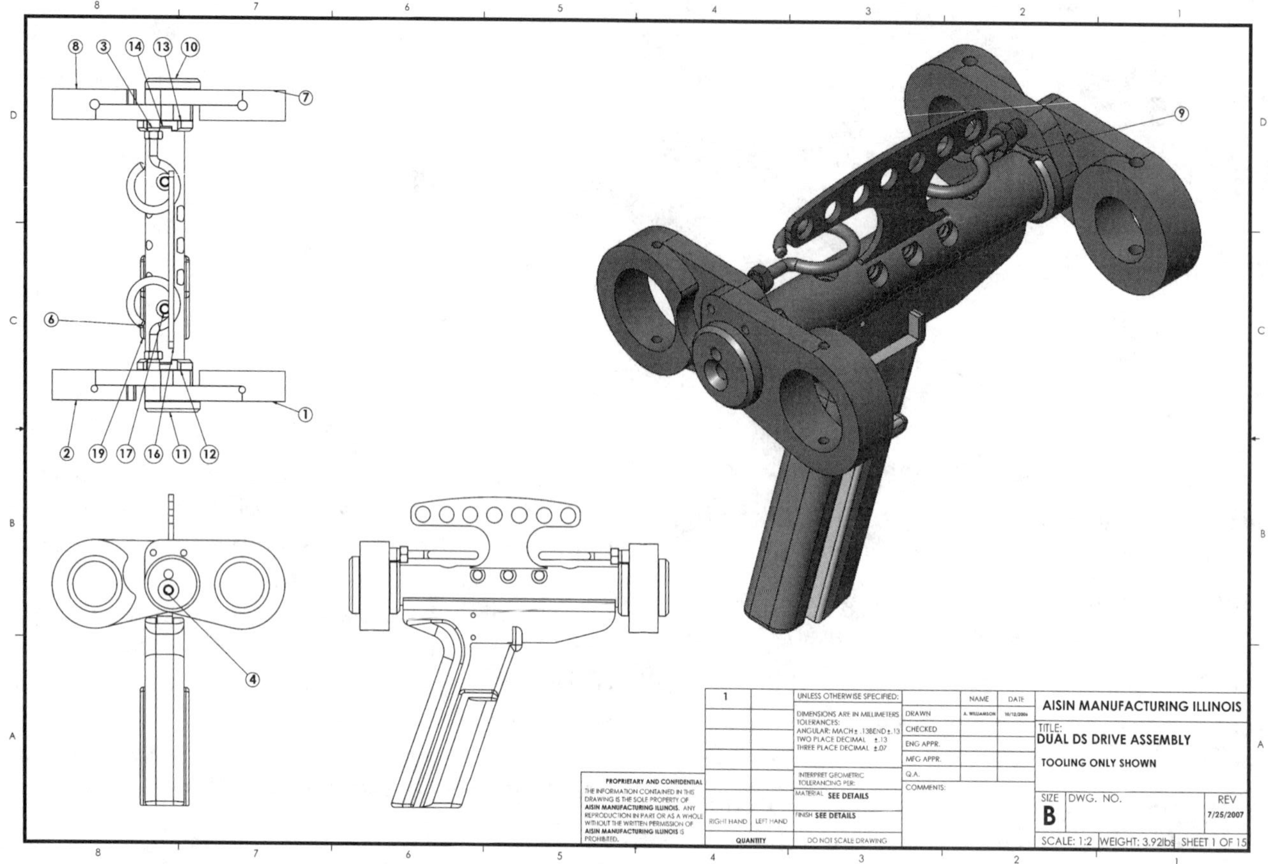


**Figure 16-1.**
Assembly drawings serve as the detail drawing of the entire product, machine, or device. (AISIN Manufacturing Illinois)

## Assembly

Assembly drawings often serve as the detail drawing of the entire product, machine, or device as a whole. See Figure 16-1. This drawing is often a multiview drawing and may be sectioned in one or more views, although one view may be sufficient to describe the assembly. The drawing contains only the hidden lines critical to explaining how the parts are assembled. This usually means few, if any, hidden lines appear on the drawing. Center lines are still appropriate. Dimensions that express the overall measurements or the range of movement for a particular part may also be applied.

## Subassembly

While an assembly drawing may be made for a complete mechanism or machine, more complex machines and structures may require several subassembly drawings for subunits. ***Subassembly*** is a term for any assembly that fits within a larger assembly. ***Subassembly drawings*** are similar in nature to other assembly drawings. They include a group of related parts composing a subunit of a larger mechanism, such as a drill press spindle assembly, automatic transmission assembly, or drive sprocket assembly.

The drawing in Figure 16-2 is an example of a subassembly drawing. In this figure, the subassembly is a multiview projection that provides information about a part composed of other parts. In essence, this subassembly drawing is like a detail drawing in that it provides identification about the group of parts as a new unit. The subassembly part number is used for inventory and process control. As with any assembly drawing, dimensions are not usually required because all the "materials" are already manufactured and inspected, although the final dimensions of the subassembly can be included.

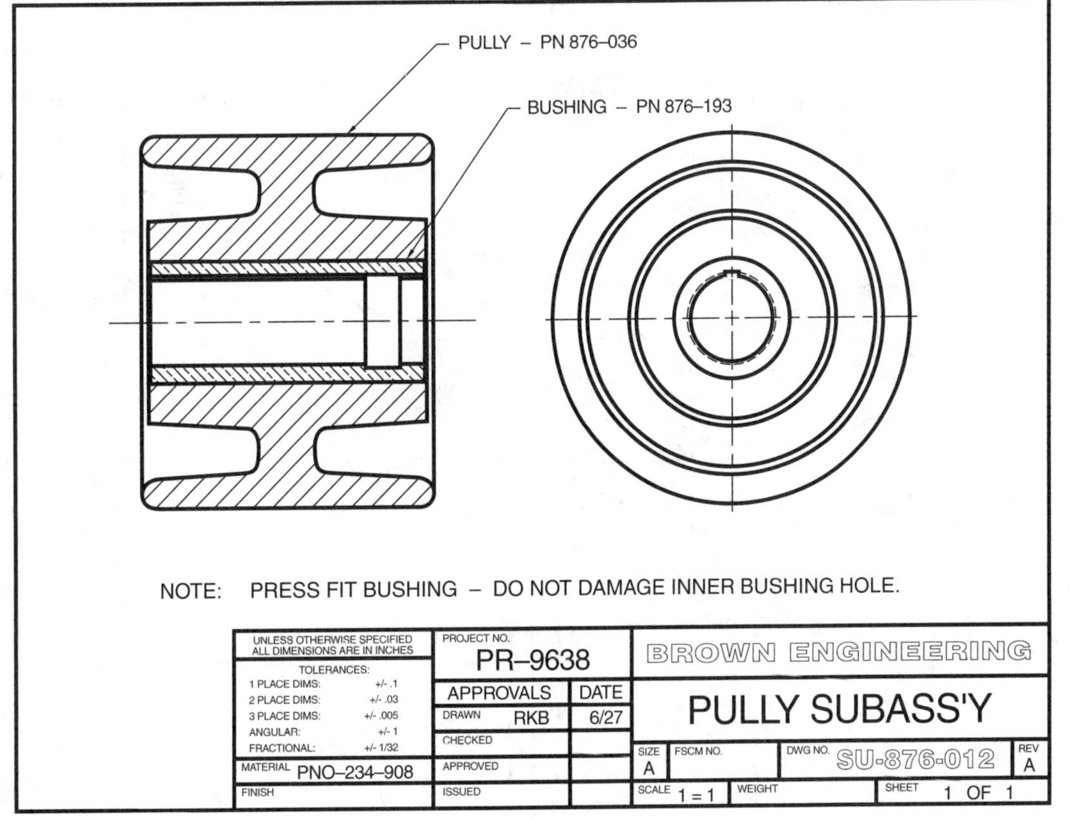


Figure 16-2.
Subassembly drawings include a related group of parts composing a subunit of a larger mechanism.

## Detail Assembly

In ASME Y14.24, the term ***detail assembly*** is applied to an assembly drawing that also contains all dimensions for all features of the individual parts within the assembly. These ***detail assembly drawings*** are useful for very simple assembly mechanisms. The drawing in Figure 16-3 is a detail assembly drawing. As shown, the company saved documentation time and expense by showing dimensions of all parts in this one drawing. However, this practice can be problematic with respect to maintaining one print for individual components as well as the assembly. Unless the pieces of this item are all manufactured and assembled before arriving in inventory, tracking can be difficult.

## Diagram Assembly

***Diagram assembly drawings*** use conventional symbols joined with single lines to show piping flow and wiring assemblies. The drawing explains to the reader how to assemble parts, but the parts are shown as symbols only. An example of a diagram assembly drawing is shown in Figure 16-4.

## Installation Assembly

***Installation assembly drawings*** provide the necessary information to install or erect a piece of equipment. For example, a dust collector system or other large piece of equipment may be shipped knocked down and erected at the customer's plant or field location. An installation assembly drawing provides the instructions for correctly installing the equipment. Installation assembly drawings are often shown in pictorial form.

## Exploded Assemblies

***Exploded assembly drawings*** show some or all of the parts separated from each other to help show the correct order or method of assembly, Figure 16-5. Exploded assembly drawings are really not a separate type of assembly drawing. Both multiview and pictorial assemblies can be drawn in an exploded manner. Exploded assemblies are easy to read and understand when analyzing a mechanism or product. Thus, they are most often used for service manuals or part request diagrams intended for the general public.

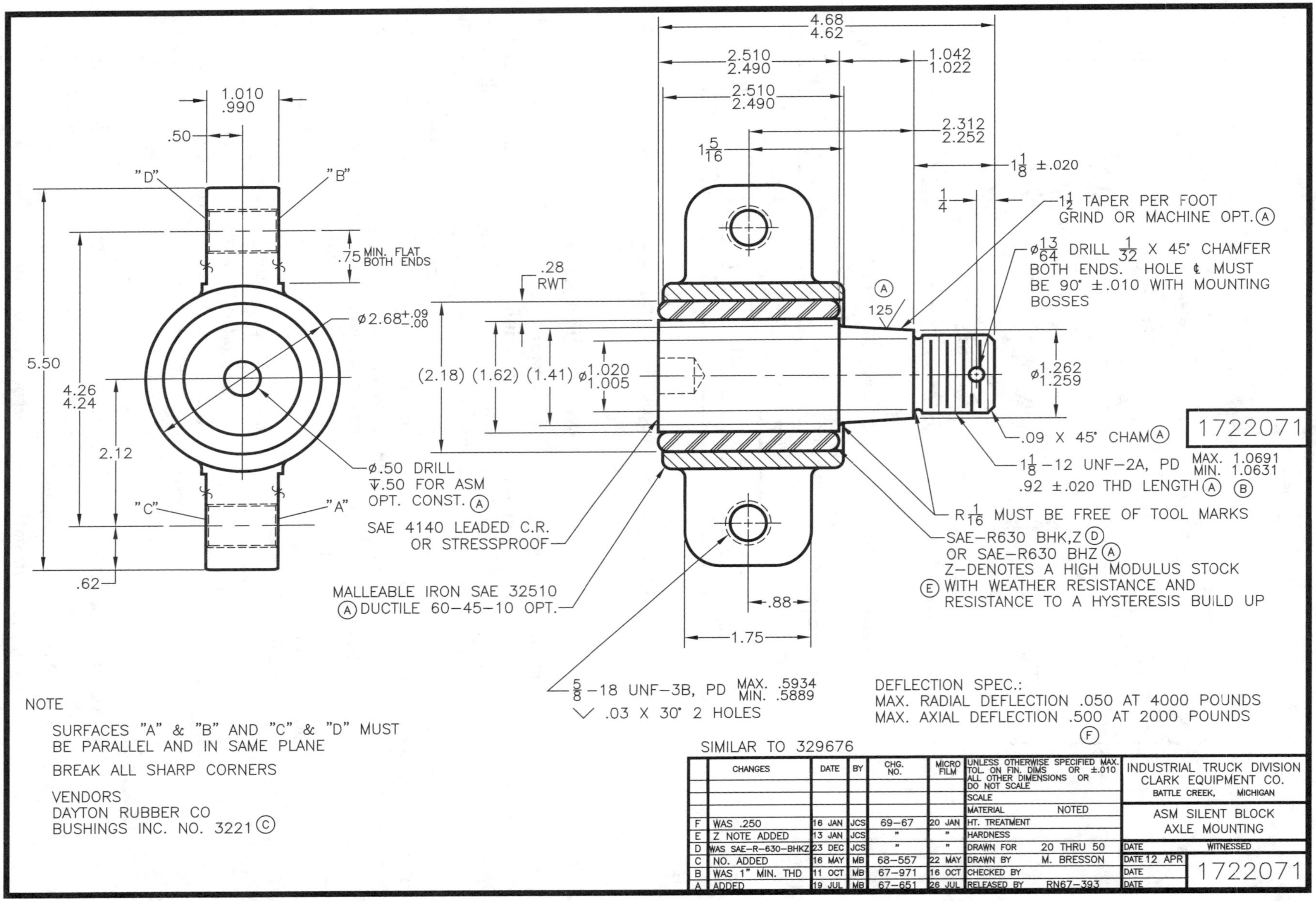

| | CHANGES | DATE | BY | CHG. NO. | MICRO FILM | UNLESS OTHERWISE SPECIFIED MAX. TOL. ON FIN. DIMS OR ±.010 ALL OTHER DIMENSIONS OR DO NOT SCALE | INDUSTRIAL TRUCK DIVISION CLARK EQUIPMENT CO. BATTLE CREEK, MICHIGAN |
|---|---|---|---|---|---|---|---|
| | | | | | | SCALE | |
| | | | | | | MATERIAL NOTED | ASM SILENT BLOCK AXLE MOUNTING |
| F | WAS .250 | 16 JAN | JCS | 69–67 | 20 JAN | HT. TREATMENT | |
| E | Z NOTE ADDED | 13 JAN | JCS | " | " | HARDNESS | |
| D | WAS SAE–R–630–BHKZ | 23 DEC | JCS | " | " | DRAWN FOR 20 THRU 50 | DATE WITNESSED |
| C | NO. ADDED | 16 MAY | MB | 68–557 | 22 MAY | DRAWN BY M. BRESSON | DATE 12 APR — 1722071 |
| B | WAS 1" MIN. THD | 11 OCT | MB | 67–971 | 16 OCT | CHECKED BY | DATE |
| A | ADDED | 19 JUL | MB | 67–651 | 26 JUL | RELEASED BY RN67–393 | DATE |

Figure 16-3.
Detail assembly drawings combine the features of detail and assembly drawings. (Clark Equipment Company)

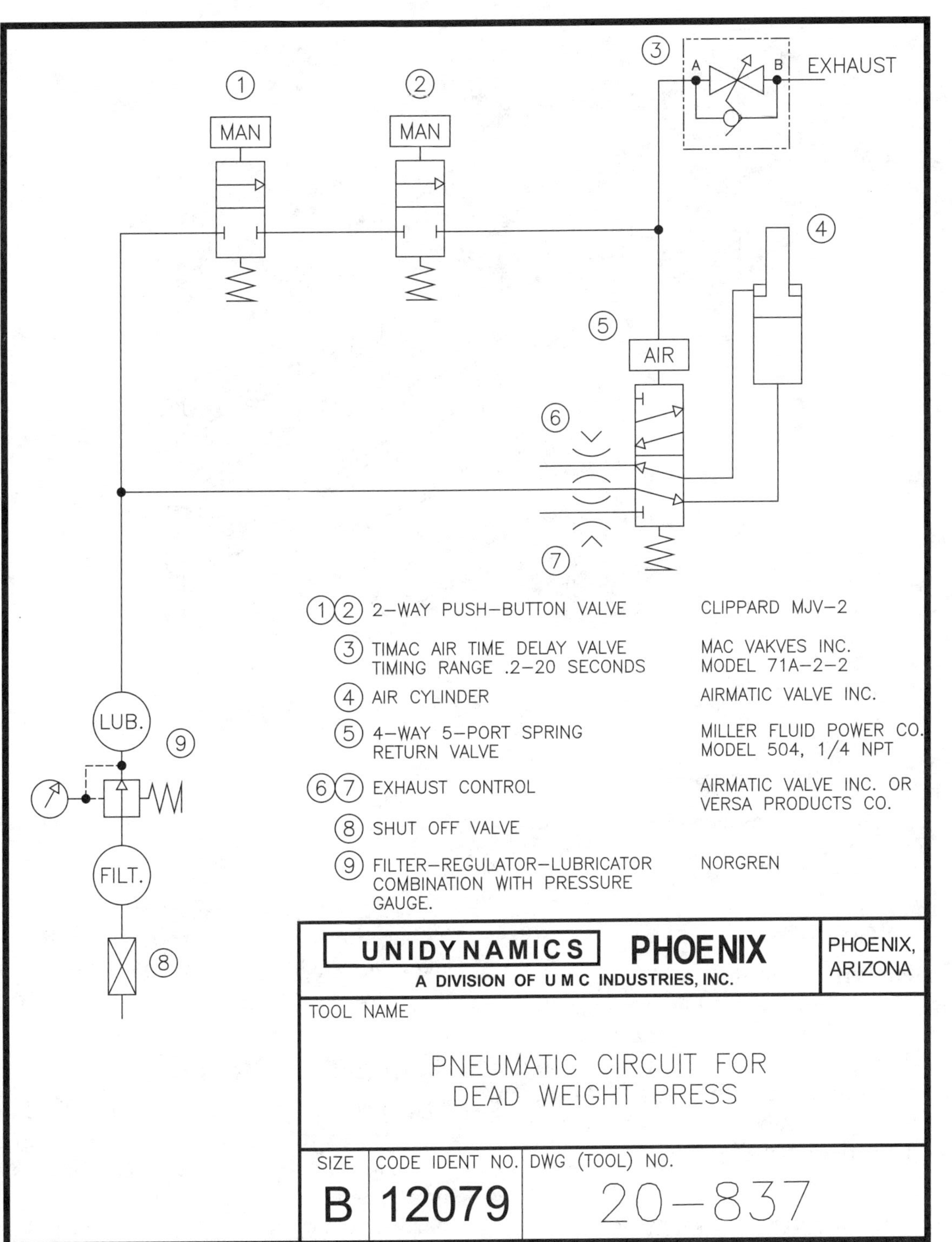


**Figure 16-4.**
Diagram assembly drawings use conventional symbols joined with single lines to show flow and wiring. (Unidynamics)

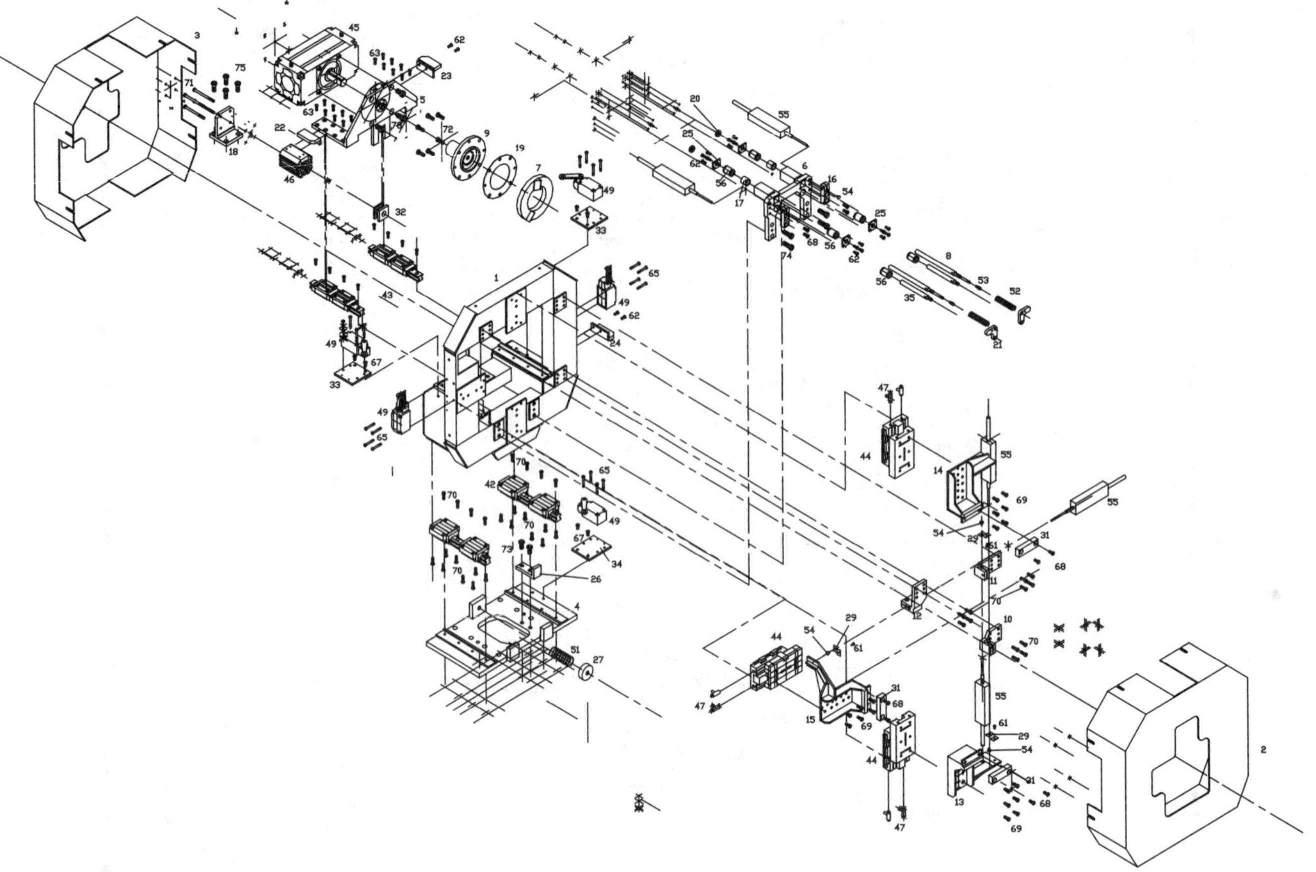

**Figure 16-5.**
Exploded assembly drawings represent parts separated from each other to help show the correct order or method of assembly. This drawing also serves as a maintenance document.

## Part Identification

A system is necessary for identifying the parts of the assembly. Most assembly or subassembly drawings use leaders to identify the parts. Many assembly drawings use balloons neatly arranged on the drawing to identify the parts. ***Balloons*** are circles containing a number or letter and are usually connected to leader lines, Figure 16-6.

## Sections in Assembly Drawings

Section lines for different parts of an assembly drawing are turned at different angles or the scale and spacing of the section lines may be different. This is important to keep in mind when reading an assembly drawing. All section lines assigned to any one part should all be the same angle and spacing. See Figure 16-7. To clarify an assembly drawing with many section lines, it is also helpful to use the section line patterns representing the part material, rather than general-purpose section lines. Thin materials such as gaskets may have the sectioned area filled in solid.

## Application Blocks

In some cases, an application block is located adjacent to the title block. An ***application block*** provides the opportunity to indicate the larger unit, subassembly, or assembly for a particular part or subassembly. As illustrated in Figure 16-8, the application block usually has a column for the part number or assembly number entitled NEXT ASSEMBLY. Another column entitled USED ON indicates a higher level "system" number, such as the model number of the product in which the part or assembly is used. Additional quantity columns can be added to indicate the number of subassemblies required at the next assembly or total quantity used in the product. The application block can also be incorporated into the parts list.

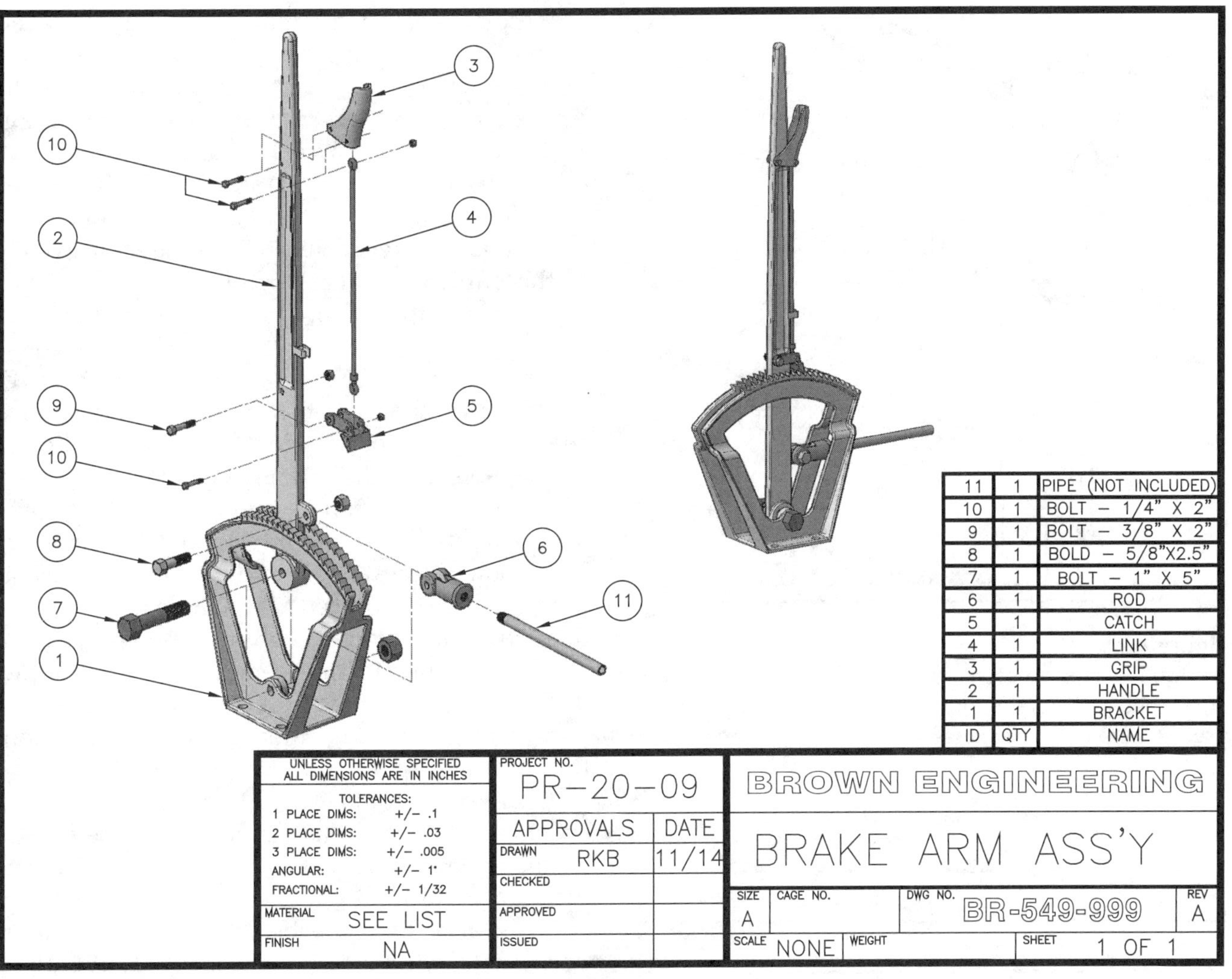


**Figure 16-6.**
Balloons contain the find number for the parts list.

## Parts List Information

As discussed in Unit 3, the ***parts list*** is a chart listing the parts within an assembly drawing, Figure 16-9. It usually appears immediately above the title block on a print. As illustrated, the following items are usually included in a basic parts list.

- Find Number: Matches the number found within a balloon on the drawing.
- Quantity Required: Indicates the number of the item needed within the assembly.
- Part Number (PIN): Indicates the identification number assigned by the company to the part.
- Nomenclature or Description: Provides an opportunity to name or describe the part.

## Additional Listings

Additional columns are often found in the parts list for assembly drawings. These columns may include information about materials, code numbers, or even application block information. See Figure 16-10 for an expanded parts list that incorporates the following columns.

The ***CAGE code*** column appears on assembly drawings that use parts with these supplier codes. See Figure 16-10A. This code is only listed if the part is purchased from an outside vendor. Cage codes are applicable to activities and parts provided for the federal government. At one time, this was referred to as the FSCM number.

The ***material*** column lists the commercial name of the material used in making the part. See Figure 16-10B. This may be a trademark name, such as Lexan, or an industry abbreviation, such as CRS (cold rolled steel).

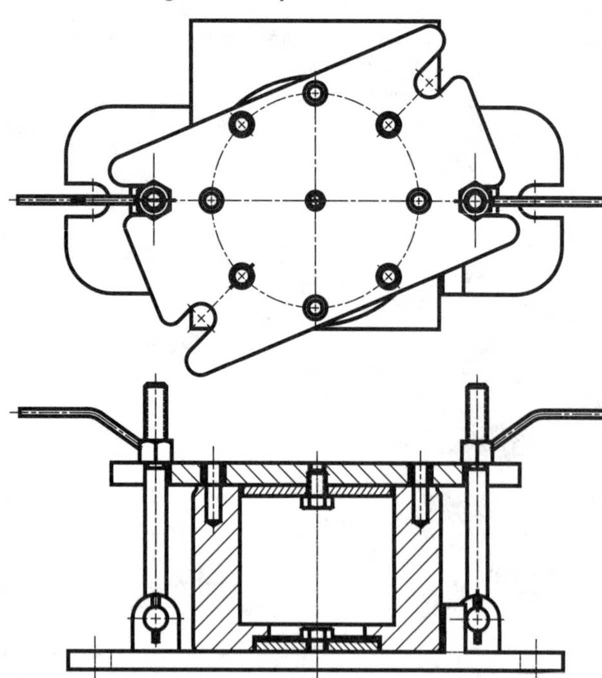

Figure 16-7.
Section lines for different parts of an assembly drawing are turned at different angles. Different spacing is also used to help delineate parts.

| | |
|---|---|
| | |
| BR–548X–4 | 548Z |
| BR–549X–4 | 549Z |
| NEXT ASSEMBLY | USED ON |
| APPLICATION | |

Figure 16-8.
Application blocks are used to indicate what assembly is next in line for the part or subassembly. They may include a number for the model or main system in which the assembly is used. Application blocks are sometimes incorporated into the parts list.

The material specification, sizes, notes, suppliers column furnishes the commercial specification and stock size of the part material. This column is frequently referred to as the ***procurement specification.*** See Figure 16-10C. The names and addresses of manufacturers of purchased parts are sometimes included in this column.

The ***unit weight*** column gives the actual weight of the part. See Figure 16-10D. This information is typically supplied only when it is required by contract.

An additional column for ***notes or remarks*** can be added to provide information related to a vendor source, general notes on the drawing, or specific conditional information about the part. See Figure 16-10E.

Notes concerning interchangeability and replaceability are sometimes placed in a column. These notes indicate parts that are interchangeable with parts in other assemblies. These notes are also used to indicate parts likely to rapidly wear and need replacement. Replacement parts is another designation for replaceability. When these notes appear on a drawing, they are often indications that additional parts must be stocked.

The ***zone*** column is used for a zone designation code. Detail parts on larger prints may be located by the zone designation code for ease in locating the parts on prints.

Part lists can also be created as a separate document. In these cases, the note SEE SEPARATE PARTS LIST should be placed on the drawing above the title block.

## Versions in Assembly Drawings

It is sometimes practical to create an assembly drawing that illustrates more than one version of the assembly. In these cases, each version is identified by a dash number following the assembly number, such as 65302-1, 65302-3, and 65302-5. When dash-numbered assembly drawings are created, columns and rows are added to accommodate all versions of the assembly.

Figure 16-11 illustrates an assembly drawing parts list configured in such a way as to define two different assembly options. The first column is for assembly number 557-2100-5, while the second column is for 557-2100-3. The third column is for shared parts, in this case the frame and slide. In this example, the -3 assembly has a base, while the -5 assembly has a channel, two glides, and two setscrews.

| FIND NO. | QTY REQD | PART NO. | NOMENCLATURE OR DESCRIPTION |
|---|---|---|---|
| 5 | 4 | 725–P45 | ¼" LOCK WASHER |
| 4 | 4 | 3452–M12 | ¼" X 1" MACHINE SCREW |
| 3 | 2 | 658–4512 | SLIDE BAR |
| 2 | 2 | 985–4567 | BASE GROOVE GUIDE |
| 1 | 1 | 985–1345 | BASE |

PARTS LIST

**Figure 16-9.**
At minimum, a parts list should contain a find number, if applicable; the quantity required; the part number; and the description of that part.

| FIND NO. | QTY REQD | CAGE CODE | PART NO. | NOMENCLATURE OR DESCRIPTION | MATERIAL | PROCUREMENT SPECIFICATIONS | UNIT WEIGHT | NOTES OR REMARKS |
|---|---|---|---|---|---|---|---|---|
| 5 | 4 | | 725–P45 | ¼" LOCK WASHER | | ALLIED FSNRS | .05 | |
| 4 | 4 | | 3452–M12 | ¼" X 1" MACHINE SCREW | | ALLIED FSNRS | .04 | SLOT HEAD ONLY |
| 3 | 2 | 81349 | 658–4512 | SLIDE BAR | AL–6061 | | .57 | |
| 2 | 2 | 81349 | 985–4567 | BASE GROOVE GUIDE | STL | | .3 | |
| 1 | 1 | 81349 | 985–1345 | BASE | CI | | 2.5 | |

PARTS LIST

A B C D E

**Figure 16-10.**
Parts lists can also include columns for (A) CAGE codes; (B) material; (C) procurement specifications; (D) unit weight; and/or (E) notes or remarks.

| NO. RQD. ASS'Y | NO. RQD. ASS'Y | NO. RQD. ASS'Y | PART OR IDENTIFYING NO. | ITEM | NOMENCLATURE OR DESCRIPTION | CODE IDENT. | MATERIAL | MATERIAL SPECIFICATION SIZES, NOTES, SUPPLIERS | UNIT WT. |
|---|---|---|---|---|---|---|---|---|---|
| 2 | | | AR5–13 | 6 | GLIDE | | ACETAL RSN | | |
| 2 | | | G84–3S | 5 | SETSCREW | | | ALLIED FSN | |
| 1 | | | 557–1845 | 4 | CHANNEL | | ALY6061–0 | 5–2X15 | |
| | 1 | | 557–1843 | 3 | BASE | | ALY6061–0 | 8–14X20 | |
| | | 2 | 557–3547 | 2 | SLIDE | | STL–4392 | | |
| | | 1 | 557–1754 | 1 | FRAME | | CI | | |
| | | | | | | | | | |
| –5 | –3 | X | 557–2100–X | | SLIDE ASS'Y | | | | |

PARTS LIST

**Figure 16-11.**
One assembly drawing can sometimes be configured to cover more than one version of the assembly, each with a dash number.

## Review Questions

*Circle the answer of choice, fill in the blank, or write a short answer.*

1. What is an inseparable assembly drawing? ____________________

____________________

____________________

____________________

2. *True or False?* Assembly drawings show the working relationship of various parts of a machine, structure, or product as they fit together.

3. Which of the following is usually *not* provided by an assembly drawing?
   A. An assembly name and number.
   B. Surface texture symbols for each part.
   C. A list of materials (parts).
   D. Overall size and location dimensions necessary for critical information.

*For each of the items below, indicate whether or not they are typically included on an assembly drawing.*

4. Hidden lines:
   Usually Drawn __________ Sometimes Drawn __________ Never Drawn __________

5. Center lines:
   Usually Drawn __________ Sometimes Drawn __________ Never Drawn __________

6. Dimensions:
   Usually Drawn __________ Sometimes Drawn __________ Never Drawn __________

7. __________ are shown on an assembly drawing to reference parts back to the list of material.

8. *True or False?* While subassemblies are common for more complex machines and devices, these assembly drawings do not usually have their own drawing or part number.

9. A(n) __________ assembly is one that contains all dimensions for all features of the individual parts within the same drawing as the assembly of the individual parts.

10. A(n) ____ assembly drawing uses symbols for the parts. An example would be a piping and wiring flow diagram where the pipes are shown as lines and components like pumps and valves are shown as symbols.
    A. pipe
    B. schematic
    C. diagram
    D. exploded

11. *True or False?* An exploded assembly simply means several or all of the parts of an assembly are shown separated from each other, yet still aligned with each other.

## Industry Print Exercise 16-1

*Refer to the print PR 16-1 and answer the questions below.*

1. How many assemblies are described by this one assembly drawing? ______________________
2. Each assembly described by this drawing uses one of two subassemblies, either part number ______________________ or part number ______________________.
3. What one part number is common to all of the assemblies described by this drawing? ______________________
4. How many different BODY options are there with respect to this drawing? ______________________
5. How will the date code be applied to the assembly? ______________________
6. What is the find number of the part that is adjusted to set the spring compression? ______________________
7. What is the basic height of this assembly as shown in the section view? ______________________
8. What process is to be performed just prior to assembly? ______________________
9. Is the main spring shown as a section? ______________________

*Review questions based on previous units:*

10. What paper size is the original version of this print? ______________________
11. Who checked this drawing? ______________________
12. What does it mean that the .875 HEX dimension value is in parentheses? ______________________
13. What is the scale of the views on the original drawing? ______________________
14. In what month was this drawing issued? ______________________
15. Are there any cutting-plane lines shown in this drawing? ______________________

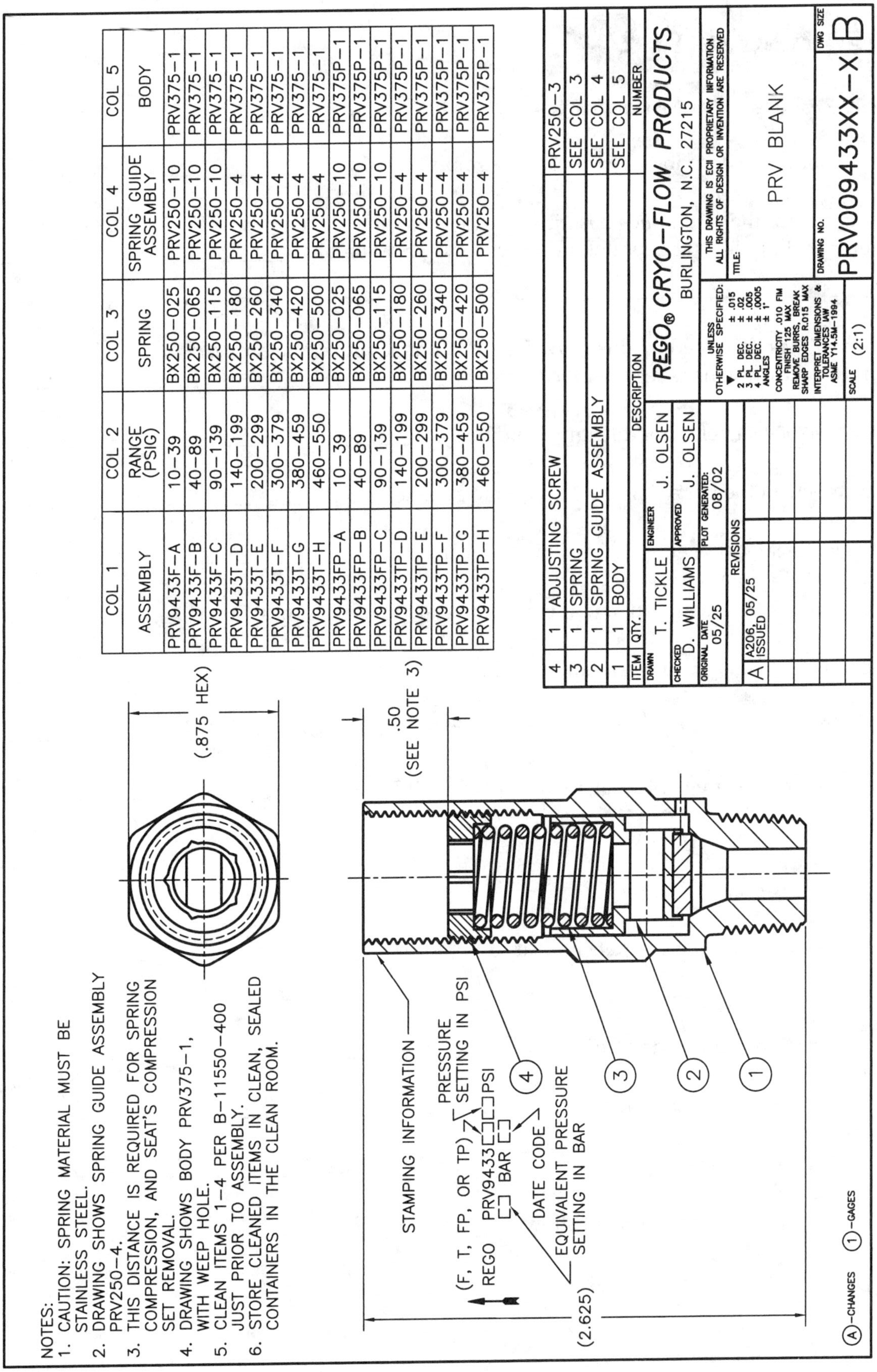

NOTES:
1. CAUTION: SPRING MATERIAL MUST BE STAINLESS STEEL.
2. DRAWING SHOWS SPRING GUIDE ASSEMBLY PRV250-4.
3. THIS DISTANCE IS REQUIRED FOR SPRING COMPRESSION, AND SEAT'S COMPRESSION SET REMOVAL.
4. DRAWING SHOWS BODY PRV375-1, WITH WEEP HOLE.
5. CLEAN ITEMS 1-4 PER B-11550-400 JUST PRIOR TO ASSEMBLY.
6. STORE CLEANED ITEMS IN CLEAN, SEALED CONTAINERS IN THE CLEAN ROOM.

(A)-CHANGES (1)-GAGES

(.875 HEX)

STAMPING INFORMATION

(F, T, FP, OR TP)
PRESSURE SETTING IN PSI
REGO PRV9433[ ][ ][ ]PSI
[ ] BAR [ ]
DATE CODE
EQUIVALENT PRESSURE SETTING IN BAR

(2.625)

.50 (SEE NOTE 3)

1 2 3 4

| COL 1 | COL 2 | COL 3 | COL 4 | COL 5 |
|---|---|---|---|---|
| ASSEMBLY | RANGE (PSIG) | SPRING | SPRING GUIDE ASSEMBLY | BODY |
| PRV9433F-A | 10-39 | BX250-025 | PRV250-10 | PRV375-1 |
| PRV9433F-B | 40-89 | BX250-065 | PRV250-10 | PRV375-1 |
| PRV9433F-C | 90-139 | BX250-115 | PRV250-10 | PRV375-1 |
| PRV9433T-D | 140-199 | BX250-180 | PRV250-4 | PRV375-1 |
| PRV9433T-E | 200-299 | BX250-260 | PRV250-4 | PRV375-1 |
| PRV9433T-F | 300-379 | BX250-340 | PRV250-4 | PRV375-1 |
| PRV9433T-G | 380-459 | BX250-420 | PRV250-4 | PRV375-1 |
| PRV9433T-H | 460-550 | BX250-500 | PRV250-4 | PRV375-1 |
| PRV9433FP-A | 10-39 | BX250-025 | PRV250-10 | PRV375P-1 |
| PRV9433FP-B | 40-89 | BX250-065 | PRV250-10 | PRV375P-1 |
| PRV9433FP-C | 90-139 | BX250-115 | PRV250-10 | PRV375P-1 |
| PRV9433TP-D | 140-199 | BX250-180 | PRV250-4 | PRV375P-1 |
| PRV9433TP-E | 200-299 | BX250-260 | PRV250-4 | PRV375P-1 |
| PRV9433TP-F | 300-379 | BX250-340 | PRV250-4 | PRV375P-1 |
| PRV9433TP-G | 380-459 | BX250-420 | PRV250-4 | PRV375P-1 |
| PRV9433TP-H | 460-550 | BX250-500 | PRV250-4 | PRV375P-1 |

| ITEM | QTY. | DESCRIPTION | NUMBER |
|---|---|---|---|
| 4 | 1 | ADJUSTING SCREW | PRV250-3 |
| 3 | 1 | SPRING | SEE COL 3 |
| 2 | 1 | SPRING GUIDE ASSEMBLY | SEE COL 4 |
| 1 | 1 | BODY | SEE COL 5 |

DRAWN T. TICKLE
ENGINEER J. OLSEN
CHECKED D. WILLIAMS
APPROVED J. OLSEN
ORIGINAL DATE 05/25
PLOT GENERATED: 08/02

REGO® CRYO-FLOW PRODUCTS
BURLINGTON, N.C. 27215

REVISIONS
A A206, 05/25 ISSUED

UNLESS OTHERWISE SPECIFIED:
▼ ± .015
2 PL DEC. ± .02
3 PL DEC. ± .005
4 PL DEC. ± .0005
ANGLES ± 1°
CONCENTRICITY .010 FIM
FINISH 125 MAX
REMOVE BURRS, BREAK SHARP EDGES R.015 MAX
INTERPRET DIMENSIONS & TOLERANCES IAW ASME Y14.5M-1994

THIS DRAWING IS ECII PROPRIETARY INFORMATION
ALL RIGHTS OF DESIGN OR INVENTION ARE RESERVED

TITLE: PRV BLANK

SCALE (2:1)
DRAWING NO. PRV009433XX-X
DWG SIZE B

**PR 16-1.**
Print supplied by RegO Cryo-Flow Products.

## Industry Print Exercise 16-2

*Refer to the print PR 16-2 and answer the questions below.*

1. How many assemblies are described by this one assembly drawing? ____________

2. The catalog number for the assemblies described by this drawing all begin with KM50-MCLN. List the unique numbers that finish the catalog numbers for each assembly: ____________

3. Some of the assemblies described by this drawing are left hand and some are right hand. Which of the two is the way the drawing is represented? ____________

4. Is STEEL BODY 224292-01 a left-hand or right-hand body? ____________

5. List the find numbers of the parts that are common to all assemblies. ____________

6. In inches, what is the largest overall dimension of this object? ____________

7. What is the part number and description of the part associated with find number 6? ____________

8. In general, what is the name of this series of assemblies? ____________

9. Give a one word description of the items that must be purchased separately. ____________

10. List the three angular measurements that are important to this assembly. ____________

*Review questions based on previous units:*

11. What do the numerical values in brackets indicate? ____________

12. What size of paper is the original drawing? ____________

13. What former value was specified for the 7° angular measurement? ____________

14. What are the initials of the person who approved adding the KM logo? ____________

15. Are the reference dimensions on this print shown in parentheses? ____________

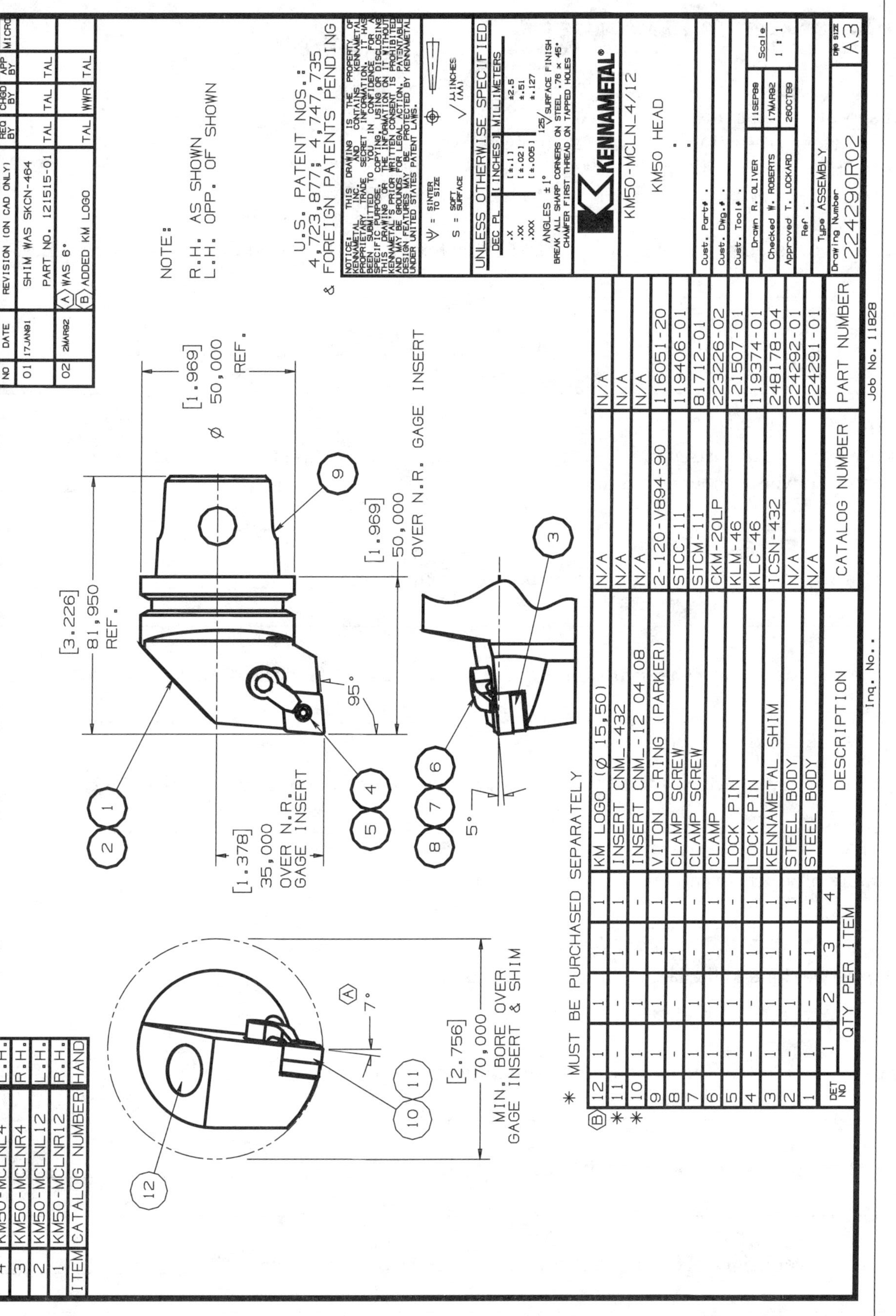


PR 16-2.
Print supplied by Kennametal.

## Bonus Print Reading Exercises

The following questions are based on various bonus prints located in the folder at the back of this textbook. Refer to the print indicated, evaluate the print, and answer the question.

**Print AP-004:**

1. Of the types of assembly drawings discussed in the text, which type best describes this drawing?
2. Name the three parts indicated on this drawing.

**Print AP-010:**

3. Carefully examine this assembly drawing. Within each identification bubble, there is an upper number and lower number. What does each number stand for?
4. Do the dimensions on this drawing apply before assembly or after?
5. On this print, the assembly is sectioned. Are there any of the parts that have *not* been sectioned?

**Print AP-016:**

6. Give the total number of parts (not number of different parts) required for this assembly.
7. How many products or models are covered by this one assembly drawing?
8. What must be done to the label before it is attached?
9. Give the part names of the two parts that are different, depending on the model number.
10. How is item 5 to be positioned in relationship to item 13?
11. What is the total height of the assembly, as shown on the drawing?

**Print AP-017:**

12. How many products are covered by this one assembly drawing? ________________

13. Why is item 5 not shown or identified in the views of the drawing? ________________

14. Are items 1 and 2 both visible in the top view? ________________

15. Why is item 4 not shown? ________________

16. If you think of the top view as a clock, there is one insert shown at 6 o'clock. How many inserts are there on the actual manufactured milling cutter? ________________

**Print AP-018:**

17. This assembly print shows one insert position at 6 o'clock in the top view. How many inserts are there on the actual manufactured product? ________________

18. What items are included in the parts list that are not shown in the views? ________________

Fasteners come in all shapes, sizes, and types. Here, two types of specialty fastener are shown.

UNIT 17

# Springs and Fasteners in Industrial Prints

*After completing this unit, you will be able to:*

**Define** terms related to springs in industrial applications.

**Define** terms related to fasteners used in industrial applications.

**Identify** types of threaded fasteners (cap screws, machine screws, etc.) shown on industrial prints.

**Identify** types of nonthreaded fasteners (pins, keys, rivets, etc.) shown on industrial prints.

**Read** specifications related to springs and fasteners specified on a print.

Within the realm of product design, designers have a vast supply of standard products from which to choose when assembling the components of a product. These objects include springs, pins, keys, and other fasteners such as nuts, screws, and bolts. Although springs are items often purchased to complete an assembly of parts, sometimes the designer needs to custom design a spring for a particular application. In this unit, springs and fasteners are presented as objects that often do not need a detail drawing. However these items are visually drawn and represented in an assembly drawing, so the well-trained print reader should be familiar with these items to better read the drawings and specifications for a product.

ASME standards for springs and fasteners focus on the complete dimensional and general data associated with these parts, including strength and accuracy specifications. In general, they do not focus on the drawing of the parts, although thread representation is covered in the ASME Y14.6, as presented in Unit 8. In general, drawings of the parts are often simplified to represent the fastener or spring and sometimes do not represent the accurate production measurements. Since these parts are available for purchase, most of them are available for CAD systems in symbol libraries provided on the Internet by fastener vendors or CAD bulletin boards. Some common ASME standards that define the size and characteristics of fasteners are:

- B17.1, *Keys and Keyseats*
- B17.2, *Woodruff Keys and Keyseats*
- B18.2.1, *Square and Hex Bolts and Screws*
- B18.3, *Socket Cap, Shoulder, and Set Screws, Hex and Spline Keys*
- B18.5, *Round-Head Bolts*
- B18.8.1, *Clevis Pins and Cotter Pins*
- B18.8.2, *Taper Pins, Dowel Pins, Straight Pins, Grooved Pins, and Spring Pins*
- B18.25, *Square and Rectangular Keys*
- B18.27, *Tapered and Reduced-Cross-Section Retaining Rings*

In summary, multiple ASME standards have been established on behalf of springs and fasteners. As listed above, the B18 series of standards includes separate documents for lock washers, round-head bolts, rivets, square-neck bolts, hex nuts, and machine screws. ASME B18.12 is entitled *Glossary of Terms for Mechanical Fasteners.* ASME Y14.3, *Mechanical Spring Representation,* was discontinued in 2006, but many companies still refer to that standard as good practice for drawing springs.

## Springs

A spring is a device that is designed to either compress or expand due to a force and, in doing so, will counteract that force with an equivalent force. Springs are made from wire material. Companies that produce springs often also create other wire forms for various applications. See Figure 17-1.

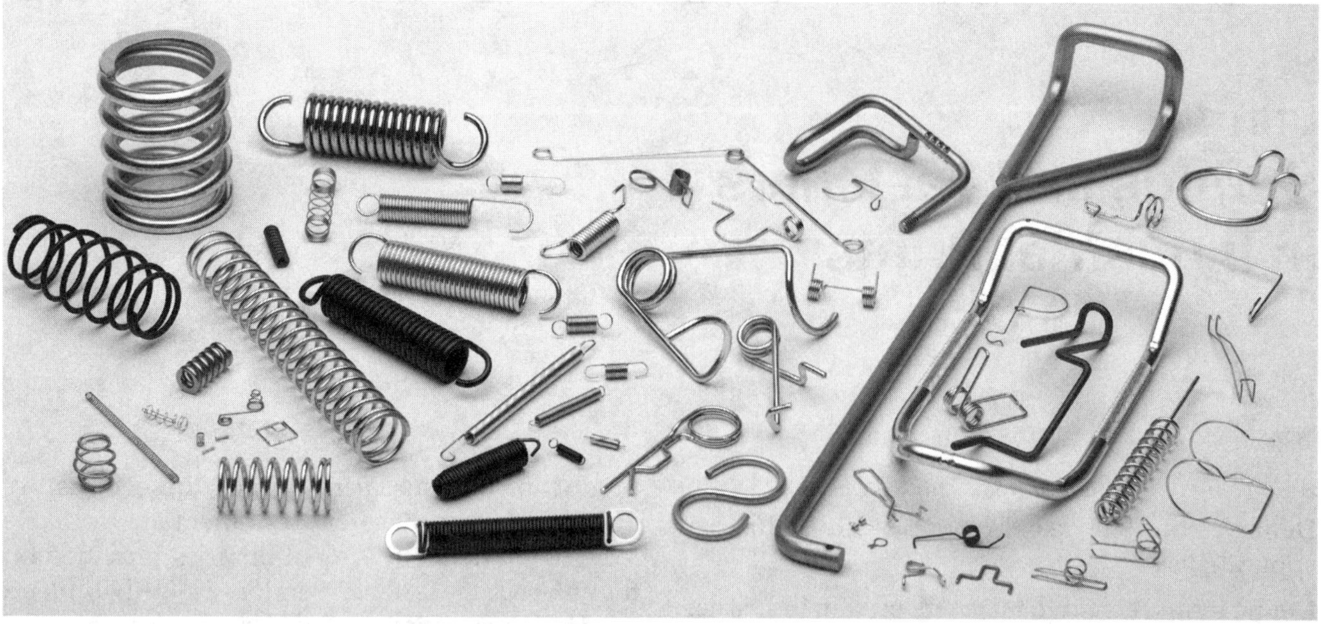

Figure 17-1.
Springs are used to help maintain or control certain relationships between other parts. This figure also shows other parts that can be made from wire or rods. (Master Spring and Wire Form Co.)

## *Types of Springs*

A common example of a spring is a helical-wound coil of wire that, if pulled, will stretch to be longer. The coil of wire will return to its original, shorter shape if the pulling stops. This type of spring is called an ***extension spring,*** Figure 17-2. The springs are characterized by coils that touch each other in the free position.

A ***compression spring*** is a coil of wire designed to hold its shape with the coils separated from each other in the free position, as if it is already stretched out. See Figure 17-3. A force is required to compress the coils and the spring exerts a force that attempts to return the coil to the longer length.

Figure 17-2.
Extension springs are used to help pull parts toward each other. (Master Spring and Wire Form Co.)

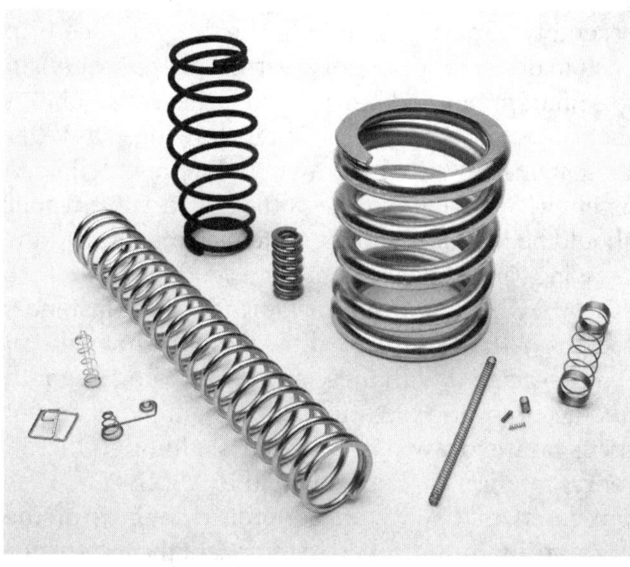

Figure 17-3.
Compression springs are used to push parts away from each other. (Master Spring and Wire Form Co.)

Extension springs and compression springs are designed to have force applied in the same direction as the axis of the spring coil. ***Torsion*** springs are designed to be used in situations where the force is at a right angle to the axis of the coil of wire. Torsion springs are wound either in a helical or spiral form and are commonly used on parts that rotate about an axis. These springs can be used to apply force to the rotation of one part about an axle or pivot pin. See Figure 17-4.

Other springs found in industrial assemblies come in the form of ***spring washers.*** These are flat springs fabricated from thin metals that can hold a bowed shape, thus exerting a force if another part attempts to flatten it. ***Garter*** springs are long extension springs designed to encompass a circular feature in a belt-like manner. See Figure 17-5.

The following terms are related to springs. Study Figure 17-6 as you examine these terms to see how each may be applied.

- **Active coils.** For a compression or extension spring, the windings in the helical shape that are subject to move when the spring is under force to stretch or compress. Some coils on the end of an extension spring are not active, but rather form the end of the spring for connecting to other objects.
- **Diameter, average.** The diameter average between ID and OD, usually the diameter measured from the center axis of the wire across the axis of the spring from side to side.

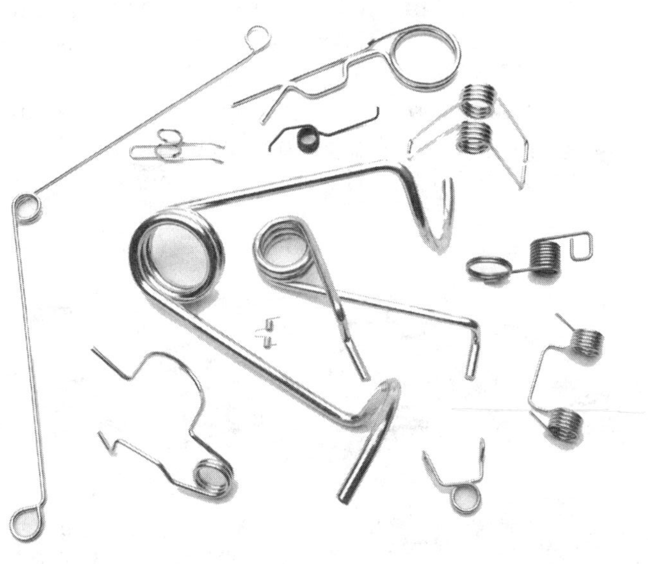

**Figure 17-4.**
Torsion springs can be used to help provide tension or force around an axis. (Master Spring and Wire Form Co.)

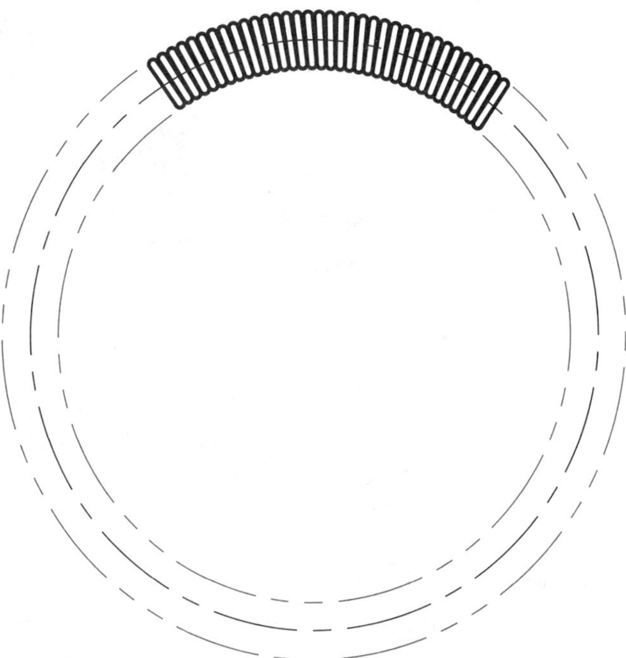

**Figure 17-5.**
Garter springs are long extension springs designed to encompass a circular feature in a belt-like manner. The ends have either hooks, one end can be tapered to fit within the other end, or ends can be connected with internal inserts.

- **Diameter, inside (ID).** The diameter on the inside of the coil of a compression or expansion spring.
- **Diameter, outside (OD).** The diameter on the outside of the coil of a compression or expansion spring.
- **Free length.** The overall distance from one end of the spring to the other when no force is being applied.
- **Handed.** The helical winding direction; either left handed (LH) or right handed (RH). This may not have an impact on the design of the spring unless other parts are inserted into the spring.
- **Load.** The force applied to the spring in a particular application.
- **Pitch.** The distance from the center of one wire to the center of the next wire in the active coils. For an extension spring, the pitch is usually equal to the wire diameter. Specifying the number of coils is more common than specifying the pitch.

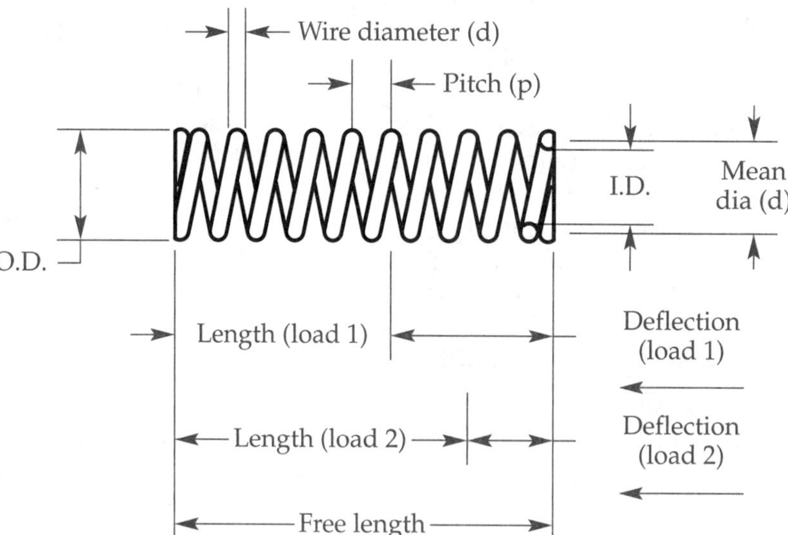


**Figure 17-6.**
Terms related to springs are illustrated in this compression spring example.

## *Drawing Representation*

Before the advent of CAD systems, it was tedious to draw every coil of a spring. Therefore, it was common to only draw each end of the spring, including one or two coils. The ends of the spring were then connected with phantom lines, as shown in Figure 17-7. Another technique is to draw the spring as a schematic single-line representation, still showing the number of coils, also shown in the figure.

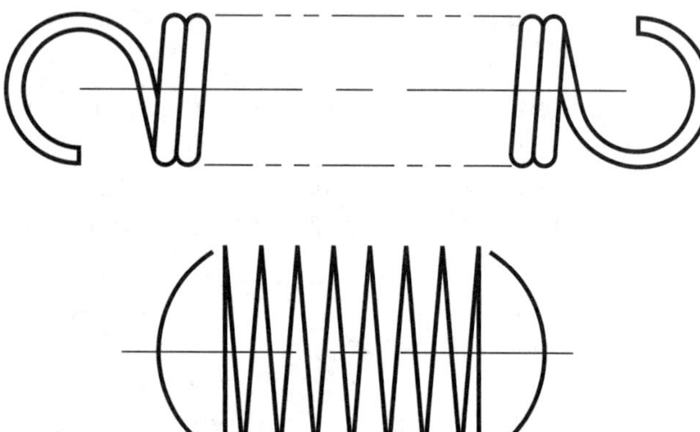

**Figure 17-7.**
In the past, phantom lines were commonly used to show repetitive detail like that found in spring drawings or springs were drawn in a schematic format. With CAD systems, these techniques are not as common.

Most CAD systems can easily duplicate entities. This has made the phantom-line method unnecessary, but it can still be used to keep the number of lines to a minimum. Garter springs use circular phantom lines to assist the drafter in showing the spring without complex details.

## *End Styles*

When drawing compression springs, the type of end finish is usually shown. Figure 17-8 shows the common end styles. The ***plain end,*** also called ***open end,*** style simply features the wire ending after a specified number of coils, spaced evenly based on the pitch. The ***square end,*** also called ***closed end,*** style changes the pitch on the last coil in such a way that the end of the spring is somewhat perpendicular to the axis without grinding. The other options for

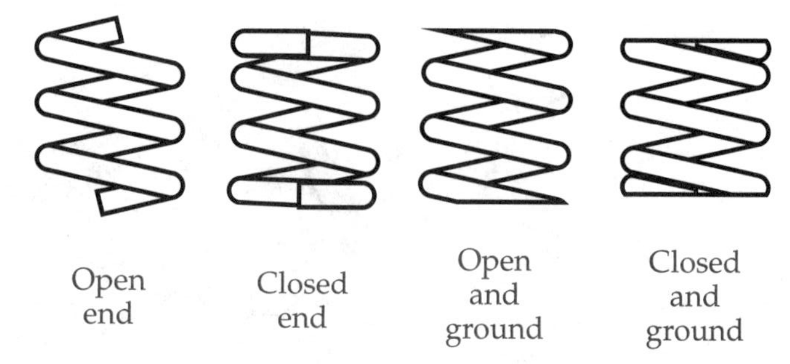


**Figure 17-8.**
Compression spring end styles include either open (plain) or closed (square), each either ground or not ground.

end styles are based on whether or not a grinding process is applied to create a more perpendicular surface for the spring. ***Closed and ground*** and ***open and ground*** are two additional end styles that can be specified when ordering springs from most spring manufacturers.

For extension springs, the hook type is often drawn at each end. As shown in Figure 17-9, hook types include machine hook, full round, full loop, double loop, loop on side, and extended hook. Garter springs may be connected end-to-end with a tapered end, a connection, or loops.

## *Typical Specifications*

Depending on the application, the typically specified spring data may include the wire material, wire diameter, outside diameter, inside diameter, number of coils, and free length. For extension springs, the free length is often clarified to be "inside hooks." The gap between the hook tip and the body length can also be specified. See Figure 17-10. For compression springs, a specification similar to "works inside a ____ diameter hole" or "works outside a ____ diameter shaft" can be indicated. For expansion springs, an initial tension force can be specified to indicate the conditions under which the spring will be at rest. For extension and compression springs, it is common to specify load lengths, such as "length at load 1 = ____", "length at load 2 = ____", etc. For compression springs, a solid height can be specified. Sometimes special requirements are specified related to temperature, chemical exposure, or finish.

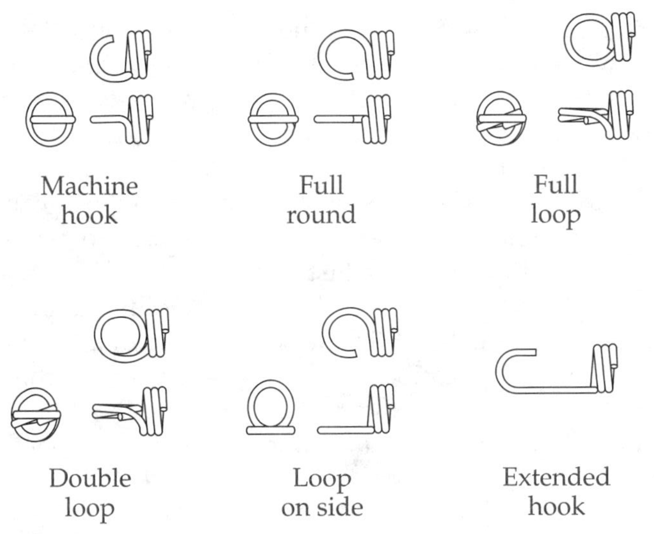


**Figure 17-9.**
Extension springs can be made with a variety of ends for attaching to other parts.

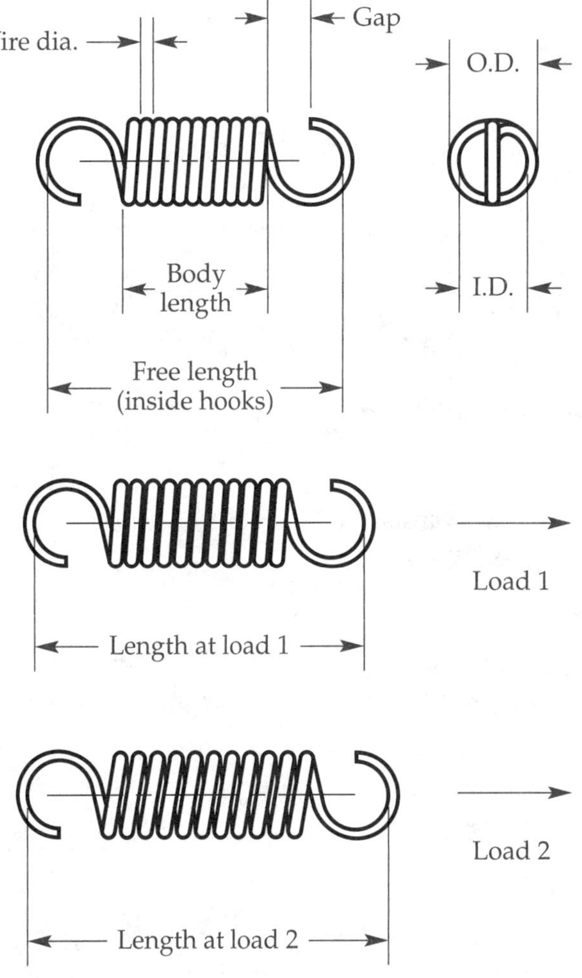


**Figure 17-10.**
Terms related to extension springs are illustrated in this example.

Torsion springs can be drawn to show various specifications. The drawing usually includes the free state leg angle between the two legs, the installed deflection, and the maximum deflection. See Figure 17-11.

***Spring washers*** are special washers designed to exert force as the threaded fasteners are tightened against them. Common styles of spring washers include the basic curved washer, wave washer, and Belleville washer. See Figure 17-12.

## Threaded Fasteners

The most common way of fastening two or more parts together is with threaded fasteners, such as nuts and bolts. This section discusses the terminology associated with threaded fasteners. Also discussed are the types of threaded fasteners.

Wire dia.
A
C
B
I.D.
O.D.
L1
L2
Body length

A = Leg angle (free state)
B = Installed deflection
C = Max. deflection
L1 = Length of leg 1
L2 = Length of leg 2

**Figure 17-11.**
Terms related to torsion springs are illustrated in this example.

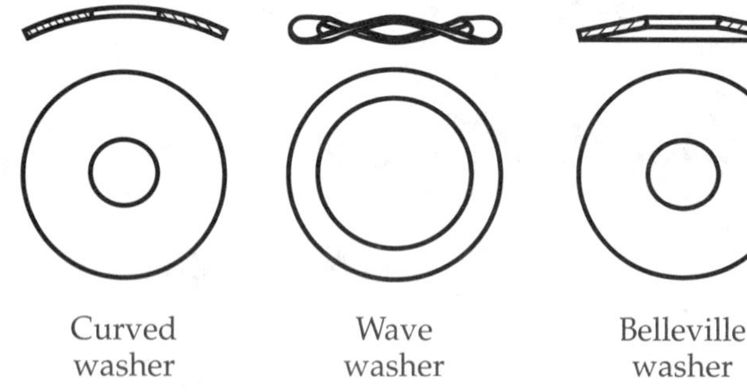


**Figure 17-12.**
Spring washer options include curved, wave, and Belleville.

## General Terminology

Terminology for threaded fasteners is not as standardized across the industry as you would expect. There are arguments for and against many definitions. Within the scope of this textbook, the term ***bolt*** is used as a general term for a uniform-diameter cylinder that has external threads on one end and a head on the other. A ***stud*** is a rod that has external threads on both ends, Figure 17-13. The threads may or may not be the same on each end. Typically, one end of a stud may be screwed into a threaded hole, leaving a threaded post sticking up that accepts a nut. In general, a ***nut*** is a device with internal threads designed to fit on the threaded end of a bolt or stud.

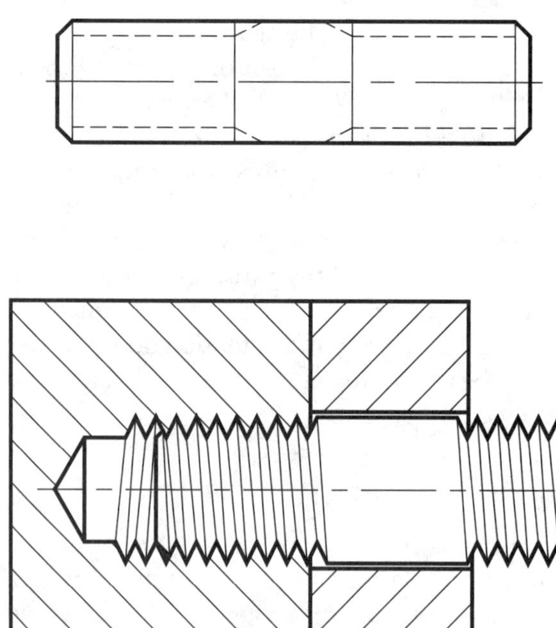

**Figure 17-13.**
A stud is a cylindrical fastener with threads on both ends, shown here in the simplified (upper) and detailed (lower) methods. In the detailed view, the stud is shown assembled in a blind tapped hole with a cover plate and clearance hole.

A universal distinction between a bolt and a screw does not exist. One of the common terms used throughout the years is ***cap screw,*** which usually applies to bolts that are finished. Cap screws may have hexagonal heads with built-in washer surfaces or they may have other head types, as discussed in the next section. At one time, the term cap screw indicated the body was completely threaded from the head to the tip. The term ***machine screw*** is similar in definition to a cap screw, but the term is usually applied to smaller fasteners of 1/4" diameter or less.

Both cap screws and machine screws can accept nuts on the end so as to hold two parts together in a "floating fastener" manner. They can also be used through a clearance hole in one part into a threaded hole in the second part in a "fixed fastener" application. Figure 17-14 illustrates, from left to right, a bolt and nut; a cap screw in a blind, tapped hole; and a machine screw in a tapped through hole.

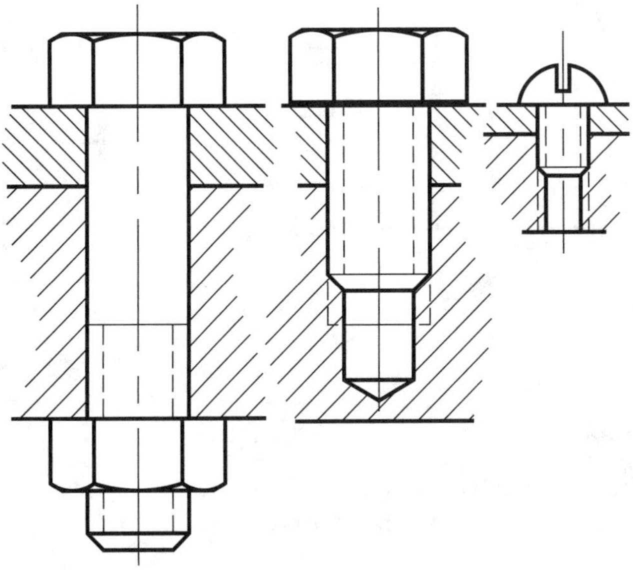

**Figure 17-14.**
Bolts typically have a head on one end. Often, a nut is threaded onto the other end. Cap screws are bolts that can also be screwed into a threaded hole. Machine screws are smaller screws that may or may not use a nut.

## Head Types

There are a wide variety of head styles for cap screws and machine screws, Figure 17-15. The most common style is the hexagon head, which may or may not include a washer surface underneath the head. Flat head and round head fasteners with a slot are shown looking through the slot. If a round view or top view is drawn, the slot is rotated 45°, even though the other view is "looking through" the slot. This is a common conventional practice that ignores true projection for the sake of clarity. Other common head styles include the fillister, hex socket, and oval.

## Nut Types

The common ***hexagon nut*** is the item that mounts on the end of a bolt to provide the "second head." By tightening the nut, the fastener remains in place. Nuts are identified by the bolt body size, not the distance across the flats of the square or hexagon head. If hexagon nuts are finished, there is a slight chamfer on one or both sides. A heavy nut is thicker than a standard nut.

In addition to the hexagon nut, there are several variations designed to function in particular ways. As illustrated in Figure 17-16, these include the flange, jam, slotted, and square. The ***flange nut*** has a built-in washer surface that provides the same function as a washer. ***Jam nuts*** can be used in situations where a regular nut is too thick, but also as a second nut that tightens against the other nut to lock it in place. ***Slotted nuts*** have slots that accommodate cotter pins that can be placed through a hole in the bolt. ***Square nuts*** are not as common as hexagon nuts, but are available in various sizes. The square nut requires an open-end wrench, but provides more bearing surface for the wrench than the hexagon nut. ***Hex coupling nuts*** and ***sleeve nuts*** are designed to connect two pieces of threaded material. The length of a coupling nut is approximately three times the nominal size.

## Screws

***Set screws*** are designed to help hold parts in place. For example, a set screw can be used to hold the hub of a pulley onto a shaft by tightening the set screw into a flat area on the axle. Figure 17-17 shows an example of a set screw used as a stop pin. While set screws are often headless, as with the common

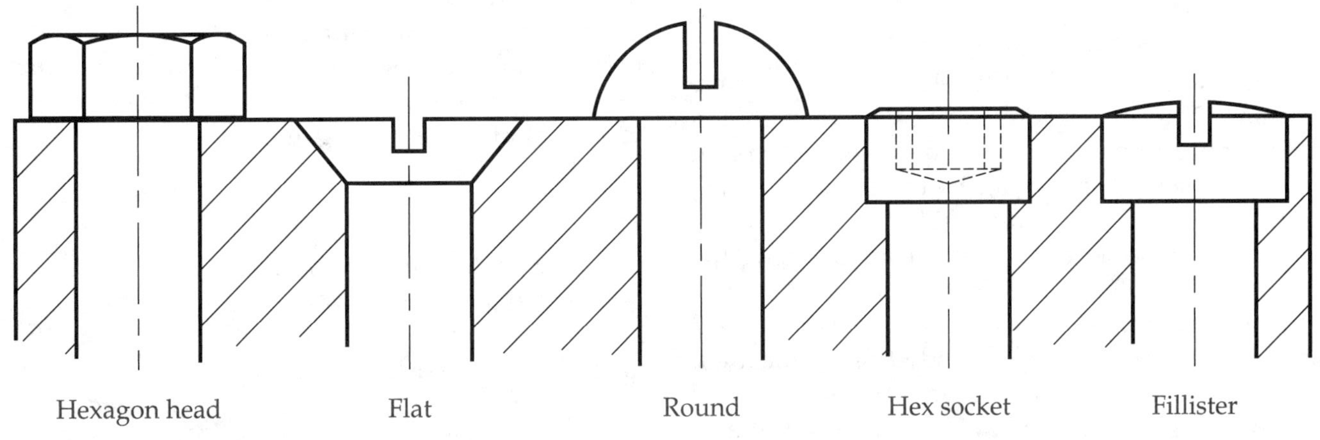


**Figure 17-15.**
Machine screws and cap screws are available in a variety of head shapes

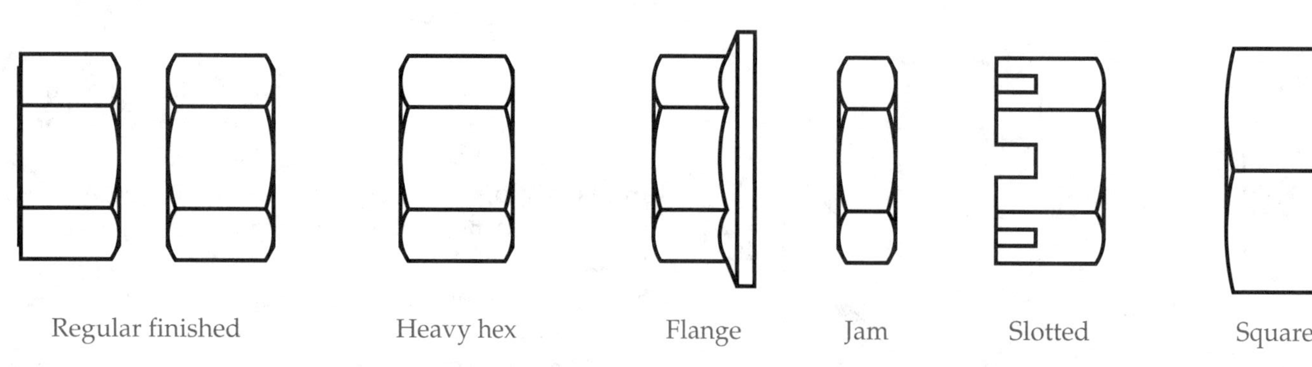


**Figure 17-16.**
Nuts are available in a variety of shapes and for a variety of purposes.

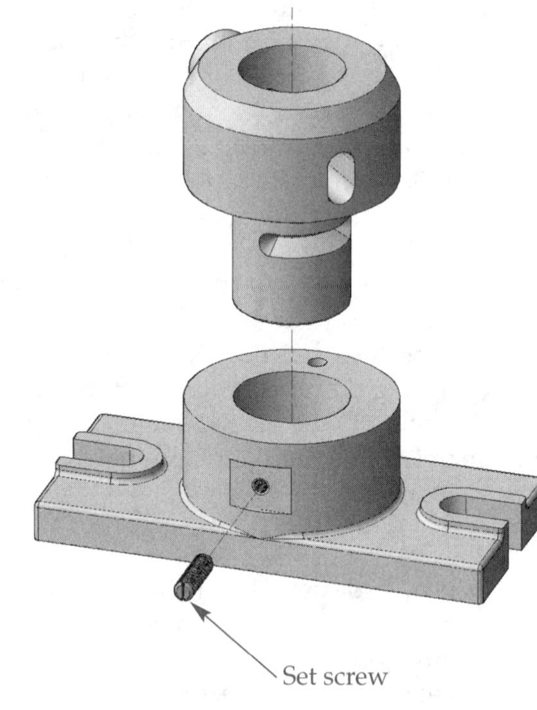


**Figure 17-17.**
Set screws can be used to help hold one part to another. In this illustration, the set screw serves as a stop pin that allows a pivoting part with a mating groove to rotate between limits.

socket head or slotted head, a set screw can also have a square head. Figure 17-18 shows four common types of head for set screws and six common point types. Set screw points can be categorized as flat, cup, oval, cone, half dog, and full dog.

There is some disagreement about using the word "screw" for a nontapered, threaded body. In some standards, the word "screw" is proposed as a definition for a tapered body that is threaded, but also designed to cut its own threads. These are sometimes referred to as ***tapping screws***, but also include such fasteners as a ***wood screw*** or a ***lag screw***, both of which might require a pilot hole.

## Nonthreaded Fasteners

The previous sections discussed threaded fasteners. This section discusses nonthreaded fasteners. Included in this category are washers, rivets, pins, keys, and retaining rings.

### Washers

***Washers*** are thin, flat cylinders that have a clearance hole for a fastener. They are usually made of metal and are designed to work in conjunction with bolts, nuts, pins, and other fasteners. See Figure 17-19. They help distribute the pressure of the tightened fastener across more surface area and also may be used to fill in space.

***Flat washers*** are simply rings of metal with a clearance hole for the fastener. Fender washers have an outside diameter larger than a typical flat washer to help spread the tightening forces even more. ***Spring lock washers*** are designed with a split in the washer and a slight helical warp to the washer. This design feature places tension against the nut as the nut is tightened, which helps prevent it from loosening on its own. Some ***tooth lock washers*** are designed with external or internal teeth that provide locking tension.

### Rivets

A ***rivet*** is a mechanical fastener designed to be permanent after installation. A ***solid rivet*** typically has one head and a smooth, cylindrical body that is malleable. After the rivet is inserted through the materials being connected, the opposite end of the rivet can be formed to also have a head. The new head of the solid rivet can be formed with a hammer

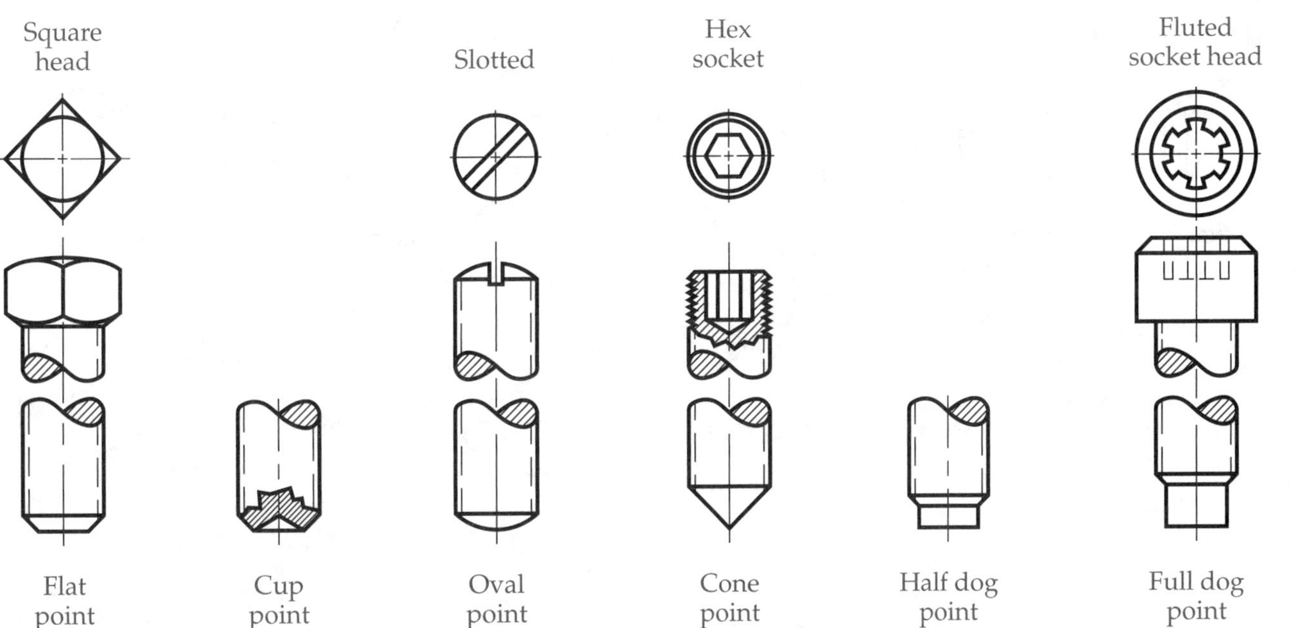


**Figure 17-18.**
Set screws are available in a variety of head and tip options.

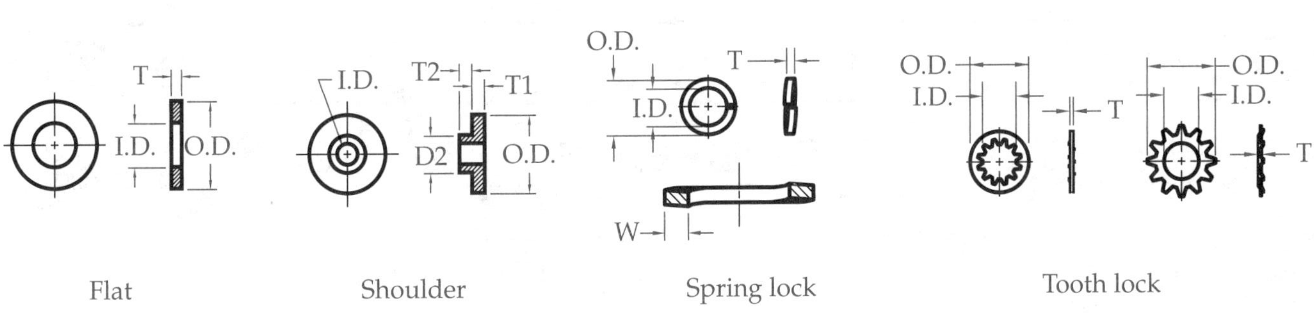


**Figure 17-19.**
Washers are available in a variety of sizes and shapes, each with its own purpose.

or compression tool. Figure 17-20 shows various solid rivet heads that are available today and the standard print expression and line techniques for indicating how rivet heads are to be finished.

***Blind rivets,*** or ***pop rivets,*** are designed to fasten thin materials when the installer can only access one side of the riveted hole. A special tool is used to pull the shaft of the rivet through a hollow center in the rivet, which is not as strong as a solid rivet, to form the second head.

## Pins

Some parts are assembled and then fastened with pins. Pins are classified in various ways, including tapered, straight, spring, grooved, clevis and cotter. See Figure 17-21. ***Tapered pins*** taper 1/4″ per foot (1:48) on the diameter and are classified by number. A #0 taper pin is the smallest and a #14 is the largest. The large end of the taper pin is constant by the number of pin, but the pin length determines the diameter of the small end. Drilling the hole in steps or using a tapered reaming tool is necessary to get a good fit.

***Straight pins*** are more difficult to assemble. Tight tolerances are required to ensure a good fit so the pin does not fall out. ***Spring pins*** are rolled and have a slot that allows for expansion and contraction, thus assuring a more secure fit. In similar fashion, ***grooved pins*** have grooves pressed into the cylindrical or tapered bodies of the keys that are designed to help the pin remain securely in place. Vendors of grooved pins usually have a wide selection of groove types, each with certain advantages. Many pins, including grooved pins, are designed to require a press fit.

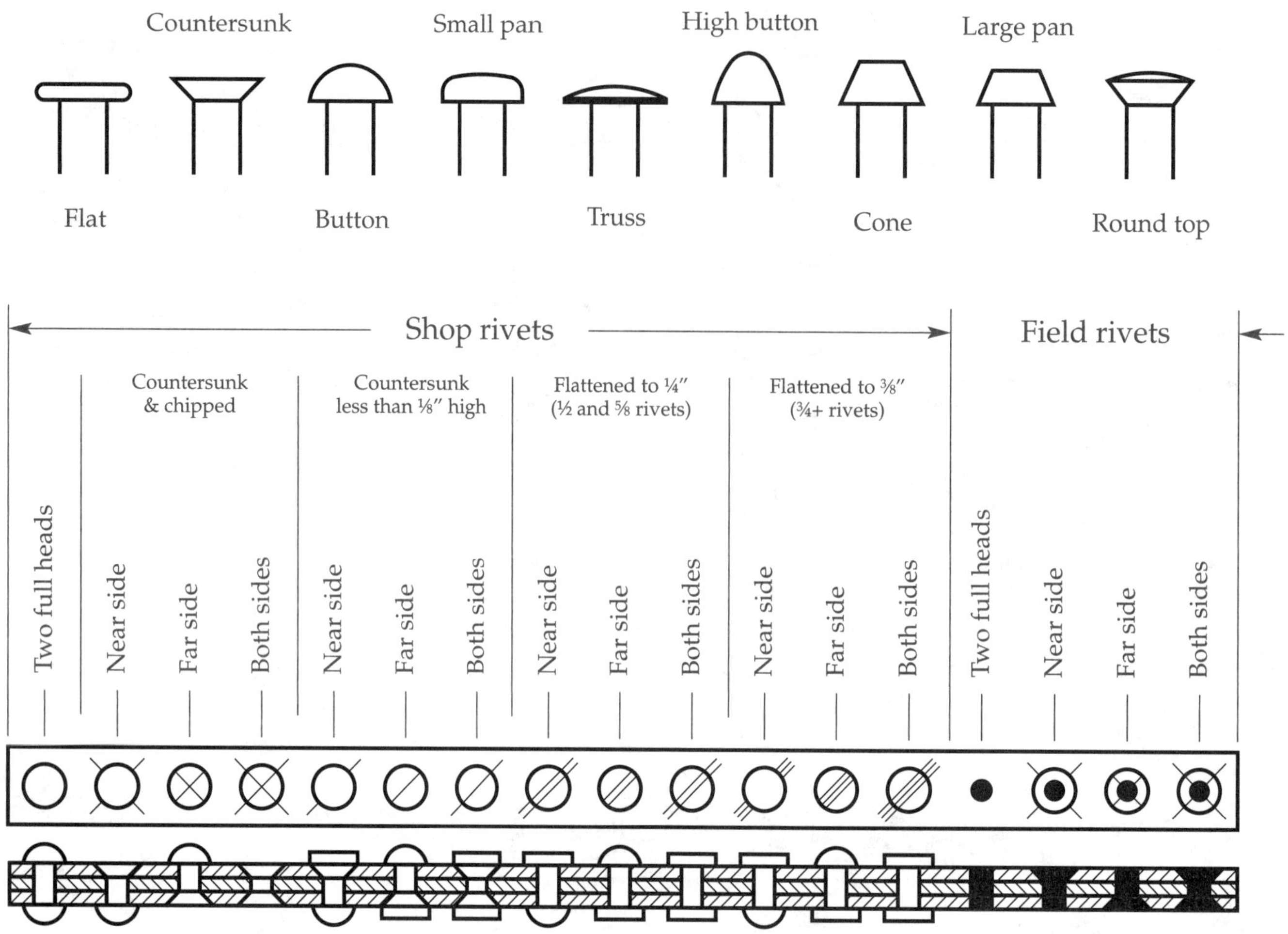

Flat, Countersunk, Button, Small pan, Truss, High button, Cone, Large pan, Round top

Shop rivets

Two full heads

Countersunk & chipped: Near side, Far side, Both sides

Countersunk less than ⅛″ high: Near side, Far side, Both sides

Flattened to ¼″ (½ and ⅝ rivets): Near side, Far side, Both sides

Flattened to ⅜″ (¾+ rivets): Near side, Far side, Both sides

Field rivets

Two full heads, Near side, Far side, Both sides

**Figure 17-20.**
Solid rivets are available in a variety of head shapes. On a print, 45° lines are used in a variety of combinations to indicate various finish techniques.

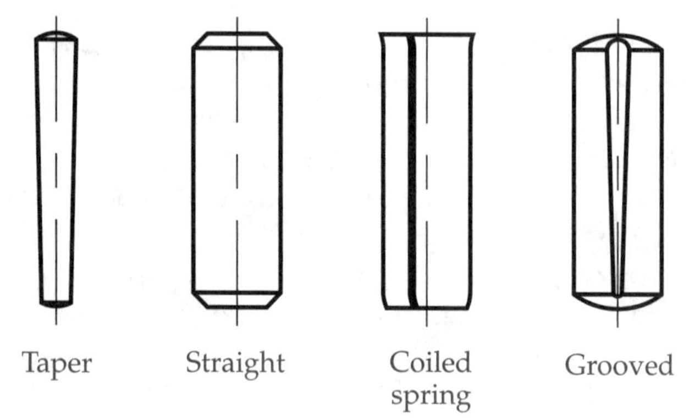


**Figure 17-21.**
Pins are available in a variety of sizes and shapes, each with its own purpose.

***Clevis pins*** are designed for quick assembly or disassembly. On one end, the clevis pin has a head and on the other end is a hole designed to accept a cotter pin. See Figure 17-22. Clevis pins come in a range of diameters and lengths. The important dimensions of a clevis pin include the main shank diameter, the distance from the head to the cotter pin hole, and the diameter of the cotter pin hole.

***Cotter pins*** are made of a folded strip of metal, which provides a split end to the pin that can be flared by separation. The folded end of the pin bulges enough to form a head. The tips of the split end are offset enough to be able to grab the ends and bend them with pliers or other tools. Cotter pins are used in conjunction with clevis pins and slotted nuts. They are also available in shapes that do not require flaring, Figure 17-23.

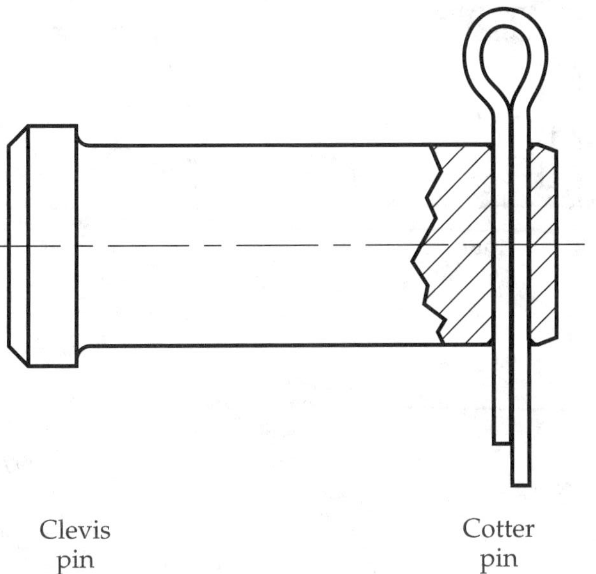


**Figure 17-22.**
The clevis pin has a head on one end and a hole for a cotter pin on the other end of the shaft.

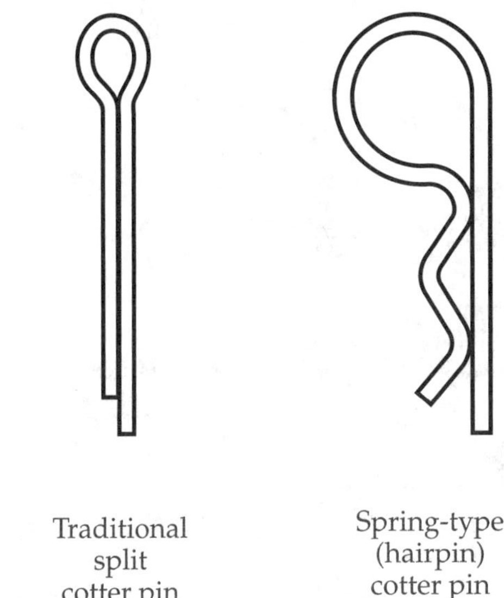


**Figure 17-23.**
A traditional cotter pin features a split end that can be flared by separation. Another type of cotter pin is the hairpin, or spring, type that can be snapped into place.

***Dowel pins*** are another type of pin common in industry. These heat-treated and precision-made pins are used more in subassemblies than as fasteners. Dowel pins are designed to be press fit into holes to create a boss or guide pin that helps parts line up for fasteners. For example, a base or container may have two dowel pins press fit into the edge that protrude to help align the cover, with clearance holes that not only fit over the pins, but additional clearance holes for threaded fasteners. The dowel pins are usually about the same diameter as the fasteners used in the other holes, and are usually designed to have twice the length of the diameter embedded into each part. Dowel pins usually have a small chamfer on each end, although some are designed with spherical ends.

## Keys

Keys are available in a variety of shapes and forms, Figure 17-24. ***Keys*** are primarily used to keep a part from rotating about the shaft to which it is mounted. For example, a wheel or pulley is often mounted to a shaft and a key is inserted into grooves designed into the hub of the wheel and the shaft. The grooves are designed to accommodate the particular key that it used. A square or rectangular key can be placed into a groove in the shaft that is long enough to accommodate the key. The slot in the hub is then aligned to slide over the key.

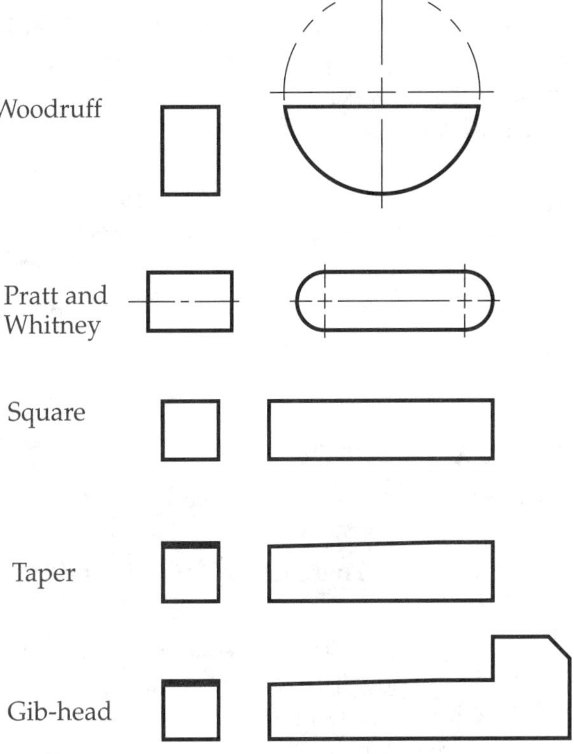


**Figure 17-24.**
Keys are available in a variety of sizes and shapes, each with different characteristics. Tapered and gib-head keys are designed to be inserted into the key groove after the hub is mounted on the shaft.

***Woodruff keys*** are somewhat semicircular, so the keyseat is cut into the shaft with a Woodruff cutter. The ***tapered key*** and ***gib-head key*** are both designed to be inserted into the keyway after the hub has been mounted over the shaft. Both of these keys taper at 1/8″ per foot. The hub length is designed to engage in the mating parts in such a way that the key can be grabbed and removed. Other key shapes are available through various vendors. These may be named with trademark names or with general descriptive names.

## Retaining Rings

***Retaining rings*** are used in assemblies to help parts stay together. They are somewhat permanent, although they are designed in such a way that they can be removed with special tools. Retaining rings typically are inserted into grooves. The groove may be external with the retaining ring fitting into a groove near the end of a shaft. The groove may also be internal with the retaining ring snapping into the groove near the end of a hole. See Figure 17-25.

The most common retaining rings contain tiny holes on the two ends to allow special tools to expand or compress the rings. Other retaining rings include snap style rings designed to function in a similar fashion, but without the ears of traditional retaining rings. E-style rings are also available and can be found in a variety of sizes.

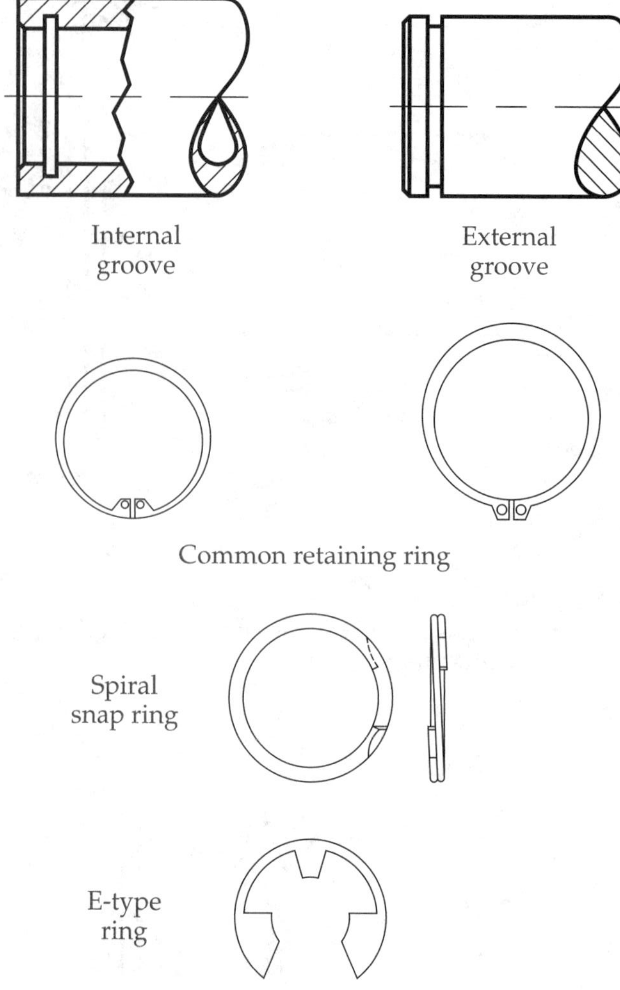


**Figure 17-25.**
Retaining rings commonly have holes for a tool, but are also available in styles that snap in and out with other techniques.

## Review Questions

*Circle the answer of choice, fill in the blank, or write a short answer.*

1. *True or False?* There is one ASME standard for all springs and fasteners.

2. Which of the following is *not* a type of spring?
   A. Extension spring.
   B. Circular spring.
   C. Torsion spring.
   D. Compression spring.

3. The overall distance from one end of a spring to the other when no force is being applied is called the ______________________.

4. How many end styles are typically available for compression springs? ____________________

____________________________________________________________

5. Which of the following typically would *not* have a mating nut?
   A. Cap screw.
   B. Machine screw.
   C. Stud.
   D. Wood screw.

6. Which of the following is *not* a head type for a cap screw?
   A. Pot.
   B. Hex socket.
   C. Flat.
   D. Oval.

7. ______________________________ nuts are thinner than regular nuts and heavy nuts.

8. Which of the following is *not* a set screw point type?
   A. Cone.
   B. Hex socket.
   C. Flat.
   D. Oval.

9. ______________________ washers are designed with teeth that provide locking tension.

10. *True or False?* Rivets come in three forms: solid, blind, and threaded.

11. Which of the following is *not* a type of pin fastener?
    A. Tapered
    B. Grooved
    C. Woodruff
    D. Spring

12. *True or False?* Clevis pins have a head on one end and a hole on the other end designed to accommodate a cotter pin.

13. What type of pin is typically used as an alignment feature rather than a fastener? ____________

____________________________________________________________

14. Which of the following is *not* a type of key?
    A. Snap.
    B. Square.
    C. Woodruff.
    D. Gib head.

15. *True or False?* One way to categorize retaining rings is external versus internal.

## Industry Print Exercise 17-1

*Refer to the print PR 17-1 and answer the questions below.*

1. What type of head does this fastener have? ______________________

2. Based on the thread specification, what is the nominal diameter of this fastener? ______________________

3. Is the across-the-flats measurement of the head 1.5 times the nominal body diameter? ______________________

4. The bearing surface of the flange on the head has a minimum diameter of ______________________.

5. What is the total maximum length of this fastener head tip–to–body tip? ______________________

6. What is to be applied to a portion of the threads of this fastener? ______________________

*Review questions based on previous units:*

7. At what scale was the original drawing created? ______________________

8. What is the part number that is replaced by this part? ______________________

9. What paper size is the original version of this print? ______________________

10. What ASME standard is referenced for drawing interpretation? ______________________

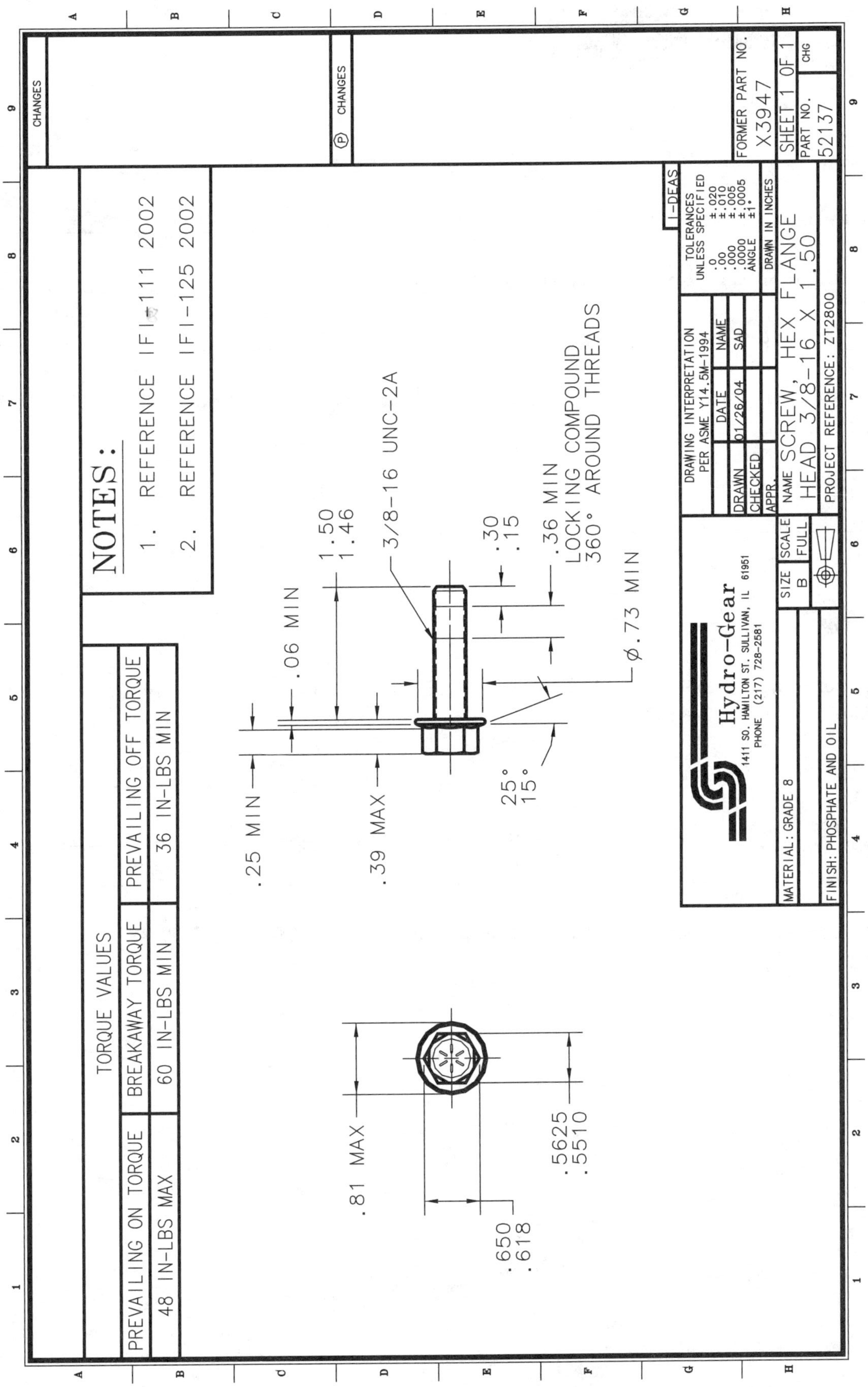

NOTES:

1. REFERENCE IFI-111 2002
2. REFERENCE IFI-125 2002

| TORQUE VALUES | | |
|---|---|---|
| PREVAILING ON TORQUE | BREAKAWAY TORQUE | PREVAILING OFF TORQUE |
| 48 IN-LBS MAX | 60 IN-LBS MIN | 36 IN-LBS MIN |

.81 MAX
.650
.618
.5625
.5510
.25 MIN
.39 MAX
.06 MIN
1.50
1.46
3/8-16 UNC-2A
.30
.15
.36 MIN
LOCKING COMPOUND
360° AROUND THREADS
Ø.73 MIN
25°
15°

Hydro-Gear
1411 SO. HAMILTON ST. SULLIVAN, IL 61951
PHONE (217) 728-2581

MATERIAL: GRADE 8
FINISH: PHOSPHATE AND OIL

DRAWING INTERPRETATION PER ASME Y14.5M-1994

| | DATE | NAME |
|---|---|---|
| DRAWN | 01/26/04 | SAD |
| CHECKED | | |
| APPR. | | |

TOLERANCES UNLESS SPECIFIED
.0 ±.020
.00 ±.010
.000 ±.005
.0000 ±.0005
ANGLE ±1°
DRAWN IN INCHES

I-DEAS

SIZE B
SCALE FULL

NAME SCREW, HEX FLANGE HEAD 3/8-16 X 1.50
PROJECT REFERENCE: ZT2800

FORMER PART NO. X3947
SHEET 1 OF 1
PART NO. 52137
CHG

CHANGES
Ⓟ CHANGES

PR 17-1.
Print supplied by Hydro-Gear.

## Industry Print Exercise PR 17-2

*Refer to the print PR 17-2 and answer the questions below.*

1. Is the locknut shown as sectioned or not sectioned? ____________________

2. How many total washers are used in this assembly? What type are they? ____________________

3. What type of spring is used in this assembly: compression, extension, or torsion? ____________________

4. Does the spring used in this assembly have a ground end? ____________________

5. Does the part associated with find number 13 fit into the part associated with find number 14? ____________________

6. What must be done to the handwheel before adding the washer and locknut? ____________________
____________________

*Review questions based on previous units:*

7. What word describes the manner in which the parts associated with find numbers 9–14 are shown in the detail, as compared to the way they are shown in the main section view? ____________________

8. How many pressure seal rings are needed? ____________________

9. What are the part numbers of the two subassemblies (*not* find numbers)? ____________________
____________________

10. What part number is applied to the lubricant used prior to assembly of the parts 7509-9 and 7509-42? ____________________

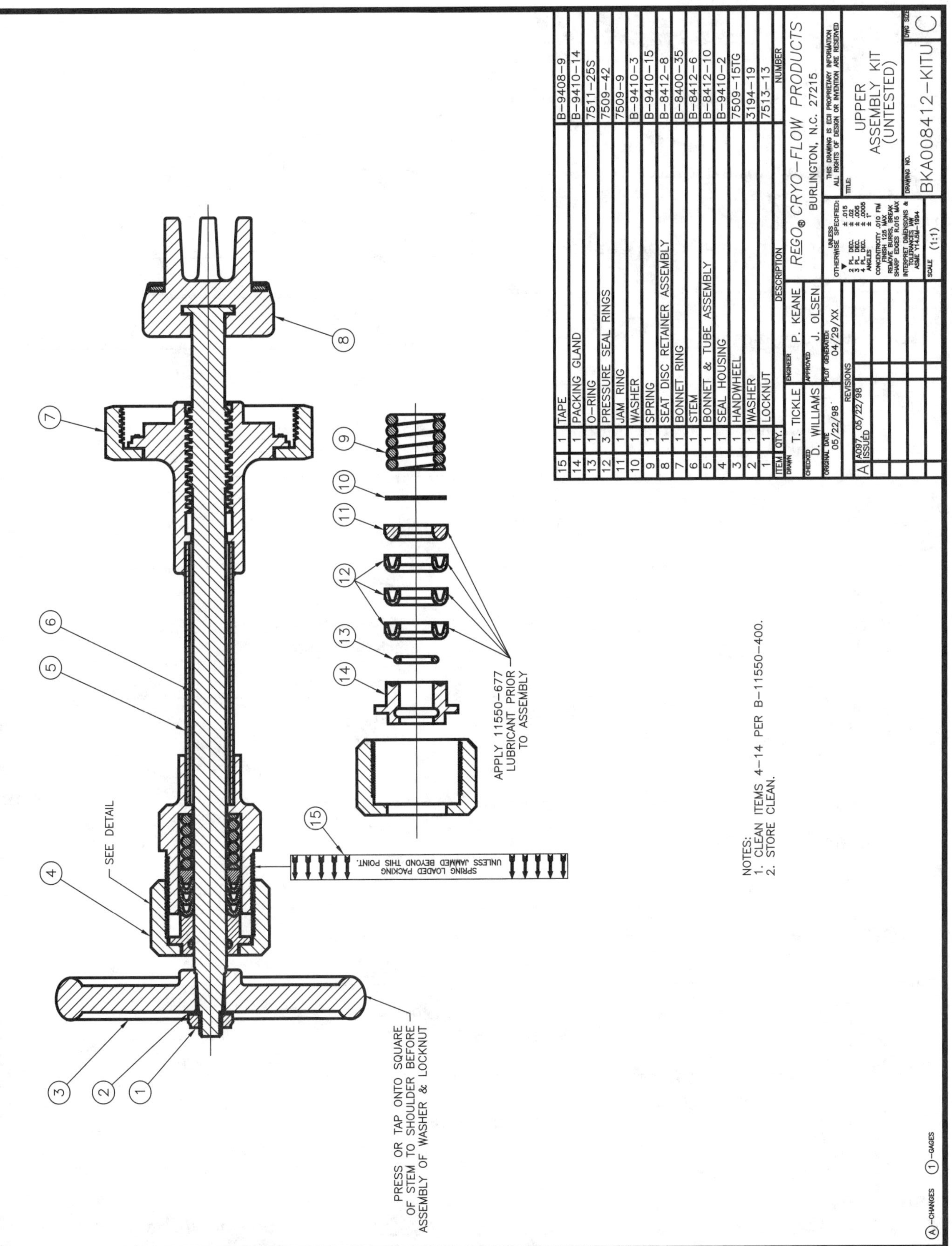


**PR 17-2.**
Print supplied by RegO Cryo-Flow Products.

## Bonus Print Reading Exercises

The following questions are based on various bonus prints located in the folder at the back of this textbook. Refer to the print indicated, evaluate the print, and answer the question.

**Print AP-010:**

1. Give the specific, complete description for item 3. ______
2. By visual identification, what type of head does item 5 feature? ______
3. What instructions are given about the $1/8 \times 1/4$ roll pin? ______
4. By visual identification, what type of tip does item 6 feature? ______

**Print AP-014:**

5. Although many fasteners may be associated with this part, what is one specific fastener identified on the print whose seat has a positional tolerance in reference to datum D? ______

**Print AP-016:**

6. How many springs are required for this assembly? ______
7. What type of spring ends are shown for item 7? ______
8. What type of head is illustrated for the fasteners that hold the bonnet to the body? ______
9. Is the lock nut (item 4) shown across the corners or across the flats? ______
10. What type of head is illustrated for the adjusting screw? ______

**Print AP-018:**

11. In this assembly, item 2 is held on by a(n) ______, which is not shown, but is listed in the parts list. ______

**Print AP-025:**

12. This part will have three fasteners pressed into place. Give the specific, complete description for these items. ____________________

____________________

Gears have many applications. Here, a gear is used with a drive chain.

UNIT 18

# Gears, Splines, and Serrations

*After completing this unit, you will be able to:*

**Identify** spur gears and their representations and specifications.

**Identify** bevel gears and their representations and specifications.

**Identify** worm gears and their representations and specifications.

**Identify** splines and serrations and their representations and specifications.

The purpose of this unit is to acquaint you with gears and the terminology associated with gears and to familiarize you with gear representation on prints. There are many types and sizes of gears used in industry today. See Figure 18-1. In this unit, three basic types of gears are discussed—spur, bevel, and worm gear. With respect to gear representation in drawings, the standards are ASME Y14.7.1, entitled *Gear Drawing Standards-Part 1: For Spur, Helical, Double Helical and Racks,* and ASME Y14.7.2, entitled *Gear and Spline Drawing Standards-Part 2: Bevel and Hypoid Gears.*

## Spur Gears

A ***spur gear*** resembles a wheel with a number of equally spaced teeth cut parallel to the axis. See Figure 18-2. It is the most commonly used type of gear. Spur gears are used for drives on mechanisms such as machine lathes and mills where the axes of the gears are parallel and the gears are in the same plane with each other. The reverse gear in a manual automotive transmission is a spur gear.

**Figure 18-1.**
Gears have many applications. These are spur gears, a type of gear most people will recognize. (Creatas)

### Spur Gear Terminology

There are many terms associated with spur gears. The terms explained in this section are commonly used and should be understood.

***Diametral pitch*** ($P_d$) is the number of teeth in a gear per inch of pitch diameter. For example, a gear having 48 teeth and a pitch diameter of three has a diametral pitch of 16 ($48 \div 3 = 16$). Mating gears must have identical diametral pitches. In industrial prints, it is common to identify gears by the diametral pitch, for example, 48 PITCH GEARS or 32 PITCH GEARS.

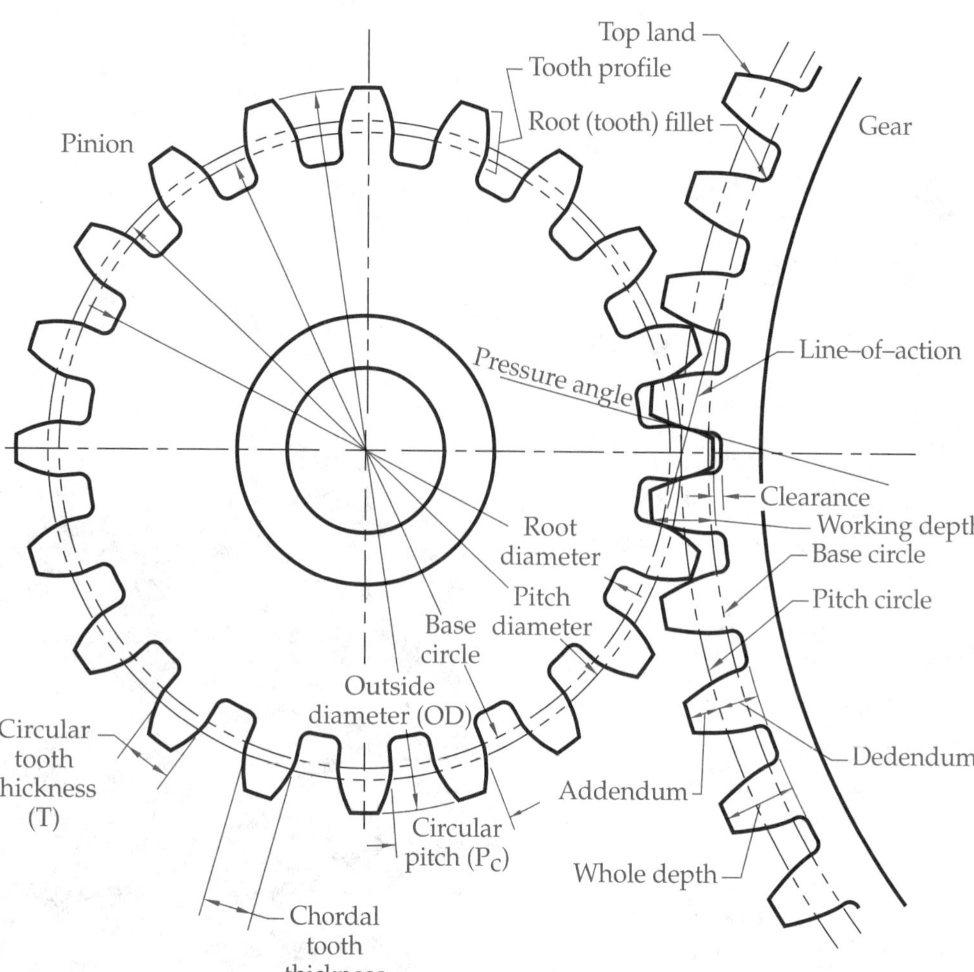


**Figure 18-2.**
The geometry and identification of spur gear parts.

The ***pitch circle*** is an imaginary circle located approximately halfway between the roots and tops of the gear teeth. A gear's pitch circle is tangent to the pitch circle of any mating gear. See Figure 18-3. The ***pitch diameter*** (D) is the diameter of the pitch circle. Pitch diameters of mating gears are compared to determine the gear ratio. As illustrated in Figure 18-3, the relationship of one pitch diameter to another is identical to the relationship of two friction rollers.

***Circular pitch*** ($P_c$) is the length of the arc along the pitch circle between the midpoint of one tooth to the midpoint of the next tooth. This value can be calculated by dividing the circumference of the pitch circle (πD) by the number of teeth on the gear. Nominally, this value is two times the circular tooth thickness, since each circular pitch distance contains one tooth and one matching tooth space.

The ***addendum*** (a) is the radial distance between the pitch circle and the top of the tooth. This distance is usually $1/P_d$ (one divided by the diametral pitch). For example, the addendum for a gear with a diametral pitch of 4 is typically 1/4″ of an inch. The ***addendum circle diameter*** is equal to the pitch circle diameter plus twice the addendum (D + 2a). The addendum circle diameter is equivalent to the outside diameter of the gear.

The ***dedendum*** (b) is the radial distance between the pitch circle and the bottom of the tooth. This distance is usually $1.25/P_d$ (1.25 divided by the diametral pitch). For example, the dedendum for a gear with a diametral pitch of 4 is typically 5/16″ of an inch. The dedendum must be larger than the addendum to allow for clearance between the mating gear teeth. The ***dedendum circle diameter*** is equal to the pitch circle minus twice the dedendum (D–2b). The dedendum circle diameter is equivalent to the root diameter of the gear.

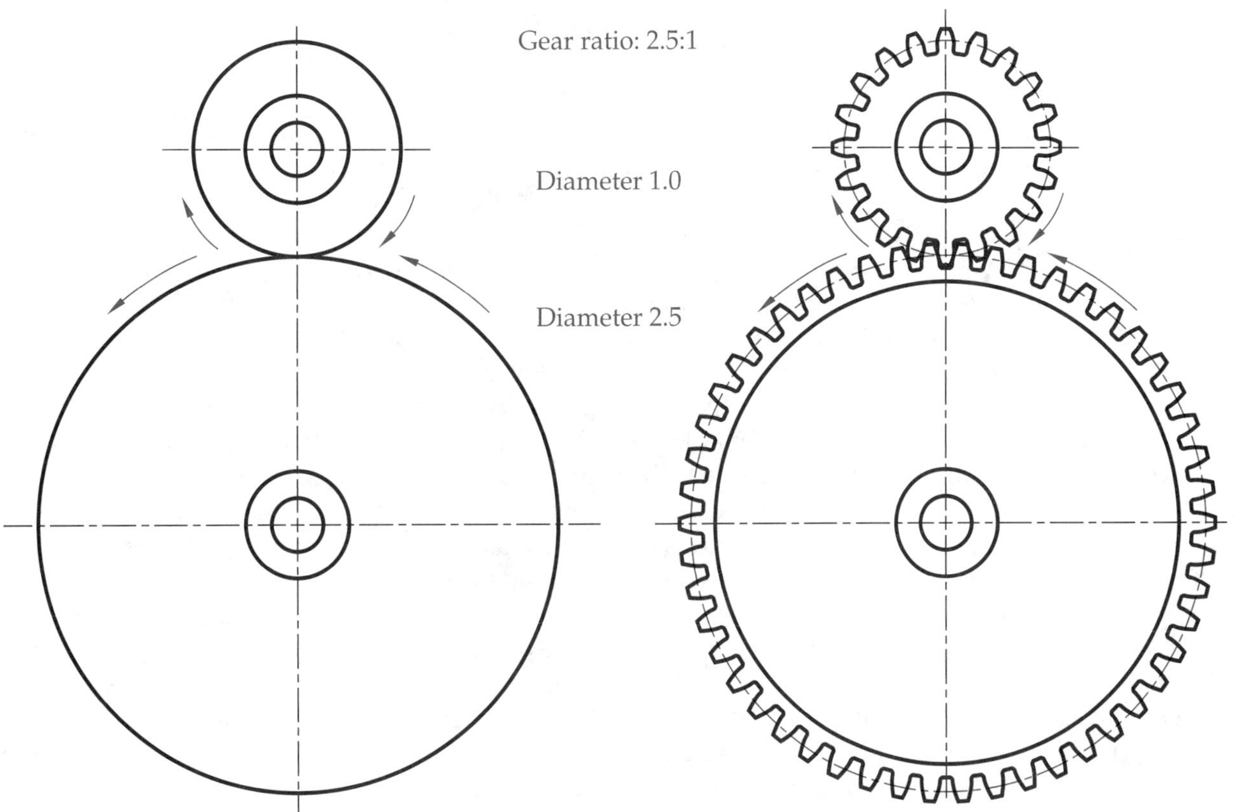


Figure 18-3.
The pitch circle is an imaginary circle located approximately halfway between the roots and tops of the gear teeth.

***Clearance*** is the radial distance between the top of a tooth on one gear and the bottom of the tooth space on the mating gear. See Figure 18-2. Clearance can be expressed as the dedendum minus the addendum (b – a).

The ***tooth face*** is the curved surface of the tooth that lies outside of the pitch circle. The ***tooth flank*** is the curved surface of a tooth that lies inside of the pitch circle. ***Tooth space*** is the distance at the pitch circle between two adjacent teeth. In general, the tooth space is the circular pitch divided by two ($P_c \div 2$). However, a calculated amount is added to allow for clearance between the tooth and the tooth space, so tooth space is slightly larger than the corresponding tooth.

The ***circular tooth thickness*** is the length of the *arc* along the pitch circle between the two sides of the tooth. This thickness is equivalent to the circular pitch divided by two ($P_c \div 2$). The ***chordal tooth thickness*** is the length of the *chord* between the intersection of the pitch circle and the two sides of the tooth. The chordal tooth thickness is the distance measured when a gear-tooth caliper is used to measure the tooth thickness at the pitch circle. See Figure 18-4. The ***chordal addendum*** is the distance from the top of the tooth to the chord at the pitch circle. It is the height dimension used in setting gear-tooth calipers to measure tooth thickness. Chordal thickness and addendum values are available in table form. The table values are often given for a diametral pitch of one, but specific distances can be found by dividing the table value by the diametral pitch for a particular gear.

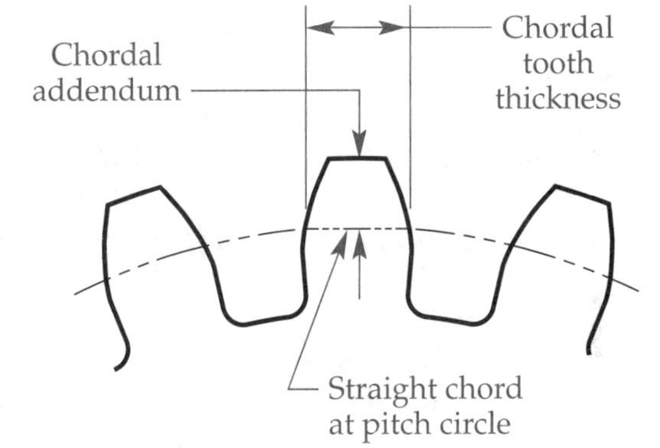


Figure 18-4.
The chordal tooth thickness is the distance measured when a gear-tooth caliper is used to measure the tooth thickness at the pitch circle.

The ***working depth*** of two mating gears is the sum of their addendums (a + a). The ***whole depth*** is the total depth of a tooth space or addendum plus dedendum (a + b).

The ***pressure angle*** is the angle between the tooth profile and a radial line at their intersection on the pitch circle. The pressure angle can also be illustrated by drawing a line through the points of contacts between engaged gear teeth. Standard values for pressure angles throughout the years have been 14.5°, 20°, and 25°. The pressure angles of mating gears must be the same. In general, the 20° pressure angle is the most common and versatile.

***Backlash*** is the amount by which the width of a tooth space exceeds the thickness of the engaging tooth on the pitch circles. While the basic calculations for a gear would create a tooth space identical to a tooth, some backlash must be allowed to compensate for manufacturing error, expansion, and lubrication.

The ***outside diameter*** is the diameter of a circle coinciding with the tops of the teeth of an external gear. This is the equivalent to the addendum circle. The ***internal diameter*** is the diameter of a circle coinciding with the tops of the teeth of an internal gear.

The ***center distance*** is the distance between the center axes of mating gears. ***Base circle*** is the diameter from which an involute tooth curve is generated or developed. Figure 4-12 in Unit 4 illustrates how the involute shape is generated. The base circle diameter is calculated by multiplying the pitch circle diameter by the cosine of the pressure angle. As a result, gear teeth with a 20° pressure angle have more involute surface length than gear teeth with a 14° pressure angle.

***Pin measurement*** is sometimes used to specify a condition or size for gears or splines. This measurement is made over pins for an external gear and between pins for an internal gear. ***Pin diameter*** is the diameter of the measuring pin or ball used between teeth.

## Spur Gear Representation and Specification

Rather than draw all tooth profiles of a gear, spur gears are often represented by showing the addendum circle (outside diameter) and the dedendum circle (root diameter) as a phantom line or a long-dash line. This practice was developed primarily to save pencil drafters from tedious work tracing involute shapes of intricate gear teeth. While CAD systems can easily create all teeth on a gear, phantom lines are still often used as the conventional standard practice. For drawings that do not show all teeth, a centerline is also drawn to show the pitch diameter.

The specifications required to machine a gear are usually given in a data block or table located on the drawing. This eliminates the need for calculations to be done in the shop, thus reducing possible errors. However, it may be necessary at times to calculate gear data. When the diametral pitch and number of teeth are known, most remaining data can be calculated with the aid of formulas common in most machinist's handbooks or gear supplier catalogs.

The American Gear Manufacturers Association (AGMA) has established quality class standards for gears. First, gears can be classified as coarse pitch or fine pitch. For coarse-pitch gears, the quality class can range from 3 (least precise) to 15 (most precise). For fine-pitch gears, the quality class can range from 5 (least precise) through 16 (most precise).

While there are several options for manufacturing gears, sometimes special cutters called ***hobs*** are used to cut gear teeth. The hobbing process may be only one of a series of processes, with another process such as grinding used to finish the teeth. Hobs are available in a variety of shapes and sizes for helping produce a wide range of gears. Information about the hobs may also appear on the print.

## Bevel Gears

A ***bevel gear*** resembles a cone with teeth on its conical side. See Figure 18-5. Like all gears, bevel gears transmit motion and power to a mating gear. Bevel gears are commonly used in industry. Most mating bevel gears have their shafts at right angles. However, the shaft angle can be other than 90°. The geometry and identification of bevel gear parts are shown in Figure 18-6.

The specifications necessary for cutting a bevel gear are usually given in the gear data block on the drawing. Formulas for cutting bevel gears vary with the type of gear and are too numerous to list in this print reading textbook. If a certain formula is needed, it may be found in a machinist's handbook.

**Figure 18-5.**
A bevel gear resembles a cone with teeth on its conical side. (Boston Gear)

## Worm Gears

A ***worm gear*** is a cylindrical gear, similar to a spur gear, with teeth cut on an angle. The worm gear is driven by a ***worm.*** The worm is similar to a screw. See Figure 18-7. Worms and worm gears are used to connect nonparallel, nonintersecting shafts. Usually, worms and worm gears are used for transmitting motion and power at a 90° angle. The worm and worm gear are also used for speed reduction. This is because one revolution of a single-thread worm is required to advance one tooth on the worm gear.

To increase the length-of-action, the worm gear is made of a throated (concave) shape. This allows the worm gear to partially wrap around the worm. On double-enveloping worms, the worm is an hourglass shape.

The specifications for cutting a worm gear and worm are usually given in the drawing data block. If additional data are needed, check the required formula in a machinist's handbook.

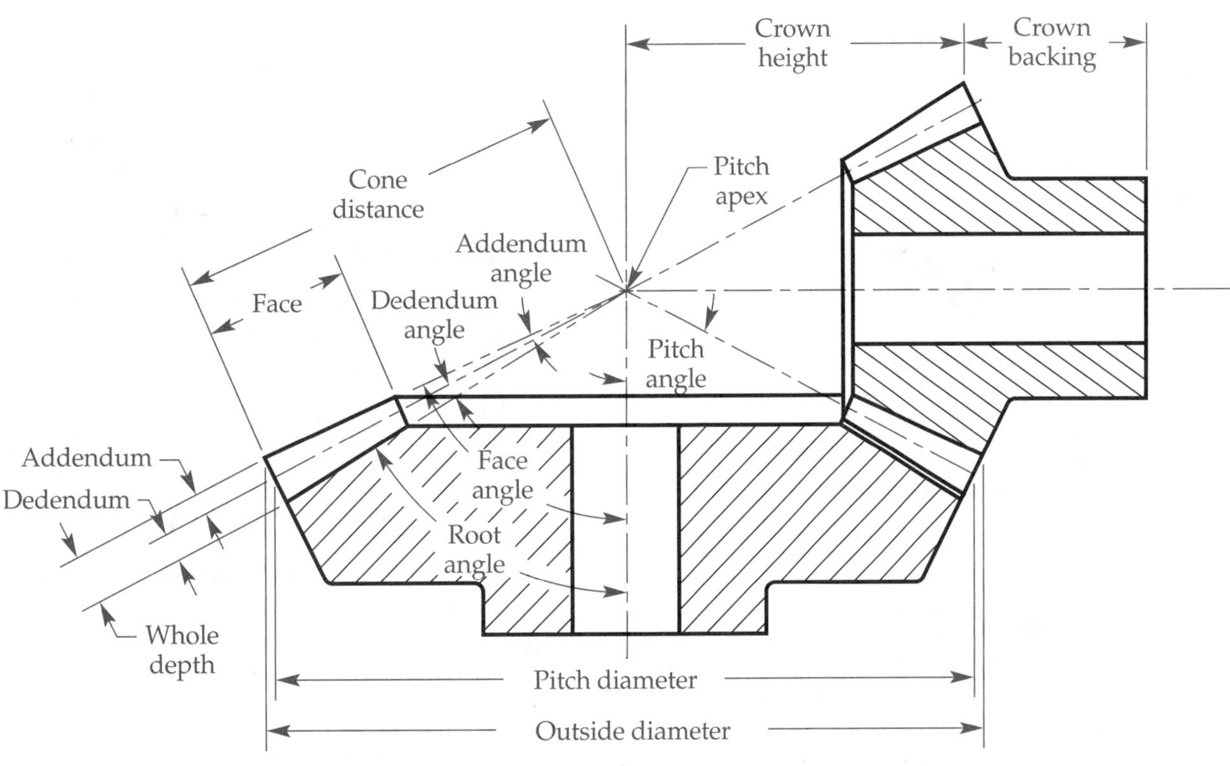


**Figure 18-6.**
The geometry and identification of bevel gear parts.

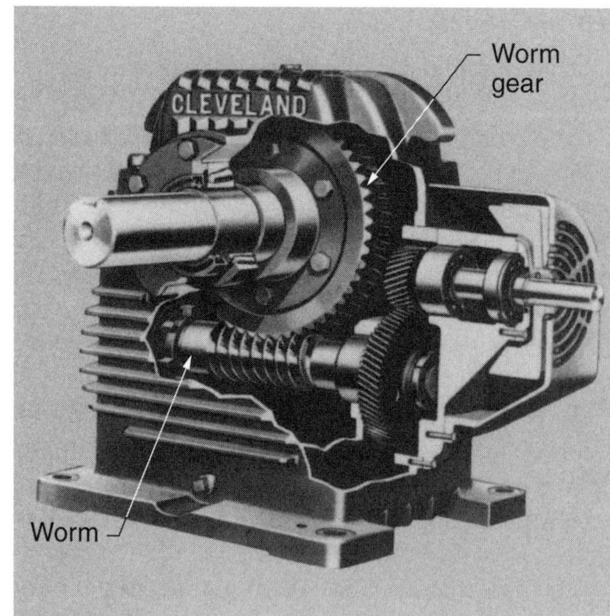


Figure 18-7.
A worm gear is a cylindrical gear, similar to a spur gear, with teeth cut on an angle and driven by a worm. (Cleveland Gear)

The terminology for worm gearing is much the same as for spur and bevel gearing. See Figure 18-8. However, there are some additional terms associated with worms and gears. These terms are discussed below.

- ***Linear pitch*** is the distance from a given point on one worm thread to the next. This distance is equal to the circular pitch of the worm gear.
- In screw thread applications, ***lead*** is the distance a thread advances in one revolution. For a single-threaded worm, the linear pitch and the lead are the same. For a double-threaded worm, the lead is twice the linear pitch; for a four-threaded worm, the lead is four times the linear pitch; and so on.
- ***Lead angle*** is the angle the lead makes with a perpendicular line to the worm axis.
- ***Throat diameter*** is the diameter of a circle coinciding with the tops of the worm gear teeth at their center plane.

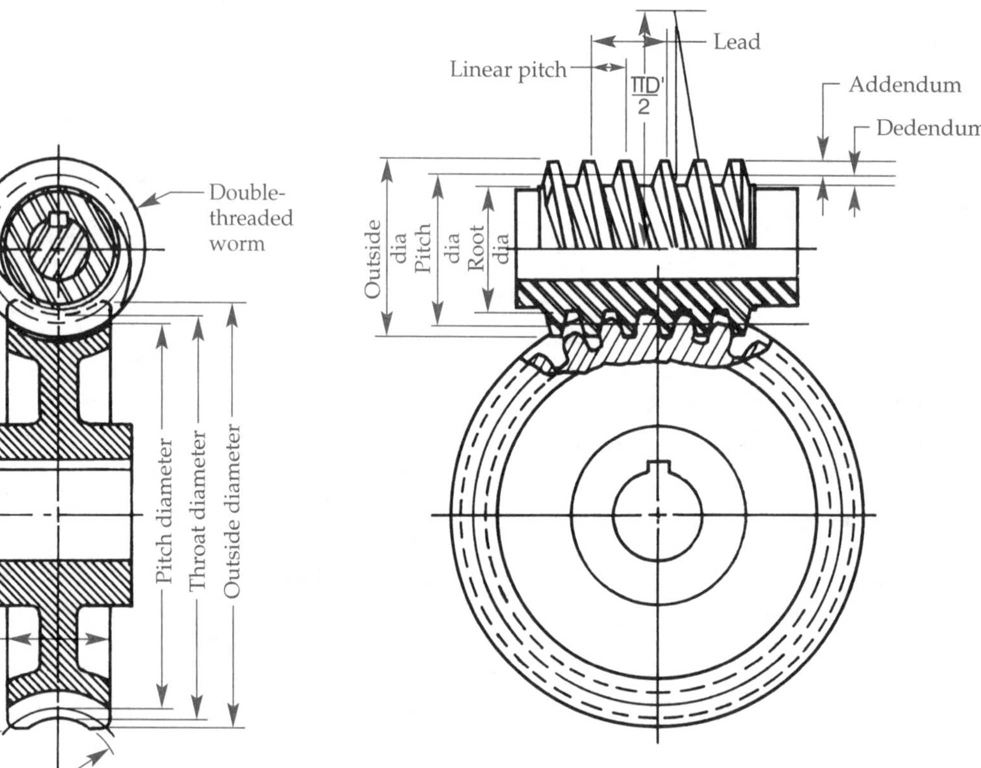


Figure 18-8.
The terminology for worm gearing is much the same as for spur and bevel gearing.

## Splines and Serrations

***Splines*** and ***serrations*** are like multiple keys or grooves on a shaft that prevent rotation between the shaft and a related member. By definition, splines are external, raised features and serrations are internal notches. As applied in industrial parts, they often appear similar to wide, small-diameter spur gear teeth. Traditionally, splines have teeth with parallel sides. However, splines with involute sides are increasing in popularity. See Figure 18-9. Involute splines are produced with the same technique and equipment as is used for gears.

There are two types of fits for splines. The first type relates to the major diameter. The second type relates to the sides of the teeth. For each type, there are three classes of fit. A class 1 fit is a sliding fit. A class 2 fit is a close fit. A class 3 fit is a press fit.

As notches, serrations are primarily used for parts permanently fitted together, perhaps by wedging. They have different tooth proportions and higher pressure angles than splines. Serrations are well adapted for use on thin wall tubing. Serrations are also common on hand tools like pliers as a type of gripping surface.

Terminology associated with involute splines and serrations is often the same as for spur gears. Usually, data for producing splines or serrations are given on the print. When a formula is needed, check a machinist's handbook.

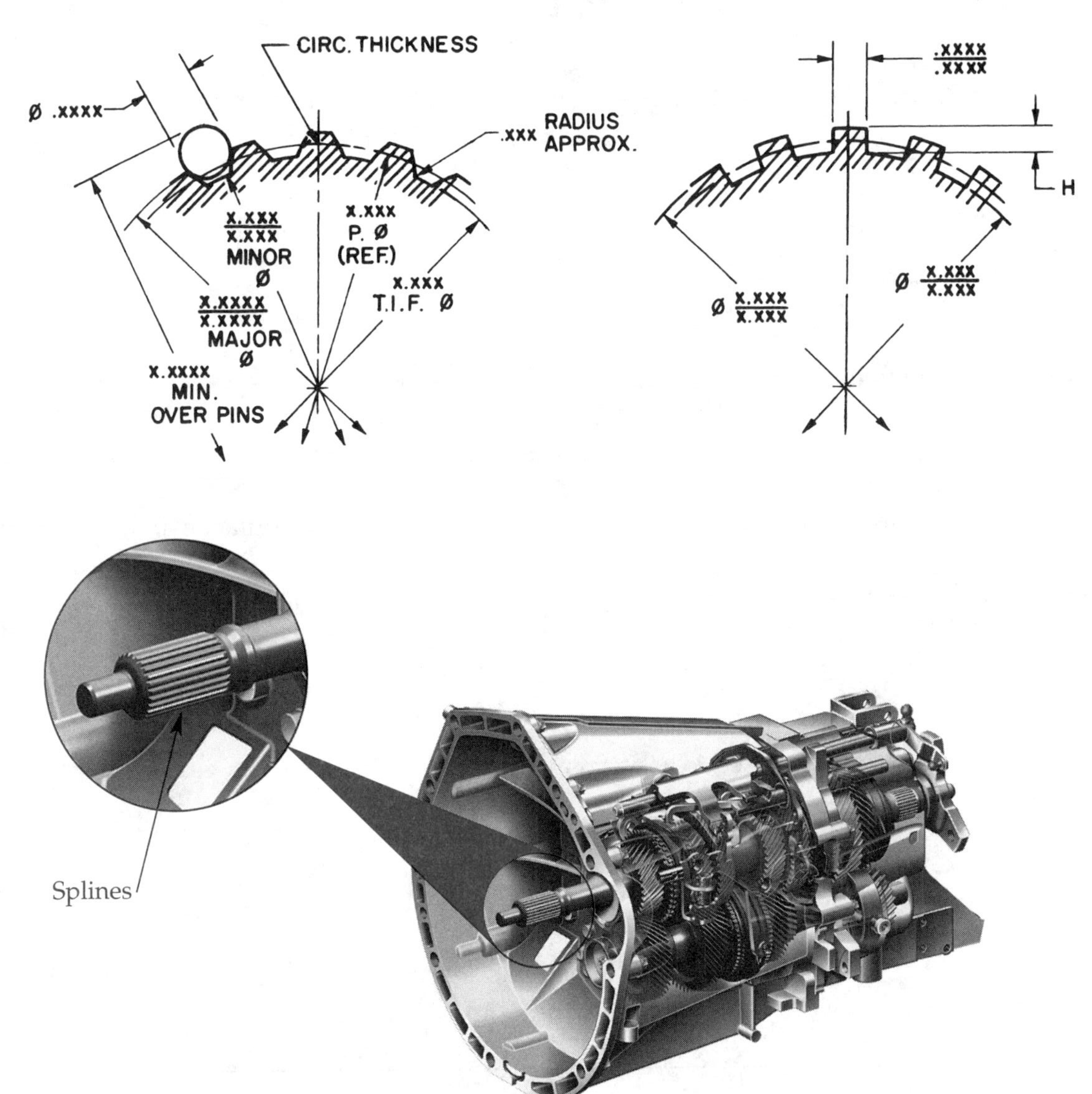


**Figure 18-9.**
Splines are like multiple keys on a shaft that prevent rotation between the shaft and its related member. The input shaft on an automotive transmission has splines to mate it to the torque converter or clutch plate. (Photo: Daimler)

## Review Questions

*Circle the answer of choice, fill in the blank, or write a short answer.*

1. A(n) ______________ gear resembles a wheel with a number of equally spaced teeth cut parallel to the axis.

2. *True or False?* The diametral pitch describes a ratio between the number of teeth on a gear and the diameter of that gear.

3. Mating gears must have the same:
   A. pitch diameter.
   B. diametral pitch.
   C. base circle diameter.
   D. pin diameter.

4. For a gear set, the relationship of one pitch diameter to the other is identical to the relationship between two ______________________________ rollers that have the same diameters.

5. A 10″ gear has 32 teeth. Which of the following is the closest to the circular pitch?
   A. 1″
   B. 2″
   C. 3″
   D. 4″

6. If the diametral pitch is 8, the addendum is ______________________________ .

7. *True or False?* The dedendum is always a little smaller than the addendum.

8. The ____________________ tooth thickness is the distance checked with gear tooth calipers.

9. Which term applies to the angle of a line drawn through the points of contact between engaged gear teeth?
   A. Base angle.
   B. Tangent angle.
   C. Working angle.
   D. Pressure angle.

10. What auxiliary objects may be used to assist in measuring the diameter of a gear?
    A. Bolts.
    B. Pins.
    C. Keys.
    D. Screws.

11. From the alphabet of lines, what line type can be used on a drawing to represent the area in which the gear teeth exist, although the gear teeth are not drawn? ______________________

______________________________________________________________

12. What does AGMA stand for?
    A. Association for Gear Machining in America
    B. Allied Gear Mechanisms Act
    C. American Gear Manufacturers Association
    D. Appendix of Gear Moving Articles

13. What type of gear transfers motion 90° by using conical gears? ______________________

______________________________________________

14. What drives a worm gear? ______________________________

______________________________________________

15. *True or False?* Splines are like multiple keys on a shaft that prevent rotation between a shaft and its related member.

## Industry Print Exercise PR 18-1

*Refer to the print PR 18-1 and answer the questions below.*

1. What is the diametral pitch ($P_d$) of this gear? ____________
2. Is the addendum $1/P_d$? ____________
3. What is the pitch diameter of this gear? ____________
4. A mating gear for this gear must have a pressure angle of ____________.
5. Is the part number to be marked on both sides of the gear? ____________
6. What size pin is specified for measuring the diameter over pins? ____________
7. After hobbing, the circular tooth thickness should be ____________, and then reduced to ____________ by grinding.
8. Is the AGMA gear class specified by the print very precise or not very precise? ____________
9. What type of line is used to indicate the outside diameter of the gear? ____________
10. What is the range of backlash specified for this gear? ____________

*Review questions based on previous units:*

11. What is the 1/2″ notch on the central hole called? ____________
12. Are the section lines indicative of the material? ____________
13. What material is used for this gear? ____________
14. What geometric tolerances are specified by a general note instead of feature control frame? ____________
____________
15. Are any dimensions specified using the limit method of tolerancing? ____________

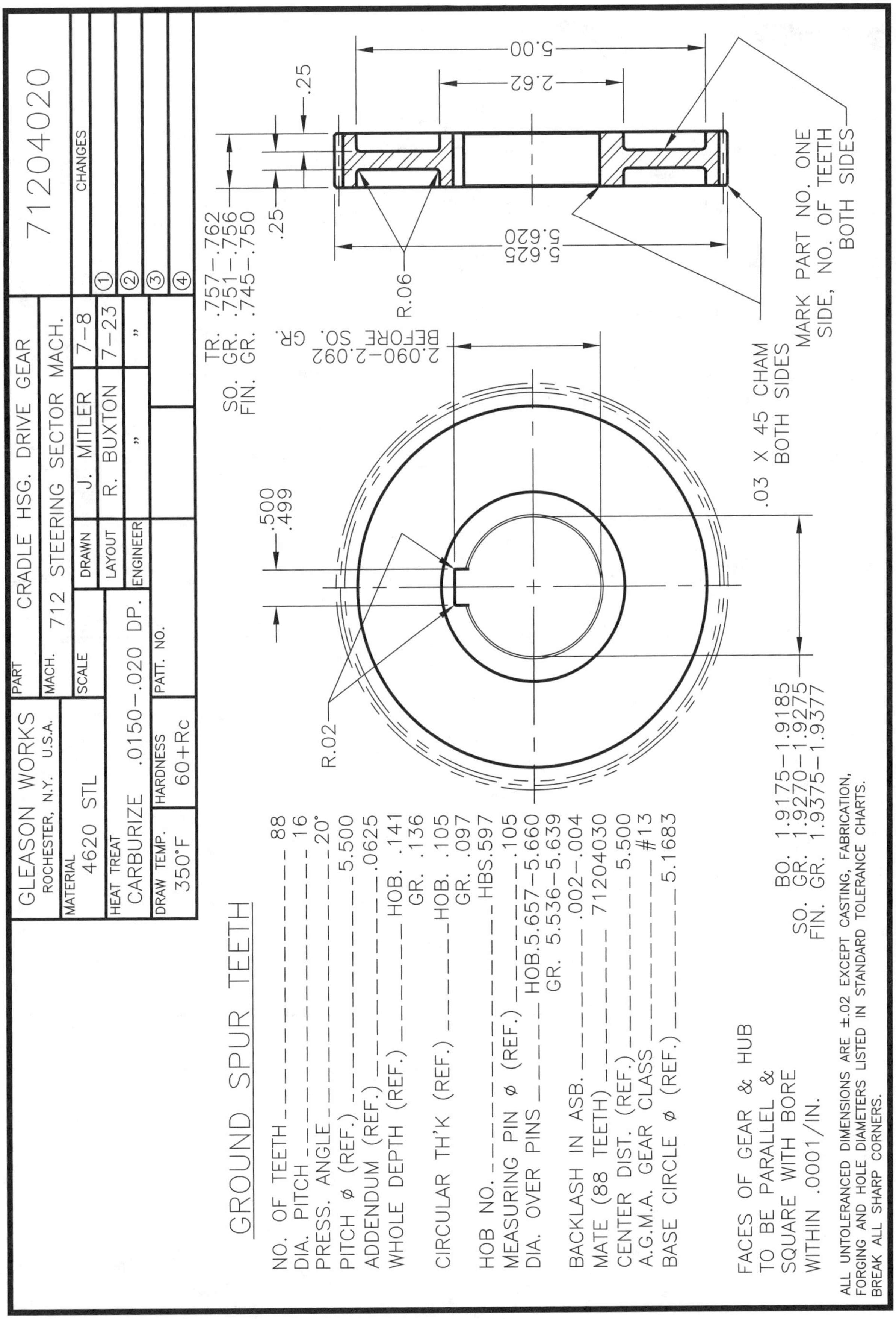

GROUND SPUR TEETH

| | |
|---|---|
| NO. OF TEETH | 88 |
| DIA. PITCH | 16 |
| PRESS. ANGLE | 20° |
| PITCH ⌀ (REF.) | 5.500 |
| ADDENDUM (REF.) | .0625 |
| WHOLE DEPTH (REF.) | HOB. .141<br>GR. .136 |
| CIRCULAR TH'K (REF.) | HOB. .105<br>GR. .097 |
| HOB NO. | HBS.597 |
| MEASURING PIN ⌀ (REF.) | .105 |
| DIA. OVER PINS | HOB.5.657–5.660<br>GR. 5.536–5.639 |
| BACKLASH IN ASB. | .002–.004 |
| MATE (88 TEETH) | 71204030 |
| CENTER DIST. (REF.) | 5.500 |
| A.G.M.A. GEAR CLASS | #13 |
| BASE CIRCLE ⌀ (REF.) | 5.1683 |

FACES OF GEAR & HUB
TO BE PARALLEL &
SQUARE WITH BORE
WITHIN .0001/IN.

BO. 1.9175–1.9185
SO. GR. 1.9270–1.9275
FIN. GR. 1.9375–1.9377

.03 X 45 CHAM
BOTH SIDES

MARK PART NO. ONE
SIDE, NO. OF TEETH
BOTH SIDES

R.02

.500
.499

2.090–2.092
BEFORE SQ. GR.

R.06

5.625
5.620

2.62

5.00

.25

.25

TR. .757–.762
SQ. GR. .751–.756
FIN. GR. .745–.750

ALL UNTOLERANCED DIMENSIONS ARE ±.02 EXCEPT CASTING, FABRICATION,
FORGING AND HOLE DIAMETERS LISTED IN STANDARD TOLERANCE CHARTS.
BREAK ALL SHARP CORNERS.

| | | |
|---|---|---|
| GLEASON WORKS<br>ROCHESTER, N.Y. U.S.A. | PART: CRADLE HSG. DRIVE GEAR | 71204020 |
| | MACH.: 712 STEERING SECTOR MACH. | |
| MATERIAL: 4620 STL | SCALE | DRAWN: J. MITLER 7–8 | CHANGES ① |
| HEAT TREAT: CARBURIZE .0150–.020 DP. | | LAYOUT: R. BUXTON 7–23 | ② |
| | | ENGINEER: " " | ③ |
| DRAW TEMP.: 350°F | HARDNESS: 60+Rc | PATT. NO. | ④ |

PR 18-1.
Print supplied by Gleason Works.

## Industry Print Exercise 18-2

*Refer to the print PR 18-2 and answer the questions below.*

1. What is the total number of gears specified by this one print? ____________
2. What is the diametral pitch ($P_d$) of the gears specified by this one print? ____________
3. How many options are there for the basic size of diameter C? ____________
4. What is the pressure angle specified for the gears of this print? ____________
5. All of the parts covered by this one print have a depth of .460", but the face width of the worm wheel is specified as a fractional distance of ____________.
6. What hole size is specified for part W5-15S? ____________
7. What three part numbers each have 90 teeth? ____________
8. What is the smallest pitch diameter specified? ____________
9. What material is specified for these gears? ____________
10. What worm is specified as the mate for this worm wheel? ____________

*Review questions based on previous units:*

11. Is Section E-E shown in alignment with other views or as a removed view? ____________
12. What scale is specified for this print? ____________
13. How many feature control frames are used on this print? ____________
14. What is the current revision for this print? ____________
15. What surface finish is specified for the parts? ____________

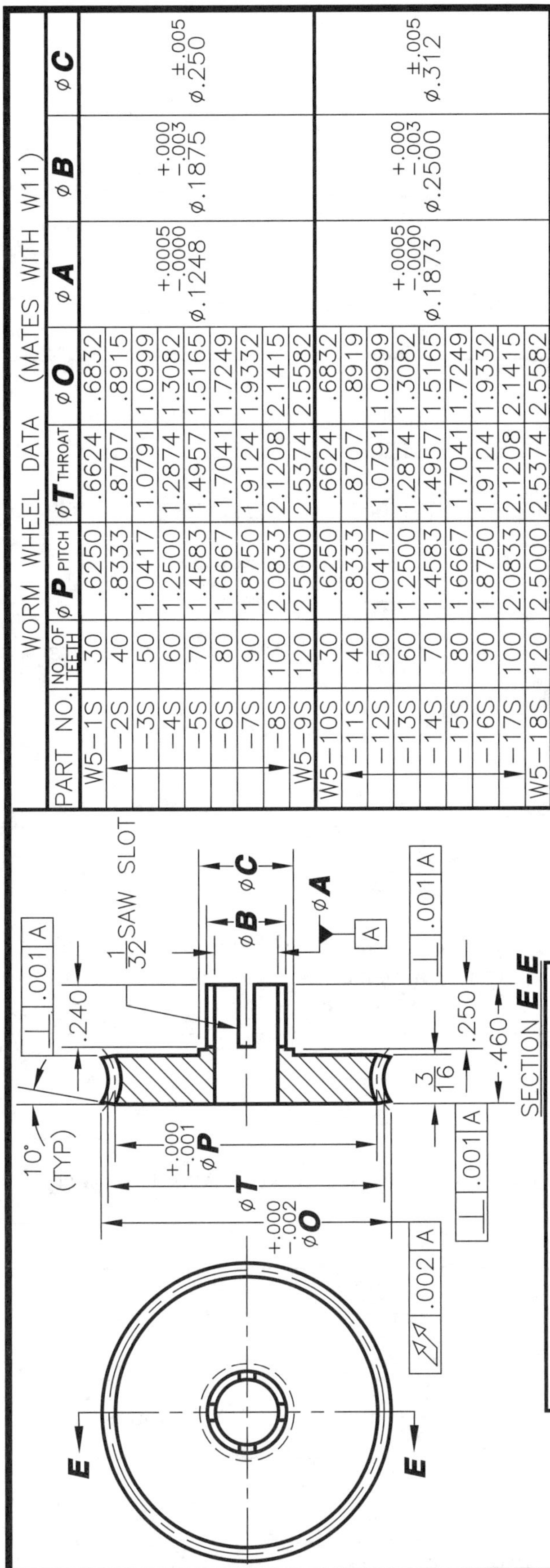

WORM WHEEL DATA (MATES WITH W11)

| PART NO. | NO. OF TEETH | ⌀ P PITCH | ⌀ T THROAT | ⌀ O | ⌀ A | ⌀ B | ⌀ C |
|---|---|---|---|---|---|---|---|
| W5–1S | 30 | .6250 | .6624 | .6832 | ⌀.1248 +.0005 −.0000 | ⌀.1875 +.000 −.003 | ⌀.250 ±.005 |
| –2S | 40 | .8333 | .8707 | .8915 | | | |
| –3S | 50 | 1.0417 | 1.0791 | 1.0999 | | | |
| –4S | 60 | 1.2500 | 1.2874 | 1.3082 | | | |
| –5S | 70 | 1.4583 | 1.4957 | 1.5165 | | | |
| –6S | 80 | 1.6667 | 1.7041 | 1.7249 | | | |
| –7S | 90 | 1.8750 | 1.9124 | 1.9332 | | | |
| –8S | 100 | 2.0833 | 2.1208 | 2.1415 | | | |
| W5–9S | 120 | 2.5000 | 2.5374 | 2.5582 | | | |
| W5–10S | 30 | .6250 | .6624 | .6832 | ⌀.1873 +.0005 −.0000 | ⌀.2500 +.000 −.003 | ⌀.312 ±.005 |
| –11S | 40 | .8333 | .8707 | .8919 | | | |
| –12S | 50 | 1.0417 | 1.0791 | 1.0999 | | | |
| –13S | 60 | 1.2500 | 1.2874 | 1.3082 | | | |
| –14S | 70 | 1.4583 | 1.4957 | 1.5165 | | | |
| –15S | 80 | 1.6667 | 1.7041 | 1.7249 | | | |
| –16S | 90 | 1.8750 | 1.9124 | 1.9332 | | | |
| –17S | 100 | 2.0833 | 2.1208 | 2.1415 | | | |
| W5–18S | 120 | 2.5000 | 2.5374 | 2.5582 | | | |
| W5–19S | 30 | .6250 | .6624 | .6832 | ⌀.2498 +.0005 −.0000 | ⌀.3125 +.000 −.003 | ⌀.375 ±.005 |
| –20S | 40 | .8333 | .8707 | .8919 | | | |
| –21S | 50 | 1.0417 | 1.0791 | 1.0999 | | | |
| –22S | 60 | 1.2500 | 1.2874 | 1.3082 | | | |
| –23S | 70 | 1.4583 | 1.4957 | 1.5165 | | | |
| –24S | 80 | 1.6667 | 1.7041 | 1.7249 | | | |
| –25S | 90 | 1.8750 | 1.9124 | 1.9332 | | | |
| –26S | 100 | 2.0833 | 2.1208 | 2.1415 | | | |
| –27S | 120 | 2.5000 | 2.5374 | 2.5582 | | | |
| W5–28S | 180 | 3.7500 | 3.7874 | 3.8082 | | | |

WORM WHEEL DATA

| | |
|---|---|
| PITCH (DIAM.) | 48 |
| THREAD | FOUR (R.H.) |
| LEAD OF WORM | .2618 |
| PRESSURE ANGLE | 25° |
| LEAD ANGLE | 14°–2' |
| WHOLE DEPTH | .0407 |
| A.G.M.A. | PREC 1 |
| TESTING PRESSURE | 20 OZ. |
| TOOTH FORM | INVOLUTE |



| REV | DATE | APP'D | CHANGE |
|---|---|---|---|
| F | 12/4 | GUY | ECO1321–2314 |
| | | | CONC. .002 WAS .0005 |
| E | 1/11 | MU | ECO1321–2119 |
| D | 10/4 | AEH | ECO1321–959 |
| C | 9/18 | AEH | ECO1321–941 |

ITEM NO. | NO. REQ'D | USED ON | REV | DATE | APP'D | CHANGE

DO NOT SCALE THIS DRAWING

REMOVE ALL BURRS AND SHARP EDGES UNLESS OTHERWISE SPEC

SURFACE FINISH 125√ UNLESS OTHERWISE SPECIFIED | WEIGHT

TOLERANCES UNLESS SPECIFIED | FRACT DIM ± 1/64 | DECIMAL DIM ± .005 | ANGULAR DIM ± -- | SCALE NONE

DRAWN BY A.E.H. | APPROVED BY

CHECKED BY A.J.C. | ENG'R

CLASS NO. | JOB NO.

HEAT TREAT ---

FINISH ---

MATERIAL BRONZE AS PER QQ-B-637, COMP. 1

STERLING PRECISION CORPORATION INSTRUMENT DIVISION NEW YORK

TITLE WORM WHEEL (CLAMP TYPE) 48 PITCH R.H. 3/16 FACE

NEXT ASSEMBLY N/A | DRAWING NUMBER W5–1S TO W5–28S | REV F

PR 18-2.
Print supplied by Sterling Precision Corporation.

## Bonus Print Reading Exercises

The following questions are based on various bonus prints located in the folder at the back of this textbook. Refer to the print indicated, evaluate the print, and answer the question.

**Print AP-008:**

1. What mean dimension represents the diameter of the cylinders that feature involute splines?
2. Given this part has two sections that feature splines, each with a relief groove, calculate the length of the splines on the right end.
3. In inches, what diameter would be the mean diameter of the splines if measured over .060″ diameter pins?
4. Are the splines to be heat treated?
5. What pressure angle is specified for the splines?

**Print AP-019:**

6. For the spur gear on this part, give the number of teeth, pitch diameter, addendum, maximum circular tooth thickness, and theoretical diameter to expect when measuring over wires.
7. For the worm gear on this part, give the lead, lead angle, working depth, and diameter to expect when measuring over wires.
8. What is the mating part number for the spur gear, how many teeth does it have, and what is the gear ratio between it and the spur gear?
9. Do either of the two gears on this part have backlash specified?

**Print AP-020:**

10. How many teeth are specified for this helical gear? ____________________

11. What pressure angle is specified for this helical gear? ____________________

12. What is the pitch diameter of this gear? ____________________

13. At the standard pitch diameter, what maximum circular tooth thickness is specified? ____________________

14. What is the maximum diameter across this gear? ____________________

Perhaps the most recognizable example of a camshaft is in an automotive engine. This Mustang engine has dual, overhead camshafts. (Ford)

# UNIT 19
# Cam Diagrams and Prints

*After completing this unit, you will be able to:*

**Identify** types of cams used in industrial applications.

**Explain** terms related to cams and followers.

**Interpret** displacement diagrams with respect to the rise, fall, and dwell of a cam follower during a cycle.

**Identify** different methods for calculating motion transition in displacement diagrams.

Some companies manufacture parts called ***cams,*** which transmit rotary motion into linear or special angular motion. Various mechanisms incorporate the principles of cams. The camshaft of an automotive engine is perhaps the most common example. See Figure 19-1. However, the history of the cam and camshaft is hundreds of years older than the internal combustion engine. Cam surfaces can also be incorporated into clamps and tool fixtures to produce quick release handles.

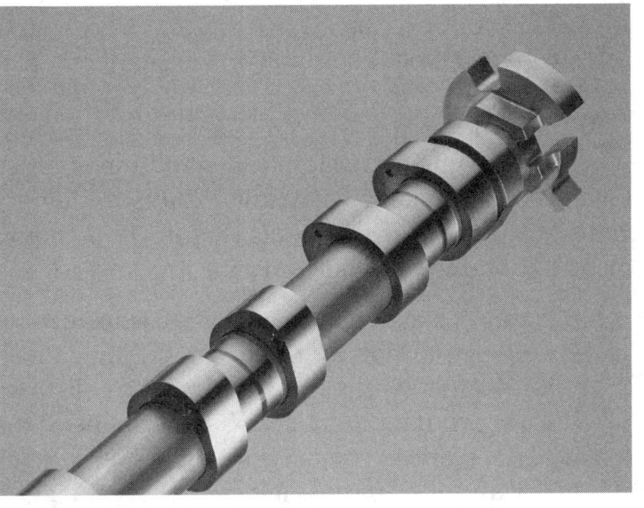

**Figure 19-1.**
The camshaft is one of the most common examples of the cam. (BMW)

Figure 19-2 features a double-cam handle used to help lock tooling into a secure position. These cam surfaces are often a simple spiral design. This unit focuses on cams that are used as mechanisms to help transfer motion in a systematic manner.

## Cam Terminology

In general, cams change uniform rotary motion into oscillating up-and-down or back-and-forth motion (linear motion). Cams are often designed in a variety of ways, including radial or plate cams, cylindrical or barrel cams, endface cams, and groove or face cams. See Figure 19-3. Even a wedge or linear slot can serve as a cam, as illustrated in Figure 19-4.

There is most likely a ***follower*** device that moves as a result of contact with the cam, but not all cams have followers. It may have a piston-type or a pivoting-type motion. The contact surface of the follower may be a knife edge, flat face, or roller. Figure 19-5 shows examples of followers. The roller provides the most flexibility. Precision-roller followers in various degrees of quality and bearing types are available from vendors.

**Figure 19-2.**
Cam surfaces can also be incorporated into clamps and tool fixtures to produce quick release handles.

Radial or plate cam
Cylindrical or barrel cam
Endface cam
Groove or face cam
Follower
Follower
Follower

**Figure 19-3.**
Types of cams include radial or plate cams, cylindrical or barrel cams, endface cams, and groove or face cams.

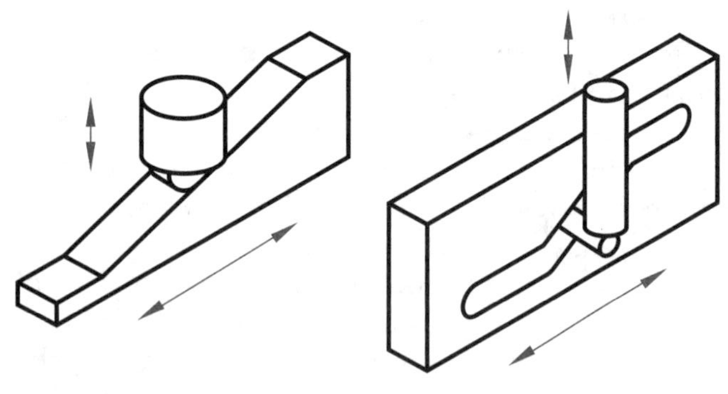

**Figure 19-4.**
Even a wedge or an angled slot can serve as a cam, causing a follower to move or oscillate.

***Radial*** or ***plate cams*** can be manufactured primarily as a "wheel" with an irregular shape, perhaps as a spline-shaped, pear-shaped, or spiral outer profile. This type of cam performs primarily as a rotating disk with a profile surface that contacts the follower.

***Cylindrical cams,*** also referred to as ***barrel cams,*** feature a follower groove cut within the cylindrical surface of a cylinder. These types of cams are often positioned with the axis of rotation vertical. Then, the rotating cam causes a follower to oscillate up and down. This type of cam may be found in certain clock mechanisms or in sewing machines. A barrel cam may be referred to as an ***endface cam*** or ***side cam*** if the cylinder has only one cam surface, allowing the follower to ride on the rim of the cylinder.

***Face cams*** are manufactured with a follower groove in the flat surface of a disc. The principle is the same as the radial cam, but with this cam the follower is nested in a groove. Some references refer to these as ***groove cams.***

A ***linear cam*** gets its name from the manner in which the cam moves in a straight line. For example, as a linear cam moves a certain horizontal distance, the follower nested within the groove is designed to move up or down as the slot dictates. These cams offer a nonelectrical solution in mechanisms that repeat motion over and over.

The following terms are used in discussions relating to cams. A basic understanding of these terms will help you when reading prints containing cams and related components. Figure 19-6 should also be examined as you study these terms.

- **Cam.** A machine element that transmits or delivers motion by rolling or sliding contact with a follower.
- **Cam profile.** For most cams, the shape of the surface upon which the follower has contact. For a face or barrel cam, this is the shape of the groove.
- **Follower.** The machine or mechanism that moves with reciprocating movement by following the cam as it rotates or oscillates.

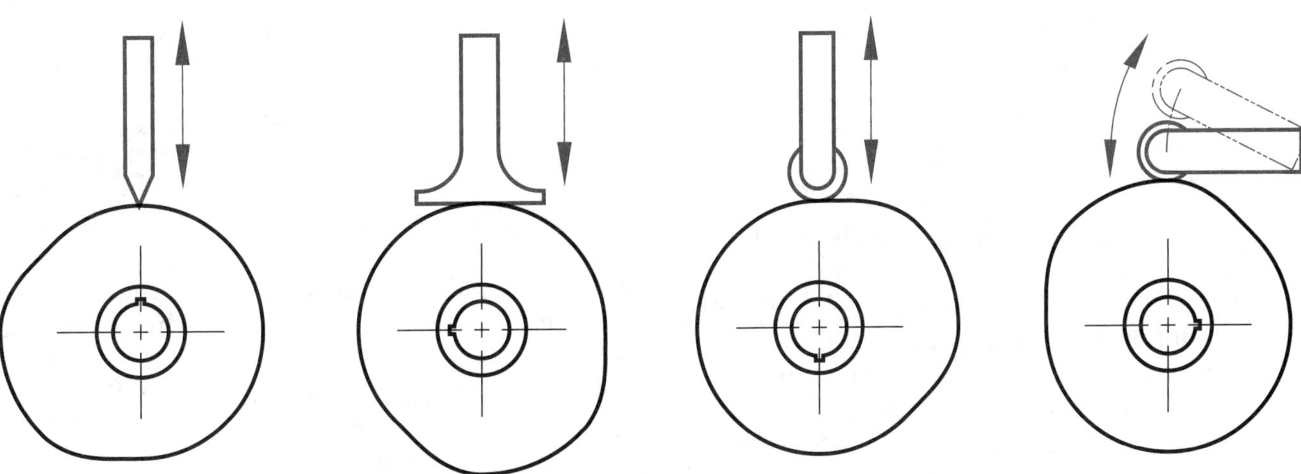

**Figure 19-5.**
Types of followers include (from left to right) knife edge, flat face, and roller. Roller followers in various degrees of quality and bearing types are available from vendors.

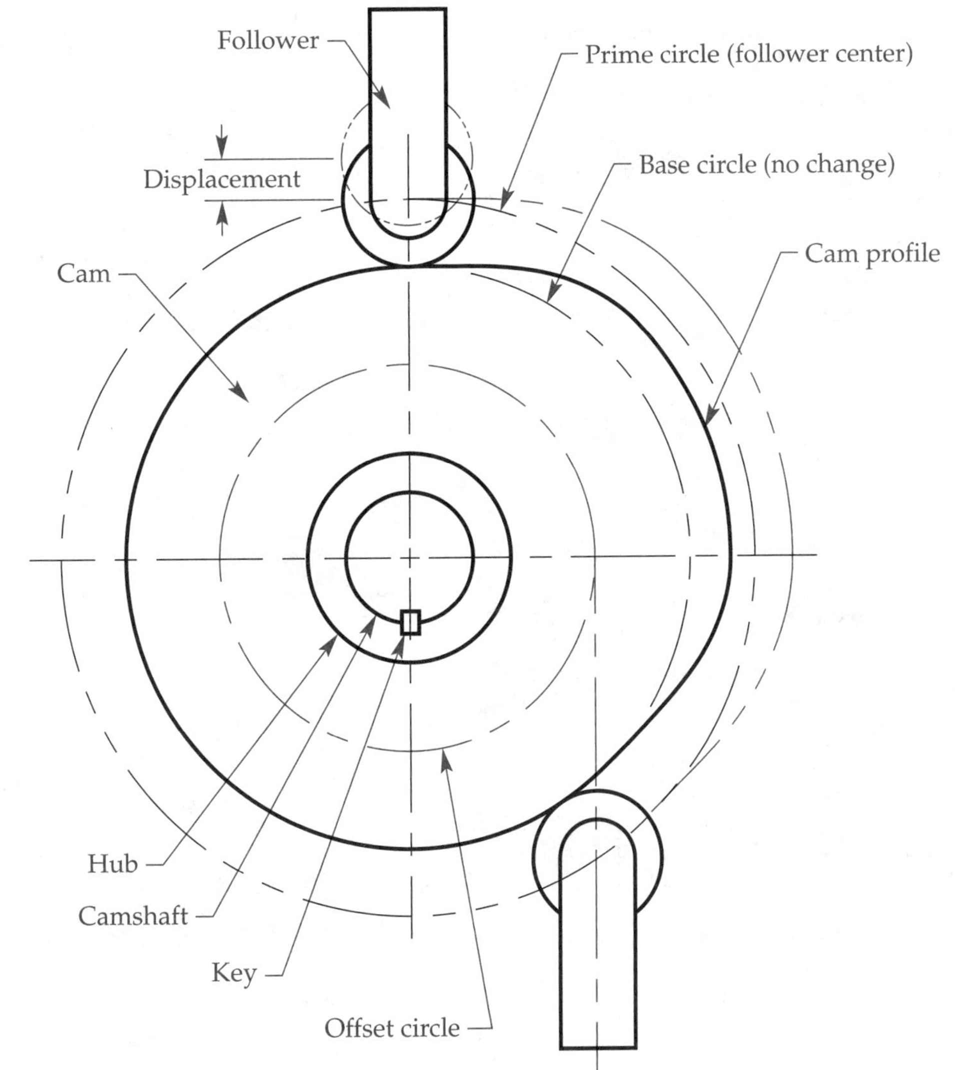


**Figure 19-6.**
This illustration features some of the terms common to radial cams.

- **Oscillate.** To move back and forth in a repeating fashion. A follower may oscillate in a linear fashion (straight up and down or in and out) or may rock back and forth about a pivot point.
- **Displacement.** The distance that a follower rises or falls during a portion of the cycle.
- **Dwell.** A period of time for which the follower does not move or change position.
- **Cycle.** In the case of round cams, one 360° revolution of the cam. In the case of a linear cam, one cycle of cam movement from a home position to other position(s) and then back. The displacement diagram usually charts one cycle.
- **Base circle.** A circle concentric with the center axis of a radial (plate) cam having a radius equivalent to the follower edge at rest or in the home position. If the cam had a base circle as its overall shape, there would be no movement of the follower.
- **Prime circle.** A circle that represents the position of a roller follower's center axis at the home position. The cam profile must take into consideration the roller wheel diameter at the prime circle position to accommodate surface-to-surface tangencies.
- **Offset circle.** A circle that represents the offset amount for a roller follower's center axis. For a roller follower offset from the center axis of the cam, the cam profile must take into consideration the offset position to accommodate surface-to-surface tangencies.

## Cam Displacement Diagrams

Cams are usually designed with the assistance of ***displacement diagrams.*** Displacement amounts can be charted over the course of one cycle and then transferred into the cam profile, which is often a spline shape not easily dimensioned. Reading a print that describes a cam may require an understanding of the displacement diagrams that are used to graphically represent the movement of the follower. Figure 19-7 shows a displacement diagram representing the rises and fall for one complete revolution of a cam.

The displacement diagram not only represents 360° of revolution, but usually represents a certain period of time. For example, the diagram illustrated in Figure 19-7 indicates the follower is to rise a given amount in one second, dwell for .5 seconds, rise an additional amount in 1.5 seconds, dwell for .5 seconds, fall back to the beginning level in two seconds, and then dwell for .5 seconds before starting the cycle again. The total cycle is, therefore, six seconds.

Depending on the accuracy of the desired cam and the level of accuracy required, the data from the displacement diagram can be transferred to the cam drawing in increments of 10°, 15°, or 30°. In terms of two-dimension drafting, CAD systems can calculate spline curves through the points that can provide useful data for the CNC machine or other manufacturing equipment.

The displacement information may also be calculated using computer programs and spreadsheets. In these cases, these data can be placed on the print in table form, as well as stored electronically. Cam design

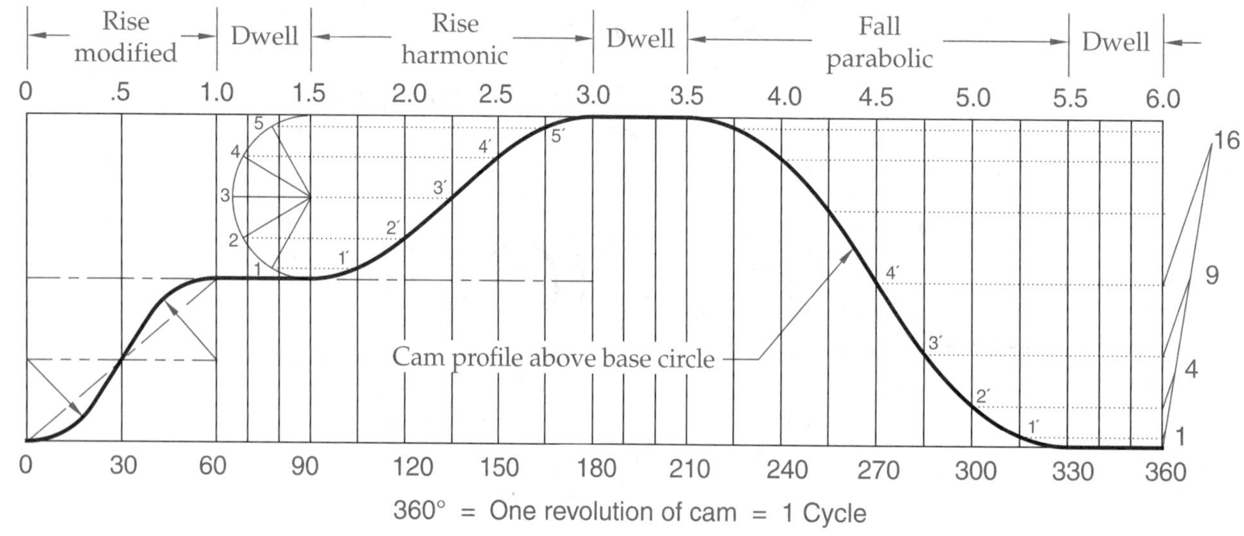


Figure 19-7.
This displacement diagram represents a 360° cycle that occurs in six seconds. It features a rise during the first second, a dwell of .5 seconds, then an additional rise equal to the first, but over the next 1.5 seconds, another dwell of .5 seconds, and then a fall over the next 2.5 seconds.

software is also available for most cam applications, allowing designers to use additional tools that automate the design process.

## Follower Motion

There are different methods of drafting the rise and fall of the follower. Some methods cause more stress on the follower as the cam forces it through various transitions. The designer must consider velocity, acceleration, and even surges or jerks that may be created by moving parts. To prevent the components from being stressed too much by the changing motion, some motion transitions are desirable. The designer understands this information and, therefore, it is seldom a concern for the print reader. In Figure 19-8, methods of creating the fall and rise displacements in the diagram include:

- **Uniform motion.** In Figure 19-8A, the rise and fall lines are simply straight-line segments. This motion is characterized by a constant rate, but may subject the follower to more sudden change.
- **Modified uniform motion.** At Figure 19-8B, a radius is added at the beginning and end of a straight line to smooth the transition.
- **Harmonic motion.** At Figure 19-8C, the shape is determined by projecting from an equally divided semicircle.
- **Parabolic motion.** At Figure 19-8D, the displacement curve is created with parabolic construction.
- **Cycloidal motion.** At Figure 19-8E, the displacement curve is generated from a cycloid.

## Reading Cam Prints

Drawings for cam devices follow standard practices. They are often created as multiview drawings, but with the cam profile dimensioned in a table or displacement diagram. However, angular dimensions, such as HARMONIC RISE 45° or DWELL 90°, may be given. See Figure 19-9. With respect to the tolerance of the geometry, profile of a surface is a common geometric control for the cam profile.

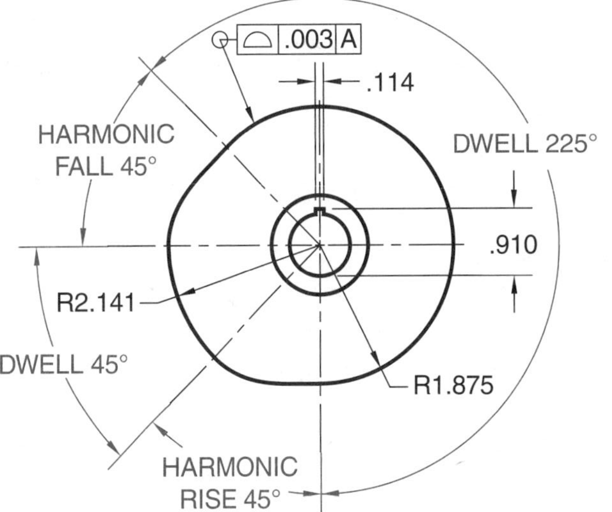


**Figure 19-9.**
While cam drawings usually follow standard practices, additional terms or techniques may be incorporated to describe the irregular profile.

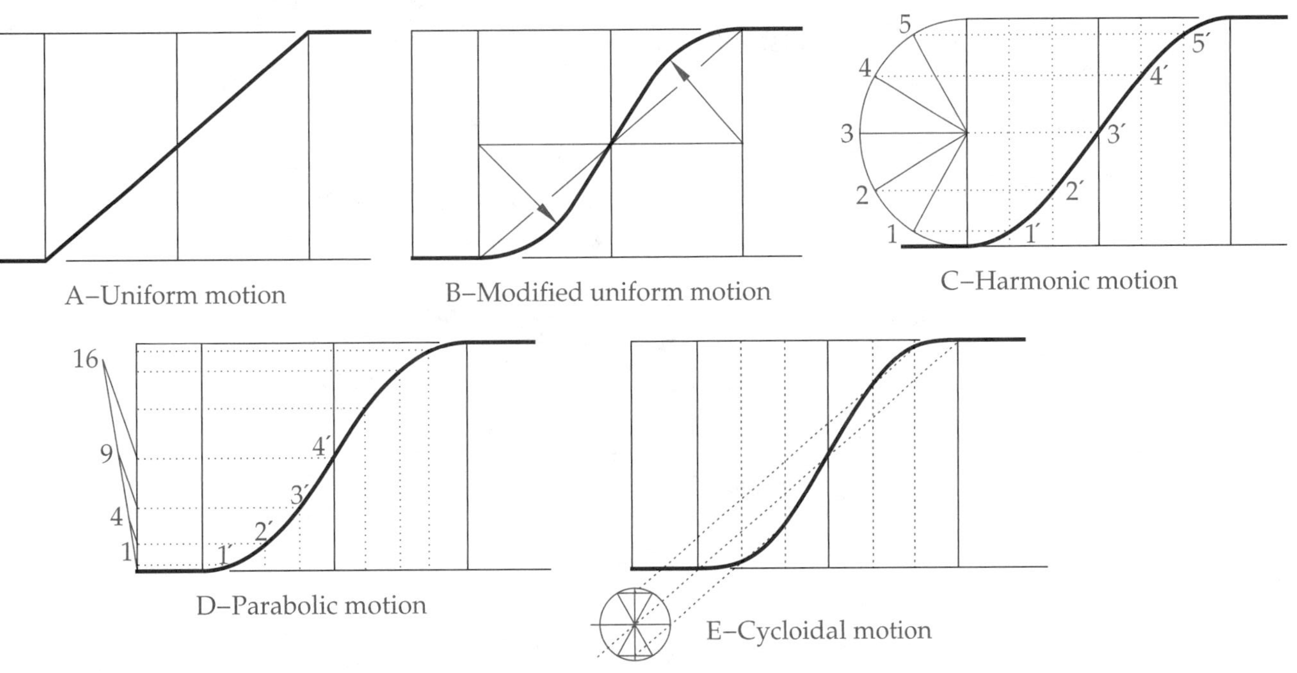


**Figure 19-8.**
Different techniques are available for designing the displacement motion as transitions occur.

## Review Questions

*Circle the answer of choice, fill in the blank, or write a short answer.*

1. A cam can be designed to transmit rotary or linear motion into ______________________, ______________________, or ______________________ motion.

2. *True or False?* Cam surfaces always mate with follower surfaces.

3. List four specific and different types of cams based on the shape. ______________________

______________________

______________________

4. List three specific types of followers based on the contact surface. ______________________

______________________

5. The term ______________________ means to move back and forth in a repeating pattern.

6. *True or False?* During a cycle, the period of time that a follower does not move, but maintains its position, is called the *freeze*.

7. For a radial cam, one cycle is usually ______________________ degrees.

8. All of the following is usually shown in a displacement diagram *except:*
   A. time.
   B. rise.
   C. fall.
   D. base circle diameter.

9. What one word is the key factor in why a designer may choose one transition motion over another?

______________________

10. Which of the following is *not* a displacement motion term?
    A. Cycloidal.
    B. Parabolic.
    C. Hyperbolic.
    D. Harmonic.

## Review Activity 19-1

*Analyze the following displacement diagrams for one cycle of the follower displacement. Then, answer the questions for each diagram.*

1. Time for one cycle ________________
   Total dwell time ________________
   Total rise after two seconds ________________
   Total rise after eight seconds ________________
   Total time to fall ________________

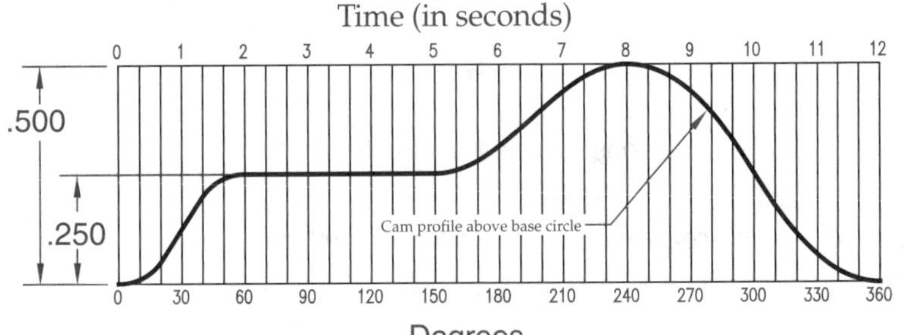


2. Degrees in one cycle ________________
   Total dwell time ________________
   Degrees in first rise ________________
   Degrees in second rise ________________
   Degrees in fall ________________

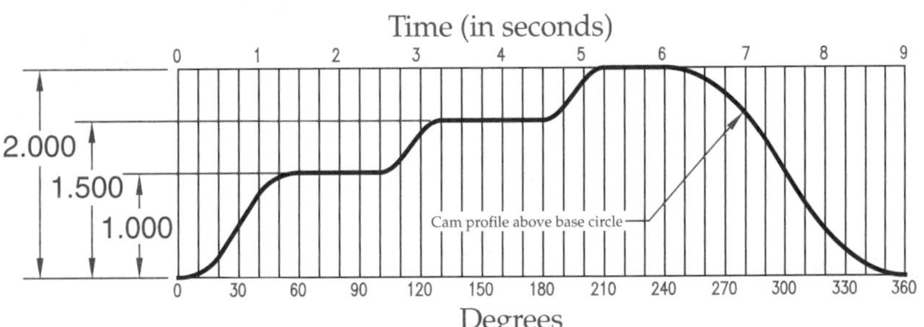


3. Time for one cycle ________________
   Total dwell time ________________
   Time to make first rise ________________
   Displacement—second rise ________________
   Total time spent falling ________________

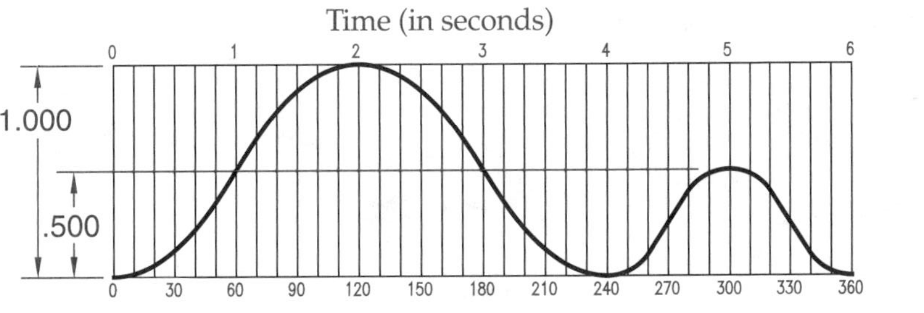

## Industry Print Exercise 19-1

*Refer to the print PR 19-1 and answer the questions below.*

1. What view of the drawing details the size measurements of the cam profile? ____________
2. Which cam type describes the cam featured on this part? ____________
3. Is there a displacement diagram featured on this print? ____________
4. As drawn, is the cam profile symmetrical above and below the horizontal center line? ____________
5. As drawn, is the cam profile symmetrical with respect to the vertical center line? ____________
6. Calculate the displacement of the follower surface if the base circle has a radius of .173″: ____________
7. How many sets of displacement measurements are given to describe the profile of the cam surface between the 9 o'clock position and the 3 o'clock position? ____________
8. How much geometric tolerance (profile of a surface tolerance) is applied to the cam profile? ____________
9. What surface hardness is specified for the cam surface? ____________
10. How "wide" is the cam surface, as specified in the main view? ____________

*Review questions based on previous units:*

11. At what scale is the original drawing of this part created? ____________
12. At what scale is Detail B created? ____________
13. What name is given to the ridges formed on the left end of this part, just to the right of Detail B? ____________
14. How deep is the center drill in the right end of this part? ____________
15. What surface quality is specified for the .465″ diameter surface down in the .142″ wide groove just to the left of datum A? ____________

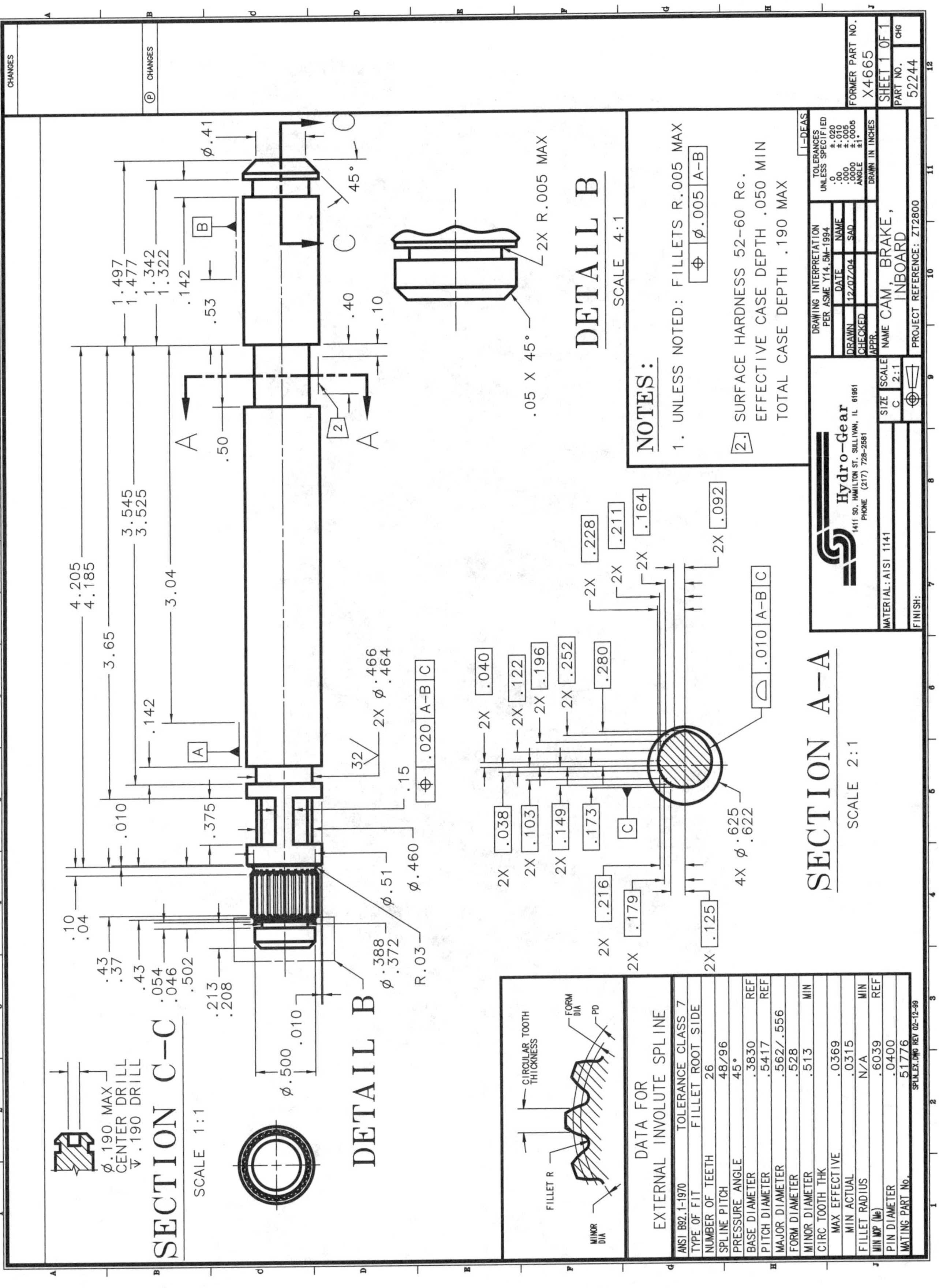


PR 19-1.
Print supplied by Hydro-Gear.

Plastic parts can be found everywhere in everyday life. Look at your cell phone and identify how many different plastic components are used in the assembly.

UNIT 20

# Plastic Parts

*After completing this unit, you will be able to:*

**List** common names and abbreviations used in the manufacture of plastic parts.

**Explain** the difference between thermoplastic and thermoset plastics.

**Explain** the processes used in the plastics industry.

**Describe** methods used to fasten and join plastics.

**Read** and **interpret** prints from the plastics industry.

***Plastic*** refers to a product of polymer chemistry—a synthetic material formed by chemical reaction rather than a substance of natural origin. Early forms of plastics had limited use as toys, kitchen utensils, and other products that do not require great strength, durability, or heat and chemical resistance. Now, plastics are an important engineering material and have a wide variety of applications. Some of these applications include electrical parts, electrical insulation, gears, bearings, packaging, automobile bodies, motor housings, and architectural products. By the latter part of the 20th century, the plastics industry represented a major segment of manufacturing and continues to do so today.

## Materials in the Plastics Industry

Plastics are broadly classified as either thermoplastics or thermoset plastics. The division is based on the effect temperature has on their properties. The next sections cover the two types of plastics.

### Thermoplastics

***Thermoplastics*** soften when heated and become solid when cooled. Thermoplastics are classified further into categories based on solvent resistance, degree of volumetric change during processing, and optical properties. ***Amorphous thermoplastics*** are typically transparent, exhibit relatively little shrinkage during processing, and have poor solvent resistance. Some common amorphous thermoplastics are polystyrene (PS), polyvinyl chloride (PVC), and polycarbonate (PC). ***Semicrystalline thermoplastics*** are typically translucent, exhibit more shrinkage than amorphous plastics, and have much higher solvent resistance than amorphous plastics. Some common semicrystalline thermoplastics are polyethylene (PE), polypropylene (PP), and nylon (polyamide [PA]). Plastics are made opaque by adding pigments. A listing of common plastics, their abbreviations, their classification, and some typical applications is given in Figure 20-1.

Thermoplastics are solid at room temperature, but become liquid when heated beyond their melting points. This "melting/freezing" cycle is the principle used to manufacture the plastic into finished parts. The initial form of the thermoplastics is either pellets or powders (solid forms). Some fabrication techniques using melted thermoplastics are injection molding, extrusion, blow molding, and rotational molding. Several plastic products made with these processes are shown in Figure 20-2. When a sheet of thermoplastic is softened to just below its melting point, it can be fabricated by thermoforming using vacuum or pressure. In their solid state, thermoplastics can be processed by standard metal-cutting techniques, such as drilling and milling.

### Thermoset Plastics

***Thermoset plastics,*** unlike thermoplastics, cannot be made liquid again by heat or solvent. Thermoset materials are chemically altered through a curing process known as *cross-linking polymerization,* which occurs during the molding process. The chemical makeup of thermoset materials causes extensive

| Thermoplastics | | |
|---|---|---|
| **Abbreviation and Type** | **Name** | **Typical Applications** |
| ABS (amorphous) | Acrylonitrile-butadiene-styrene | Appliance housings, refrigerator liners |
| PA (semicrystalline) | Polyamide (Nylon®) | Bearings, gears, housings for tools |
| PC (amorphous) | Polycarbonate | Helmets, headlights, compact discs |
| PE (semicrystalline) | Polyethylene | Bottles, bags, tubing |
| PMMA (amorphous) | Polymethylene methacrylate | Skylights, stoplights |
| POM (semicrystalline) | Polyoxymethylene (Acetal) | Gears, plumbing fixtures |
| PP (semicrystalline) | Polypropylene | Suitcases, Tupperware® |
| PS (amorphous) | Polystyrene | Cutlery, packaging, insulation |
| PVC (amorphous) | Polyvinyl chloride | Pipes, window frames, siding, hoses |

| Thermosets | | |
|---|---|---|
| **Abbreviation** | **Name** | **Typical Applications** |
| EP | Epoxy | Adhesive, aerospace panels |
| MF | Melamine-formaldehyde | Heat-resistant surfaces, dishes |
| PF | Phenol-formaldehyde | Cooking pan handles, breaker boxes |
| UP | Unsaturated polyester | Tubs and showers, car panels, boats |

**Figure 20-1.**
This table shows common plastics, their abbreviations, and typical applications.

bonding between molecules, which results in the formation of a rigid mass. Curing can be initiated by the addition of an initiator chemical (catalysis), the application of heat, ultraviolet light, or by the simple mixing of two components, depending on the type of thermoset.

The most common thermoset plastics are phenolics, unsaturated polyesters, epoxies, and melamine. Some processes used to make products from thermoset materials include compression molding, transfer molding, and fiberglass lay-up. The initial form of thermoset materials may be granules, pastes, or liquids. To obtain the better product qualities, such as higher strength and durability, molding compounds made from these resins always contain additional fillers and reinforcing agents.

## Processes in the Plastics Industry

There are many processes applied in industry to form plastics into useful products. Those most widely used are discussed here to define terms that might appear on prints.

***Molding*** is the process of forming a plastic object by forcing molten plastic material into a hollow mold using pressure. Examples of molding processes are injection molding and blow molding. Injection molding is the most widely used process to create products of great complexity with the lowest cost, Figure 20-3. The majority of the prints that involve plastic parts are for injection molding. Blow molding is typically used to make plastic bottles, Figure 20-4.

***Thermoforming*** is the process of forming a plastic object by forcing a softened sheet of thermoplastic onto a mold using vacuum or positive pressure. Figure 20-5 illustrates a part that was formed about a mold. Many large sheet-type parts, such as motorcycle windscreens, are made by thermoforming.

***Extrusion*** is a continuous process of melting thermoplastics, then pumping the molten plastic through a heated die. The shape of the die exit forms the product. Once the molten plastic passes through the die, it is cooled and finally sized by downstream equipment. The types of products made from extrusion depend on the die. Some products made from extrusion include:

- Films and sheets used in single or multiple layers or as coatings for cloth and paper.
- Profile shapes such as rod, pipe, or channels, Figure 20-6.
- Coatings around wire and cable.

***Casting*** is the process of forming a plastic object by pouring molten material into an open mold. No pressure is used in casting. In the case of a thermoset, the cast material polymerizes, or cross-links, in the mold. ***Rotational molding*** is a casting process that converts powdered thermoplastics into hollow objects by rotating the powder in a closed, heated mold until the powder melts and fuses into a solid mass.

Machining, finishing, decorating, and assembly for plastic products are the same as in other fabricating industries. Therefore, the drawings and prints providing the design and manufacturing specifications of plastic parts and assemblies are also similar.

Figure 20-2.
Examples of thermoplastics made using various processes can be found in everyday life. (Reed Spectrum)

## Fastening and Joining Plastics

There are four principal techniques to join plastics to each other and to other materials:

- **Mechanical fasteners.** Thread-cutting screws, rivets, and press-in inserts.
- **Mechanical fits.** Press-in and snap-on fits.
- **Welding.** Ultrasonic welding, heat welding, and spin welding.
- **Adhesives.** Elastomers, epoxies, and solvents.

An ultrasonic welding machine is shown in Figure 20-7.

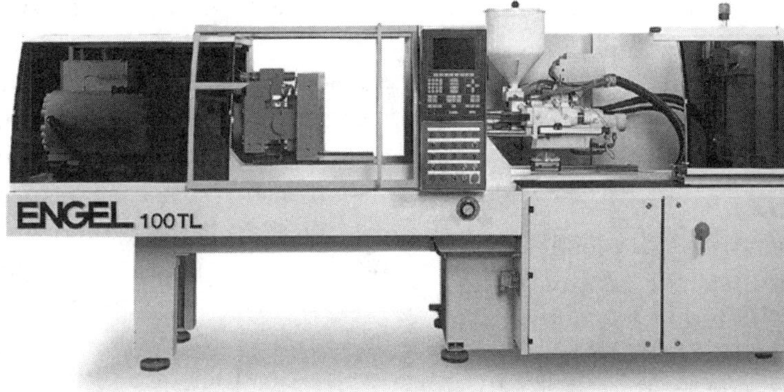


Figure 20-3.
This machine is a typical injection molding machine. (Engel, Inc.)

Figure 20-4.
These plastic bottles are made using a blow molding machine. (Uniloy)

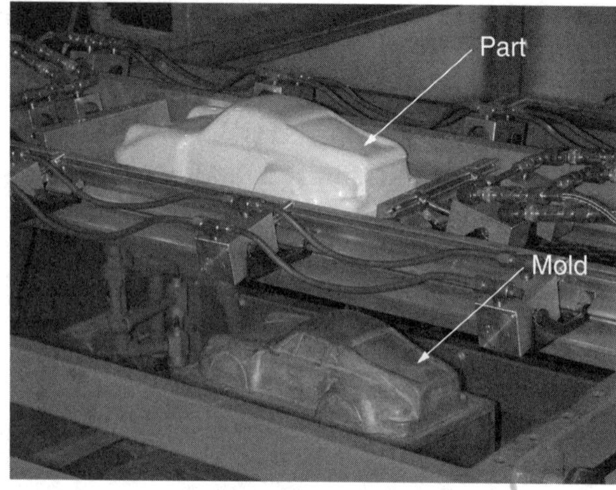


Figure 20-5.
This plastic car model was created by thermoforming. (MAAC Machinery)

## Plastics Terminology

The terminology associated with plastics technology differs somewhat from that of metalworking. You must understand plastics terms and features in order to read prints for the plastics industry. Many of the terms discussed in this section frequently appear on prints related to manufacturing with plastics. Other general terms are also discussed.

Figure 20-6.
Plastic pipes and tubing are created using the extrusion process. (Polygon Company)

A ***monomer*** is a molecule of low molecular weight capable of reacting with identical or different monomers to form a polymer. A ***polymer*** is the basic molecular structure of a plastic material. It is the product of polymerization. ***Polymerization*** is the joining of two or more molecules to form a new and more complex molecule whose physical properties are different. ***Resin*** is the basic organic material from which plastics are formed. The term is also used to describe plastic not yet converted into a product. An ***elastomer*** is a material having similar characteristics to those of rubber. Elastomers stretch to at least twice their "normal" length. In addition, they rapidly return to the original length. Thermoplastic elastomers developed in recent years have some advantages over rubber.

***Molding process*** is the placing of a plastic material, in a heated or liquid state, into a hollow mold under pressure to form a plastic part. The ***mold*** is a hollow form into which a plastic material is placed to give the material its final shape. A ***cavity*** is the depression, or set of matching depressions, in a mold used for plastic forming. The cavity shapes the surfaces of the case or molded article. This term also refers to the stationary part of a mold used for injection molding. The ***parting line*** is the edge of a mold cavity where mold halves come together.

An ***ejector pin*** is a pin designed to push a plastic product out of a mold. The ejector pin is typically made of hardened steel and its location is often indicated on the print for the associated part. A ***knockout*** is any part of the mechanism of a mold used to eject the molded articles.

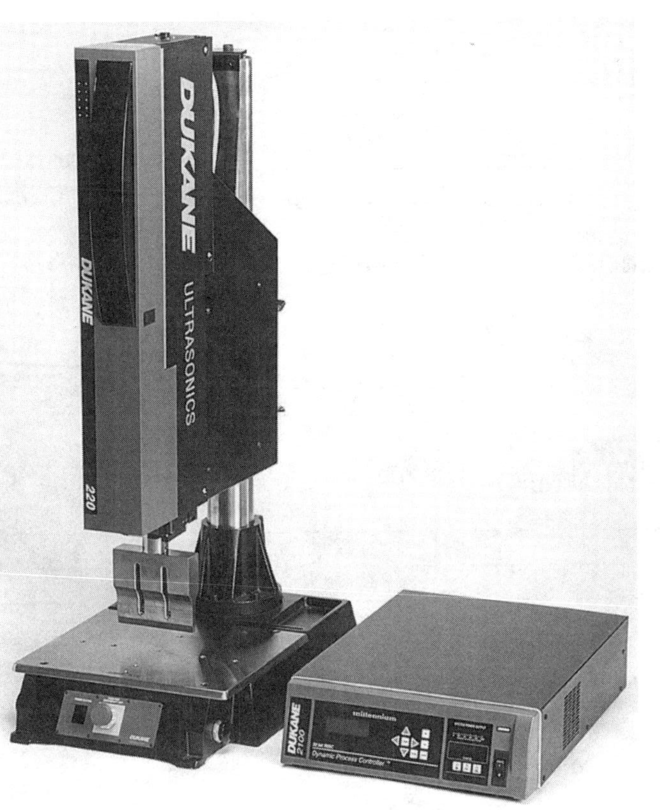


Figure 20-7.
This is an ultrasonic plastic welding machine. (Dukane)

The term *core* can have three meanings. First, a core is part of a mold that hollows out a section of a part. In this usage, a core is also called ***core pin.*** Second, a core is the movable portion of a mold used for injection molding through which the ejector pins typically pass. Third, a core is a channel in a mold for circulation of heat-transfer media.

A ***draft*** is a slight outward taper in a mold wall designed to facilitate removal of the molded object. A ***back draft*** is a taper in the opposite direction. This tends to impede removal of the object, or hold it in place.

In injection molding and transfer molding, the ***sprue*** is the main feed channel that connects the mold orifice to the runners leading to each cavity gate. The term is also given to the piece of plastic material formed from this channel. A ***runner*** is the usually circular channel that connects the sprue with the gate of the mold cavity. The term is also given to the piece of plastic material formed from this channel. A ***gate*** is the channel connecting the runner to the part cavity. The gate may have the same diameter or cross section as the runner, but is often restricted to a diameter of 1/8″ or less. ***Gate vestige*** is the residual material that remains after a gate is torn or cut away from a molded part.

***Molding shrinkage*** is the immediate shrinkage of the plastic part after it has been removed from the mold. Both the mold and molded part are measured at normal room temperature. ***Warpage*** is the distortion of the plastic part caused by nonuniform shrinkage. A ***sink mark,*** or ***sink,*** is a shallow depression or dimple on the surface of an injection-molded article caused by a short shot or local internal shrinkage after the gate seals.

***Flash*** is the thin web of excess material that is forced into the crevices between mating mold surfaces. Flash remains attached to the part when removed from the mold. Often, flash is removed in a later finishing step.

A ***boss*** is a circular protrusion from the plastic body. A boss is used to anchor a threaded fastener into the plastic. A ***gusset*** is an angular piece of material added to strengthen two adjoining walls. An ***insert*** is a material placed into the mold cavity prior to molding that is then surrounded by the molding material. ***Overmold*** is the process of injection molding a plastic over an existing plastic part. This term also refers to the material that forms the overmolding.

## Reading Prints from the Plastics Industry

Once you are familiar with plastic materials, processes, and terminology, reading prints for plastic parts and assemblies is much like reading prints for other manufacturing industries. The drawing layout, title block, notes, and dimensioning procedures are the same.

## Review Questions

*Circle the answer of choice, fill in the blank, or write a short answer.*

1. Plastic is a general term that indicates a(n) ________________ material formed by chemical reaction, rather than a substance of ________________ origin.

2. *True or False?* In the industry, some use the term thermoplastics and some use the term thermoset plastics, but they both mean the same thing.

3. The initial form of thermoplastics is often ________________ or ________________, in both cases solid form.

4. For each of the following, indicate the corresponding acronym.
   A. Polystyrene ________________
   B. Polyvinyl chloride ________________
   C. Polycarbonate ________________
   D. Polyethylene ________________
   E. Polypropylene ________________
   F. Nylon ________________

5. *True or False?* Thermoset plastics cannot be melted once they are formed.

6. Which of the following processes is *not* used with plastic materials?
   A. Thermoforming.
   B. Extruding.
   C. Casting.
   D. All of the above could be a plastic material process.

7. Which of the following processes is *not* used to join plastic materials together?
   A. Soldering.
   B. Adhesives.
   C. Welding.
   D. Mechanical fits, such as snap-on.

8. A(n) ________________ is a plastic material having similar characteristics to those of rubber.

9. What is one of the most common plastic processes, wherein liquid plastic is forced into a hollow mold with great pressure? ________________

   ________________

10. On a print, what term is used to indicate the location on a molded object where the two halves of the mold meet?
    A. Point of demarcation.
    B. Core datum.
    C. Ejection edge.
    D. Parting line.

11. Sometimes on a print, the location of the ________________ that push the part out of the mold will be indicated.

12. *True or False?* With respect to plastic molding, the word *core* has more than one meaning.

13. For parts that are molded, the print reader should expect to see information about the taper of a surface that helps the part come out of the mold more easily. This is called:
    A. allowance.
    B. shrink.
    C. draft.
    D. flash.

14. What term applies to the thin web of excess material that may be on a molded plastic part that just came out of the mold? ______________________________

______________________________________________

15. *True or False?* Sometimes molded plastic parts have metal inserts placed into the mold cavity or additional plastic is molded over the top of existing plastic.

## Industry Print Exercise 20-1

*Refer to the print PR 20-1 and answer the questions below.*

1. How much draft, if any, is specified for this part? ____________________

2. The material specified for this part is a common plastic. What is its trade name? __________
3. What is the maximum allowable depth specified for a knockout pin mark? __________
4. What datum surface is referred to with respect to flash and gate breakoff? __________
5. To prevent sharp edges on the final molded part, what radius is allowed on all edges unless otherwise specified? __________

*Review questions based on previous units:*

6. What is the name of this part? __________
7. What scale are views on the original print? __________
8. Are there any revisions made to this drawing that were *not* done in the CAD system? __________
9. What is the diameter of the feature that is used to establish datum axis C? __________
10. What type of dimensions are those that have boxes around the values? __________
11. What scale are the isometric pictorial views? __________
12. What maximum material condition is specified for the diameter of each of the three posts? __________
13. How much do the three posts stick out above the main body of the part? Do not apply tolerances. __________
14. What revision issue is this drawing? __________
15. Issue A of this drawing was in conjunction with ECN 328945. What does ECN most likely stand for?

____________________

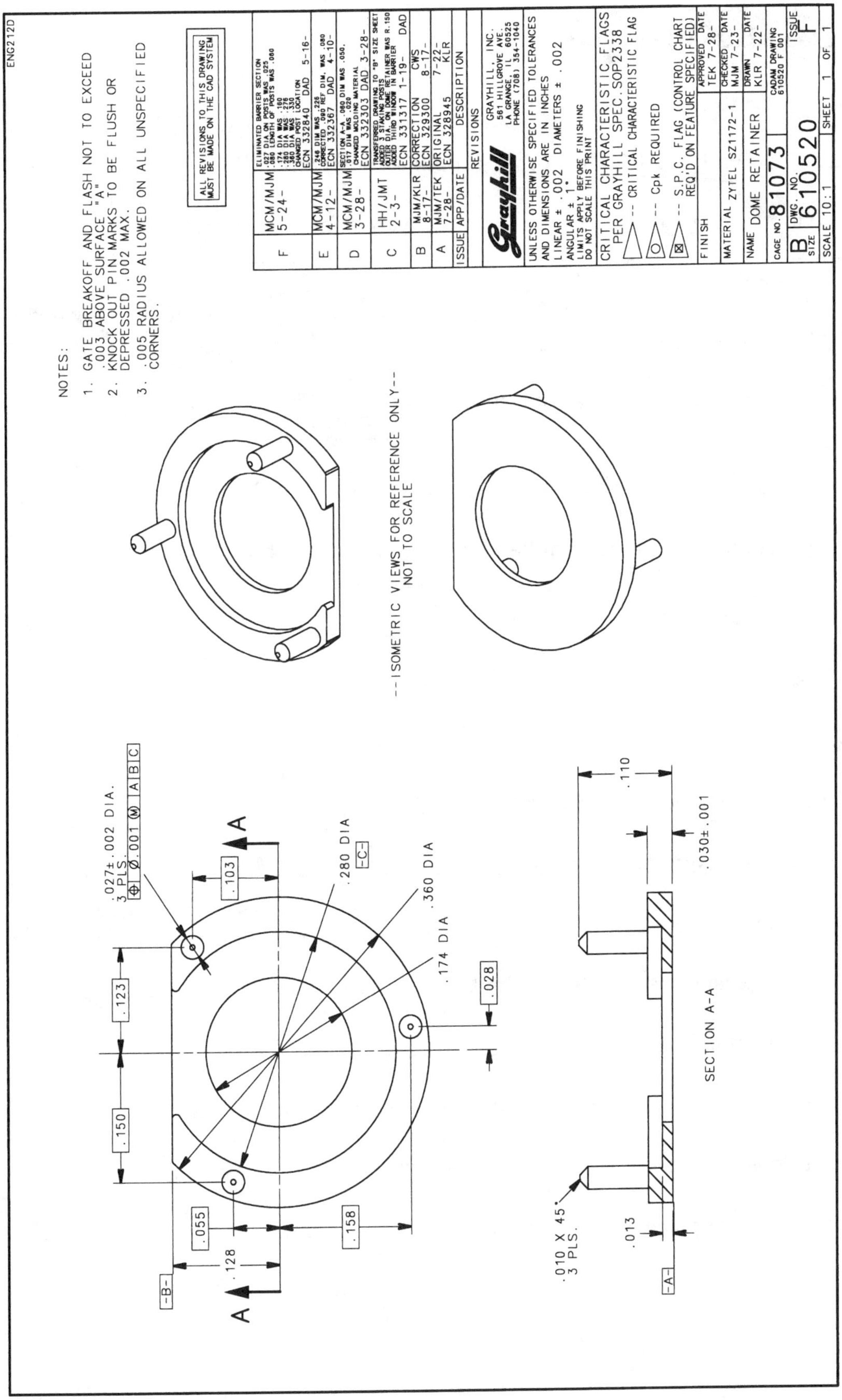

NOTES:

1. GATE BREAKOFF AND FLASH NOT TO EXCEED .003 ABOVE SURFACE "A"
2. KNOCK OUT PIN MARKS TO BE FLUSH OR DEPRESSED .002 MAX.
3. .005 RADIUS ALLOWED ON ALL UNSPECIFIED CORNERS.

ALL REVISIONS TO THIS DRAWING MUST BE MADE ON THE CAD SYSTEM

| ISSUE | APP/DATE | DESCRIPTION |
|---|---|---|
| F | MCM/MJM 5-24- | ELIMINATED BARRIER SECTION .027 DIA ON POSTS WAS .025 .080 LENGTH OF POSTS WAS .060 .174 DIA WAS .160 .280 DIA WAS .276 .360 DIA WAS .330 CHANGED POST LOCATION ECN 332840 DAD 5-16- |
| E | MCM/MJM 4-12- | .246 DIM WAS .226 CORRECTED .090 REF DIM. WAS .080 ECN 332367 DAD 4-10- |
| D | MCM/MJM 3-28- | SECTION A-A .080 DIM WAS .050. .017 DIM WAS .020 CHANGED MOLDING MATERIAL ECN 332303 DAD 3-28- |
| C | HH/JMT 2-3- | TRANSFERRED DRAWING TO "B" SIZE SHEET ADDED 3 STAKING POSTS OUTER DIA. ON DOME RETAINER WAS R.150 ADDED THIRD WINDOW IN BARRIER ECN 331317 1-19- DAD |
| B | MJM/KLR 8-17- | CORRECTION CWS ECN 329300 8-17- |
| A | MJM/TEK 7-28- | ORIGINAL 7-22- ECN 328945 KLR |

REVISIONS

Grayhill

GRAYHILL, INC.
561 HILLGROVE AVE.
LA GRANGE, IL 60525
PHONE (708) 354-1040

UNLESS OTHERWISE SPECIFIED TOLERANCES AND DIMENSIONS ARE IN INCHES
LINEAR ± .002 DIAMETERS ± .002
ANGULAR ± 1°
LIMITS APPLY BEFORE FINISHING
DO NOT SCALE THIS PRINT

CRITICAL CHARACTERISTIC FLAGS PER GRAYHILL SPEC.SOP2338
-- CRITICAL CHARACTERISTIC FLAG
-- Cpk REQUIRED
-- S.P.C. FLAG (CONTROL CHART REQ'D ON FEATURE SPECIFIED)

FINISH — APPROVED DATE TEK 7-28-
MATERIAL ZYTEL SZ1172-1 — CHECKED DATE MJM 7-23-
NAME DOME RETAINER — DRAWN DATE KLR 7-22-
CAGE NO. 81073 — CADAM DRAWING 610520 F 001
SIZE B DWG. NO. 610520 ISSUE F
SCALE 10:1 SHEET 1 OF 1

--ISOMETRIC VIEWS FOR REFERENCE ONLY--
NOT TO SCALE

.027±.002 DIA.
3 PLS.
⌖ Ø.001 Ⓜ A B C

A
.103
.280 DIA
-C-
.360 DIA
.174 DIA
.123
.028
.150
.055
.158
.128
-B-
A

.110
.030±.001
SECTION A-A
.010 X 45°
3 PLS.
.013
-A-

ENG212D

PR 20-1.
Print supplied by Grayhill, Inc.

## Industry Print Exercise 20-2

*Refer to the print PR 20-2 and answer the questions below.*

1. The largest diameter of this part is .470″. What draft is applied to that dimension? ______
2. What draft is applied to the six ridges of the knob that are 60° apart, as shown in the section view? ______
3. The small central cavity with a diameter of .165″ simply indicates +DFT. According to drawing notes, what draft is allowed for that feature? ______
4. The material used for this part is a blended plastic developed by GE. What is its trade name? ______
5. In total, what is the angular tolerance for angles, unless otherwise specified? ______
6. How thick is the material of the part at the right-hand side of the part, as shown in the section view? ______

*Review questions based on previous units:*

7. What is the part number for this print? ______
8. Are there any leader lines shown on this print? ______
9. What scale are the views on the original drawing? ______
10. Who checked this drawing? ______
11. Does this object have any dimensions that use the limit method of tolerancing? ______
12. How many cutting planes are indicated? ______
13. What tolerance is to be applied to decimal values unless otherwise specified? ______
14. For some of the dimensions, there is a TYP note. What does that stand for? ______
15. How many changes were made as part of revision A? ______

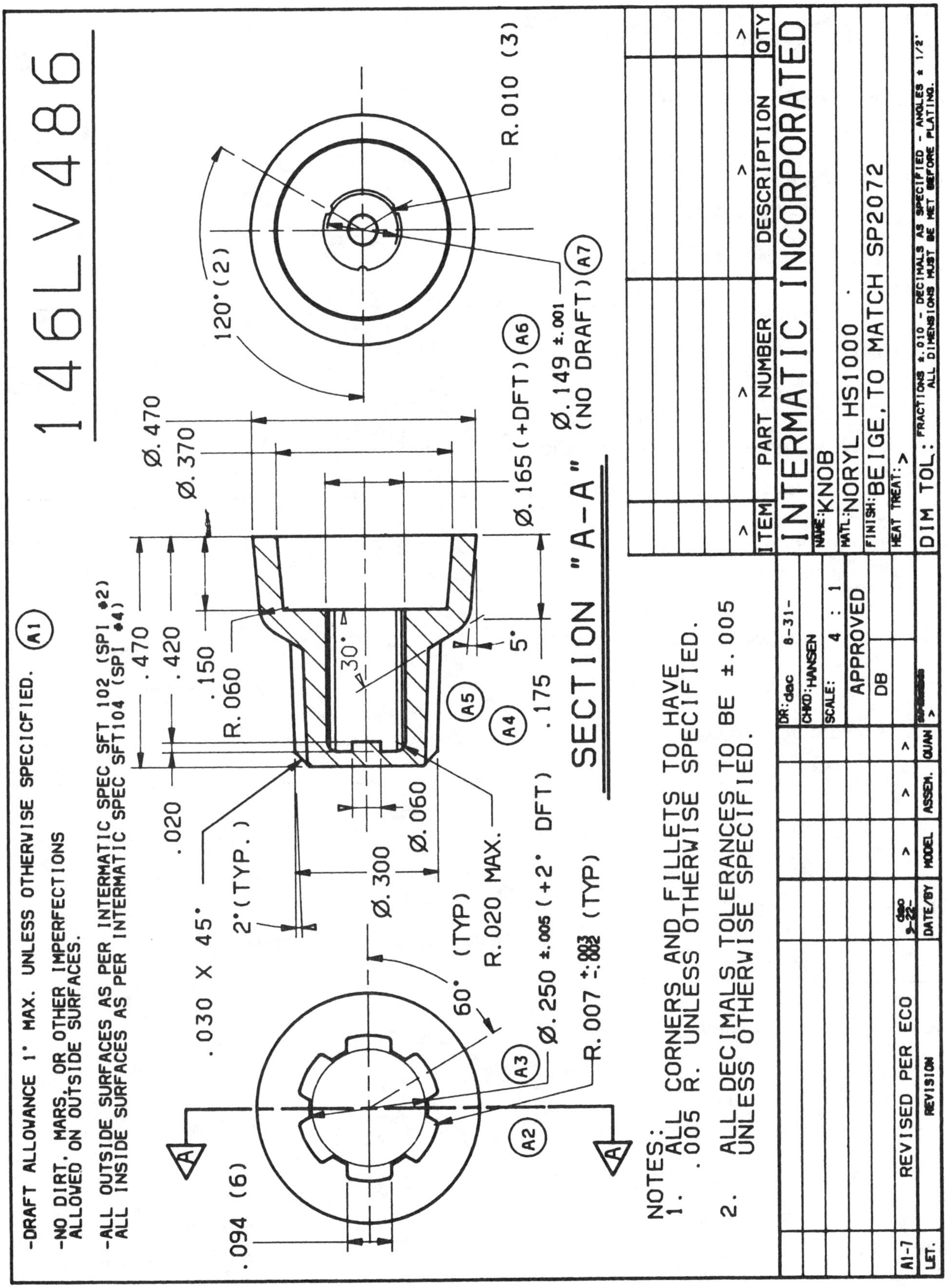

-DRAFT ALLOWANCE 1° MAX. UNLESS OTHERWISE SPECICFIED. (A1)

-NO DIRT, MARS, OR OTHER IMPERFECTIONS
ALLOWED ON OUTSIDE SURFACES.

-ALL OUTSIDE SURFACES AS PER INTERMATIC SPEC SFT 102 (SPI #2)
ALL INSIDE SURFACES AS PER INTERMATIC SPEC SFT104 (SPI #4)

146LV486

.470
.020
.420
Ø.470
.030 X 45°
.150
Ø.370
.094 (6)
2°(TYP.)
R.060
120°(2)
30°
Ø.300
Ø.060
60° (TYP)
R.020 MAX.
(A5)
(A4)
5°
Ø.165(+DFT) (A6)
R.010 (3)
(A2) (A3)
Ø.250 ±.005 (+2° DFT)
.175
Ø.149 ±.001
(NO DRAFT) (A7)
R.007 +.003 -.002 (TYP)
SECTION "A-A"
A
A

NOTES:
1. ALL CORNERS AND FILLETS TO HAVE .005 R. UNLESS OTHERWISE SPECIFIED.
2. ALL DECIMALS TOLERANCES TO BE ±.005 UNLESS OTHERWISE SPECIFIED.

| ITEM | PART NUMBER | DESCRIPTION | QTY |
|---|---|---|---|
| > | > | > | > |

INTERMATIC INCORPORATED

NAME: KNOB
MATL: NORYL HS1000
FINISH: BEIGE, TO MATCH SP2072
HEAT TREAT: >
DIM TOL: FRACTIONS ±.010 - DECIMALS AS SPECIFIED - ANGLES ± 1/2° ALL DIMENSIONS MUST BE MET BEFORE PLATING.

DR: dac 8-31-
CHKD: HANSEN
SCALE: 4 : 1
APPROVED
DB

| LET. | REVISION | DATE/BY | MODEL | ASSEM. | QUAN |
|---|---|---|---|---|---|
| A1-7 | REVISED PER ECO | dac 9-22- | > | > | > |

PR 20-2.
Print supplied by Intermatic, Inc.

## Bonus Print Reading Exercises

The following questions are based on various bonus prints located in the folder at the back of this textbook. Refer to the print indicated, evaluate the print, and answer the question.

**Print AP-021:**

1. What material is specified for this part? ____________________
2. In which view is the parting line located? ____________________
3. Does the cavity number appear inside of the part or outside? ____________________
4. What type of mold release *cannot* be used? ____________________
5. In the view where the cutting-plane line is shown, what do the circles represent? ____________________
6. Give the maximum allowed amount of flash. ____________________
7. What is the typical draft on the outer sides of the part? ____________________

**Print AP-022:**

8. What surface finish is specified for this part and what document is referenced for that? ____________________
9. What type of plastic is specified for this part? ____________________
10. How much flashing or mismatch is permitted at the parting line? ____________________
11. In Section A-A, a .352″ diameter hole is specified. Is the other end of that hole larger, smaller, or the same diameter? ____________________
12. What is the maximum amount of draft that can be applied to any feature of this part? ____________________

UNIT 21

# Precision Sheet Metal Parts

*After completing this unit, you will be able to:*

**Explain** the design implications of bending sheet metal to form a part.

**Define** various terms related to sheet metal bends.

**Use** a setback chart to calculate bend radii for precision sheet metal.

**Use** the proper formula to calculate bend radii for precision sheet metal.

**Explain** proper layout and fold procedures.

**Read** and interpret precision sheet metal prints.

Growth in the instrumentation and electronic industries in the latter part of the 20th century brought about the need for a special type of sheet metal work. Precision sheet metal ***parts*** are those manufactured from thin metals to machine shop tolerances. The metal is machined in the flat position and then folded or assembled. The calculations and layout must be exact so the relationship and location of the resulting planes and features of the folded part are within specified tolerances. See Figure 21-1. Geometric dimensioning and tolerancing controls may also be applied to the flat pattern as well as the final folded part.

The thickness of sheet metal is referred to as the ***gage*** of the material, which is abbreviated GA. Sometimes the spelling *gauge* is used. The higher the gage, the thinner the material is. The gage for sheet metal is standardized in a manner that is based on the weight of the material. For example, 16 gage cold rolled steel (CRS) is 2.5 pounds per square foot, while 18 gage is 2.0 pounds per square foot and 20 gage is 1.5 pounds per square foot. As a result of this method, not all materials have the same thickness value per gage value. For example, 15 gage standard steel is .0673″ thick, while 15 gage aluminum is .0571″ thick.

## Bending Terms

While terms associated with sheet metal bending are not necessarily standardized, there are some common terms that will help the print reader develop an understanding of how flat patterns are developed. Figure 21-2 illustrates some of the terms.

- ***Bend:*** a uniformly curved section of material that serves as the edge between two sides or legs.
- ***Bend allowance:*** the amount of metal required to create the desired bend, as based on a given bend angle, material thickness, and bend radius.
- ***Bend angle:*** the number of degrees between flat surfaces on each side of a bend. This angle can be described as acute (< 90°), square (= 90°), or obtuse (> 90°).
- ***Bend axis:*** the theoretical center axis of the bend.
- ***Bend deduction (setback):*** the amount of material to deduct from the blank length to accommodate a bend, rather than using a sharp corner, between two legs. The setback method of calculating is often used as an alternative method of calculating the total stock length, rather than calculating the leg lengths plus bend allowance.
- ***Bend line:*** the line of transition between the flat surface leg and the curved bend.
- ***Bend radius:*** the desired amount of the curvature, usually measured on the inside of the bend.
- ***Thickness:*** the stock thickness of the sheet metal, usually identified as T.
- ***Leg:*** the straight segment of stock on either side of a bend.
- ***Neutral plane:*** the area within a flat blank that will neither compress nor stretch as bending occurs. This plane is not necessarily in the exact central portion of the material. After bending, this plane is characterized by a neutral radius curve. By definition, bend allowance can be determined by calculating the circumference of the neutral plane.

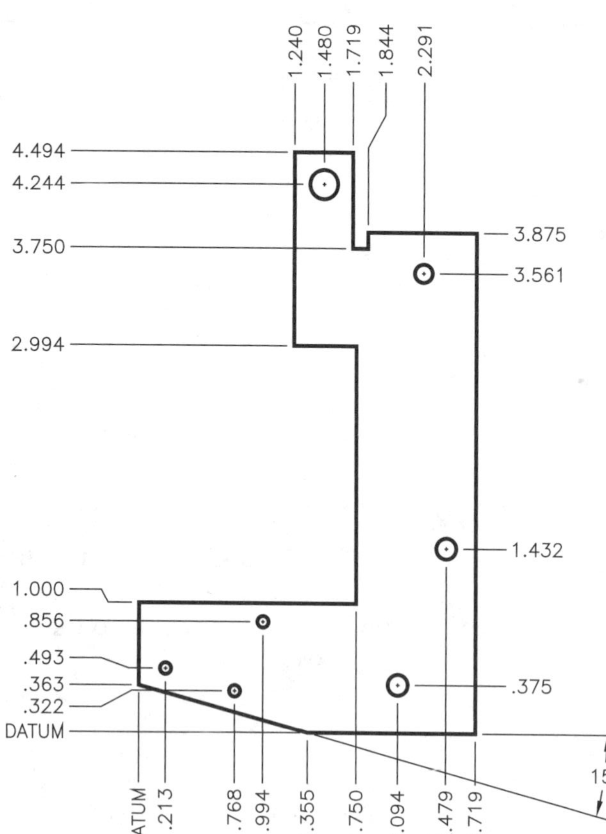


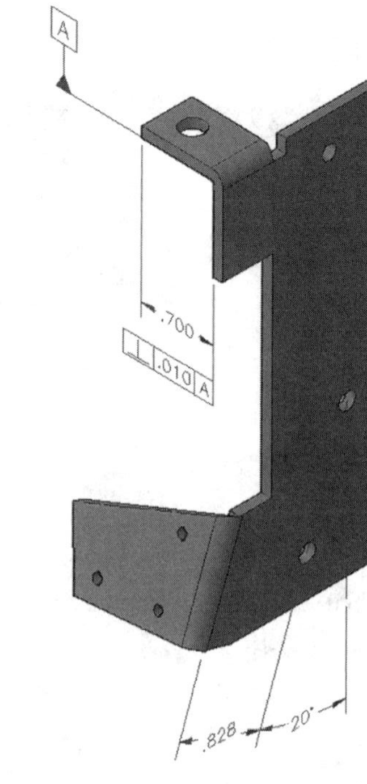


**Figure 21-1.**
A folded or "bent" part begins as a flat blank of stock material, as shown on the left.

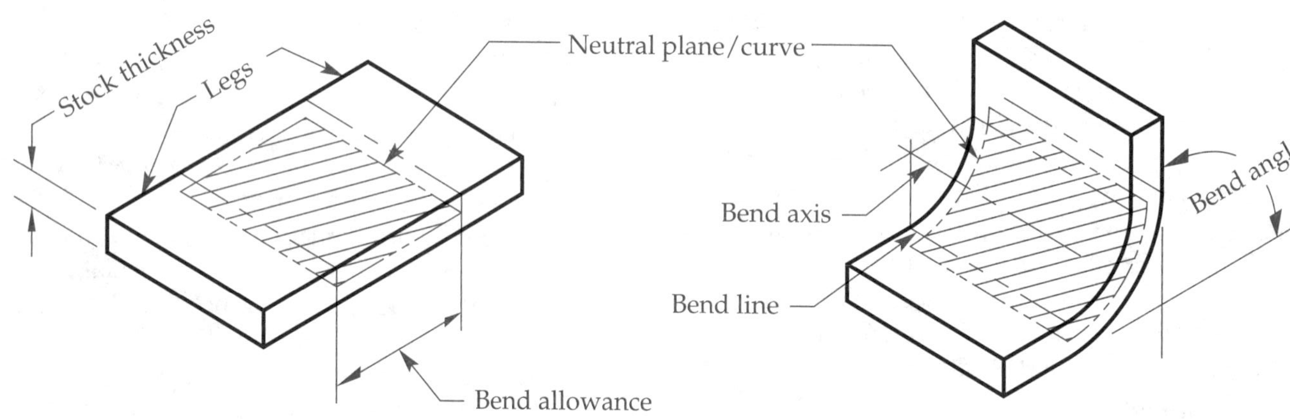


**Figure 21-2.**
Bend terminology common to most handbooks and references is illustrated here.

- ***K-factor:*** a factor used in formulas to allow for the neutral plane offset. The K-factor varies and may be impacted by bend angle sharpness, bend tools, and material properties. It is reasonable to assume that the K-factor will be between .25 and .50, which means the neutral plane falls between 1/4T and 1/2T on the axis side of the center plane of the bend. See Figure 21-3.
- ***Neutral arc bend radius:*** the bend radius plus a percentage (K-factor) of the metal thickness.
- ***Blank length:*** the total distance across all legs and the bend allowance.
- ***Springback:*** opening of a bend after forming due to the elasticity of the material, usually specified in degrees or a percentage.
- ***Overbending:*** bending an extra increment to compensate for springback.

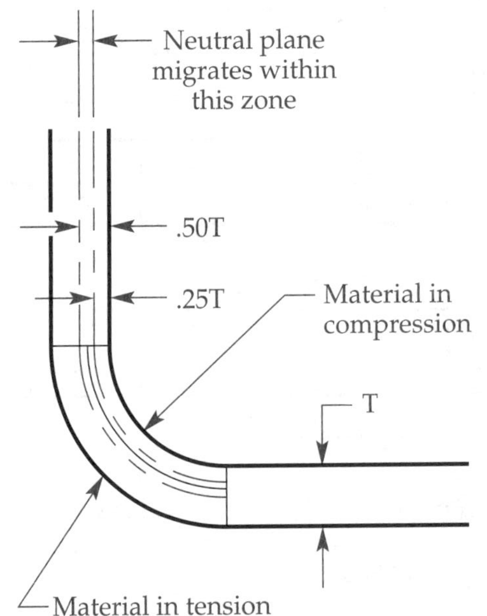


**Figure 21-3.**
Bend calculations require an understanding of the K-factor, which approximates the offset position of the neutral plane.

## Calculations and Layouts

Prints are often dimensioned with the part shown flat with a pictorial (3D) view available on the print. This unit examines those situations wherein manual calculations are made and is designed to assist the print reader in understanding the transition from a folded 3D part to a flat pattern. The ***developed length*** is the dimensions of the flat part that result in the correct dimensions for the folded part. There are primarily two methods of calculating the developed length of a part. One is to use a setback chart and the other is to use bend allowance calculations. Various bend allowance values can also be placed in a chart for quick reference. While these calculations are seldom the responsibility of the print reader, it is useful to know how the dimensions are calculated to double-check accuracy in the shop or field. Folding procedures and notes are sometimes added to the print, either in local note or chart form.

## Setback Charts

One method of calculating bends is to use a bend deduction value as found in a machinist's handbook chart or table known as a setback chart. ***Setback*** is the amount of distance subtracted from the total length of two sides that would otherwise form a sharp square corner. In other words, setback is the amount "deducted" from the distance of the square corner when "cutting across the corner" in a curved manner. Setback charts are available in most precision sheet metal industries. These charts save time and eliminate calculation errors. In some handbooks, more detailed charts are available that allow for a wider range of material thickness, bend angles, and bend radii.

A small setback chart for 90° bends is shown in Figure 21-4. The setback is found by first locating the bend radius in the left-most vertical column. Then, locate the metal thickness column using the headings along the top. Follow the column down until it meets the radius row. As illustrated by the colored entries in Figure 21-4, a 1/8″ bend radius with a metal thickness of .040″ requires a setback, or bend deduction, of .106″.

Figure 21-5 and the formula below illustrate the application of the setback example given above followed through to calculate the developed length of a precision sheet metal part. The formula for calculating the developed length of a precision sheet metal part is:

Developed length = X + Y – Z

In the formula above, X = the outside overall distance of one side (as if the corner is square), Y = the outside overall distance of the other side, and Z = the setback deduction for the bend, as indicated by the chart.

Developed length = X + Y – Z

Developed length = 1.00 + .750 – .106

Developed length = 1.750 – .106

Developed length = 1.644

PRECISION SHEET METAL SETBACK CHART

| 90° Bend Radius | Material Thickness .016 | .020 | .025 | .032 | .040 | .051 | .064 |
|---|---|---|---|---|---|---|---|
| 1/32 | .034 | .039 | .046 | .055 | .065 | .081 | .097 |
| 3/64 | .041 | .046 | .053 | .062 | .072 | .086 | .104 |
| 1/16 | .048 | .053 | .059 | .068 | .079 | .093 | .110 |
| 5/64 | .054 | .060 | .066 | .075 | .086 | .100 | .117 |
| 3/32 | .061 | .066 | .073 | .082 | .092 | .107 | .124 |
| 7/64 | .068 | .073 | .080 | .089 | .099 | .113 | .130 |
| 1/8 | .075 | .080 | .086 | .095 | .106 | .120 | .137 |
| 9/64 | .081 | .087 | .093 | .102 | .113 | .127 | .144 |
| 5/32 | .088 | .093 | .100 | .109 | .119 | .134 | .150 |
| 11/64 | .095 | .100 | .107 | .116 | .126 | .140 | .157 |
| 3/16 | .102 | .107 | .113 | .122 | .133 | .147 | .164 |

**Figure 21-4.**
A bend radius of 1/8″ on metal with a thickness of .040″ requires a setback of .106″, as shown in color.

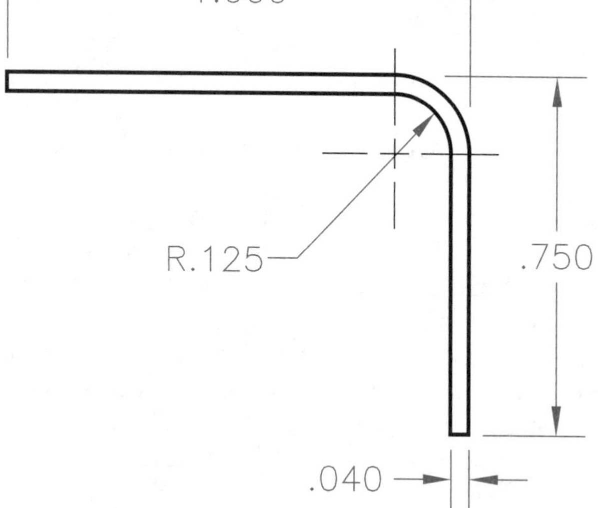


Figure 21-5.
When calculating the developed length with a setback chart, use the measurements of the sides as if there was a sharp corner and then subtract the amount given in the chart.

## Bend Allowance

When sheet metal is bent, a ***bend allowance*** must be calculated based on how the material stretches or compresses. If you think of the process of bending 1/8″ thick sheet metal about a 1/4″ radius bar to form a 90° rounded corner, the surface on the inside of the corner is compressed and the surface on the outside of the corner is stretched. Between those surfaces is a neutral plane, which is not stretched or compressed, that is not quite at the center of the material.

The calculations for bends may be obtained in a variety of ways. Each company may have its own system or charts developed from formula. Of course, computer-aided design software is now able to create the flat patterns directly from the CAD model using standard formulas within the software, but the software may require the designer to designate a formula method. In any case, the print reader must be able to interpret the drawings, usually shown in a flat pattern, but often accompanied by a pictorial or multiview that may also show finished measurements *after* the bends.

### Bend Allowance Charts

Other bend allowance charts are available that express calculations in a different manner. These charts are *additive* in nature. Figure 21-6 shows a portion of a bend allowance chart that provides the user with the amount of material to add to the

| Bend Allowance for 90° Bends | | | | |
|---|---|---|---|---|
| Radius → / Thickness ↓ | .031 | .063 | .094 | .125 |
| .016 | .060 | .110 | .159 | .208 |
| .020 | .062 | .113 | .161 | .210 |
| .025 | .066 | .116 | .165 | .214 |
| .032 | .071 | .121 | .170 | .218 |
| .040 | .077 | .127 | .176 | .224 |
| .051 | | .134 | .183 | .232 |
| .064 | | .144 | .183 | .241 |

Figure 21-6.
This chart is an example of a bend allowance chart for 90° bends that have been for various material thicknesses and bend radii.

straight sides (legs) of a part. In this manner, the distances that are straight on the finished product are added together. The distance provided on the chart is added to the flat pattern development. In summary, this chart requires the designer to calculate the leg distances by subtracting R + T from the side length. For example, with the part illustrated in Figure 21-5, the finished leg lengths are calculated:

$$\text{Leg one} = X - (R + T) = 1.000 - (.125 + .040) = .835$$

$$\text{Leg two} = Y - (R + T) = .750 - (.125 + .040) = .585$$

The bend allowance from the table indicates .224 should be added for the bend to determine the total developed length:
.835 + .585 + .224 = 1.644

### Formulas

Without charts, the bend allowance can be calculated using a formula. Using a formula to calculate the bend allowance is simply determining the circumference, or lineal length, of the bend. The circumference of a complete circle is $\pi D$, so the circumference of a 90° arc is $1/4\pi D$ or $1/2\pi R$. However, for bend allowance, the radius value should be based on the bend radius of the neutral arc. This can be determined by adding the K-factor distance to the inside bend radius.

The various means of determining the K-factor are beyond the scope of this textbook. There are several factors, including material properties, bending tools, and the material thickness–to–bend radius ratio. Some references recommend reverse engineering the K-factor using a sample piece of material bent with the intended bending tool.

In the following examples, two different K-factors are given for calculating the bend allowance. Many companies have developed their own formulas or guidelines for determining the K-factor. These two examples are representative of the types of rules a company may use. Refer to Figure 21-7 to assist you in the examples below. In these examples, each formula is calculating the bend allowance of a 90° bend. If the angle of bend is not 90°, a different fractional amount will need to replace 1/2 in the formula:

## Example Rule A

When the inside radius is *less* than twice the material thickness and the bend is 90°, use this formula, which incorporates a K-factor of .4:

$$A = 1/2\pi\ (R + .4T)$$

In the formula above, A = the bend allowance, R = the inside radius, and T = the material thickness.

As an example, to find the lineal length of a 90° bend when the stock thickness is .064″ and the inside radius is 1/16″ (.0625″), apply the formula as follows.

$$A = 1/2\pi\ (R + .4T)$$

$$A = 1/2 \times 3.1416 \times (.0625 + .4 \times .064)$$

$$A = 1.5708 \times (.0625 + .0256)$$

$$A = 1.5708 \times .0881$$

$$A = .1384$$

The bend allowance for this example is .1384″.

## Example Rule B

When the inside radius is *more* than twice the material thickness and the bend is 90°, use this formula, which incorporates a K-factor of .5:

$$A = 1/2\pi\ (R + .5T)$$

As an example, to find the lineal length of a 90° bend when the stock thickness is .032″ and the inside radius is 1/16″ (.0625″), apply the formula as follows.

$$A = 1/2\pi\ (R + .5T)$$

$$A = 1/2 \times 3.1416 \times (.032 + .5 \times .032)$$

$$A = 1.5708 \times (.032 + .016)$$

$$A = 1.5708 \times .048$$

$$A = .0754$$

The bend allowance for this example is .0754″.

## Developed Length

After the bend allowance (A) has been found, the developed length of the entire part can be calculated. In the earlier example, each leg length was calculated by subtracting R + T. The following formula incorporates the subtraction of R + T from both legs:

$$\text{Developed length} = X + Y + A - (2R + 2T)$$

In the formula above, X = the outside distance of one side (as if the corner was square), Y = the outside distance of the other side, A = the bend allowance, R = the inside radius, and T = stock thickness.

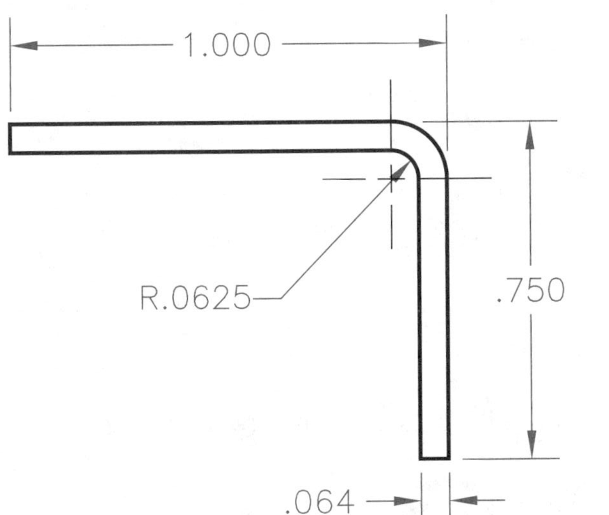


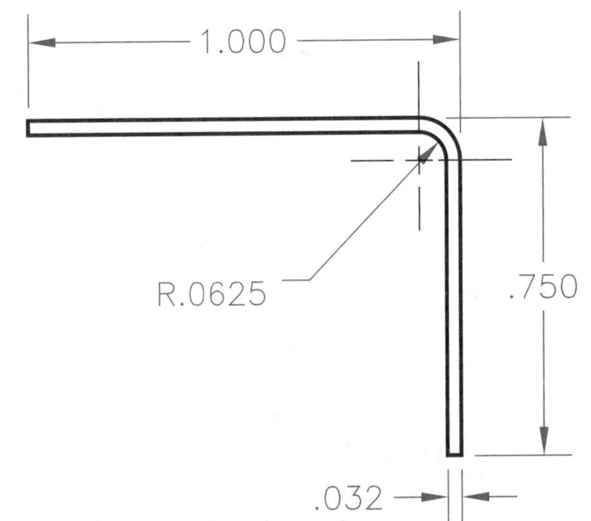


**Figure 21-7.**
These examples are explained in the text. The bend allowance for the thicker material was calculated using a K-factor of .4 and the thinner material with a K-factor of .5.

The .064″ thick material in the example above using Rule A results in a developed length of 1.6354″:

$$\text{Developed length} = 1.00 + .75 + .1384 - (.125 + .128)$$
$$\text{Developed length} = 1.8884 - .253$$
$$\text{Developed length} = 1.6354$$

The .032″ thick material in the example above using Rule B resulted in a developed length of 1.6364″:

$$\text{Developed length} = 1.00 + .75 + .0754 - (.125 + .064)$$
$$\text{Developed length} = 1.8254 - .189$$
$$\text{Developed length} = 1.6364$$

## Layout and Fold Procedure

The bend allowance formulas and setback charts for precision sheet metal parts provide approximate data. The coefficient of expansion for various materials can vary. Therefore, it is a good idea to have a duplicate flat blank for reference when making adjustments. Once all overall measurements and feature locations meet specified tolerances, the duplicate blank can be discarded or used. The procedure to lay out and fold precision sheet metal parts is:

1. Calculate and lay out the developed length and feature locations.
2. Make two identical flat blanks to the overall dimensions.
3. With the forming die, form one piece and check measurements.
4. If the folded part is not within tolerances, make machine adjustments as needed.
5. Make two more identical blanks.
6. With the forming die, form one piece and check measurements.
7. If the folded part is not within tolerances, make machine adjustments as needed. Repeat steps 5 and 6 until the folded part is within tolerances.
8. Once the folded part is within tolerances, the identical blank is the pattern to be used for building all blanks.

## Reading Precision Sheet Metal Prints

While prints for folded sheet metal parts are not that much different than other parts, the material thickness is one element unique to sheet metal parts. For a flat pattern development, often the key description is a single-view drawing of the pattern with the depth, or material thickness, not shown. Often, the print will include the thickness of the material as a reference value, usually in the material block.

In summary, prints for precision sheet metal parts are read in the same manner as other prints in the metal machining and fabrication industries. Some prints are fully dimensioned, as exemplified by print PR 21-1 at the end of this unit. The ordinate method of dimensioning is common, as shown on that print, since the values can be easily incorporated into CNC programming. In former practice, many sheet metal prints contained few, if any, dimensions, but were accompanied by a print on polyester film. The film was dimensionally stable and the toolmaker worked directly from the film to make a flat pattern. Parts were then made using the flat pattern.

## Review Questions

*Circle the answer of choice, fill in the blank, or write a short answer.*

1. Which of the following describes how sheet metal behaves as it is bent to form a rounded edge?
   A. The metal compresses about the curve of the bend.
   B. The metal stretches around the bend.
   C. There is both compression and stretching (tension) occurring at the bend.
   D. The metal neither compresses nor stretches, but remains neutral.

2. The amount of metal required to create the desired bend is called ______________________.

3. *True or False?* Bend deduction is just another way of saying bend allowance.

4. Which term defines the line of transition between a flat surface leg and the curved bend?
   A. Tangent line.
   B. Neutral line.
   C. Springback line.
   D. Bend line.

5. *True or False?* The bend radius is usually measured on the inside of the bend.

6. The ______________________ is the straight segment of stock on either side of a bend.

7. The ______________________ plane is the portion of the bend allowance that will neither compress nor stretch as bending occurs.

8. *True or False?* The K-factor is used to define the approximate position of the neutral plane, usually at a location between 25% and 50% of the thickness of the material.

9. Bending an extra increment to compensate for ______________________ of the sheet metal is referred to as ______________________.

10. The ______________________ length is the total length of the flat pattern for a folded part.

11. Which of the following is used as a deduction when calculating the length of a folded part?
   A. Bend allowance.
   B. Setback.
   C. Side one leg distance.
   D. Side two leg distance.

12. For a simple L-shape fold, when calculating the leg length for either straight side segment, what two values are subtracted from the total side length to determine the leg length? ______________________

______________________________________________

13. *True or False?* Since the circumference of a circle is $\pi D$, the circumference of a 90° bend can be calculated as $1/2\pi R$.

14. How many blanks should be made before folding a pattern into the folded shape part? ____________

______________________________________________

15. The thickness of sheet metal is referred to as ____.
   A. gage
   B. depth
   C. stock
   D. rating

## Industry Print Exercise 21-1

*Refer to the print PR 21-1 and answer the questions below.*

1. What gage of material is specified for this part? ____________

2. How many bend steps are indicated with notes? ____________

3. Of the bends indicated in question 2, how many are 90° bends? ____________

4. The overall length of the folded part is 1128.19 mm [44.42″]; how long is the developed length? ____________

5. What type of line is used on the development to indicate a bend line? ____________

6. Along the top edge of the development, approximately 1/2″ of material will be folded down using a bend radius of ____________.

7. What is the total bend angle specified for the small flap that has three small (Ø7.50 mm) holes? ____________

8. What is the total height of this part in its folded form? ____________

9. How wide is the clearance slot approximately 26.5″ to the right of the left end of this object? ____________

*Review questions based on previous units:*

10. Which type of dimensioning is used on this drawing? ____________

11. At what scale was the original drawing created? ____________

12. How many pictorial views are shown, if any? ____________

13. For the hole or holes that require a countersink, what countersink diameter is specified? ____________

14. What is the weight of this part? ____________

15. What finish is indicated for this part? ____________

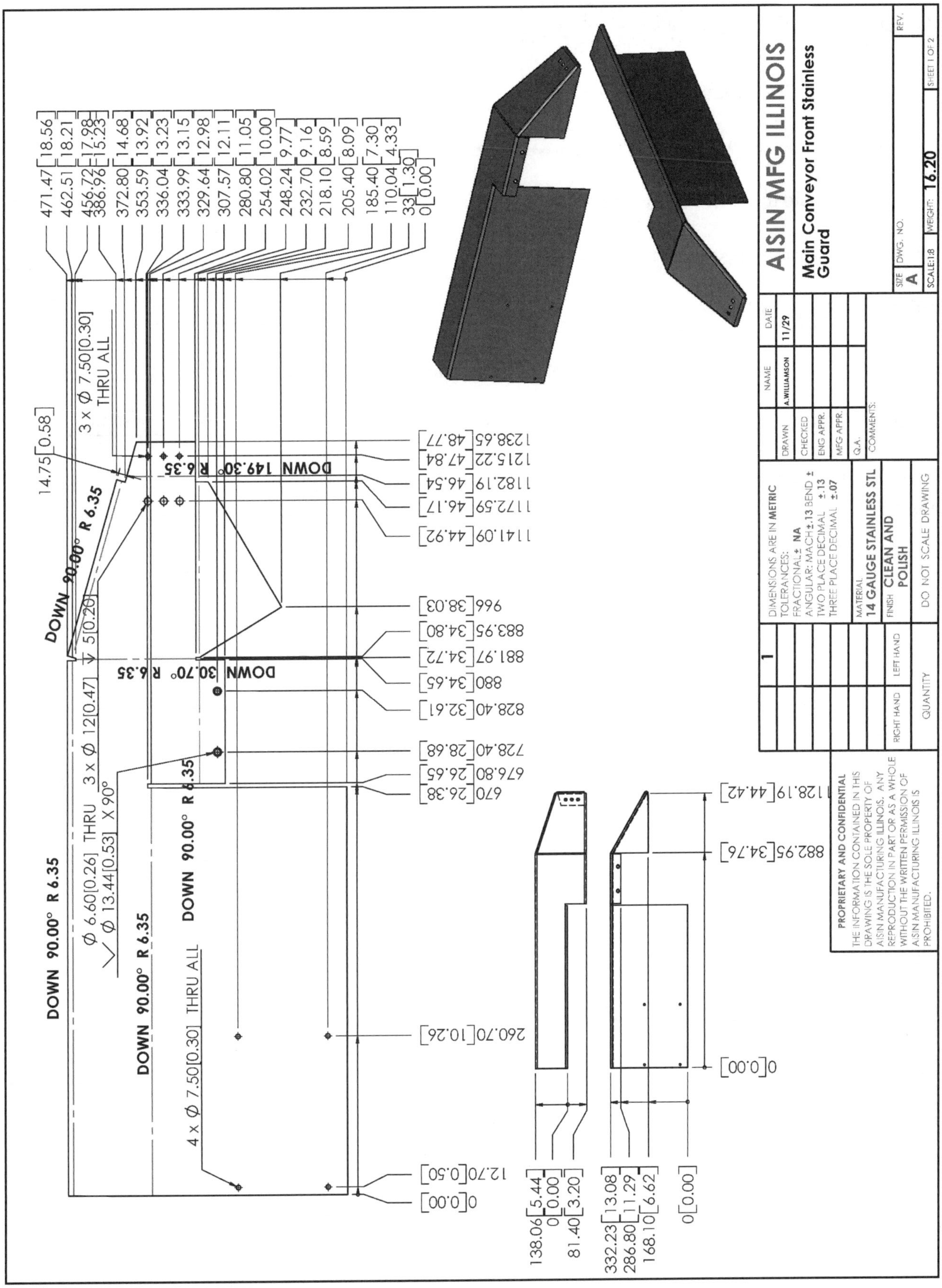

DOWN 90.00° R 6.35

Ø 6.60[0.26] THRU 3 x Ø 12[0.47] ↓ 5[0.20]

⌵ Ø 13.44[0.53] X 90°

DOWN 90.00° R 6.35

DOWN 90.00° R 6.35

4 x Ø 7.50[0.30] THRU ALL

3 x Ø 7.50[0.30] THRU ALL

DOWN 90.00° R 6.35

DOWN 30.70° R 6.35

DOWN 149.30° R 6.35

**AISIN MFG ILLINOIS**

**Main Conveyor Front Stainless Guard**

| | NAME | DATE |
|---|---|---|
| DRAWN | A.WILLIAMSON | 11/29 |
| CHECKED | | |
| ENG APPR. | | |
| MFG APPR. | | |
| Q.A. | | |

COMMENTS:

DIMENSIONS ARE IN **METRIC**
TOLERANCES:
FRACTIONAL± **NA**
ANGULAR: MACH±.13 BEND ±
TWO PLACE DECIMAL ±.13
THREE PLACE DECIMAL ±.07

MATERIAL
**14 GAUGE STAINLESS STL**

FINISH **CLEAN AND POLISH**

DO NOT SCALE DRAWING

QUANTITY: RIGHT HAND | LEFT HAND — 1

SIZE **A** DWG. NO. REV.

SCALE:1:8 WEIGHT: **16.20** SHEET 1 OF 2

**PROPRIETARY AND CONFIDENTIAL**
THE INFORMATION CONTAINED IN THIS DRAWING IS THE SOLE PROPERTY OF AISIN MANUFACTURING ILLINOIS. ANY REPRODUCTION IN PART OR AS A WHOLE WITHOUT THE WRITTEN PERMISSION OF AISIN MANUFACTURING ILLINOIS IS PROHIBITED.

PR 21-1.
Print supplied by AISIN Manufacturing Illinois.

## Industry Print Exercise 21-2

*Refer to the print PR 21-2 and answer the questions below.*

1. How many bends does this part have? ____________
2. What is the basic thickness of the metal for this part? ____________
3. What bend deduction (setback) was used in calculating the flat pattern? ____________
4. What is the bend radius for this part? ____________
5. In the formed view, is the 10.574″ dimension to the inside or outside of the bends? ____________
6. What size are the slotted holes of this part? ____________
7. Even though the finished part is 5.5″ tall, what is the developed length in that direction? ____________
8. How wide are the clearance notches on each side of the top "flap"? ____________
9. What is the radius for breaking sharp edges? ____________
10. What amount of tolerance is permitted for the dimensions shown in the formed view? ____________

*Review questions based on previous units:*

11. How many holes on this part are square holes? ____________
12. What is the part number for this part? ____________
13. What part number will this part be used on? ____________
14. Are there any hidden lines shown on this drawing? ____________
15. At what scale was this original drawing created? ____________

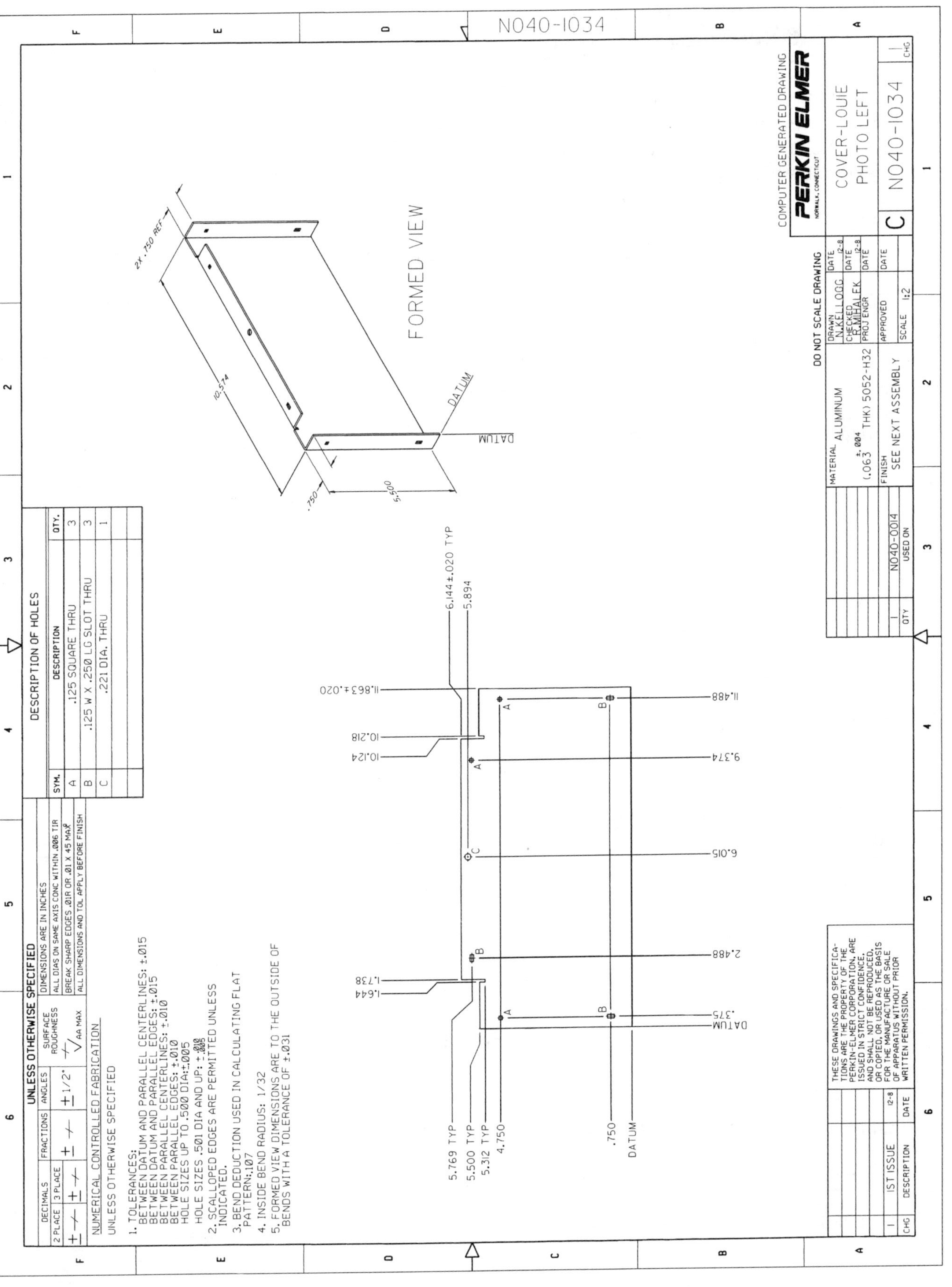


PR 21-2.
Print supplied by Perkin Elmer.

## Bonus Print Reading Exercises

The following questions are based on various bonus prints located in the folder at the back of this textbook. Refer to the print indicated, evaluate the print, and answer the question.

**Print AP-023:**

1. Which corner represents the origin (machine zero) for the NC punch press that makes this part?
2. What is the size of the holes marked B?
3. Calculate the size of the rectangular slot.
4. What do the lines drawn as hidden lines represent?

**Print AP-024:**

5. What thickness of sheet metal is used for this part, in gage and inches?
6. In its final form, what are the three overall dimensions of this part?
7. What part number (indicated in parentheses) protrudes up from the sheet metal and is located 7.260″ from the left?
8. How many 3″ long slots are featured on this part?

**Print AP-025:**

9. What term is used on this print to refer to the pictorial illustration of the part in its final shape?
10. What is the overall size of the flat pattern for this part?
11. What inside bend radius is specified for this part?
12. What bend reduction was applied in calculating the flat pattern dimensions?

UNIT 22
# Welding Prints

*After completing this unit, you will be able to:*

**Identify** a welding symbol.

**List** the elements of a welding symbol.

**Explain** the meaning of a welding symbol.

**Identify** the basic *weld* symbols used in *welding* symbols.

Reading welding drawings is quite similar to reading other drawings, except for the symbols involved. This unit explains the symbols found on welding drawings. The American Welding Society (AWS) has developed and adopted standard procedures for using symbols to indicate the exact location, size, strength, geometry, and other information needed to describe the weld required. The symbols in this unit are based on AWS A2.4, entitled *Standard Symbols for Welding, Brazing, and Nondestructive Examination.* As noted by the title, this standard also provides a means of specifying brazing operations as well as nondestructive testing methods, including the frequency, extent, and/or examination method. This unit deals with welding and the symbols used in that process. Additional information that may provide insight to the print reader can been found in AWS A3.0, entitled *Standard Welding Terms and Definitions.*

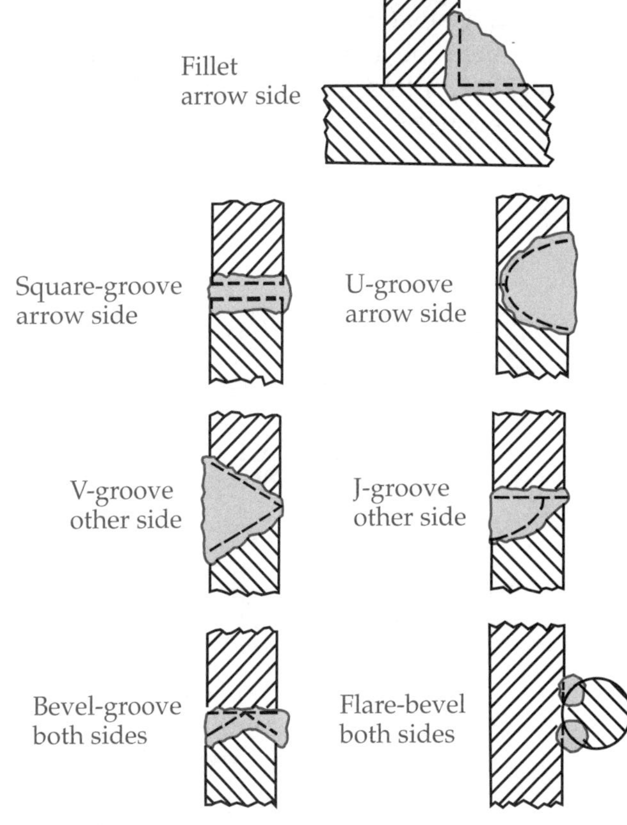


**Figure 22-1.**
A weld can take many forms.

## Welding Defined

***Welding*** is a fabrication process that usually involves joining metal pieces by melting them along a joint, with a filler material being added in some cases. As the pieces cool, the junction becomes a strong joint and is referred to as a ***weld.*** The different types of weld are named by the basic design of the joint itself, Figure 22-1. The selection of a weld is determined by three basic factors: 1) the thickness of the metals to be joined; 2) the depth of penetration of the weld required to meet the strength specifications; and 3) the type of metal or metals being joined.

Different energy sources can be used for welding, but these are not necessarily indicated on the print. Welding can be accomplished with gas flame, electric arc, lasers, and ultrasound power supplies. ***Brazing*** and ***soldering*** are fabrication processes that melt a joint material at a lower melting point than the materials being joined, so the pieces themselves are not actually melted.

## Welding Drawings

As discussed in Units 15 and 16, welding drawings are considered to be detail drawings, although as inseparable assemblies they share characteristics of other assembly drawings. The print usually serves the same purpose as a detail drawing for one part, even though the object is comprised of individual pieces shown assembled. A parts list is often included. The joints or seams between pieces to be welded together are shown on the drawing. Welding symbols are used to indicate the type and size of the weld to be applied.

### Welding Symbol

A ***welding symbol*** is a symbolic notation used to indicate the type and size of weld to be applied at a seam, joint, or location on the assembled pieces. The symbol consists of a leader line with arrow, a reference line, and a tail section. See Figure 22-2.

The ***reference line*** is the horizontal line portion of the welding symbol. Attached to one end is a leader line with an arrow. Attached to the other end is an optional tail section. This line may appear vertically on some prints.

The ***leader line with arrow*** is used to connect the welding symbol reference line to one side of the joint. This side of the joint is considered the ***arrow side.*** The side opposite of the arrow is considered the ***other side*** of the joint. The point of the arrow is usually attached to a seam or joint between two pieces to be welded.

The ***tail*** section, shown in Figure 22-3, is used for designating the welding specification, process, or other reference such as an industry specification. If a reference is not given, the tail can be omitted from the reference line. You may need to consult company standards or other resources to determine the welding process indicated. There are a multitude of specifications, including those shown in the following table.

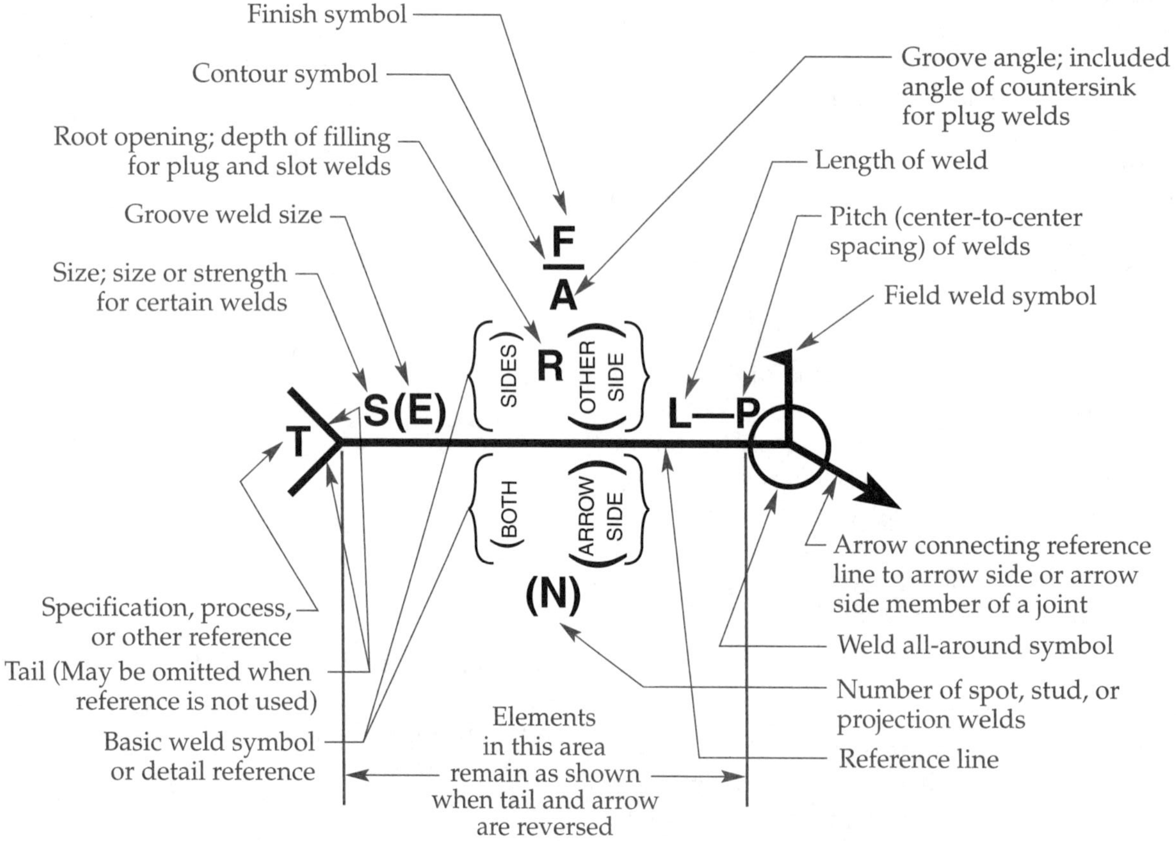


**Figure 22-2.**
The welding symbol provides much information to the reader.

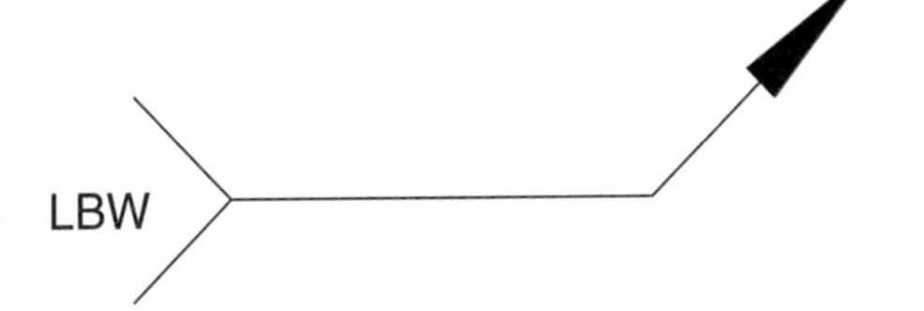


**Figure 22-3.**
The tail section of the welding symbol designates the welding specification, process, or other reference.

| Abbreviation | Meaning |
|---|---|
| EBW | Electron Beam Welding |
| FCAW | Flux Cored Wire |
| GMAW | Gas Metal Arc Welding (MIG/MAG) |
| GTAW | Gas Tungsten Arc Welding (TIG) |
| LBW | Laser Beam Welding |
| OFW | Oxyfuel Gas Welding |
| PAW | Plasma Arc Welding |
| RSEW | Resistance Seam Welding |
| SMAW | Shielded Metal Arc Welding (MMA) |
| SAW | Submerged Arc Welding |

## *Weld Symbols*

A ***weld symbol*** indicates the type of weld only. Basic weld symbols for various types of welds are shown in Figure 22-4. The weld symbol is attached to the welding symbol reference line. In cases such as the fillet weld, where the weld can be applied on both sides of a joint, it is appropriate to show the weld symbol on both sides of the reference line. In other situations, such as a spot weld or seam weld, it may be appropriate to show the symbol centered with the reference line.

In summary, the ***weld location*** is specified by where the weld symbol appears in relation to the reference line. When the symbol is placed on the side of the reference line nearest to the reader (below the reference line), the weld is to be on the *arrow side* of the joint. See Figure 22-5. When the symbol is on the side of the reference away from the reader (above the reference line), the weld is to be on the *other side* of the joint. If the weld symbol appears on both sides of the reference line, the weld is to be on *both sides* of the joint.

The ***weld dimension,*** or size, is shown on the same side of the reference line as the weld symbol. See Figure 22-6A. When the dimensions are covered by a general note, such as ALL FILLET WELDS 3/8″ IN SIZE UNLESS OTHERWISE NOTED, the welding symbol does not need to be dimensioned, but the weld symbol is still required. See Figure 22-6B. When both welds have the same dimensions, either or both welds may be dimensioned. See Figure 22-6C. The pitch, or distance between, of staggered intermittent welds is shown to the right of the weld length. See Figure 22-6D.

## Supplementary Symbols

***Supplementary symbols*** are used with the welding symbol to further specify the characteristics of the weld. These supplementary symbols include the contour, groove, spot weld, weld-all-around, field weld, melt-through, and finish symbols.

The ***contour symbol*** is a supplementary symbol that can be shown next to the weld symbol to indicate if the weld surface is to be flat-faced, convex, or concave. Figure 22-7 shows how this symbol can be used with a fillet weld. The symbol can also be used with other weld symbols.

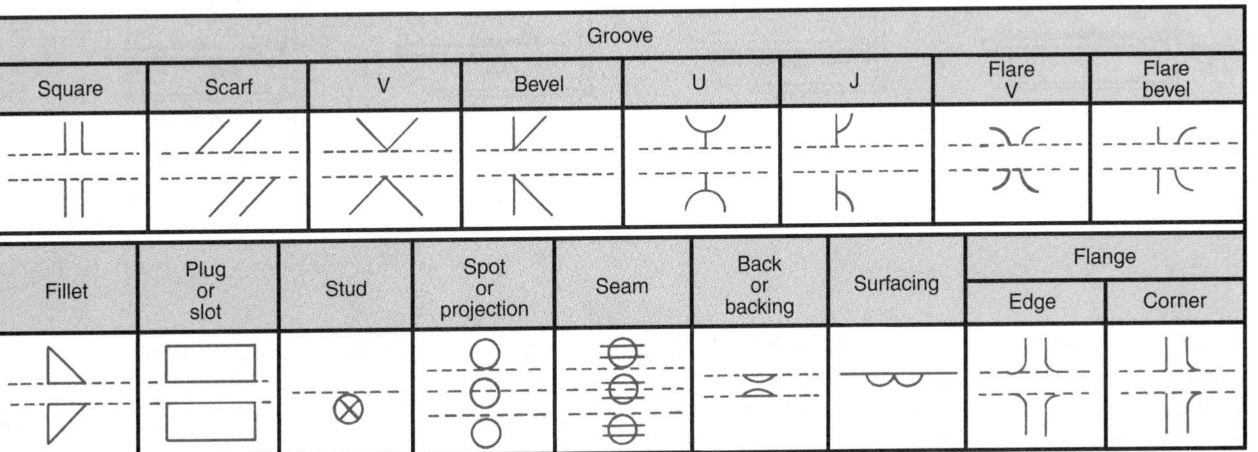


**Figure 22-4.**
Basic weld symbols for various types of welds. (The American Welding Society)

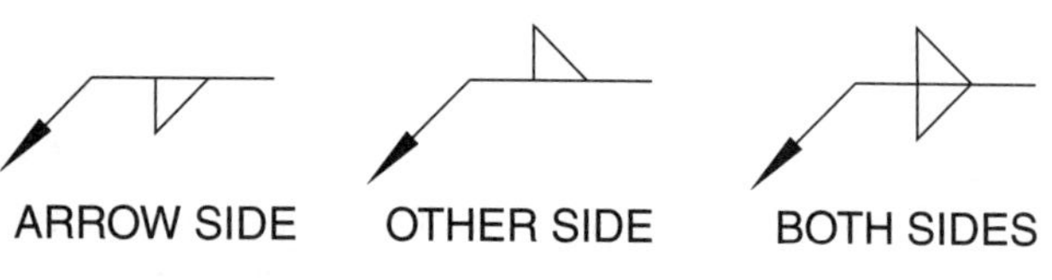


Figure 22-5.
The weld location is indicated by where the weld symbol appears in relation to the reference line.

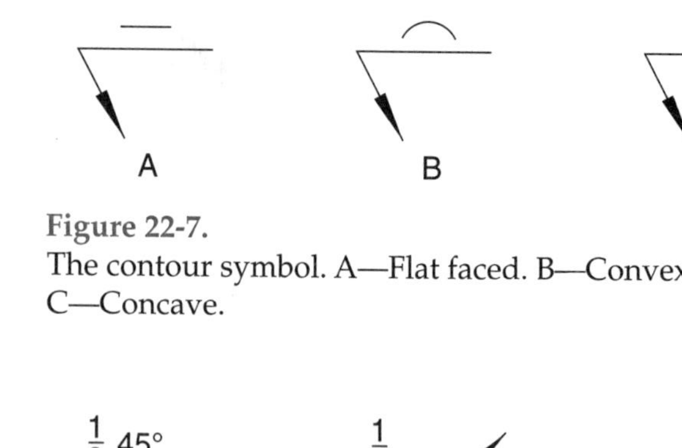


Figure 22-7.
The contour symbol. A—Flat faced. B—Convex. C—Concave.

Groove welds are often used if larger pieces of metal are to be welded. Various specifications about the size of the grooves can be incorporated into the welding symbol. While there are too many scenarios to cover within the scope of this unit, Figure 22-8 shows some examples of using a ***groove angle symbol.*** The additional symbols should be shown on the same side of the reference line as the weld symbol. In Figure 22-8A, the depth of the groove weld is shown as 1/8″ on the left of the weld symbol and the angle of the groove is 45°. The angle specification is the included angle between the surfaces of both pieces as positioned for assembly, not the individual slope of either part alone. As illustrated in Figure 22-8B, the depth on the arrow side is different than the depth on the other side. Since the angle is not specified within the welding symbol, perhaps it is covered by a general note. An empty space between two pieces to be welded is called the ***root opening.*** The root opening size of the groove weld can be specified by placing the value inside of the weld symbol, as shown in Figure 22-8C.

1/8 45°

A

1/4 3/8

B

1/16

C

Figure 22-8.
The size is shown on the same side of the reference line as the weld symbol.

The ***spot weld symbol*** specifies a spot weld. It calls out the spot weld diameter (Figure 22-9A), strength in pounds (Figure 22-9B), center-to-center pitch (Figure 22-9C), and number of spot welds (Figure 22-9D). In Figure 22-9C, the spot weld symbol is centered with the reference line, indicating a "weld through."

The ***weld-all-around symbol*** indicates the weld extends completely around a joint. See Figure 22-10.

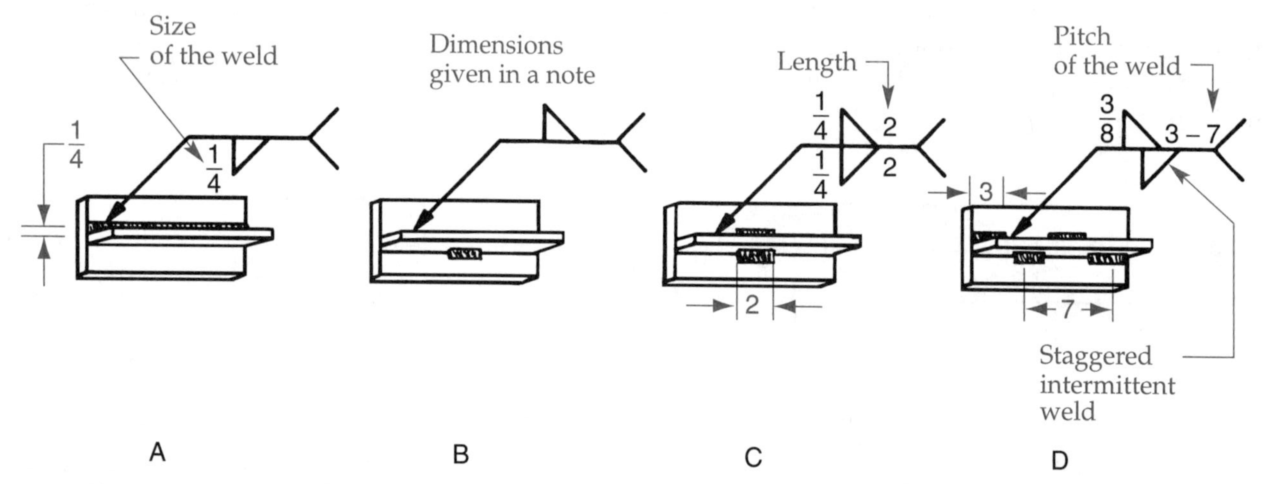


Figure 22-6.
A—The weld dimension is shown on the same side of the reference line as the weld symbol. B—When the dimensions are covered by a general note, the welding symbol does not need to be dimensioned. C—When both welds have the same dimensions, either or both welds may be dimensioned. D—The pitch of staggered intermittent welds is shown to the right of the weld length.

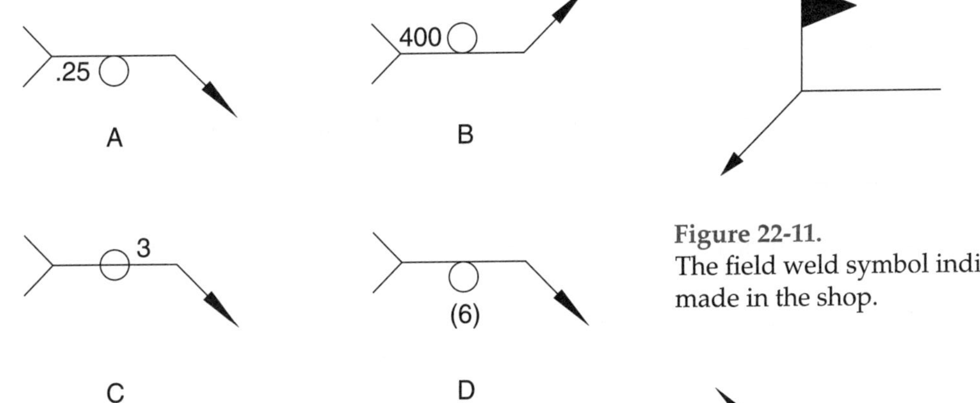


**Figure 22-9.**
The spot weld symbol can be annotated in many ways.

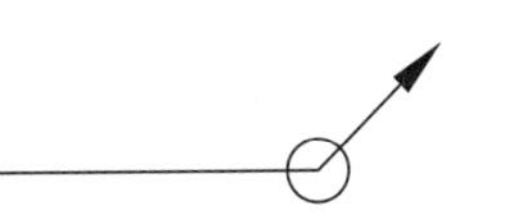

**Figure 22-10.**
The weld-all-around symbol indicates the weld extends completely around a joint.

The ***field weld symbol*** indicates a weld is not to be made in the shop or place of initial construction. See Figure 22-11. For example, the components of a large storage tank may be manufactured at a machine shop. The components must then be shipped to the job site and welded together. The assembly drawings for the tank would make use of the field weld symbol for all welds that would be done "in the field" instead of in the shop.

The ***melt-through symbol*** indicates welds where 100% joint penetration plus reinforcement is required when made from one side. See Figure 22-12A. When melt-through welds are to be finished by machining or some other process, a contour symbol is added. See Figure 22-12B.

The ***finish symbol*** indicates the method of finishing, not the surface texture. The letter C = chipping, G = grinding, M = machining, R = rolling, and H = hammering.

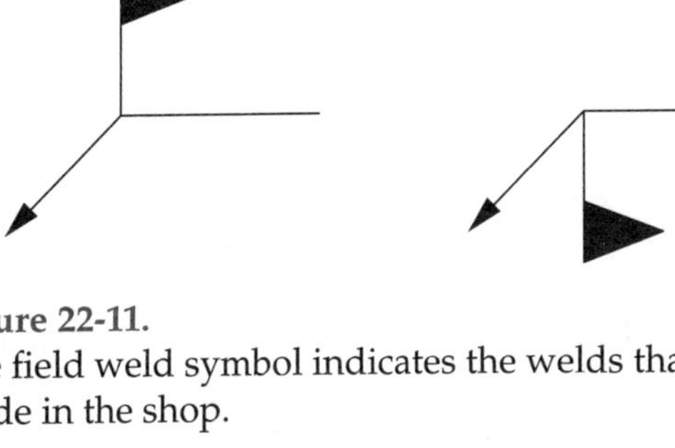

**Figure 22-11.**
The field weld symbol indicates the welds that are not made in the shop.

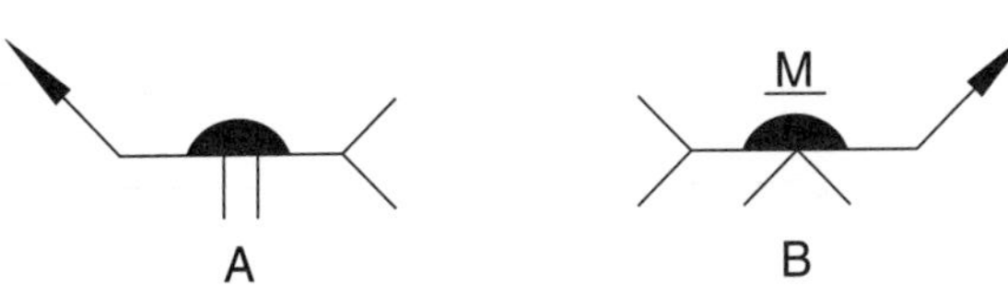


**Figure 22-12.**
The melt-through symbol indicates welds with 100% joint penetration.

## Review Questions

*Circle the answer of choice, fill in the blank, or write a short answer.*

1. What is the name and acronym for the organization that determines standards for welding?

   ______________________________

2. *True or False?* When the weld symbol appears below the reference line, the weld is on the "other side."

3. Write brief definitions for the word *weld*:
   A. Noun:

   ______________________________

   ______________________________

   B. Verb:

   ______________________________

   ______________________________

4. *True or False?* A complete welding symbol will contain three items: 1) an arrow; 2) a reference line; and 3) a tail.

5. The ____________ line is the portion of the welding symbol to where the weld symbol is attached.

6. For one T-joint with a weld on each side, how many leader arrows are incorporated into the welding symbol to connect it to the joint on the print? ______________________________

   ______________________________

7. Which of the following is *not* a type of weld that has its own weld symbol?
   A. Fillet.
   B. Spot.
   C. Flat.
   D. Plug.

8. If the reference line is horizontal and the weld symbol is placed below the reference line, then the weld is to occur on the ______________________________ side of the joint.

9. *True or False?* The weld dimension (i.e., the size of the fillet weld) can be given in the welding symbol, but the length of the weld must be dimensioned outside of the welding symbol as a linear dimension or general note.

10. Which of the following does *not* have a supplementary graphic symbol, but is accommodated by placing a letter next to the weld symbol?
    A. Finish method.
    B. Weld all around.
    C. Field weld.
    D. Melt through.

11. Which of the following is *not* a way to further describe the surface contour of a fillet weld?
    A. Concave.
    B. Flat.
    C. Round.
    D. Convex.

12. What type of weld may also require some numeric values for strength, center-to-center spacing, or quantity? ______________________________

    ______________________________

13. If a(n) ________________ is in the middle of the reference line, it is the symbol for a spot weld, but if it is located at the elbow of a reference line and the leader line, it means "all around."

14. What does field weld mean? ______________________________

    ______________________________

15. The finish symbol C means:
    A. Circular finish.
    B. Weld all around.
    C. Chipping.
    D. Convex.

## Industry Print Exercise 22-1

*Refer to the print PR 22-1 and answer the questions below.*

1. What is the total number of parts required to make the welded assembly? ____________
2. How many welding symbols are on the print? ____________
3. How many different types of welds are specified on the print and what are they? ____________
4. For the weld between parts TP1 and TP2, what does the 3 mean? ____________
5. What standard should be used to interpret the weld symbols on this print? ____________
6. As indicated by the welding symbols, are the welds to be made on the near side, far side, or both sides? ____________
7. What note is found in the tail of the welding symbol? ____________
8. Are any field welds indicated? ____________

*Review questions based on previous units:*

9. What is the hole diameter for the small hole located on part AAA316FTP2? ____________
10. What is the largest diameter on this assembly? ____________
11. What geometric tolerancing controls are used on this print? ____________

    ____________
12. What is the name of this part? ____________
13. What is the smoothest surface finish specified on this print? ____________
14. In the local note for the holes marked M6, what does B.C. stand for? ____________
15. As shown in the sectional side view, what is the total depth of this welded part? ____________

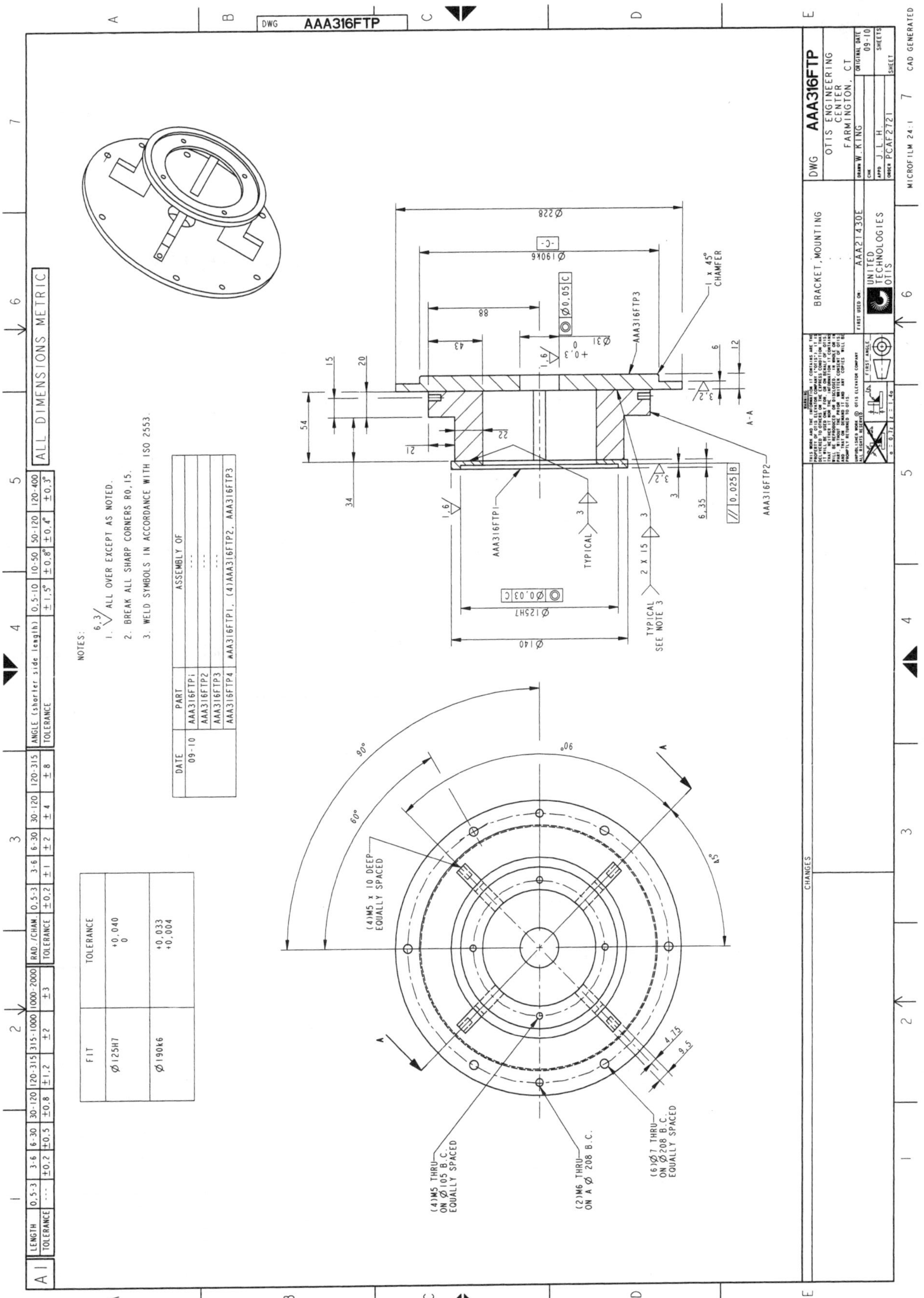
DWG AAA316FTP
BRACKET, MOUNTING
OTIS ENGINEERING CENTER FARMINGTON, CT
DRAWN W. KING
APPD J. L. H.
ORDER PCAF2721
ORIGINAL DATE 09-10
FIRST USED ON: AAA21430E
UNITED TECHNOLOGIES OTIS
MICROFILM 24:1
CAD GENERATED
CHANGES
ALL DIMENSIONS METRIC
NOTES:
1. 6,3 ALL OVER EXCEPT AS NOTED.
2. BREAK ALL SHARP CORNERS R0,15.
3. WELD SYMBOLS IN ACCORDANCE WITH ISO 2553.
DATE | PART | ASSEMBLY OF
09-10 | AAA316FTP1 | ---
AAA316FTP2 | ---
AAA316FTP3 | ---
AAA316FTP4 | AAA316FTP1, (4)AAA316FTP2, AAA316FTP3
FIT | TOLERANCE
Ø125H7 | +0,040 0
Ø190k6 | +0,033 +0,004
LENGTH | 0,5-3 | 3-6 | 6-30 | 30-120 | 120-315 | 315-1000 | 1000-2000
TOLERANCE | --- | ±0,2 | ±0,5 | ±0,8 | ±1,2 | ±2 | ±3
RAD./CHAM. | 0,5-3 | 3-6 | 6-30 | 30-120 | 120-315
TOLERANCE | ±0,2 | ±1 | ±2 | ±4 | ±8
ANGLE (shorter side length) | 0,5-10 | 10-50 | 50-120 | 120-400
TOLERANCE | ±1,5° | ±0,8° | ±0,4° | ±0,3°
A-A
1 x 45° CHAMFER
AAA316FTP1
AAA316FTP2
AAA316FTP3
TYPICAL
TYPICAL SEE NOTE 3
2 X 15
(4)M5 x 10 DEEP EQUALLY SPACED
(4)M5 THRU ON Ø105 B.C. EQUALLY SPACED
(2)M6 THRU ON A Ø 208 B.C.
(6)Ø7 THRU ON Ø208 B.C. EQUALLY SPACED
Ø228
Ø190k6
Ø140
Ø125H7
Ø31

PR 22-1.
Print supplied by United Technologies Otis.

## Industry Print Exercise 22-2

*Refer to the print PR 22-2 and answer the questions below.*

1. What is the total number of parts (not part numbers) required for this welded assembly? ____________

2. What is the name and number of the part identified as item 3? ____________

3. According to note 3, what section of the AWS D1.1 standard should be referenced for the 100% weld penetration? ____________

4. What type of weld should be used to "tack" the parts associated with item 4 in place? ____________

5. What process can be used after welding to blend the weld with the parts? ____________

6. For the process in question 5, does the welding symbol indicate flat, convex, or concave?

   ____________

7. A couple of the welding symbols have a circle on the symbol elbow. What does that mean?

   ____________

8. Are there complete dimensions on the drawing detailing all measurements for the individual parts?

   ____________

9. Unless specified, all welds are to be class ____________.

*Review questions based on previous units:*

10. Are there cutting-plane lines shown on this print? ____________

11. What scale are the main views on the original drawing? ____________

12. How much does this part weigh after welded together? ____________

13. What revision status was given to this print? ____________

14. What is the length of each of the six spacers? ____________

15. At what scale is the enlarged detail? ____________

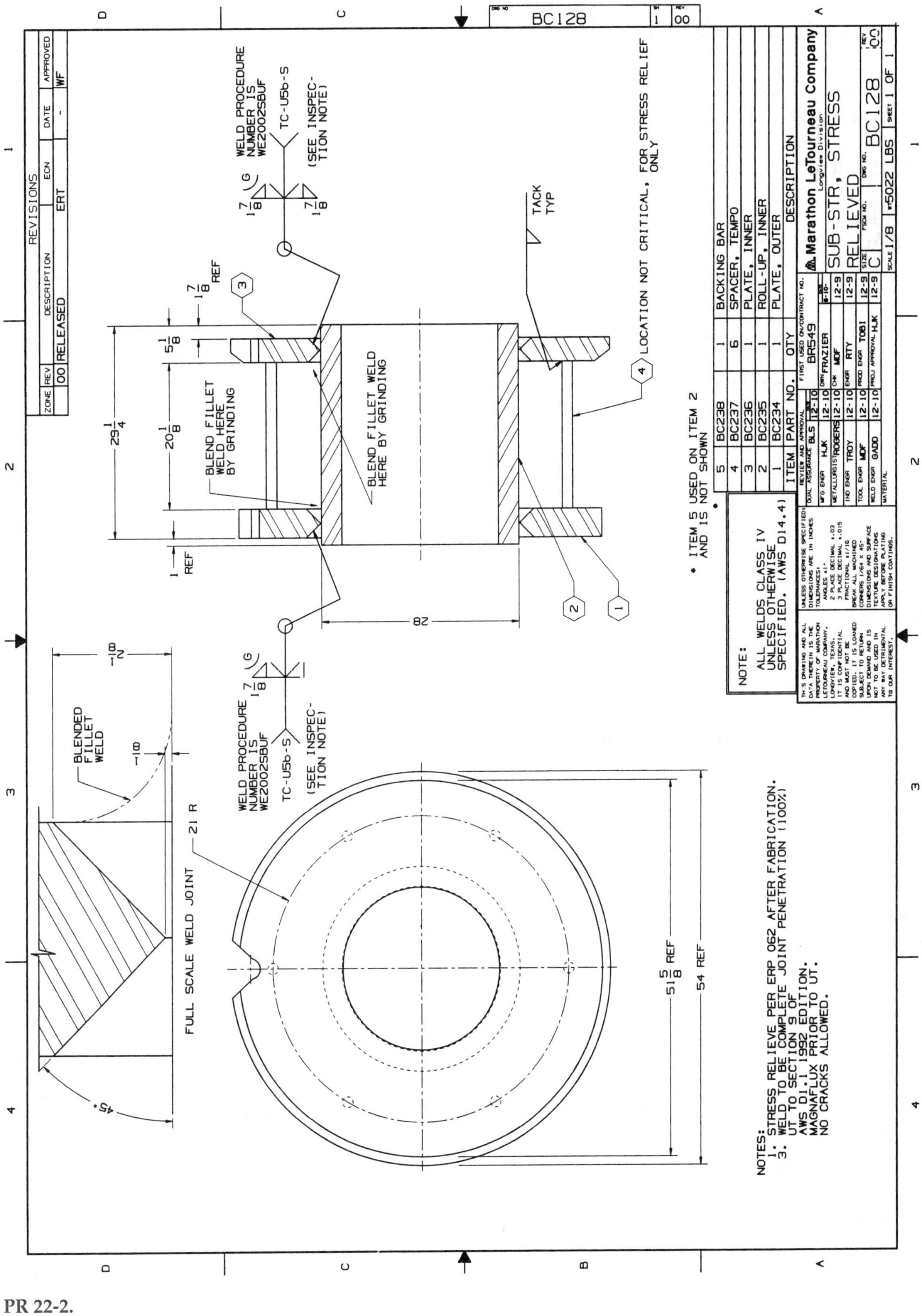

| ZONE | REV | DESCRIPTION | ECN | DATE | APPROVED |
|---|---|---|---|---|---|
| | 00 | RELEASED | ERT | - | WF |

REVISIONS

LOCATION NOT CRITICAL, FOR STRESS RELIEF ONLY

TACK TYP

WELD PROCEDURE NUMBER IS WE2002SBUF

TC-U5b-S

(SEE INSPECTION NOTE)

BLEND FILLET WELD HERE BY GRINDING

• ITEM 5 USED ON ITEM 2 AND IS NOT SHOWN

29 1/4 — 20 1/8 — 5 1/8 — 1 7/8 REF — 1 REF — 28

FULL SCALE WELD JOINT

BLENDED FILLET WELD

1/8 — 2 1/8 — 45° — 21 R — 51 5/8 REF — 54 REF

NOTE:
ALL WELDS CLASS IV UNLESS OTHERWISE SPECIFIED. (AWS D14.4)

NOTES:
1. STRESS RELIEVE PER ERP 062 AFTER FABRICATION.
3. WELD TO BE COMPLETE JOINT PENETRATION (100%) UT TO SECTION 9 OF AWS D1.1 1992 EDITION. MAGNAFLUX PRIOR TO UT. NO CRACKS ALLOWED.

| ITEM | PART NO. | QTY | DESCRIPTION |
|---|---|---|---|
| 5 | BC238 | 1 | BACKING BAR |
| 4 | BC237 | 6 | SPACER, TEMPO |
| 3 | BC236 | 1 | PLATE, INNER |
| 2 | BC235 | 1 | ROLL-UP, INNER |
| 1 | BC234 | 1 | PLATE, OUTER |

THIS DRAWING AND ALL DATA THEREIN IS THE PROPERTY OF MARATHON LETOURNEAU COMPANY, LONGVIEW, TEXAS. IT IS CONFIDENTIAL AND MUST NOT BE COPIED. IT IS LOANED SUBJECT TO RETURN UPON DEMAND AND IS NOT TO BE USED IN ANY WAY DETRIMENTAL TO OUR INTEREST.

UNLESS OTHERWISE SPECIFIED: DIMENSIONS ARE IN INCHES. TOLERANCES: ANGLES ±1°, 2 PLACE DECIMAL ±.03, 3 PLACE DECIMAL ±.015, FRACTIONAL ±1/16. BREAK ALL MACHINED CORNERS 1/64 X 45°. DIMENSIONS AND SURFACE TEXTURE DESIGNATIONS APPLY BEFORE PLATING OR FINISH COATINGS.

REVIEW AND APPROVAL: QUAL ASSURANCE BLS 12-10; MFG ENGR HJK 12-10; METALLURGIST ROGERS 12-10; IND ENGR TROY 12-10; TOOL ENGR MDF 12-10; WELD ENGR GADD 12-10; MATERIAL

FIRST USED ON/CONTRACT NO. BR549; DRN FRAZIER 8-10-; CHK MDF 12-9; ENGR RTY 12-9; PROD ENGR TOBI 12-9; PROJ APPROVAL HJK 12-9

Marathon LeTourneau Company, Longview Division

SUB-STR, STRESS RELIEVED

SIZE C — DWG NO. BC128 — REV 00

SCALE 1/8 — #5022 LBS — SHEET 1 OF 1

DWG NO BC128 — SH 1 — REV 00

PR 22-2.
Print supplied by Marathon LeTourneau Company.

## Bonus Print Reading Exercises

The following questions are based on various bonus prints located in the folder at the back of this textbook. Refer to the print indicated, evaluate the print, and answer the question.

**Print AP-010:**

1. What type(s) of weld is specified on this print? ______
2. How much applied pressure must the specified welds withstand? ______

**Print AP-012:**

3. What type(s) of weld is specified for this part? ______
4. Do any of the welding symbols indicate weld all around? ______

**Print AP-026:**

5. What does the 2X mean in the tail of the welding symbols on this print? ______
6. How many welding symbols appear on this print? ______
7. How many different types of welds are specified on this print? ______
8. What welding method is used for the welds on this assembly? ______
9. How many times does the bevel-groove weld symbol appear on this print? ______
10. What type of weld is specified between part T150618 and part T150620 along the bottom edge of T150618? ______
11. List the weld or welds necessary for welding part T150619 to T150618: ______

UNIT 23

# Instrumentation and Control Drawings

*After completing this unit, you will be able to:*

**Describe** the types of control diagrams.

**Identify** symbols used on fluid power diagrams.

**Recognize** supplementary information accompanying circuit diagrams.

**Read** a graphic diagram.

**Interpret** graphic diagrams for fluid control devices.

This unit introduces the basic symbols and diagrams used in instrumentation and control circuits. This knowledge will assist you in reading and interpreting graphic diagrams.

Several types of diagrams or drawings are used to show instrumentation and control circuits. These include graphic, pictorial, cutaway, and combination diagrams. The emphasis in this unit is on graphic diagrams. Graphic symbols are standardized and graphic diagrams are the most widely used presentation in industry for instrumentation and control circuits.

## Diagram Types

There are four basic types of diagrams. The types are explained in this section. After learning the basic symbols and sequence of operations used in instrumentation and control circuits, you will have little difficulty reading any of these four types of diagrams.

A ***graphic diagram*** consists of graphic symbols joined by lines. These diagrams provide an easy method of emphasizing functions of the circuit and its components. See Figure 23-1. A ***pictorial diagram*** shows components in a more picture-like manner and the piping as lines. See Figure 23-2. A ***cutaway diagram*** consists of cutaway symbols of components and piping. This emphasizes component function. See Figure 23-3. A ***combination diagram*** shows components and piping as graphic symbols in pictorial or as cutaways. See Figure 23-4. The type of illustration used is the one that best suits or describes the purpose of the diagram.

Many instrumentation and control circuits for machine tools involve fluid power, either liquid or gas. The symbols in Figure 23-5 are commonly found on graphic diagrams for fluid power circuits.

## Instrumentation and Control Diagram Terminology

You must become familiar with the terms used for instrumentation and control diagrams. The following terms often appear on diagrams. These terms may also be used to describe diagrams.

- ***Accumulator:*** A container in which fluid is stored under pressure as a source of fluid power.
- ***Actuate:*** To put devices and circuits into action.
- ***Actuator:*** A device for converting hydraulic or pneumatic energy into mechanical energy. Examples of an actuator are motors and cylinders.
- ***Circuit:*** The complete path of flow in a fluid power system, including the flow-generating device (pump).
- ***Component:*** A single hydraulic or pneumatic unit.
- ***Control:*** A device used to regulate the function of a component or system.
- ***Enclosure:*** A housing for components. A phantom line rectangle drawn around components indicates the limits, or enclosure, of an assembly.
- ***Fluid:*** A liquid or gas.

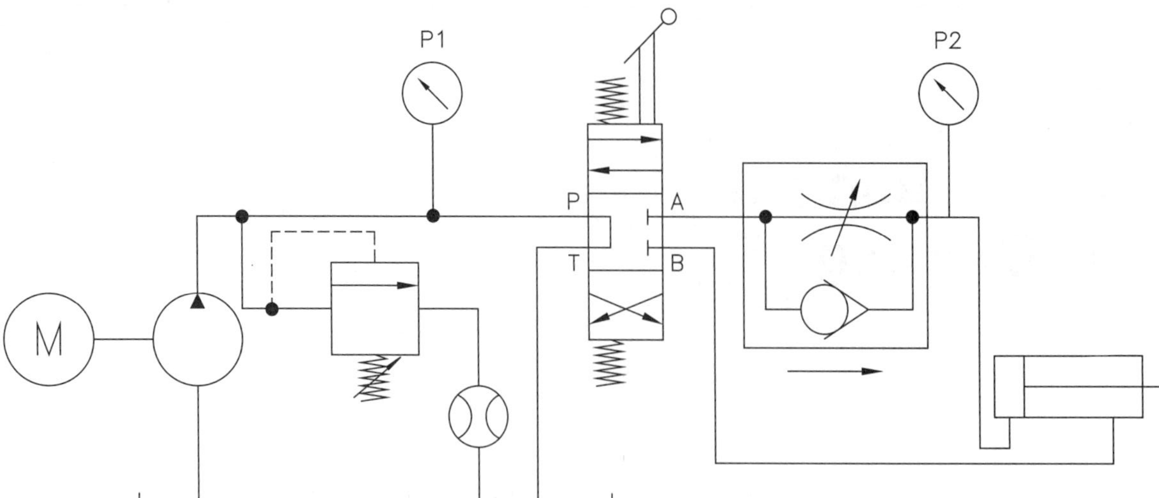


**Figure 23-1.**
A graphic diagram shows components and piping as schematic symbols.

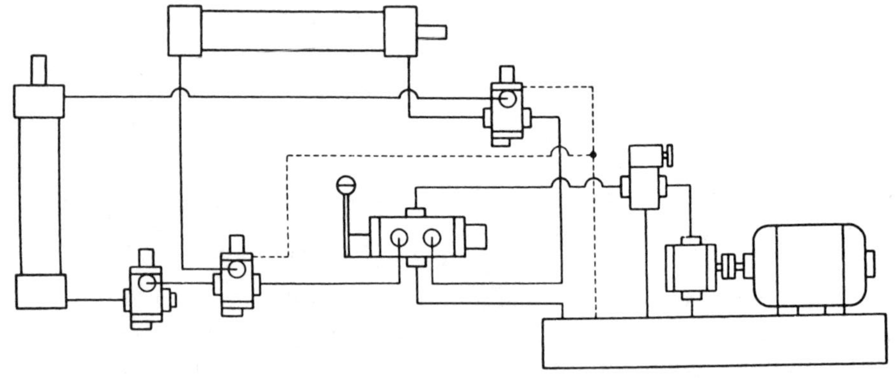

**Figure 23-2.**
A pictorial diagram shows components in pictorial and piping as lines.

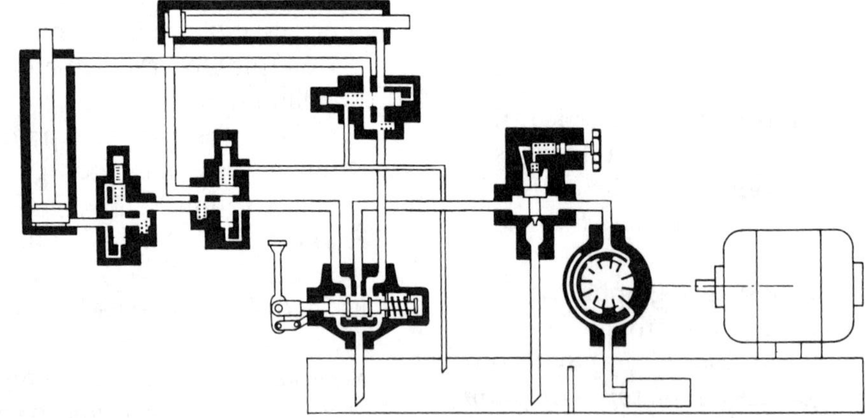

**Figure 23-3.**
A cutaway diagram emphasizes the function of components by showing them as cutaways.

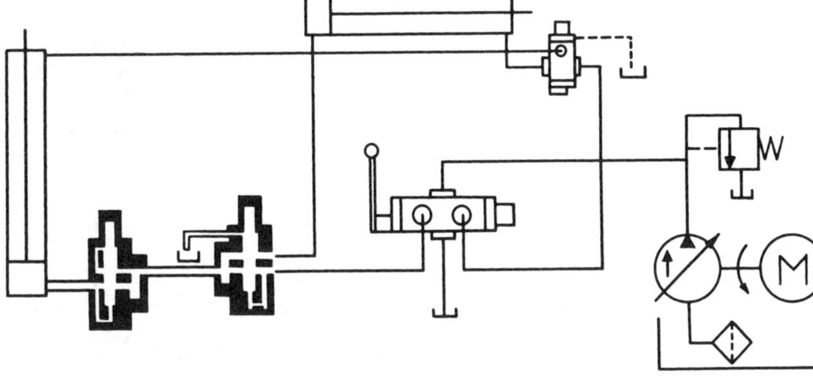


Figure 23-4.
A combination diagram shows components and piping as symbols in pictorial or as cutaways.

- *Port:* The end of a passage in a component, either internal or external.
- *Reservoir:* A container for storage of liquid in a fluid power system; not typically under pressure.
- *Restriction:* An area in a line or passage with a reduced cross section that produces a pressure drop; sometimes called an orifice.

## Supplementary Information to Accompany Circuit Diagrams

Once you are familiar with the symbols shown in Figure 23-5, you will be able to read and understand graphic diagrams. In addition to the graphic diagram, prints of circuit diagrams usually include a listing of the sequence of operations, solenoid chart, and components used. All of these items help explain the function and purpose of the circuit and its components.

The ***sequence of operations*** is an explanation of the various functions of the circuit in the order of occurrence. Each phase of the operation is numbered or lettered. A brief description is given of the initiating and resulting action. The sequence of operations is usually located in the upper part of the sheet or on an attached sheet.

***Solenoid charts*** are used if the circuit contains solenoids. ***Solenoids*** are electrical switches incorporating a metal rod that can be extended or retracted based on the application of electricity. The chart explains the operation of the electrically controlled components and is normally located in the lower-left corner of the print. Solenoids are assigned a letter on the diagrams. The chart shows whether the solenoids are energized (+) or de-energized (–) at each phase of system operation.

A ***component list,*** sometimes called a parts list or bill of materials, includes the name, model number, quantity, and supplier or manufacturer of components in the circuit. Each component in the diagram is numbered and keyed to the component list. For the most part, this is similar to other parts lists found in assembly drawings. In diagrams, where two or more identical components are used, the same key number is used, followed by a letter to help identify each instance. However, some companies prefer that each duplicate component have a new number. The component list may appear in the upper-right corner of the print.

## Reading a Graphic Diagram

Familiarize yourself with the pneumatic diagram shown in Figure 23-6. Match the numbers in the component list with their graphic symbols in the diagram. Refer to the list of graphic symbols in Figure 23-5 for further clarification. Then, try reading the pneumatic graphic circuit diagram by following the sequence of operations:

1. Manual valve (5) is depressed to start the cycle.
2. After valve (5) is shifted, follow the airflow through valve (7A) on to where it shifts valve (4). Valve (4), in turn, directs air to head end of cylinder (10).
3. The piston in cylinder (10) extends to contact the work and perform the riveting operation. The cylinder also depresses valve (7B).
4. When the tonnage control valve (6) reaches the set point of 80 pounds per square inch, it opens and reverses valve (5).

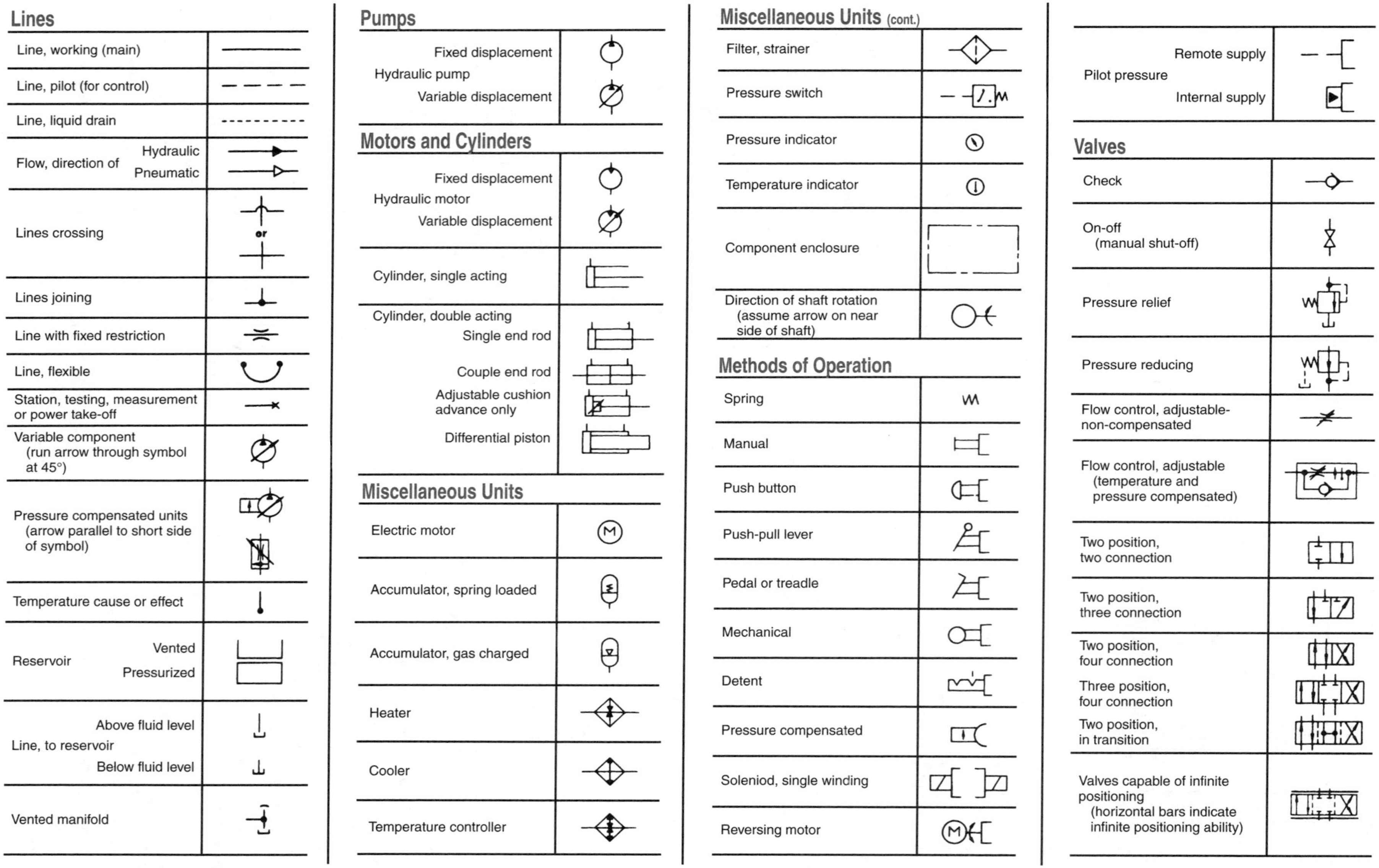

## Lines

- Line, working (main)
- Line, pilot (for control)
- Line, liquid drain
- Flow, direction of
  - Hydraulic
  - Pneumatic
- Lines crossing (or)
- Lines joining
- Line with fixed restriction
- Line, flexible
- Station, testing, measurement or power take-off
- Variable component (run arrow through symbol at 45°)
- Pressure compensated units (arrow parallel to short side of symbol)
- Temperature cause or effect
- Reservoir
  - Vented
  - Pressurized
- Line, to reservoir
  - Above fluid level
  - Below fluid level
- Vented manifold

## Pumps

- Hydraulic pump
  - Fixed displacement
  - Variable displacement

## Motors and Cylinders

- Hydraulic motor
  - Fixed displacement
  - Variable displacement
- Cylinder, single acting
- Cylinder, double acting
  - Single end rod
  - Couple end rod
  - Adjustable cushion advance only
  - Differential piston

## Miscellaneous Units

- Electric motor
- Accumulator, spring loaded
- Accumulator, gas charged
- Heater
- Cooler
- Temperature controller

## Miscellaneous Units (cont.)

- Filter, strainer
- Pressure switch
- Pressure indicator
- Temperature indicator
- Component enclosure
- Direction of shaft rotation (assume arrow on near side of shaft)

## Methods of Operation

- Spring
- Manual
- Push button
- Push-pull lever
- Pedal or treadle
- Mechanical
- Detent
- Pressure compensated
- Soleniod, single winding
- Reversing motor
- Pilot pressure
  - Remote supply
  - Internal supply

## Valves

- Check
- On-off (manual shut-off)
- Pressure relief
- Pressure reducing
- Flow control, adjustable-non-compensated
- Flow control, adjustable (temperature and pressure compensated)
- Two position, two connection
- Two position, three connection
- Two position, four connection
- Three position, four connection
- Two position, in transition
- Valves capable of infinite positioning (horizontal bars indicate infinite positioning ability)

Figure 23-5. Symbols used on fluid power diagrams are standardized.

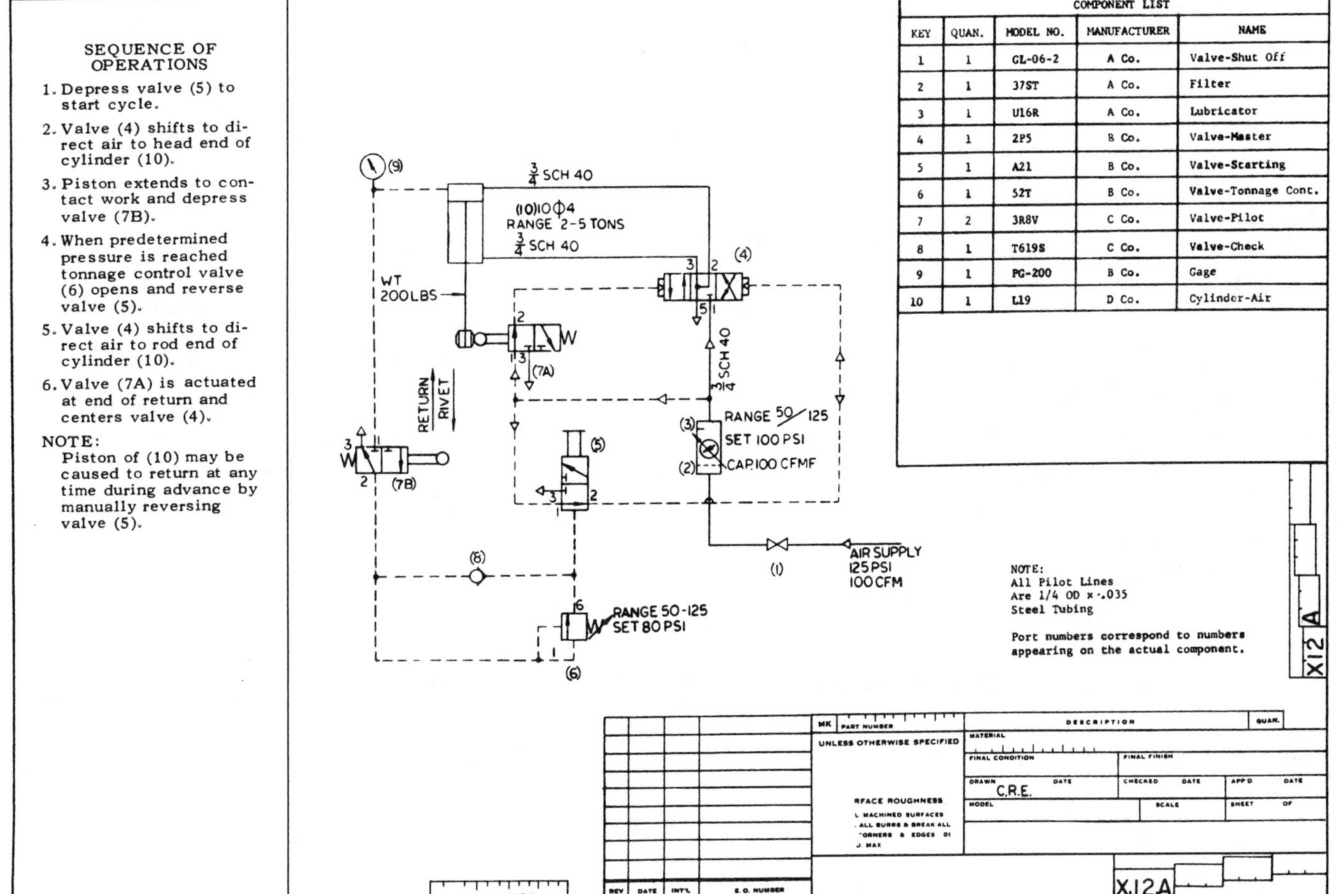


COMPONENT LIST

| KEY | QUAN. | MODEL NO. | MANUFACTURER | NAME |
|---|---|---|---|---|
| 1 | 1 | GL-06-2 | A Co. | Valve-Shut Off |
| 2 | 1 | 37ST | A Co. | Filter |
| 3 | 1 | U16R | A Co. | Lubricator |
| 4 | 1 | 2P5 | B Co. | Valve-Master |
| 5 | 1 | A21 | B Co. | Valve-Starting |
| 6 | 1 | 52T | B Co. | Valve-Tonnage Cont. |
| 7 | 2 | 3R8V | C Co. | Valve-Pilot |
| 8 | 1 | T619S | C Co. | Valve-Check |
| 9 | 1 | PG-200 | B Co. | Gage |
| 10 | 1 | L19 | D Co. | Cylinder-Air |

**Figure 23-6.**
Refer to the text for a step-by-step sequence of operations of this circuit.

5. The reversing of valve (5) shifts the airflow and shifts valve (4) to direct the airflow to the rod end of cylinder (10). Air pressure on the head of the cylinder is also vented by the shifting of valve (4).
6. At the end of the piston rod return, valve (7A) is actuated. This permits the air pressure to flow through valve (7A) and centers valve (4), neutralizing the pressure on the rod and head ends of cylinder (10).
7. The cycle can begin again by manually activating valve (5).

## Review Questions

*Circle the answer of choice, fill in the blank, or write a short answer.*

1. List the four basic types of diagrams. ______________________________________________

______________________________________________________________________________

2. *True or False?* The type of diagram consisting of symbols that represent components connected by single lines is the graphic diagram.

3. A(n) ______________________ is a device for converting hydraulic or pneumatic energy into mechanical energy.

4. A ____ is the complete path of flow for an entire fluid power system, including the pump.
   A. chain
   B. schematic
   C. path
   D. circuit

5. A phantom line around components on the diagram indicates those items are in a(n) ______________.

6. What term applies to the end of a passage in a component (either internal or external)? __________

______________________________________________________________________________

7. *True or False?* The container for storage of liquid in a fluid power system, not typically under pressure, is called the tank.

8. An explanation of the various functions of the circuit in the order of occurrence is often referred to as the ______________________ of ______________________.

9. A ____ is an electrical switch that can be either energized or de-energized.
   A. solenoid
   B. cylinder
   C. transformer
   D. diode

10. In a diagram component list, if two parts are identical, one may be noted as 16A, while the other is identified as ______________________________________________.

## Industry Print Exercise 23-1

*Refer to print PR 23-1. The sequence of operations for this print are given below. Answer the questions following the sequence of operations.*

**Sequence of operations:**

1. The operator actuates the two push-button valves (1) and (2) until the weight starts to descend.
2. Air passes through valves (1) and (2) to the time delay valve (3) and the pilot of valve (5).
3. The four-way valve (5) is shifted, venting the head end of cylinder (4) and pressurizing the rod end.
4. The weight of the ram descends.
5. The speed of descent is controlled by the exhaust control (7).
6. The time delay valve (3) continues to time until the preset interval has passed.
7. The time delay valve (3) vents air pressure from the line between valve (2) and the pilot of valve (5).
8. With the pilot vented, the spring return of valve (5) shifts the valve to its normal position.
9. The rod end of cylinder (4) is vented and the head end is pressurized.
10. The weight ascends.
11. The speed of ascent is controlled by the exhaust control (6).

**Questions:**

1. What is the name of the circuit? ____________________

2. Give the drawing number. ____________________

3. How many different component parts are listed? ____________________

4. What starts the sequence of operation? ____________________

5. What causes valve (5) to shift? ____________________

6. When valve (5) shifts, what happens to cylinder (4)? ____________________

7. What part controls the speed of the cylinder descent? ____________________

8. What causes cylinder (4) to reverse? ______

9. What does the phantom line around valve (3) indicate? What component(s) are included? ______

10. Why is it not necessary to hold valves (1) and (2) down during the entire sequence? ______

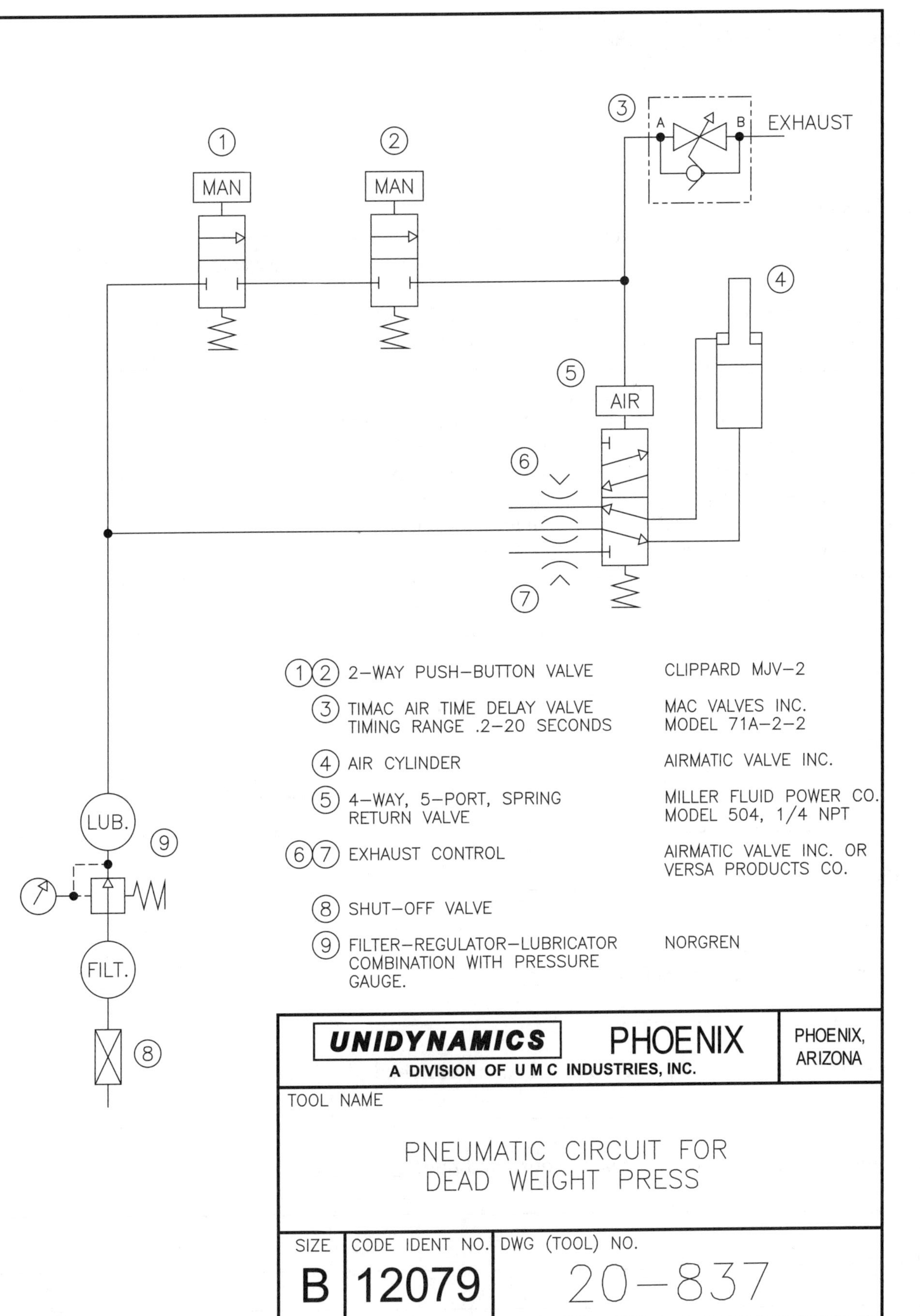


**PR 23-1.**
Print supplied by Unidynamics.

## Industry Print Exercise 23-2

*Refer to print PR 23-2. The sequence of operations for this print are given below. Answer the questions following the sequence of operations.*

**Sequence of operations:**

Description: This is a system to control a 90-ton hydraulic, explosive-compaction press. The press ram descends at a controlled rate, slowly compacting the explosive charge to avoid detonation due to a sudden shock.

1. The cycle is started by energizing solenoid A, which shifts control valve (3) and supplies pressure to the head end of the hydraulic cylinder (7). This operates the ram.
2. The cylinder (7) extends at a slow rate controlled by the fluid in the rod end of the cylinder, which is vented through the flow control valve (5).
3. When the cylinder is fully extended and the pressing pressure is reached, the pressure switch (6) de-energizes solenoid A and energizes solenoid B. This shifts control valve (3) to retract the cylinder (7). Control valve (5) has an integral check valve that permits the free flow of fluid in the opposite direction, causing cylinder (7) to rapidly retract.
4. The retraction of cylinder (7) actuates the limit switch (8), which de-energizes solenoid B. Control valve (3) returns to the neutral position. This completes the cycle. This circuit has safety features. On electrical power outage, both solenoid A and B are de-energized. This allows valve (3) to return to the center position, thus stopping ram travel. Also, valve (4) provides low pressure relief on the ram upstroke; valve (2) provides high pressure relief on ram downstroke.

**Questions:**

1. What is the name of the circuit? ______________________________

______________________________

2. Give the drawing number. ______________________________

______________________________

3. What starts the sequence of operations? ______________________________

______________________________

4. When valve (3) shifts on energizing solenoid A, to which end of cylinder (7) is pressure supplied?

______________________________

5. What causes the ram to extend at a slow, controlled rate? ______________________________

______________________________

6. What causes the ram to retract? ______________________________

______________________________

______________________________

______________________________

7. Why does cylinder (7) retract rapidly? ______________________________

8. What completes the cycle? ______________________________

9. What happens in the event of electrical power outage? ______________________________

10. Give the function of valves (2) and (4). ______________________________

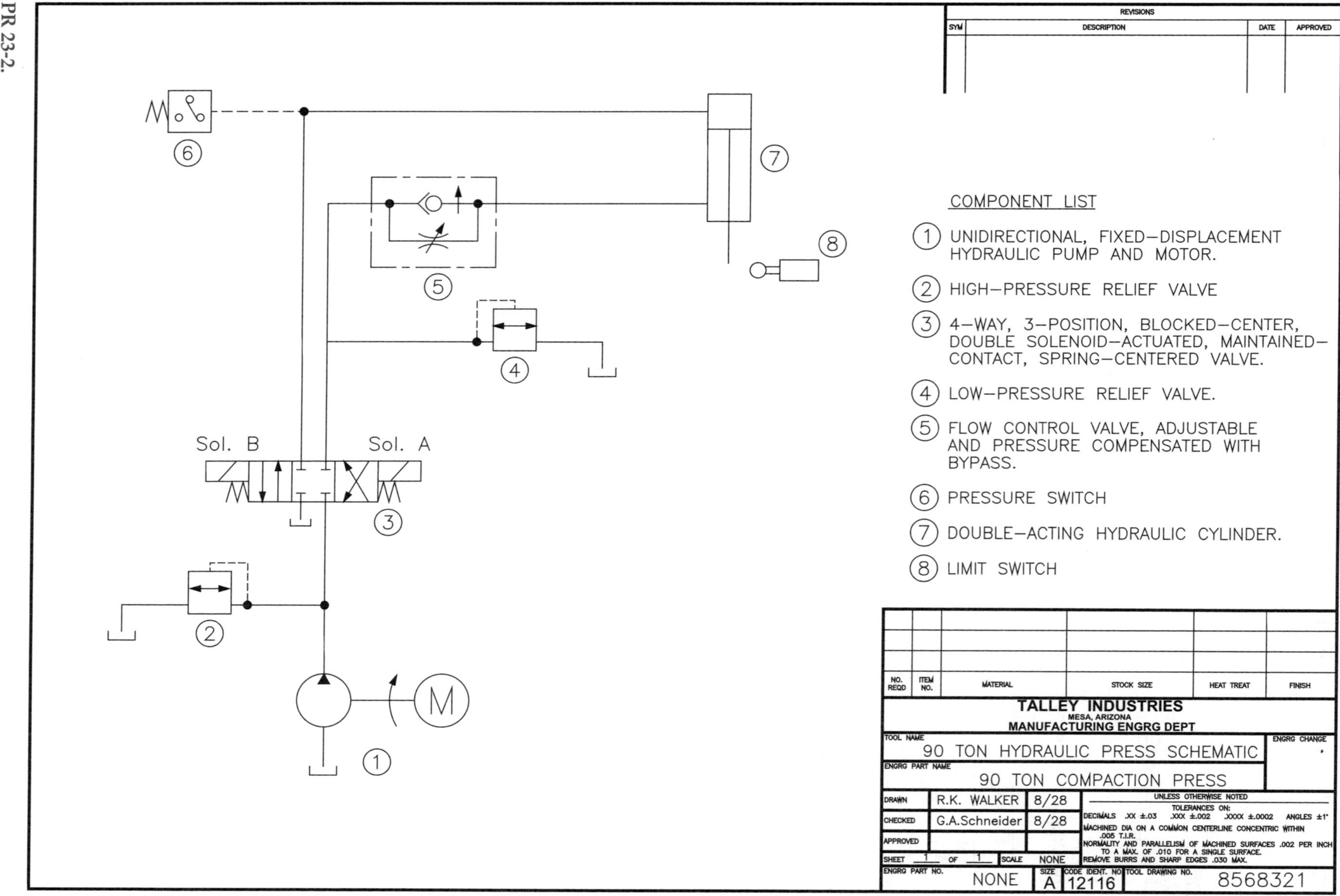


PR 23-2.
Print supplied by Talley Industries.

## Bonus Print Reading Exercises

The following questions are based on various bonus prints located in the folder at the back of this textbook. Refer to the print indicated, evaluate the print, and answer the question.

**Print AP-027:**
**Sequence of Operations 25-086**
**Description:** This machine drills four 1/8″ diameter holes 90° apart through a cylindrical shell. The shell is approximately 3″ diameter and 12″ long. After drilling, the shell is transferred vertically, where four 1/8″ diameter pins are inserted into the holes. The machine then resets itself for unloading, reloading, and the next cycle.

1. The machine is loaded and the shell is clamped in place.
2. The automatic cycle is initiated by manually depressing the button on valve 1.
3. A pulse of line pressure opens valves 2 and 3, assuring a rest position.
4. When valve 2 shifts, transfer cylinders 4 are pressurized to lower the shell into the drilling (down) position.
5. When the shell is fully in the down position, it actuates valve 5.
6. Line pressure through valve 5 starts Timac timer 6. Line pressure also passes through valve 7 to start the four drill motors.
7. The drill motors have their own internal pneumatic controls to start and stop their rotation and to retract on reaching a predetermined depth.
8. Timer 6 times out during the drill cycle. This resets valve 7 and shifts valve 3 to allow pressure from valves 8 to pass when available.
9. When all drills are retracted, they mechanically shift valves 8, which allow pressure to pass to valve 3.
10. Pressure passes through valve 3, through check-speed control valve T-1, and shifts valve 2. This allows line pressure to move transfer cylinders 4 to the up position.
11. At the full-up position of transfer cylinders 4, valve 9 is actuated allowing pressure to start timer 10. Pressure also passes through valve 11 and valve 18 to shift the two valves 12. This extends the four pin cylinders.
12. After timer 10 times out, pressure passes through the timer to shift valve 11. This vents the pressure on the two valve 12 controls.
13. Upon full extension of pin cylinders seating pins to a predetermined depth, microswitches 13 actuate solenoid valve 14. This allows pressure to shift the two valves 12. Shifting the valves 12 allows pressure to retract the pin cylinders.
14. Valve 15 is used as an emergency interrupt for the transfer cylinders. When manually actuated, the transfer cylinders retract.
15. Valves 16 and 17 are manual overrides for operating the pin cylinders. Valve 16 manually extends the pin cylinders. Valve 18 is used as a selector valve only when valve 16 is manually actuated. Valve 17 manually retracts the pin cylinders.

**Bill of Materials 25-086**

1. Air valve, 3-way, push button, spring return.
2. Air valve, 4-way, pilot operated, both ways.
3. Air valve, 3-way, pilot operated, both ways.
4. Air cylinder, double action, 2-1/2″ diameter × 4″ stroke.
5. Air valve, 3-way, lever operated, spring return.
6. Air timer, Timac, .2–20 sec. adjustable.
7. Air valve, 3-way, pilot operated, spring return.
8. Air valve, 3-way, lever operated, spring return.

9. Air valve, 3-way, lever operated, spring return.
10. Air valve, Timac, .2–20 sec. adjustable.
11. Air valve, 3-way, lever operated, spring return.
12. Air valve, 4-way, pilot operated, both ways.
13. Microswitch, V3-26.
14. Air valve, 2-way, solenoid operated, spring return, skimmer V53DB2-150.
15. Air valve, 3-way, manually operated.
16. Air valve, 3-way, manually operated, spring return.
17. Air valve, 3-way, manually operated.
18. Air valve, 3-way, pilot operated, both ways. Air drills, 2800 rpm, Rockwell 1/8″ dashpot. Air cylinders, double action, double rod, 2-1/2″ diameter × 4″ stroke.

**Questions:**

1. What is the name of the print? ____________________

2. What is the drawing number? ____________________

3. Briefly describe the machine's function. ____________________

4. How is the automatic cycle started? ____________________

5. When the transfer cylinders are in their down position, what starts the drill motors? ____________________

6. While the drilling cycle is in process, what happens to valve 7 and valve 3? ____________________

7. What activates the four valves 8? ____________________

8. When the four valves 8 are shifted, what takes place? ____________________

9. What happens when transfer cylinders 4 are in the up position? ______________________

10. What causes the pin cylinders to automatically retract? ______________________

11. When is valve 15 used? ______________________

**Print AP-028:**

**Sequence of Operations: Case Assembly Machine Hydraulic Schematic**

**Description:** This machine uses a fluid power system to perform two dimpling operations on a cylindrical assembly. The part is loaded into a fixture, dimpled, rotated 180°, and dimpled again. This is accomplished by the following sequence.

1. The workpiece is loaded into the fixture.
2. When the workpiece is in position, the operator depresses valve 7. This extends cylinder 8 and clamps the piece into position.
3. The operator depresses and holds the push buttons on valves 2 and 3. These valves allows hydraulic fluid to flow to the pilot of valve 5, thus shifting the valve. Cylinder 6 extends and dimples the workpiece.

Note: This step cannot be accomplished unless the operator depresses both palm buttons. This safety feature is to ensure that the operator is clear of the mechanism.

4. By releasing the push button on either valve 2 or 3, the fluid is exhausted from the pilot of valve 5, thus shifting the valve. Cylinder 6 retracts.
5. Once cylinder 6 is retracted, the operator manually actuates valve 10. Fluid pressure rotates rotary actuator 11 by 180°.
6. When the workpiece is rotated, steps 3 and 4 are repeated to make another dimple 180° from the first.
7. On completion of step 6, the operator releases valve 10. Rotary actuator 11 returns to its original position.
8. When the part has returned to its original position, the operator releases valve 7. Cylinder 8 retracts to unclamp the workpiece from the fixture. The cycle is complete.

Note: This circuit is a dual pressure circuit. Cylinders 6 and 8 extend under high-pressure fluid. This is accomplished by pressure-relief valve 1. Retraction of these cylinders is done under low pressure, accomplished by pressure-relief valve 7.

**Questions:**

12. What is the name of the drawing? ______________________

13. Briefly describe the purpose of the machine. ______________________

14. What actuates the workpiece clamping cylinder? ____________________

15. When the push button on valve 2 is depressed, is valve 5 actuated? Why or why not? ____________________

16. What happens when cylinder 6 is extended and valve 3 is released? ____________________

17. What causes the workpiece to rotate 180° for the second part of the operation? ____________________

18. Describe valve 10. ____________________

19. Is cylinder 6 extended or retracted by high-pressure fluid? ____________________

20. What is the normal position (as shown) for cylinder 8? ____________________

21. Does rotary actuator 11 operate on high-pressure or low-pressure fluid? ____________________

# Appendix A
# Applied Mathematics

*After completing this unit, you will be able to:*

**Identify** common fractions.

**Add, subtract, multiply,** and **divide** common fractions.

**Identify** decimal fractions.

**Add, subtract, multiply,** and **divide** decimal fractions.

**Identify** the metric system and metric numbers.

**Add, subtract, multiply,** and **divide** metric numbers.

Engineers and technicians frequently need to make calculations involving common fractions and decimals in connection with print reading. If you are in need of a quick review of the fundamentals of this area of mathematics, this unit will assist you in your study.

## Common Fractions

Common fractions are written with one number over the other, as:

$$\frac{11}{16}$$

The number on the bottom is the ***denominator.*** It indicates the number of equal parts into which a unit is divided. In the above example, the unit is divided into 16 equal parts. The number on top is the ***numerator.*** It indicates how many equal parts are represented by the fraction. In the above example, the fraction represents 11 of 16 equal parts. See Figure A-1.

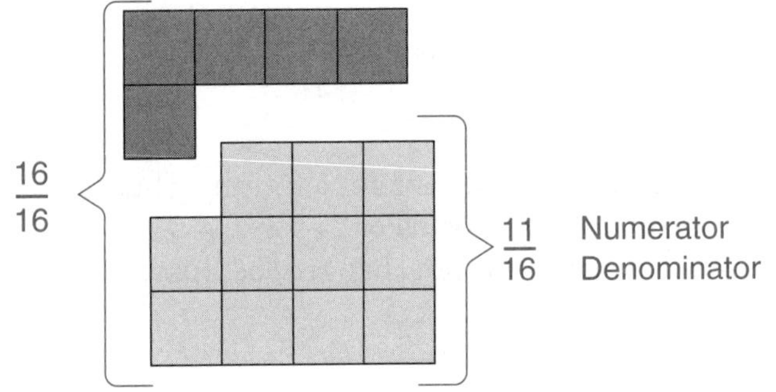


**Figure A-1.**
A unit is divided into 16 equal parts. The denominator is 16. The numerator is the number of parts selected, in this case 11.

A ***proper fraction*** is a fraction where the numerator is less than the denominator. For example:

$$\frac{7}{16} \text{ and } \frac{3}{4}$$

are proper fractions. An ***improper fraction*** is a fraction where the numerator is greater than its denominator. For example:

$$\frac{5}{4} \text{ and } \frac{19}{16}$$

are improper fractions. A ***mixed number*** is a number consisting of a whole number and a fraction. For example:

$$2\,\frac{3}{4} \text{ and } 5\,\frac{1}{8}$$

are mixed numbers.

## Fundamental Steps in the Use of Fractions

There are several fundamental principles for working with fractions. These are described in the fundamental steps below.

1. Whole numbers can be changed to fractions by multiplying the numerator and denominator by the same number. For example, to change the whole number 6 into a fraction with 4 as the denominator, place the whole number over 1 and multiply the numerator and denominator by 4 as follows.

   $$\frac{6}{1} \times \frac{4}{4} = \frac{24}{4}$$

   Each whole unit contains 4 fourths. Thus, 6 units contain 6 × 4 fourths or 24 fourths. The value of the number has not changed, because:

   $$\frac{24}{4} = 6$$

2. Mixed numbers can be changed to fractions by changing the whole number to a fraction and adding the two fractions. The two fractions must have the same denominator. For example, to change 3 5/8 to a fraction, place the whole number over 1, multiply that numerator and denominator by 8, and add the resulting fraction to 5/8 as follows.

   $$3\frac{5}{8} = \frac{3}{1} \times \frac{8}{8} + \frac{5}{8}$$
   $$= \frac{24}{8} + \frac{5}{8}$$
   $$= \frac{29}{8}$$

   Each whole unit contains 8 eighths. Thus, 3 units contain 3 × 8 eighths or 24 eighths. Adding the 5/8 part of the mixed number provides the final answer.

3. Improper fractions can be reduced to a whole or mixed number by dividing the numerator by the denominator. For example, to reduce 17/4, divide 17 by 4 as follows.

   $$\frac{17}{4} = 17 \div 4 = 4\frac{1}{4}$$

   The number 4 goes into 17 four times (16) with a remainder of 1. Place the 1 over 4 to get 1/4. Combining the 4 and the 1/4 provides the mixed number 4 1/4.

4. Fractions can be reduced to the lowest form by dividing the numerator and denominator by the same number. For example, to reduce 6/8 to the lowest form, divide the numerator and denominator by 2 as follows.

   $$\frac{6}{8} = \frac{6 \div 2}{8 \div 2}$$
   $$= \frac{3}{4}$$

   Since both the numerator and denominator are divided by the same number (2), the value of the fraction is not changed.

   In print reading, it can be helpful in analyzing a fractional measurement to divide by 2, even if the numerator is an odd number! For example, 15/32 can be thought of as 7 1/2 sixteenths. Sometimes it is hard to envision what 32nds of an inch are since they are so small, but converting to sixteenths may help the reader visualize the distance.

5. Fractions can be changed to a higher form by multiplying the numerator and denominator by the same number. For example, to change 5/8 to a higher form (16), multiply the numerator and denominator by the same number (2) as follows.

   $$\frac{5}{8} = \frac{5 \times 2}{8 \times 2}$$
   $$= \frac{10}{16}$$

   The value of the fraction is not changed by multiplying both 5 and 8 by 2. This is because multiplying by 2/2 is the same as multiplying by 1 (2 ÷ 2 = 1).

## Addition of Fractions

To add common fractions, the denominators must all be the same. The numerators are then simply added together. For example, to add 5/16, 3/8, and 11/32, the ***lowest common denominator (LCD)*** must be found. In this example, 32 is the

LCD. All fractions must be changed to have 32 as the denominator. Change fractions to a higher form (32) using fundamental step number 5 as follows.

$$\frac{5 \times 2}{16 \times 2} = \frac{10}{32}$$

$$\frac{3 \times 4}{8 \times 4} = \frac{12}{32}$$

All fractions now have a common denominator of 32 and can be added as follows.

$$\frac{10}{32} + \frac{12}{32} + \frac{11}{32} = \frac{33}{32}$$

The improper fraction 33/32 can be reduced using fundamental step number 3 as follows.

$$\frac{33}{32} = 33 \div 32$$

Dividing 33 by 32 results in 1 with a remainder of 1. Place the 1 over 32, to get:

$$\frac{1}{32}$$

Combining the 1 and the 1/32 results in the mixed number:

$$1\frac{1}{32}$$

## Subtraction of Fractions

To subtract common fractions, the denominators must all be the same. The numerators can then simply be subtracted. For example, to subtract 5/16 from 3/4, the lowest common denominator (LCD) must be found. In this example, 16 is the LCD. All fractions must be changed to have 16 as the denominator. Change fractions to a higher form (16) using fundamental step number 5 as follows.

$$\frac{3 \times 4}{4 \times 4} = \frac{12}{16}$$

The fractions can be subtracted by subtracting the numerators as follows.

$$\frac{12}{16} - \frac{5}{16} = \frac{7}{16}$$

## Multiplication of Fractions

To multiply common fractions, first change all mixed numbers to improper fractions. Then, multiply all numerators and multiply all denominators. Reduce the resulting fraction to the lowest form. For example, to multiply 1/2, 3 1/8, and 4, the mixed number must be changed to an improper fraction. In this example, 3 1/8 must be changed to an improper fraction and then multiplied with the other fractions as follows.

$$\frac{1}{2} \times 3\frac{1}{8} \times 4 = ?$$

$$\frac{1}{2} \times \frac{25}{8} \times \frac{4}{1} = \frac{100}{16}$$

The improper fraction 100/16 can be reduced using fundamental step number 3 as follows.

$$\frac{100}{16} = 100 \div 16$$

$$= 6\frac{4}{16}$$

Dividing 100 by 16 results in 6 with a remainder of 4. Place the 4 over 16 to get:

$$\frac{4}{16}$$

The fraction 4/16 can be further reduced using fundamental step number 4 as follows.

$$\frac{4 \div 4}{16 \div 4} = \frac{1}{4}$$

Combining the 6 and the 1/4 provides the mixed number of:

$$6\frac{1}{4}$$

The improper fraction of 100/16 reduced to the lowest form is 6 1/4.

## Division of Fractions

To divide common fractions, change all mixed numbers to improper fractions and invert the divisor. Inverting the divisor is reversing the numerator and denominator of the divisor. The divisor is the number by which you are dividing. Then, multiply all numerators and multiply all denominators. Reduce the resulting fraction to the lowest form. For example, to divide 5 1/4 by 1 1/2, the mixed numbers must be changed to improper fractions. The divisor is inverted and multiplied with the other fraction as follows.

$$5\frac{1}{4} \div 1\frac{1}{2} = ?$$

$$\frac{21}{4} \div \frac{3}{2} =$$

$$\frac{21}{4} \times \frac{2}{3} = \frac{42}{12}$$

The improper fraction 42/12 can be reduced using fundamental step number 3 as follows.

$$\frac{42}{12} = 42 \div 12$$

$$= 3\frac{6}{12}$$

Dividing 42 by 12 results in 3 with a remainder of 6. Place the 6 over 12 to get 6/12. The fraction 6/12 can be further reduced using the fundamental step number 4 as follows.

$$\frac{6 \div 6}{12 \div 6} = \frac{1}{2}$$

Combining the 3 and the 1/2 gives us the mixed number of:

$$3\frac{1}{2}$$

The improper fraction of 42/12 reduced to the lowest form is 3 1/2.

## Decimal Fractions

The denominator in decimal fractions is 10 or a multiple of 10. When writing decimal fractions, the denominator is omitted and a decimal point is placed in front of the numerator. For example, 1/10 is written as .1 (one-tenth) and 1/100 as .01 (one-hundredth). Other examples are:

3/10 is written as .3 (three-tenths)
87/100 is written as .87 (eighty-seven hundredths)
375/1000 is written as .375 (three hundred seventy-five thousandths).
4375/10000 is written as .4375 (four thousand three hundred seventy-five ten thousandths).

Whole numbers are written to the left of the decimal point and fractional parts are to the right. For example:

1 1/10 is written as 1.1 (one and one-tenth).
4 35/100 is written as 4.35 (four and thirty-five hundredths).
5 253/1000 is written as 5.253 (five and two hundred fifty-three thousandths).

### *Addition and Subtraction of Decimals*

Decimal fractions are added and subtracted in the same manner as whole numbers. In adding and subtracting decimals, the figures must be written so the decimal points vertically align. For example, when adding 7.3125, 1.25, .625, and 3.375, the numbers are added as follows.

```
  7.3125
  1.25
   .625
+ 3.375
-------
 12.5625
```

Additional zeros can be placed to the right so each decimal fraction has the same number of digits to the right of the decimal, but this is not required. This does not change the value since 1/10 = 10/100 = 100/1000. For example:

```
  7.3125
  1.2500
   .6250
+ 3.3750
-------
 12.5625
```

When subtracting decimal fractions, such as 2.25 from 8.625, the decimal points must vertically align in the same manner as addition.

```
  8.625
- 2.25
------
  6.375
```

This can also be written as:

```
  8.625
- 2.250
------
  6.375
```

Notice the decimal point in the answer for each example is directly below the decimal points in the problem.

### Multiplication of Decimals

Decimal fractions are multiplied in the same manner as whole numbers. Vertical alignment of decimal points is not necessary. To find the position of the decimal point in the answer, count the number of places to the right of the decimal point in each number being multiplied. Add them together to get the total number of places. In the answer, place the decimal point by counting off the total number of places from the right. For example, multiply 6.25 by 1.5 as follows.

```
   6.25  (two decimal places)
×   1.5  (one decimal place)
   3125
   6250
  9.375  (three decimal places)
```

### Division of Decimals

Decimal fractions are divided in a manner similar to whole numbers. The answer is called the quotient. However, the number of decimal places must be counted similar to multiplication of decimal fractions. To place the decimal point in the quotient (answer), count the number of places to the right of the decimal point in the divisor. Then, count this number of places to the right of the decimal point in the dividend (the number being divided). Place the decimal point directly above in the quotient. For example, to divide 36.5032 (dividend) by 4.12 (divisor), the decimal point is moved two places to the right and division is performed as follows.

```
            8.86
4.12^ | 36.50^32
        32 96
         3 543
         3 296
           2472
           2472
              0
```

## Metric System

The metric system of numbers works in the same manner as the decimal-inch system. Both systems are base-ten number systems. Only the size of the units and terms differ. This makes it easy to shift from one multiple or sub-multiple to another.

### Addition and Subtraction in the Metric System

Numbers in the metric system are added and subtracted in the same manner as they are for decimal fractions. Vertically align the decimal points and then add or subtract as needed. For example, 38.35 millimeters, 20.666 millimeters, and 116.59 millimeters are added as follows.

```
   38.35
   20.666
+ 116.59
  175.606
```

As another example, to subtract 118.06 from 107.902, align the decimal points and subtract the numbers as follows.

```
  118.06
– 107.902
   10.158
```

Notice the decimal point in the answer for each example is directly below the decimal points in the problem.

### Multiplication and Division in the Metric System

Numbers are multiplied and divided in the metric system in the same manner as for decimal fractions in the decimal-inch system. For example, multiply 81.6 millimeters by 3 as follows.

```
  81.6
×    3
 244.8
```

To divide 103.42 millimeters in four equal parts, divide as follows.

```
     25.850
4 | 103.420
     8
     23
     20
      34
      32
       22
       20
        20
        20
         0
```

## Review Questions

*Circle the answer of choice, fill in the blank, or write a short answer.*

1. What are the two components of a proper fraction? ____________________

____________________________________________

2. A(n) ____________________ number contains a whole number and a fraction.

3. A(n) ____________________ fraction has a numerator less than the denominator.

4. A(n) ____________________ fraction has a numerator more than the denominator.

5. *True or False?* A fraction can be reduced to the lowest form by dividing the denominator by the numerator.

6. *True or False?* To add fractions, the denominators must all be the same.

7. LCD stands for lowest ____ denominator.
   A. constant
   B. contrasting
   C. common
   D. created

8. To divide one fraction by another, change any mixed numbers to improper fractions, invert the ____________________, and then multiply.

9. *True or False?* When adding decimal number values, vertically align the decimals and then you must add zeros to each decimal fraction so that each has the same number of digits on the right of the decimal.

10. *True or False?* Numbers in the metric system are multiplied and divided in a different manner than in the decimal-inch system.

## Review Activity A-1

### Adding Fractions

Solve the following problems. Reduce answers to the lowest form. Show all work in the space provided.

1. $\frac{3}{4} + \frac{1}{8} + \frac{1}{2} =$

2. $\frac{7}{8} + \frac{3}{16} =$

3. $\frac{5}{12} + \frac{3}{8} + \frac{3}{4} =$

4. $\frac{3}{10} + \frac{9}{10} + \frac{1}{4} =$

5. $\frac{7}{16} + \frac{3}{32} + \frac{1}{4} =$

6. $1\frac{3}{4} + \frac{7}{8} + 1\frac{1}{16} =$

7. $\frac{5}{32} + \frac{7}{64} + \frac{7}{8} =$

8. $1\frac{3}{8} + \frac{3}{32} + \frac{7}{16} =$

9. $3\frac{1}{16} + \frac{9}{16} + \frac{1}{2} =$

10. $5\frac{1}{5} + 2\frac{3}{10} + 8\frac{1}{2} =$

11. $4\frac{5}{8} + 20\frac{7}{32} =$

12. $\frac{3}{8} + \frac{7}{64} + \frac{9}{16} =$

13. $12\frac{7}{8} + 25\frac{3}{8} =$

14. $\frac{21}{32} + \frac{9}{64} + \frac{1}{4} =$

15. $\frac{3}{8} + 1\frac{1}{2} + \frac{7}{16} + \frac{7}{8} =$

16. $2\frac{1}{4} + \frac{5}{8} + \frac{5}{16} + \frac{17}{32} =$

# Review Activity A-2

## Subtracting Fractions

Solve the following problems. Reduce answers to the lowest form. Show all work in the space provided.

1. $\frac{3}{8} - \frac{1}{4} =$

2. $\frac{3}{4} - \frac{5}{16} =$

3. $1\frac{7}{8} - \frac{3}{16} =$

4. $3\frac{1}{2} - \frac{9}{16} =$

   Note: Change $3\frac{1}{2}$ to $2\frac{24}{16}$

5. $10\frac{3}{8} - 7\frac{3}{32} =$

6. $5 - 2\frac{3}{8} =$

7. $12\frac{1}{16} - 8\frac{1}{2} =$

8. $4\frac{1}{4} - 3\frac{1}{16} =$

9. $20\frac{7}{8} - 11\frac{3}{64} =$

10. $15\frac{5}{8} - 5\frac{1}{2} =$

## Review Activity A-3

### Multiplying Fractions

Solve the following problems. Reduce answers to the lowest form. Show all work in the space provided.

1. $\frac{3}{4} \times \frac{1}{2} =$

2. $2\frac{5}{8} \times \frac{1}{4} =$

3. $\frac{7}{8} \times 5 =$

4. $6\frac{3}{4} \times \frac{1}{3} =$

5. $12\frac{1}{2} \times \frac{1}{2} =$

6. $4\frac{3}{4} \times \frac{1}{2} \times \frac{1}{8} =$

7. $16 \times \frac{3}{4} =$

8. $9\frac{5}{8} \times \frac{1}{2} =$

9. $10 \times \frac{4}{5} =$

10. $\frac{14}{3} \times 6 =$

## Review Activity A-4

### Dividing Fractions

Solve the following problems. Reduce answers to the lowest form. Show all work in the space provided.

1. $2\frac{3}{4} \div 6 =$
2. $12 \div \frac{3}{4} =$
3. $16\frac{1}{8} \div 2 =$
4. $8\frac{2}{3} \div \frac{1}{3} =$
5. $16\frac{1}{4} \div 20 =$
6. $\frac{7}{8} \div \frac{7}{16} =$
7. $15 \div 1\frac{1}{4} =$
8. $21\frac{3}{8} \div 3\frac{1}{8} =$
9. $5\frac{1}{4} \div \frac{3}{8} =$
10. $3\frac{5}{8} \div 2 =$

## Review Activity A-5

**Adding and Subtracting Decimals**

Solve the following problems. Show all work in the space provided.

1.
$$\begin{array}{r} 4.5625 \\ .875 \\ 2.75 \\ +\ 5.8137 \\ \hline \end{array}$$

2. 7.0625 + .125 + 8.0 =

3. .832 + .4375 + .27 =

4.
$$\begin{array}{r} 27.9375 \\ -\ 16.937 \\ \hline \end{array}$$

5. 4.0 – .0625 =

6. 2.25 – 1.125 =

## Review Activity A-6

### Multiplying Decimals

Solve the following problems. Show all work in the space provided.

1. $4.825 \times 1.75 =$

2. $167 \times .25 =$

3. $65.96 \times .37 =$

4. $4.95 \times 1.35 =$

5. $93.18 \times .07 =$

## Review Activity A-7

**Dividing Decimals**

Solve the following problems. Show all work in the space provided.

1. $9.45 \div 2.7 =$

2. $654.5 \div 35 =$

3. $1386.0 \div 1.65 =$

4. $331.266 \div 80.6 =$

5. $4401.25 \div 503 =$

# Review Activity A-8

## Adding and Subtracting Metric Numbers

Solve the following problems. Show all work in the space provided.

1. $$\begin{array}{r} 66.67 \\ 1.42 \\ 3.76 \\ +\quad 1.24 \\ \hline \end{array}$$

2. 41.88 + 89.112 + 8.38 =

3. 4.19 + 49.25 + 2.6 =

4. $$\begin{array}{r} 66.68 \\ -\quad 41.88 \\ \hline \end{array}$$

5. 26.97 – 7.1 =

6. 102.85 – 16.302 =

## Review Activity A-9

**Multiplying and Dividing Metric Numbers**

Solve the following problems. Show all work in the space provided.

1. Find the total length of six sections, each 72.5 mm long.

2. What is the total thickness of five spacers, each 1.22 mm thick?

3. Find the length, in millimeters, of each part when a 108 mm long rod is divided into seven equal parts.

# Appendix B
# Measurement Tools

*After completing this unit, you will be able to:*

**Identify** the importance of the steel rule.

**Read** a fractional steel rule, decimal steel rule, and metric steel rule.

**Identify** the common scale devices available to the drafter and print reader.

**Read** and **use** the fractional scale found on the architect's scale.

**Read** and **use** common reduction scales found on the architect's scale.

**Read** and **use** the decimal scale found on a civil or mechanical engineer's scale.

**Read** and **use** the scales found on a civil engineer's scale for reduction scales.

**Read** and **use** a fractional scale designed for reduction scales.

**Discuss** the common terms used to identify units within the metric system.

**Read** an inch micrometer or a metric micrometer.

**Identify** common dial or digital calipers as used in industrial applications.

The ability to make accurate measurements is a fundamental skill needed by all who read and use prints. This unit reviews the basic principles of reading many scales, including the steel rule, architect's scale, civil engineer's scale, and a micrometer. These measuring instruments or devices are commonly used by drafters, machinists, and others who work with prints on a regular basis.

## Principle of Scale Measurements

A scaled drawing is created by specifying a unit on the drawing, such as 1/2″, equal to a unit on the actual object, such as 1″. In this case, rather than stating 1/2″ = 1″, this particular scale can appear on the print as SCALE = 1:2, SCALE: 1 = 2, or simply HALF SCALE. Expressions should always be read as "paper" equals "real." In other words, if the scale is expressed as 1 = 4, then one unit on the paper equals four units in real life. If the scale is 1/4″ = 1′-0″, then 1/4″ on the paper equals one foot in real life.

***Scale*** is a term with multiple meanings. As a noun, a scale is a measuring instrument or device. As a verb, scale means the process of proportionately reducing or enlarging an object. For example, an object is "scaled down" to decrease its size. The term "scale" is used synonymously for the word "size." For example, a plastic model car may be a "1/25th scale" model, which also means it is "1/25th size."

## Reading a Steel Rule

The ***steel rule*** is a common tool for a machinist. It is very compact, fits nicely in a pocket, and provides very accurate measurements. To work with this print reading text, your instructor may want you to have and be able to read a 6″ steel rule. With this measuring tool, you would be able to read fractionally in 64ths of an inch, in a decimal format in 100ths (decimal) of an inch, or in the metric units of millimeters. In Figure B-1, an enlarged portion of a rule is shown with one edge divided into 64 parts to the inch and the other edge with 100 parts to the inch.

Steel rules can be purchased in more than one format. Some have a decimal edge in addition to the fractional edge, while others have a metric edge. If your particular industry has a lot of metric and inch conversion, you may wish to have a scale with fractional inches along one edge and millimeters along the other edge. Some machine tools and equipment are specified in fractional measurements

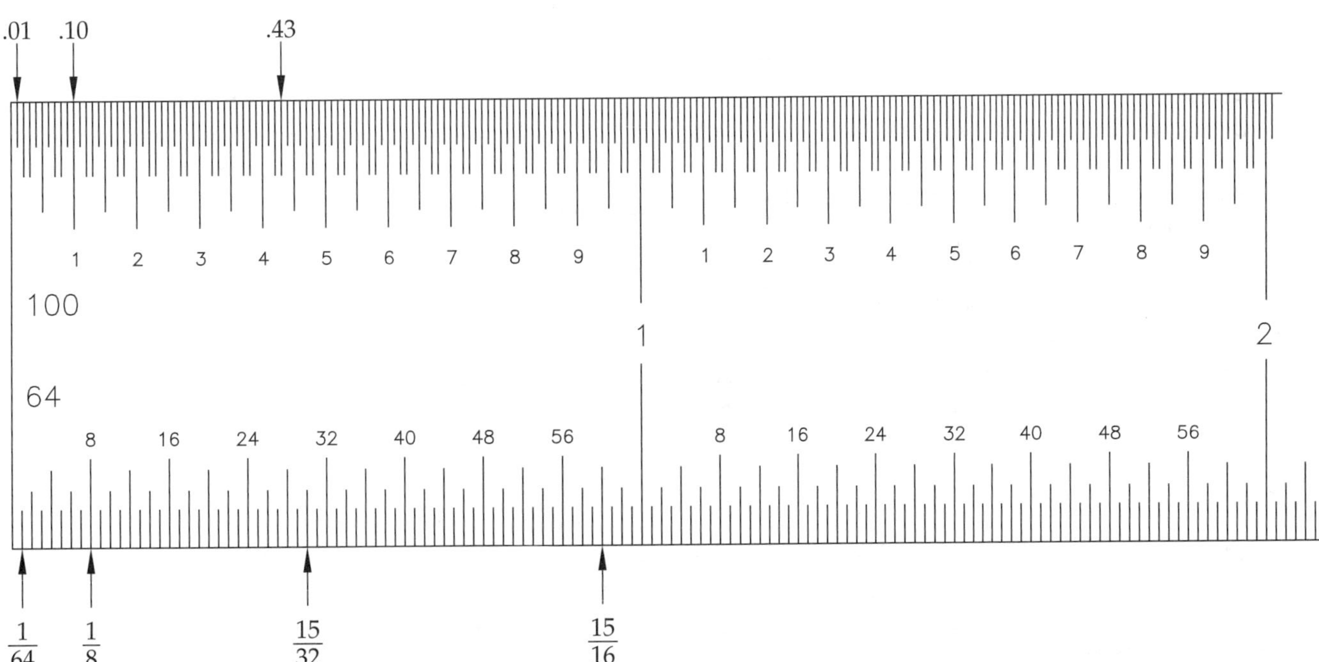


**Figure B-1.**
A typical steel rule has one edge divided into 64 parts to the inch and the other edge with 100 parts to the inch. On some steel rules, one edge is metric.

(1/2, 1/4, 1/8, 1/16, etc.). Some standard stock sizes or screw threads still are specified with fractions. However, engineering designs and tolerances (variations in allowed size) are usually provided in decimal inches. Therefore, a decimal-equivalent chart is also a handy tool to have when reading a print.

## Fractional Rule

The number 64 printed on the end of the rule means each inch is divided into 64 equal parts, Figure B-2. Therefore, each small division on the rule is 1/64″. To read a fractional rule, use the edge divided into 64ths and follow these principles to help you determine a measurement:

- Reduce the fractional reading to its smallest denominator. For example, if you are reading 26, then convert 26/64 to 13/32. If you are reading 48, then convert 48/64 to 24/32, then to 12/16, then to 6/8, and finally to 3/4. The previous unit on applied mathematics is available if you need to review these principles.
- The major divisions of each inch are numbered in increments of 8 (8, 16, 24, etc.). There are eight of these major divisions; therefore, each one is equal to 1/8 of an inch. Think of these major divisions as 1/8 (8/64ths), 2/8 (16/64ths), 3/8 (24/64ths), and so on. Of course, 2/8 can be reduced to 1/4.

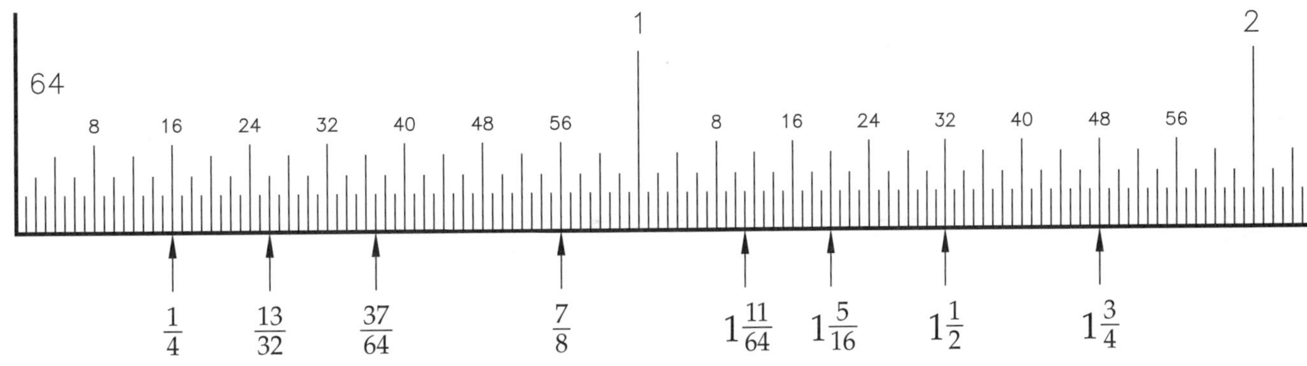


**Figure B-2.**
One edge of a steel rule may be divided into 64 parts to the inch.

- Each of these numbered divisions is also divided into 8 smaller divisions. The middle division mark is longer than the others within that division and is 1/16 (4/64) of an inch away from the labeled marks. For example, the 32 mark is 1/2″ (32/64). Therefore, halfway toward the 40 mark is 9/16″. Halfway toward the 24 mark is 7/16″.
- Have a decimal equivalent chart or a calculator handy. Some fractional readings make more sense as a decimal. For example, 39/64 may be more meaningful when converted to .609″.

## Decimal Rule

Many decimal rules have 100 printed on the end of the rule, Figure B-3. This indicates each inch is divided into 100 parts. Therefore, each small division is 1/100″ (.01″). If the scale has a 50 on the end, there are 50 increments to the inch on that scale. To read the decimal rule divided into 100ths, follow these principles to help you determine a measurement:

- Each inch has 100 major divisions. Each major division is divided into groups of ten. Therefore, subdivisions represent 1/10 (10/100), 2/10 (20/100), 3/10 (30/100), and so on.
- Each 1/10 inch division contains 10/100ths of an inch. Therefore, these major divisions can be read .20, .30, .40, and so on. Between each major division is a "half" mark that is longer than the others representing .25, .35, .45, and so on.
- Each division mark represents .01″. For example, a measurement of .43″ is three small divisions beyond the 4 mark. The smallest divisions have marks of varying lengths to help prevent the lines from all blending together.

## Metric Rule

Metric rules are used when reading prints dimensioned in centimeters or millimeters. Machined parts are usually smaller than one meter. One meter is 100 centimeters or 1000 millimeters. Metric dimensions for industrial parts are usually given in millimeters.

Conversion tables for millimeters and inches are shown in Appendix E. A metric rule is shown in Figure B-4. A metric rule is read in the same manner as the decimal rule. To read a metric rule, use the edge labeled "mm" and follow these principles to help you determine a measurement:

- Each number represents a centimeter, which is equivalent to 10 millimeters.

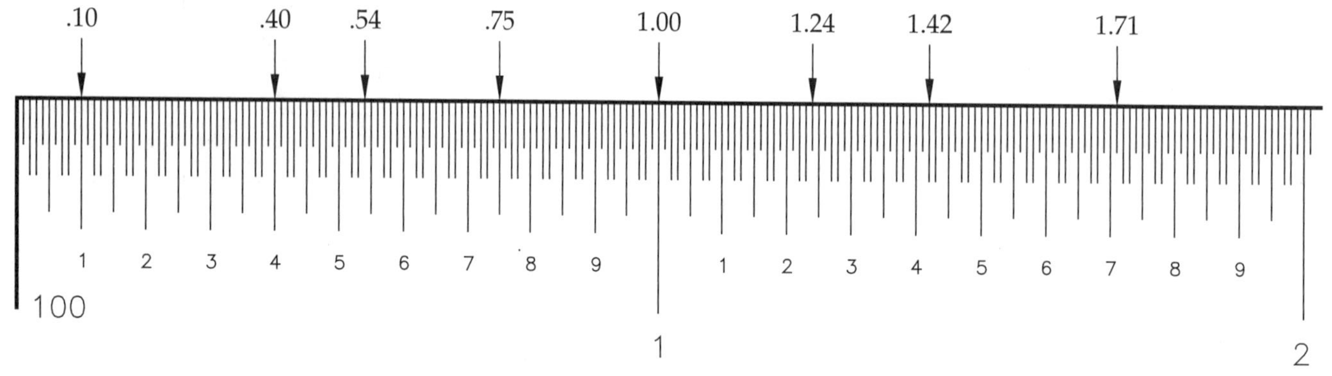


**Figure B-3.**
One edge of a steel rule may be divided into 100 parts to the inch.

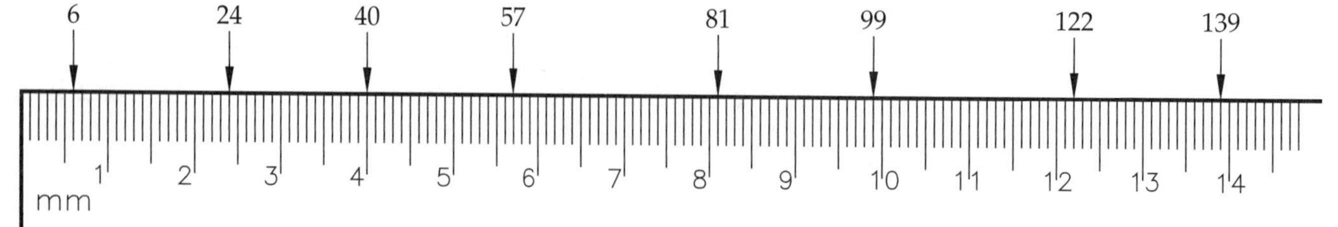


**Figure B-4.**
One edge of a steel rule may be divided into millimeters.

- Each small division is 1/10 of a centimeter, or one millimeter. The measurement is expressed as a whole number, such as 1 mm, 9 mm, 17 mm, and so on.
- Between each centimeter division is a "half" mark that is longer than the others. These marks represent 5 mm, 15 mm, 25 mm, etc.

## Common Scales Used by Drafters

Traditional instrument drawings have been created by drafters throughout the years, often executed with a variety of scales, depending on the scale of the final drawing. A drafter's ruler is called a "scale" because it has several edges designed for scaling a large object down to paper size. The most common measuring tool of the drafter's collection of instruments has always been the triangular ***architect's scale,*** Figure B-5. For objects drawn actual size (scale 1 = 1), the architect's scale has an edge of the scale that is fully divided. This "full" edge of the architect's scale is typically twelve inches, with each inch divided into 16ths. Of course, any common ruler would be fine for the drafter to use, but the architect's scale is usually manufactured more precisely than some rulers. While most architect's scales are triangular, there are some flat versions available.

While one edge of the triangular architect's scale has a fully divided inch scale for full-size measurements, the other five edges each contain two scales that share tick marks. Each scale on a given edge originates from a different end of the scale. This provides the drafter with 10 scales for reducing houses and other structures to paper size. Typically, these are expressed as 3/32″ = 1′-0″, 1/8″ = 1′-0″, 3/16″ = 1′-0″, 1/4″ = 1′-0″, 3/8″ = 1′-0″, 1/2″ = 1′-0″, 3/4″ = 1′-0″, 1″ = 1′-0″, 1-1/2″ = 1′-0″, and 3″ = 1′-0″. Architectural scales are designed for feet and inch measurements, such as room sizes of 14′-2″ or 12′-3 1/2″.

With the proper math understanding, some of the edges of the architect's scale can also be used for mechanical design applications. For example, since 3″ = 1′-0″ is a one-to-four ratio, that edge can be used for creating or measuring 1/4 size (1:4 scale) drawings.

Another common scale used by instrumental drafters is the ***civil engineer's scale,*** often simply referred to as an *engineer's scale.* See Figure B-6. Civil engineers and drafters working on roadway and property drawings often use the civil engineer's scale because it is decimal, with each inch divided into 10ths. Like the architect's scale, the engineer's scale is most commonly manufactured as a triangular device. The full-size edge of a engineer's scale is marked off in divisions of 10 parts to the inch. This scale can be used to lay out drawings based on decimal dimensions. However, the marks can also be used for reduction scales, such as 1″ = 10′, 1″ = 100′, or 1″ = 1000′. In addition, the scale can be

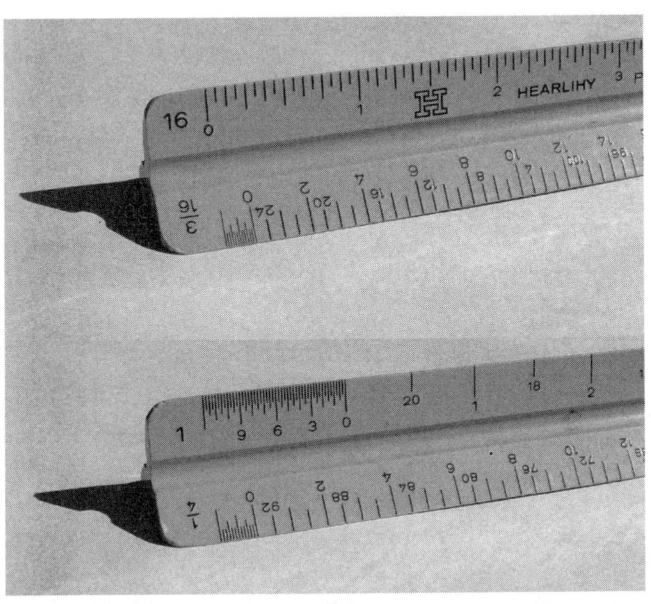


**Figure B-5.**
An architect's scale is often triangular to allow up to twelve different scales on one instrument.

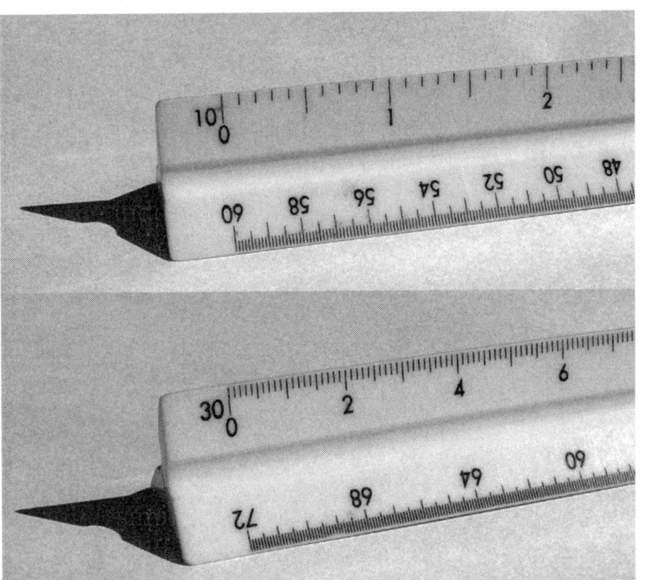

**Figure B-6.**
Civil engineering scales are often triangular, which allows for six different scales on one instrument. Each edge can also have a multiple of scales based on a factor of 10 (i.e., 1″ = 2′, 1″ = 20′, 1″ = 200′, etc.).

used for enlarging drawings at scales such as 1″ = .1″. The other five edges of the triangular civil engineer's scale are divided into 20, 30, 40, 50, and 60 parts and are marked as such. These edges are appropriate for reduction ratios such as 1″ = 20′, 1″ = 30′, 1″ = 40′, 1″ = 50′, and 1″ = 60′. Of course, with the decimal system, the same five edges can be used for 1″ = 200′, 1″ = 300′, and so on. With an understanding of math and decimal places, you can also use the 50 edge of the scale for 1″ = .5″ and, thus, have a DOUBLE SCALE. The 20 edge can be used for 1″ = 2″ and, thus, a HALF SCALE. The exercise section of this unit provides practice for some of these options.

For many mechanical applications, such as machine part drawings, the scales most frequently applied are quarter size (1/4″ = 1″), half size (1/2″ = 1″), and full size (1″ = 1″), which can also be stated as 1:4, 1:2, and 1:1. While there are scales (measuring instruments) available for these scales, they are not as common. If measuring instruments are produced in this fashion, they are often referred to as ***mechanical engineer's scales.*** See Figure B-7. Also, some smaller parts of an assembly may need to be enlarged on the drawing for clarity in shape and size. Typical scales for enlargement are twice size (2″ = 1″), triple size (3″ = 1″), and 5X (5″ = 1″). The 5X is read as "five times." For general measurements, a civil engineering scale could be used to measure these drawings using the appropriate edge, although the civil engineering scale is decimal, not fractional.

***Metric scales*** are also triangular in shape. These metric scales are configured in such a way as to provide for reduction or enlargement along each edge. See Figure B-8. In metric applications, just as in civil engineering, the divisions are decimal, so 1:10, 1:100, and 1:1000 can all use the same edge.

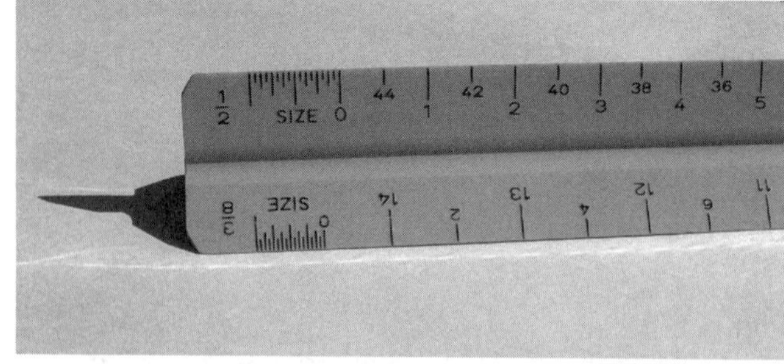

Figure B-7.
This triangular scale contains a half-size scale typically found on a mechanical engineer's scale.

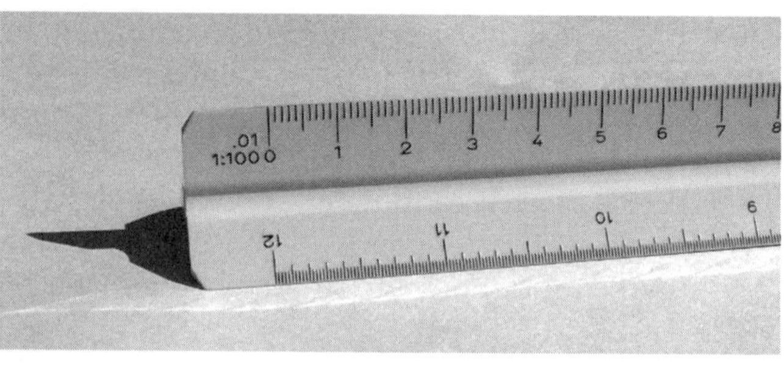

Figure B-8.
This metric scale is commonly used if large metric measurements are to be reduced at various scales.

Combination scales are also available in a variety of configurations. On these scales, common architectural, civil, mechanical, and metric scales may be combine in a variety of ways.

While CAD drafters create drawings using real world size measurements, regardless of the paper size, the views are printed at a common and standard scale. The final product is the same as drawings created by pencil drafters. Measuring instruments can be used to extract information from the print, if allowed. Remember, however, measuring a print is not typically allowed.

## Reading Common Scales

With an understanding of how scales are used, you must now learn how to read various scales. This section explains how to read the different scales you may use to read prints.

### Reading a Fractional-Inch Scale

Study Figure B-9. This figure shows the full-size fractional scale on a typical architect's scale. Each inch is divided into 16 parts. The large tick marks indicate whole inches. The next smaller tick marks are 1/2 inches. The next smaller tick marks are 1/4 inches. The next smaller tick marks are 1/8 inches. The smallest tick marks are 1/16 inches.

### Reading a Decimal-Inch Scale

Study Figure B-10. This figure shows the full-size decimal scale on a typical civil engineer's scale. The tick marks between the whole inches are all the same length and represent 10ths of an inch. To read a decimal dimension of 2.125 inches on the civil engineer's scale, start at 0 and move to the right past 2. Continue past the first tenth (.10) to one-fourth (.025) of the next tenth (judged by eye). This represents the decimal .125.

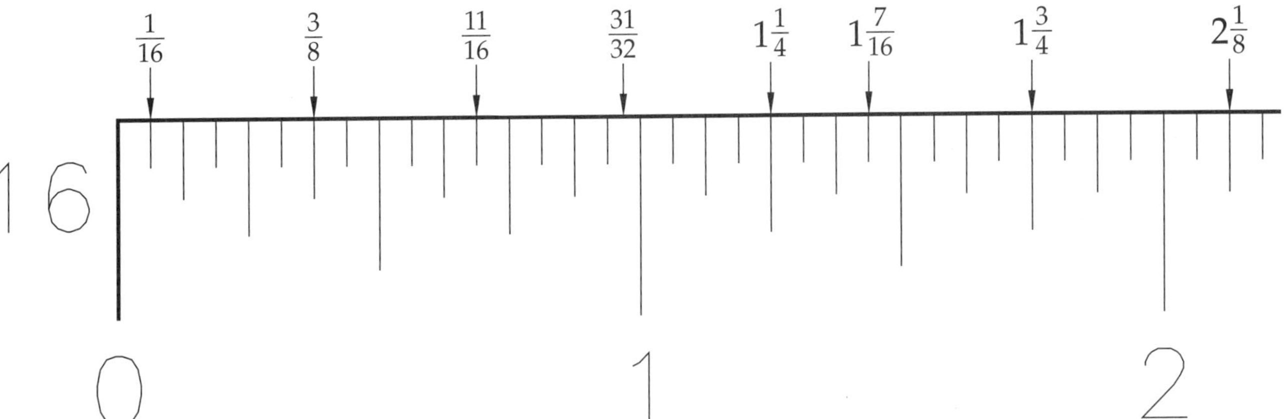


**Figure B-9.**
This full-size fractional scale on a typical architect's scale is divided into 16 parts per inch.

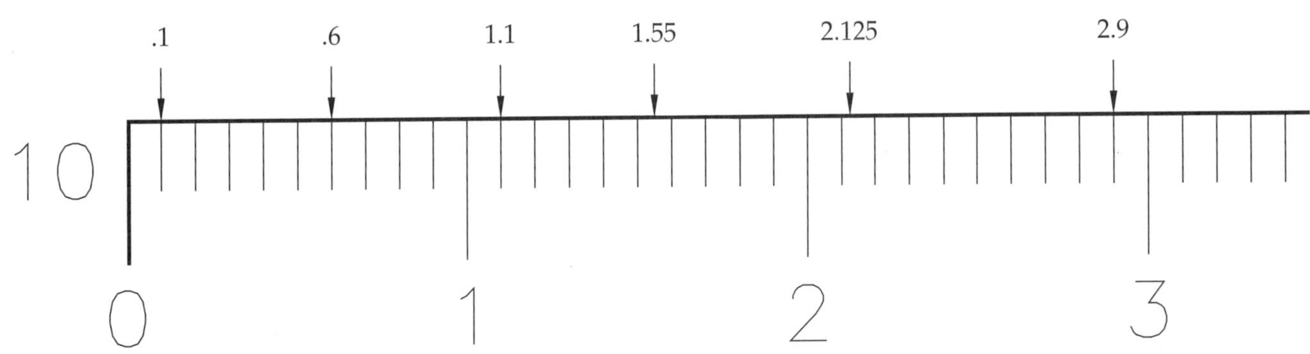


**Figure B-10.**
The full-size edge of a civil engineering scale is marked off in divisions of 10 parts to the inch.

A full-size mechanical engineer's scale typically has 50 parts to the inch. This is designed for reading in a decimal format, also. See Figure B-11. Notice how 2.125 inches is measured along this scale compared to the previous example.

## *Reading an Architect's Reduction Scale*

The ten architectural scales are designed for measuring feet and inches. Since industrial prints and topics covered in this textbook are not in feet and inch units, these scales will not be discussed in much detail.

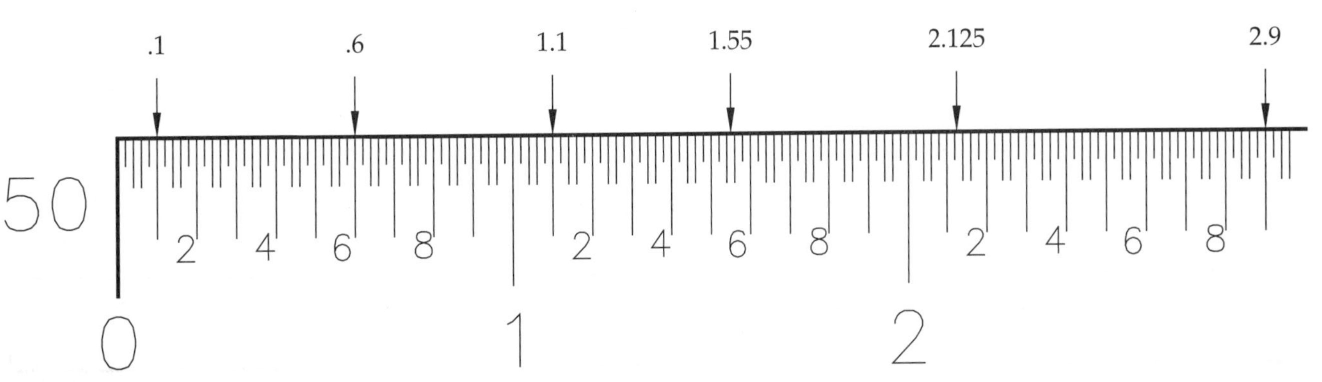


**Figure B-11.**
The full-size decimal scale on a mechanical engineer's scale is divided into 50 parts to the inch. Notice how 2.125 inches is measured along this 50 scale compared to the example in the previous illustration.

Study Figure B-12. This figure shows the edge for 1/8″ = 1′-0″ in combination with the 1/4″ = 1′-0″ on a typical architect's scale. Notice that on the "outside" side of the zero (0) is a twelve-inch ruler shrunk down to 1/8″ and on the other end of that edge on the outside side of the zero (0) is a twelve-inch ruler shrunk down to 1/4″. This principle of placing a twelve-inch ruler at each end of each edge of the architect's scale is consistent on all the edges except for the full-size edge.

On the inside side of the zero (0) are tick marks for measuring feet. Since the scale at one end is either one-half or double of the scale on the other end, these foot marks can be shared. Some of the numerical labels are for feet increments from the "far" end, while others are for the "near" end. The small "twelve-inch" rulers are divided into fractional units, depending on how much room there is for the tick marks to be readable. For example, on the 1/8″ = 1′-0″ end of the scale, there is only room for six tick marks, one for every two inches. On the 1/4″ = 1′-0″ end of the scale, there is room for twelve tick marks, one for each inch, so distances can be measured more precisely.

## Reading a Mechanical Engineer's Reduction Scale

Study Figure B-13. This figure shows the half-size scale on a mechanical engineer's scale. On instruments that have scales shrunk down for reduced-size drawings, the tick marks become much closer together. Sometimes in these cases, the more-precise tick marks are only placed on one end of the scale. To read a decimal dimension of 2.125 inches on the mechanical engineer's scale in Figure B-13, locate the 2 on the right side of the zero. Then, move the instrument an additional 1/8″ (.125″) on the scale to obtain the total measurement. The actual distance on the drawing is 1.0625″ (2.125″ at 1/2 scale).

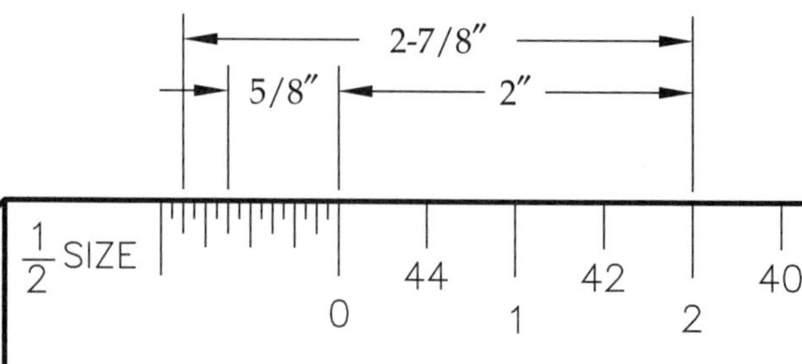


Figure B-13.
Similar to the architect's scale, this 1:2 scale is not fully divided. The whole inches are measured on one side of the zero and the fraction is measured on the other side of the zero.

## Reading a Civil Engineer's Reduction Scale

Study Figure B-14. This figure shows the 20 scale on a civil engineer's scale. The 2 is located one actual inch from zero. Therefore, this edge can be used for any reduction of 1 = 2 units, or 20 units, or 200 units. It can be used for 1″ = 2″ or 1″ = 2 cubits. Other scales include 10, 30, 40, 50, and 60.

To read a 1/2 scale (half scale) decimal dimension of 3.75″ on the civil engineer's scale, locate the 3.7 and 3.8 tick marks. Then, estimate a point halfway between those tick marks. Remember, the actual distance on the drawing is 1.875 (3.75 at 1/2 scale).

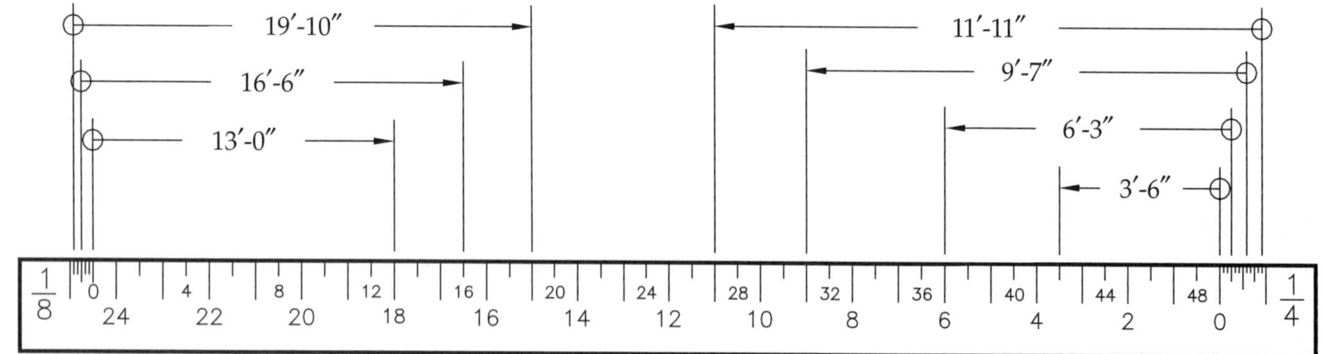


Figure B-12.
The architect's scale has five edges, each with two scales that share foot marks. There is also a small 12-inch ruler at the end of each edge for that reduction scale.

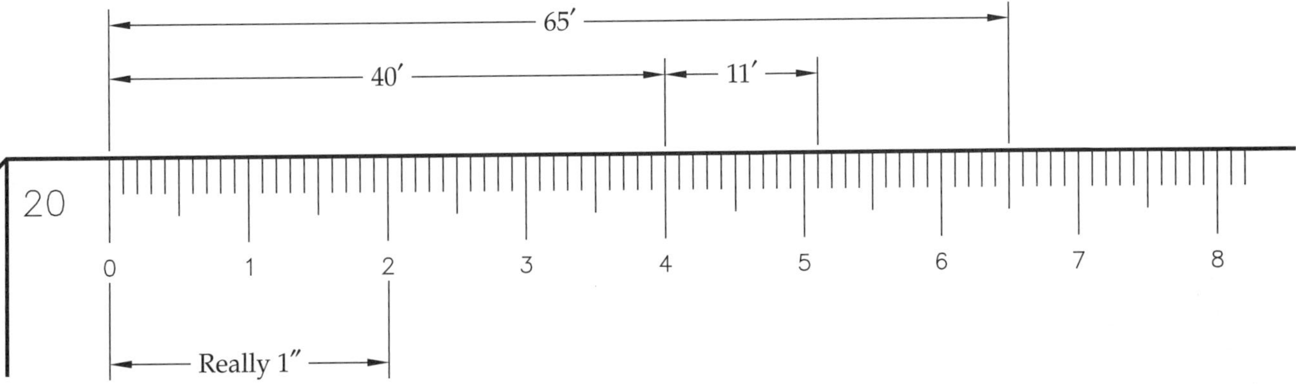


Figure B-14.
In this example, the civil engineer's scale is shown being used for measuring at a scale of 1″ = 20′.

While the scope of this textbook does not include civil engineering applications, learning to read this scale can be very useful in your professional career, especially if you are reading and measuring large prints of facility layouts or site drawings. For example, assume the 2, which is really one inch from 0, represents 20 feet. Use the tick marks to read measurements just as you would on a full-size scale. The same 2 that represents 20 feet in the example above could have represented 2 feet or 200 feet. The measurements you make simply assume the tick marks are for feet, based on the value given to the numeral 2.

## Metric Scales

In the workplace, you may encounter both the metric and decimal-inch US customary systems of measurement. Many industries use a system of dual dimensioning where both decimal-inch and metric dimensions are shown on a print. Common International System (SI) metric units are shown in Figure B-15. The metric system is somewhat similar to the decimal-inch system. Only the size of the units and terms vary. Both are base-ten number systems, which makes it easy to shift from one multiple or submultiple to another.

The most important aspect to understand in this section is how to measure millimeters. This is the most common unit for mechanical and industrial parts as discussed throughout this textbook. If you are using a triangular metric reduction scale, the 1:100 edge of the scale can be used as a full-size scale when dimensions are in millimeters, since the smallest divisions on the scale are spaced one millimeter apart.

| Quantity | Name of Unit | Symbol |
|---|---|---|
| Length | meter | m |
| Mass (weight) | kilogram | kg |
| Time | second | s |
| Electric current | ampere | A |
| Temperature | kelvin | K |
| Luminous intensity | candela | cd |
| Amount of substance | mole | mol |
| | **Supplementary Units** | |
| **Unit** | **Name of Unit** | **Symbol** |
| Plane angle | radian | rad |
| Solid angle | steradian | sr |

Figure B-15.
This chart shows common SI metric base units.

Anyone using the metric system should learn all common units and how to convert between units. Unit conversion is actually a simple shift of the decimal point. See Figure B-16.

### *Metric Units Used on Industrial Prints*

The base unit of the metric system is the meter. Other units derived from the meter are used in various industrial fields as the official unit of measure on metric drawings. See Figure B-17. For example, on precision drawings in manufacturing, the millimeter is the standard unit of length. Some typical metric scale designations for drawings are:

| Reductions | | Enlargements |
|---|---|---|
| 1:2 | 1:50 | 2:1 |
| 1:2.5 | 1:100 | 5:4 |
| 1:5 | 1:200 | 10:1 |
| 1:10 | 1:500 | 20:1 |
| 1:20 | 1:1000 | 50:1 |

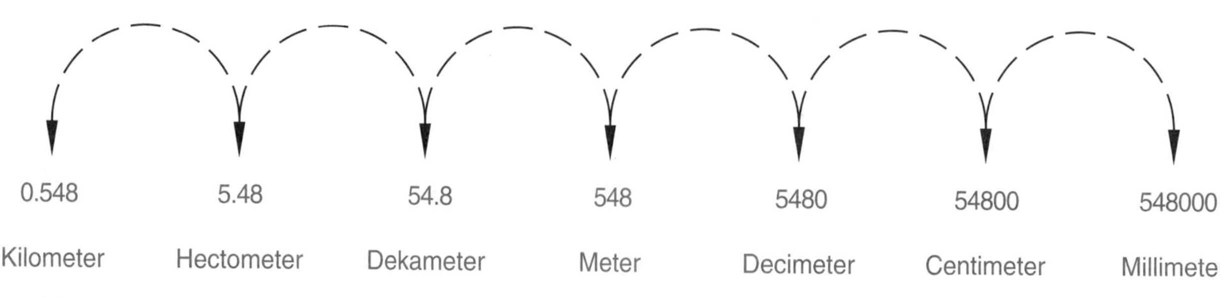


**Figure B-16.**
Converting between metric units involves an understanding of the different prefix terms used in the metric system, but otherwise is as simple as moving the decimal place.

| Industry | Unit of Measure | International Symbol | Multiple Factor |
|---|---|---|---|
| Topographical | kilometer | km | $10^3$ = 1 000 m |
| Building, Construction | meter | m | $10^0$ = 1 m |
| Lumber, Cabinet | centimeter | cm | $10^{-2}$ = 0.01 m |
| Mechanical Design, Manufacturing | millimeter | mm | $10^{-3}$ = 0.001 m |

**Figure B-17.**
Various industries use different base units of measure.

## Micrometer

The ***micrometer*** (mike) is a precision-measurement instrument. It consists of a spindle that rotates in a fixed nut, thus opening or closing the distance between an anvil and spindle. See Figure B-18. The spindle screw on an inch micrometer has 40 threads per inch. The spindle is attached to a thimble. As the thimble is turned one revolution, the spindle moves precisely 1/40 of an inch (.025″). The screw on the metric micrometer advances 1/2 millimeter per revolution of the thimble. Therefore, two revolutions move the spindle 1 mm.

### Reading an Inch Micrometer

Notice in Figure B-18 the line extending the length of the sleeve. This line is called the ***index line*** and is divided into 40 equal parts by vertical lines. Therefore, each vertical line on the sleeve designates 1/40 of an inch (.025″). Two of these lines or spaces are equal to 1/20 of an inch (.025″ + .025″ = .050″) of an inch. Every fourth line is marked with a number (1, 2, 3, etc.) and designates hundred thousandths (4 × .025″ = .100″). The line marked 1 represents .100 inch, the line marked 2 represents .200 inch, and so on.

The beveled edge of the thimble is divided around its circumference into 25 equal parts. Each line is consecutively numbered. One revolution of the thimble moves the spindle .025″. Moving from one line on the beveled edge of the thimble to the next moves the thimble 1/25 of a revolution. Therefore, the spindle is moved 1/25 of .025″, or .001″. Rotating two divisions or lines on the thimble moves the spindle .002″ and so on. Twenty-five divisions indicate a complete revolution of the thimble, or .025″, which is one division on the index line.

To read the micrometer in thousandths, the reading is taken at the edge of the thimble on the index line as follows.

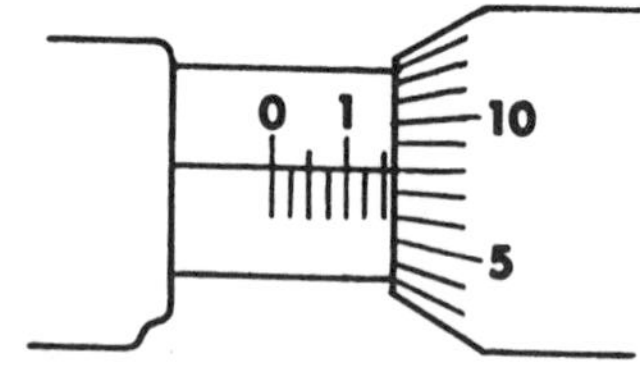


1. Line 1 on the sleeve is visible, representing .100″.
2. Two additional lines are visible representing .025″ each.
3. Line 8 on the thimble is aligned with the index line, representing .008″.
4. The micrometer reading is .158″:

```
  .100
  .050  (2 × .025″ = .050″)
+ .008
  ----
  .158
```

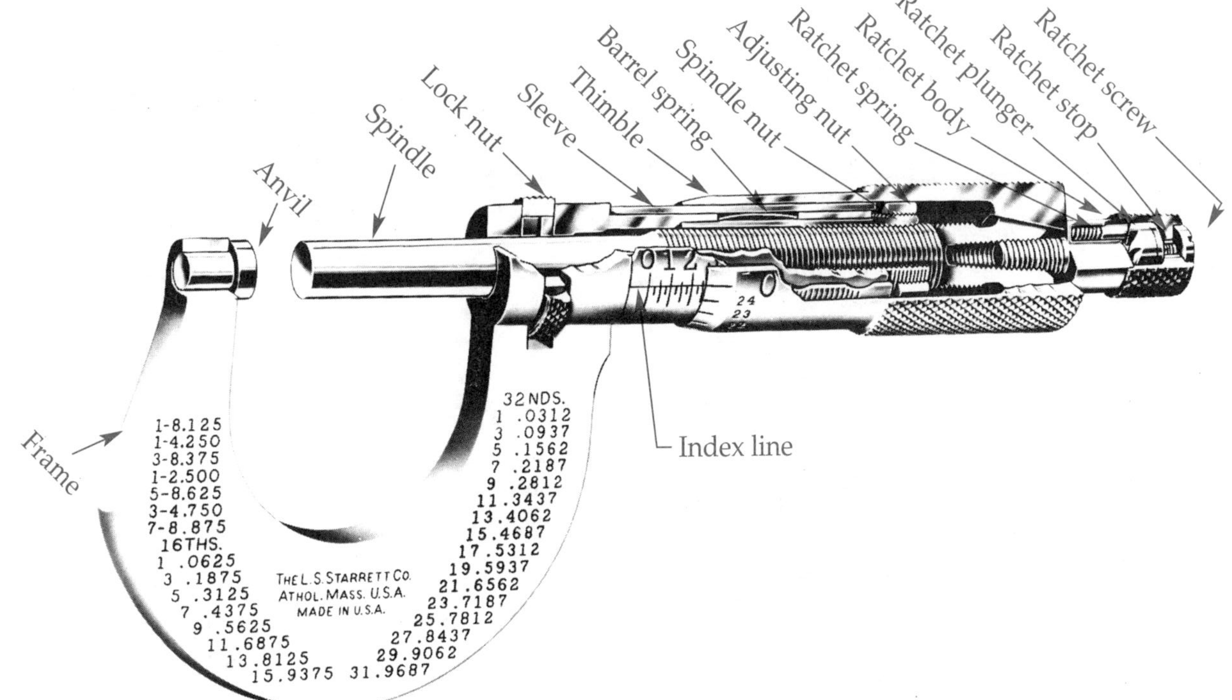


Figure B-18.
A micrometer is used to make precise measurements. This is an inch micrometer.

Try another reading:

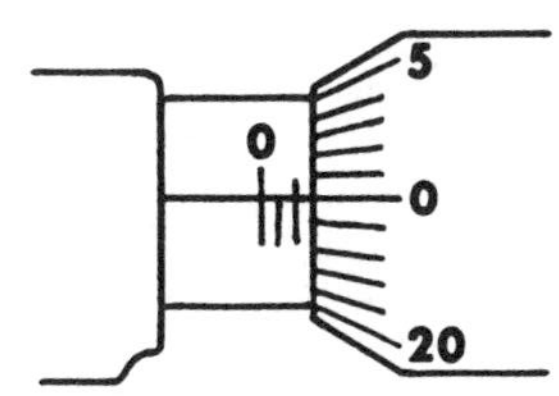


1. No numbered marks are visible.
2. Three lines are visible representing .025″ each (the thimble edge is on the third mark).
3. Line 0 on the thimble is aligned with the index line, so there is nothing to add.
4. The micrometer reading is .075″:

.000 (No numbered marks)
.075 (3 × .025″ = .075)
+ .000 (thimble reads 0)
.075

## Tricks of the Trade

An easy way to remember the values of the various micrometer divisions is to think of them as money. Count the figures on the sleeve (1, 2, etc.) as dollars. The extra vertical lines on the sleeve are then quarters. Finally, the divisions on the thimble are pennies. Add up your money and move the decimal point in place of the dollar sign. This is the micrometer reading. In the first example above, line 1 is visible on the sleeve. This is one dollar. Two additional lines are visible. This is two quarters, or fifty cents. Finally, line 8 on the thimble is aligned with the index line. This is eight pennies. Add this currency to get $1.58. Move the decimal point in place of the dollar sign to get a micrometer reading of .158″.

## Reading a Metric Micrometer

A *metric micrometer* measures in hundredths of a millimeter (0.01 mm). Reading this micrometer is quite similar to reading an inch micrometer. The basic difference is the units on the scale are in millimeters.

The sleeve of the micrometer is graduated in millimeters below the index line and in half millimeters above the line. See Figure B-19. The thimble is marked in fifty divisions around its circumference. Each small division represents 0.01 mm. Turning the thimble 10 tick marks equals 0.1 mm. Therefore, one complete turn equals 0.5 mm. If you turn the thimble a second

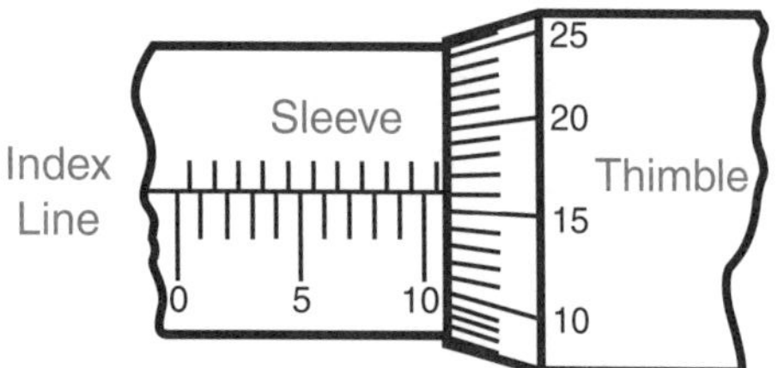


Figure B-19.
A metric micrometer is read in a manner similar to reading an inch micrometer.

full turn, you have added another 0.5 mm. To read the metric micrometer, use the following steps.

1. Note the number of whole millimeter divisions (below index line) on the sleeve. These count as whole millimeters.
2. See if a half millimeter division (above index line) is visible between the whole millimeter division and the thimble. If the thimble is beyond the half mark, add .50 millimeters.
3. Finally, read the thimble for additional hundredths (division on thimble aligning with index line).

The reading on the metric micrometer shown in Figure B-19 is as follows.

1. Ten whole-millimeter divisions.
2. One-half millimeter division.
3. Sixteen is the reading on thimble.
4. The micrometer reading is 10.66 mm:

| | |
|---|---|
| 10.00 | (10 × 1 mm) |
| 0.50 mm | (1 × 0.50 mm) |
| + 0.16 | (16 × 0.01 mm) |
| 10.66 mm | |

## Calipers

Calipers can make internal (inside), external (outside), and depth measurements. A ***caliper*** is a measuring instrument consisting of a main scale with a fixed jaw and a sliding jaw. The sliding jaw has an attached scale, either in the form of a dial or a digital readout. See Figure B-20. Each revolution of the dial is one-tenth of an inch, with 100 increments

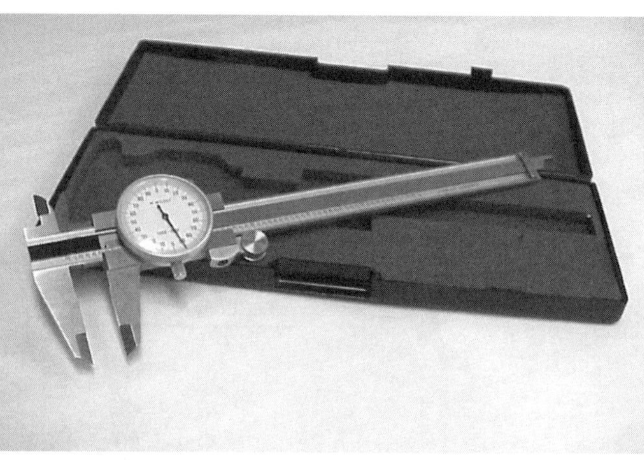

Figure B-20.
Calipers are used to make precise measurements. These are dial calipers.

around the dial. This allows readings to the nearest 1/1000th of an inch. Tick marks along the bar indicate the measurement to the nearest tenth of an inch (for example, between 1.2 and 1.3). The additional thousandths are read from the dial. The two readings are added to get the total measurement. A caliper with a dial indicator may have to be periodically reset to ensure the zero reading is correct when the jaws are closed at a zero distance.

Most state-of-the-art calipers have a digital readout (LED). See Figure B-21. Instead of reading the nearest 1/1000th of an inch on a dial, the digital readout displays the *total* measurement. It is also important when using a digital readout caliper to reset the value to zero when the calipers are closed.

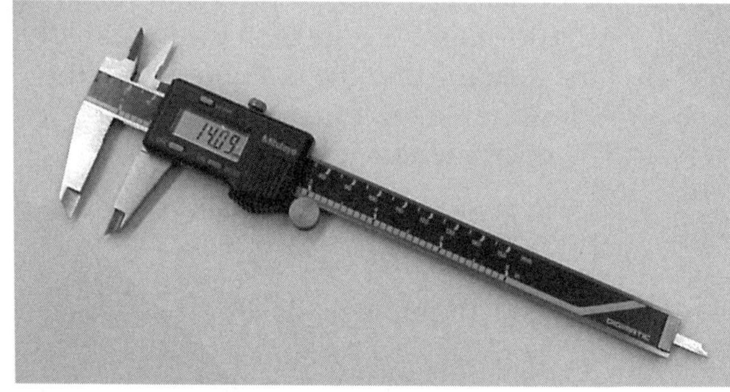

Figure B-21.
Calipers may come with digital readouts. These calipers are often referred to as digital calipers.

## Review Questions

*Circle the answer of choice, fill in the blank, or write a short answer.*

1. Using words, not numbers, fill in the blanks of this sentence: As specified in the title block, the scale is given as ________________ = ________________.

2. *True or False?* A steel rule is the most common 12-inch measuring device available to a machinist.

3. Of the measuring tools discussed in this unit, which of the following is *not* a common number of divisions for an inch?
   A. 16
   B. 25
   C. 64
   D. 100

4. The full-divided edge of an architect's scale is defined as such because every inch is fully divided into ________________ parts.

5. Besides the full-divided edge of the architect's scale, there are ________________ reduction scales on each of the other five edges of an architect's scale.

6. On each end and for each edge of the fully divided civil engineer's scale, there is a number to indicate how many divisions there are to the inch. Which of the following is *not* one of those numbers?
   A. 10
   B. 30
   C. 50
   D. 70

7. *True or False?* The 20 scale on a civil engineer's scale can be used for reducing measurements at a scale of 1″ = 2″ or 1″ = 2′-0″.

8. Which of the following is *not* a micrometer part?
   A. Spindle.
   B. Throttle.
   C. Anvil.
   D. Index line.

9. An inch micrometer is designed to move the spindle ________________ in one revolution of the thimble.

10. *True or False?* Calipers are measuring devices that have a fixed jaw as well as a sliding jaw, with readouts that may include tick marks, dial needle readings, or LED readouts. Unfortunately, as with micrometers, they can only read outside measurements and cannot be used to read internal measurements.

## Review Activity B-1

### Reading a Fractional-64 Steel Rule

Complete the readings indicated below. Write your answers in the spaces provided as fractional values.

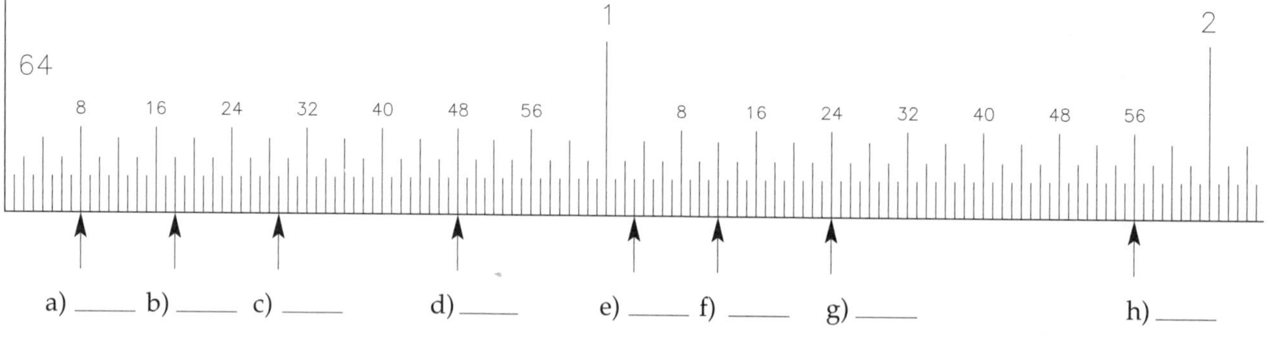


## Review Activity B-2

### Reading the Decimal-100 Steel Rule

Complete the readings indicated below. Write your answers in the spaces provided as two-place decimals.

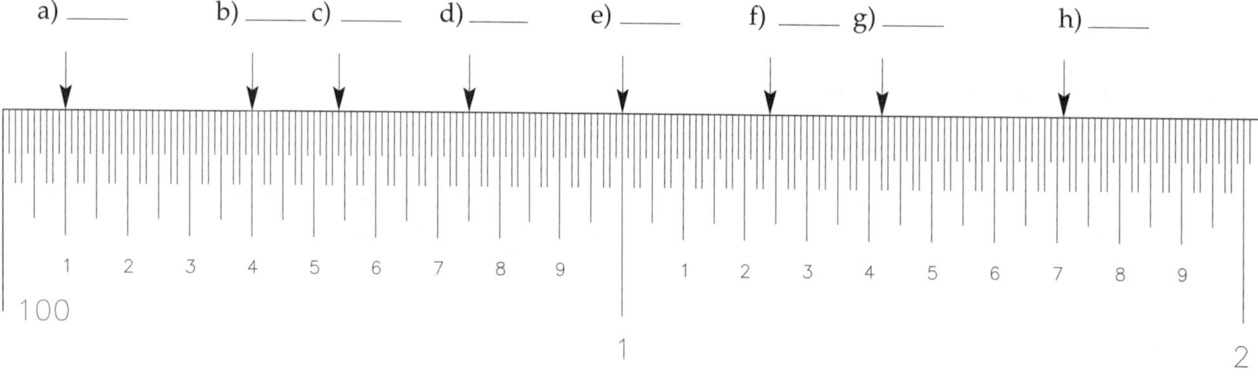

## Review Activity B-3

### Reading the Metric Steel Rule

Complete the readings indicated below. Write your answers in the spaces provided as millimeter measurements.

a) ____ b) ____ c) ____ d) ____ e) ____ f) ____ g) ____ h) ____

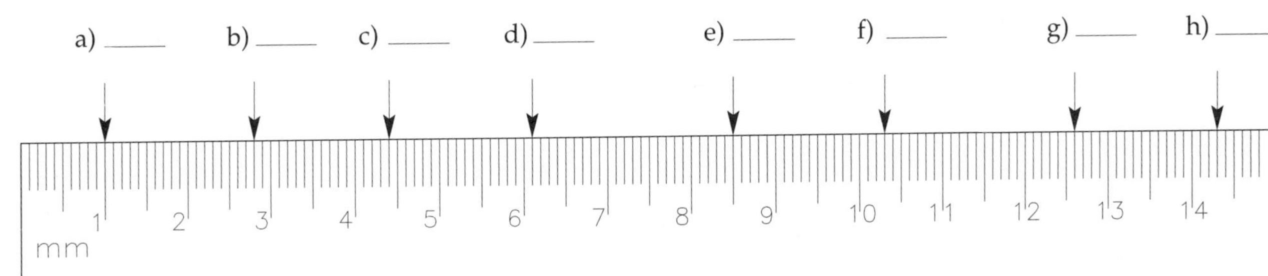

# Review Activity B-4

### Using a Fractional-Inch (16) Full Scale

Measure the length of the following lines using a fractional scale. Dividers can also be used to transfer distances to the scale on the page. Place your reading in the blank as a full scale reading.

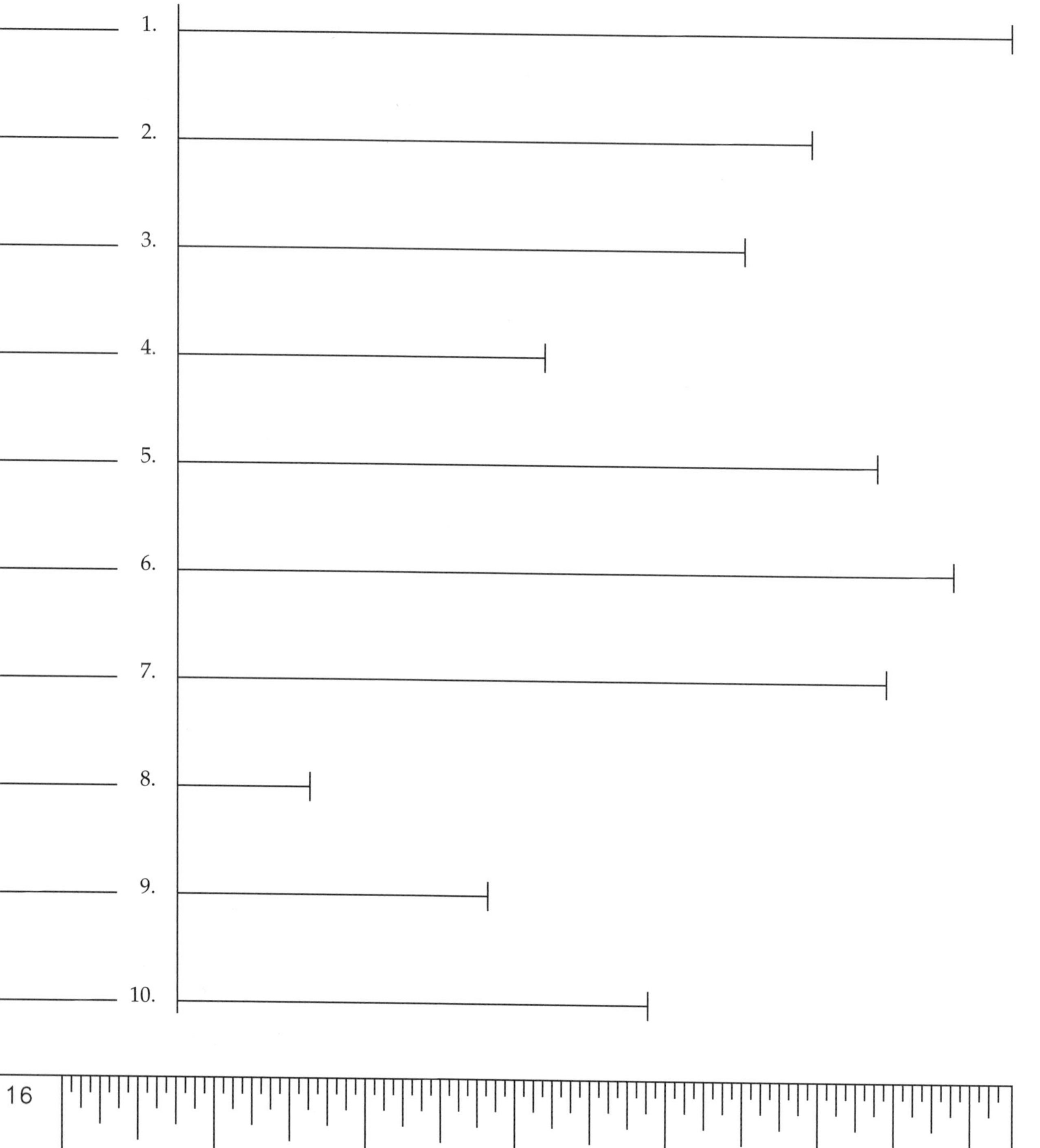

## Review Activity B-5

### Using a Decimal-Inch (10) Full Scale

Measure the length of the following lines using a decimal scale. Dividers can also be used to transfer distances to the scale on the page. Place your reading in the blank as a full scale reading.

________ 1.

________ 2.

________ 3.

________ 4.

________ 5.

________ 6.

________ 7.

________ 8.

________ 9.

________ 10.

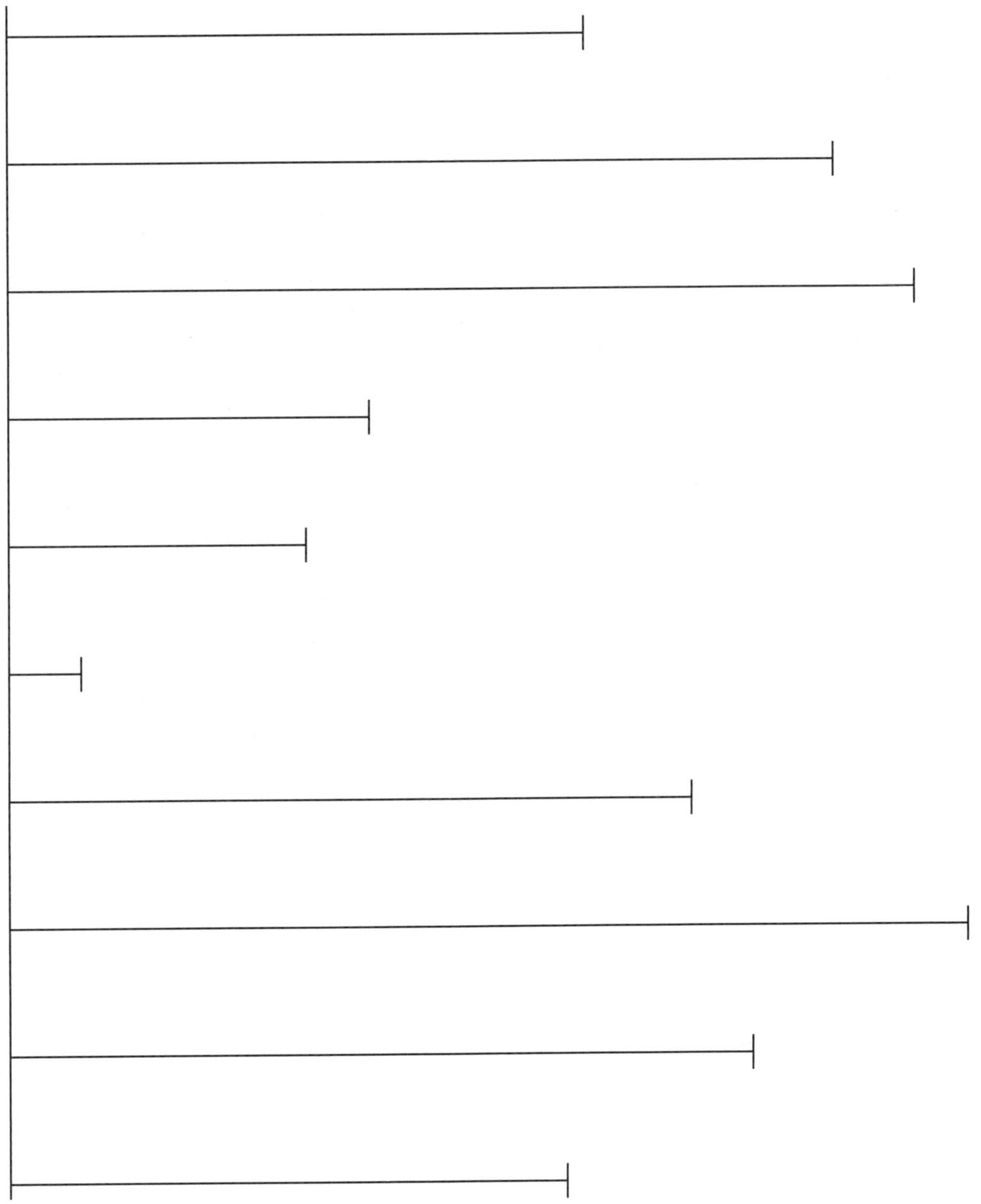

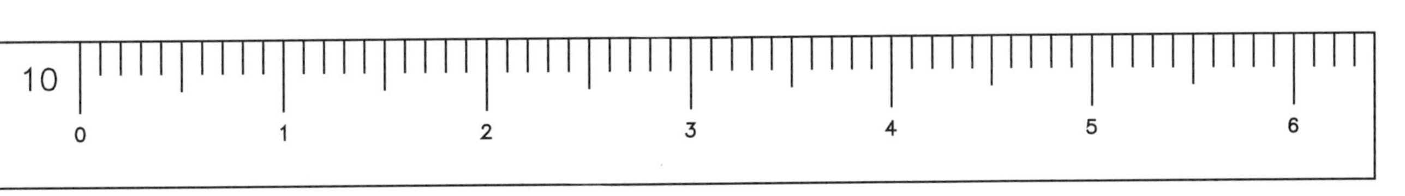

## Review Activity B-6

### Using an Architect's Reduction Scale

Shown are some dimensions along different architect's scales. Locate the zero point on each scale. Inches are to one side of zero and feet to the other side. Evaluate each dimension according to the scale illustrated. Remember, the scale is indicated on the end of the architect's scale. Place your reading in the blank provided.

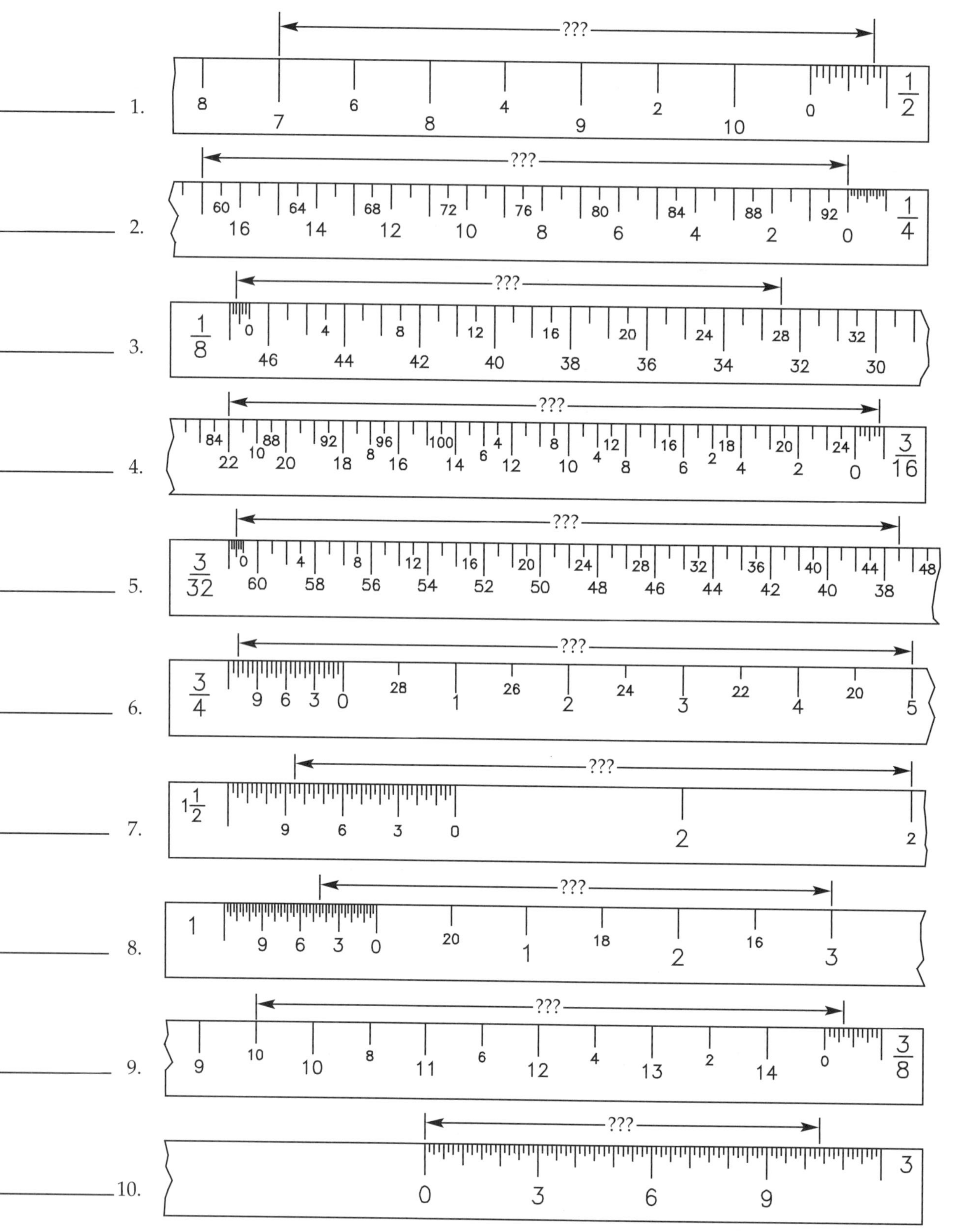

## Review Activity B-7

### Using a Civil Engineer's Reduction Scale

Shown are some dimensions along different edges of a civil engineering scale. Evaluate each dimension according to the scale indicated. Place your reading in the blank provided.

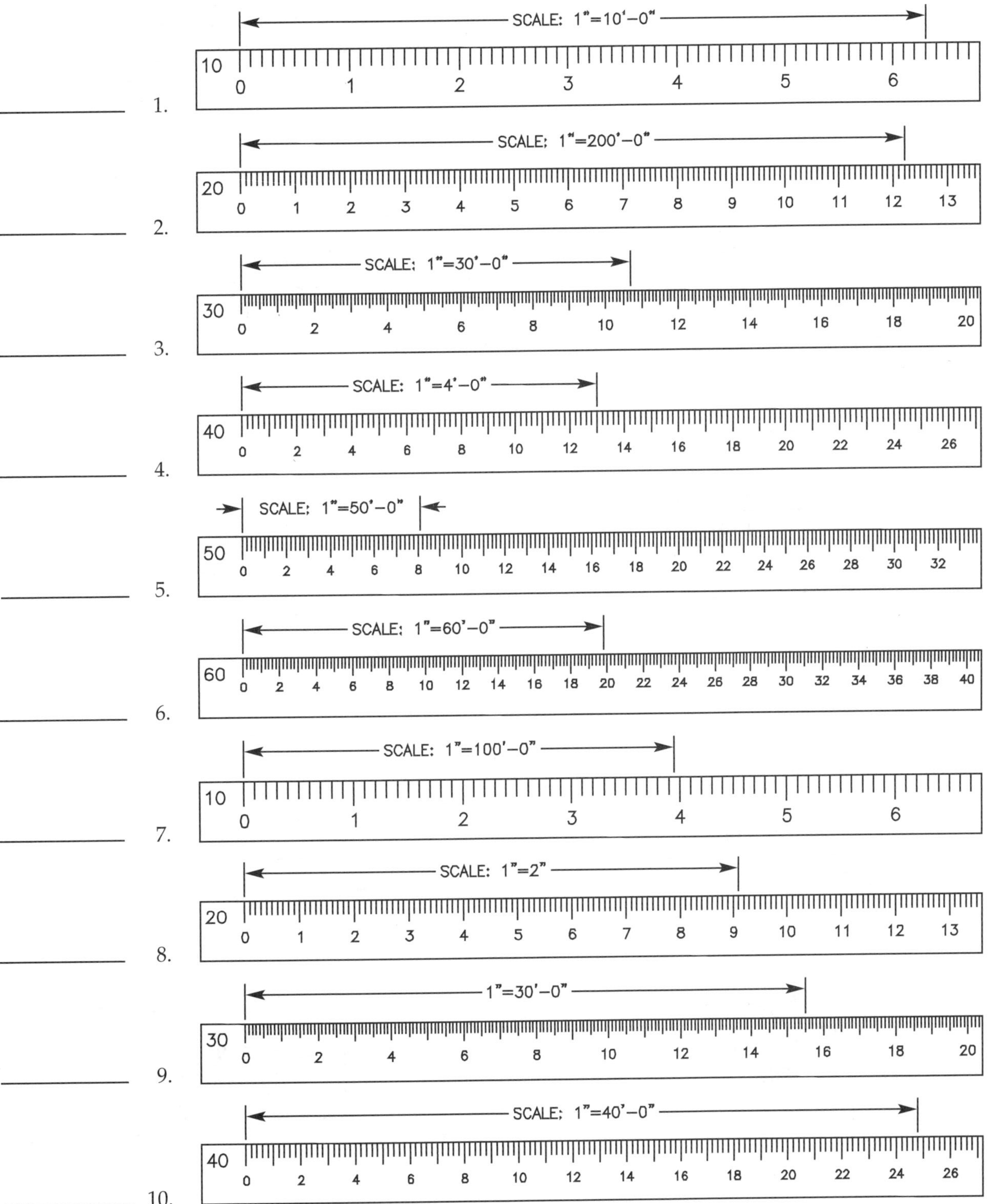


______ 1.

______ 2.

______ 3.

______ 4.

______ 5.

______ 6.

______ 7.

______ 8.

______ 9.

______ 10.

## Review Activity B-8

### Reading an Inch Micrometer

Record the readings in the space provided for the following micrometer settings.

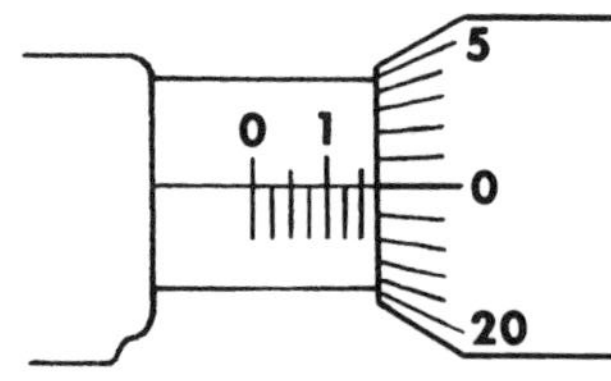


1. ______________________

6. ______________________

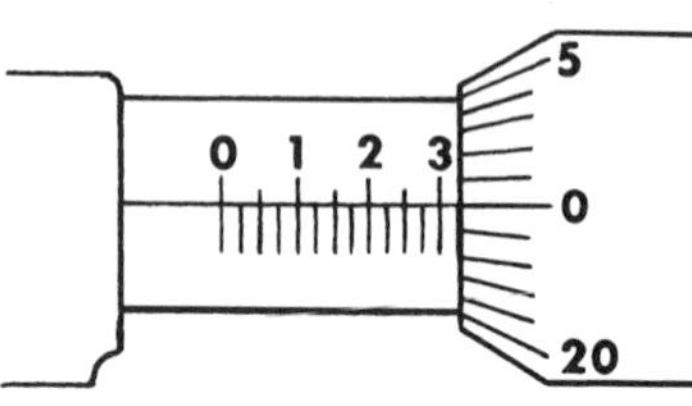


2. ______________________

7. ______________________

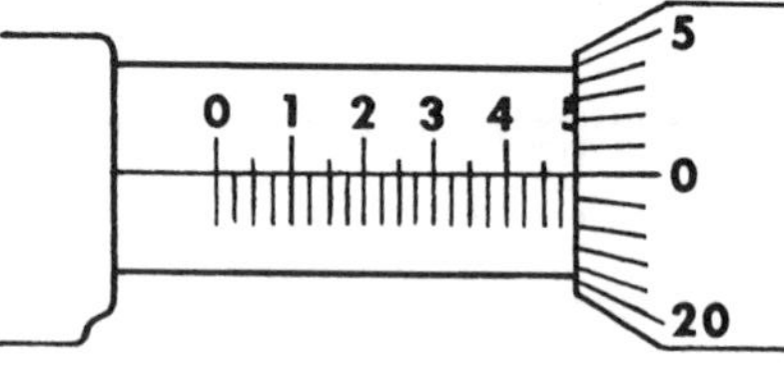


3. ______________________

8. ______________________

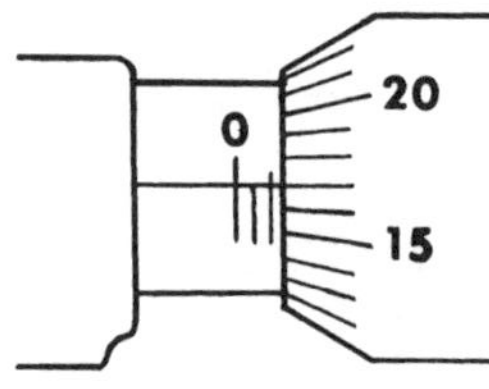


4. ______________________

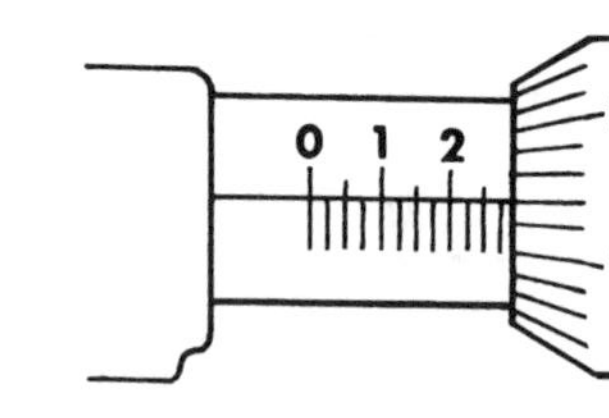


9. ______________________

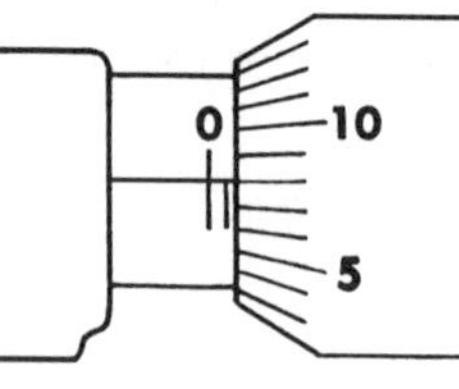


5. ______________________

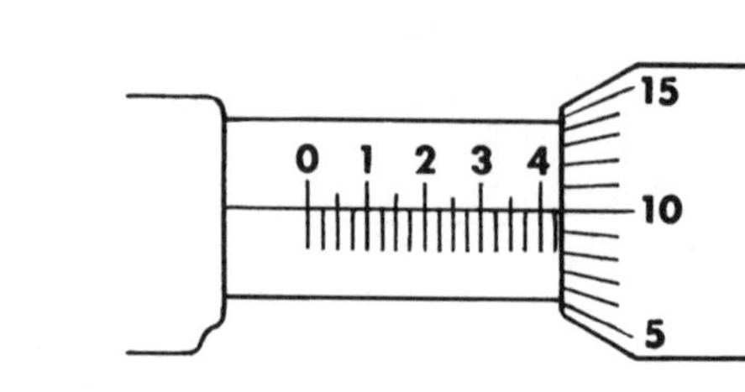


10. ______________________

## Review Activity B-9

**Reading a Metric Micrometer**

Record the readings in the space provided for the following micrometer settings.

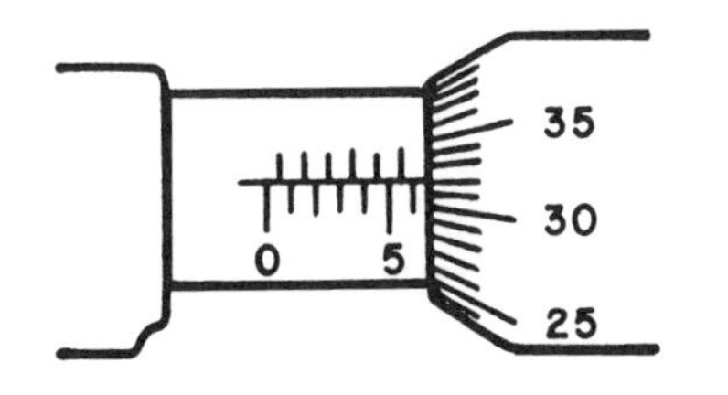


1. ______________________

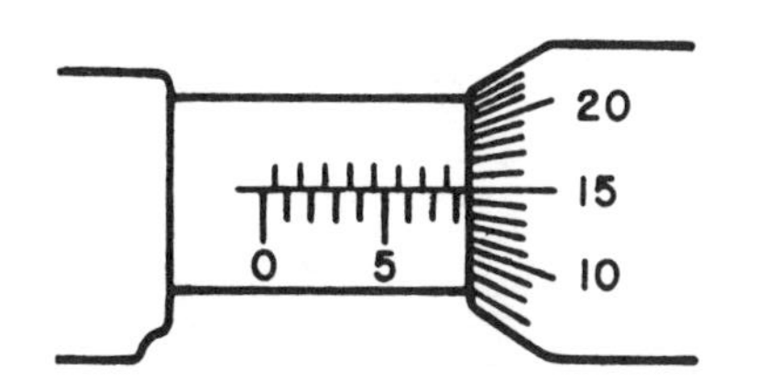


2. ______________________

0
45
40
0

3. ______________________

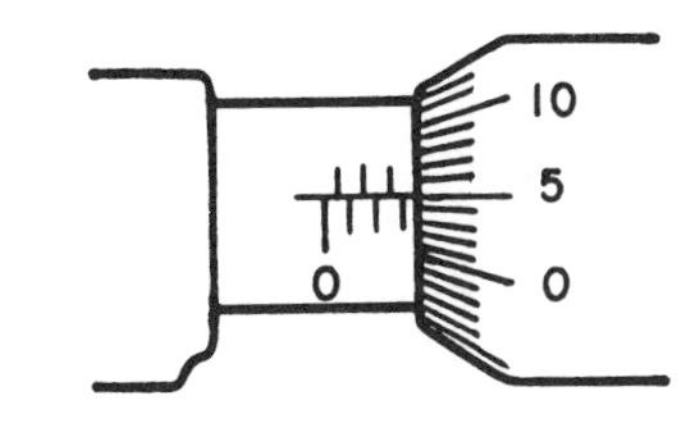


4. ______________________

5
0
45
0 5

5. ______________________

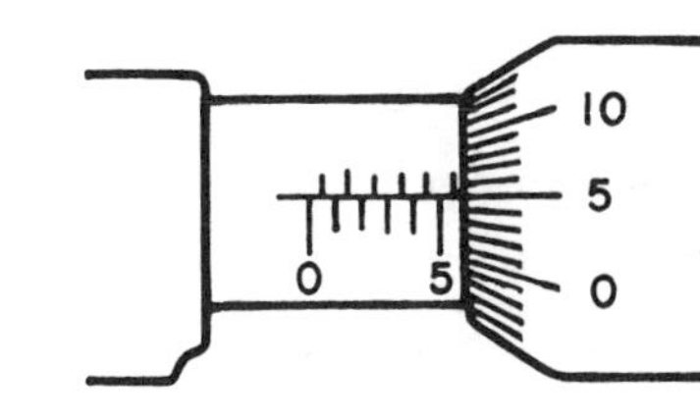


6. ______________________

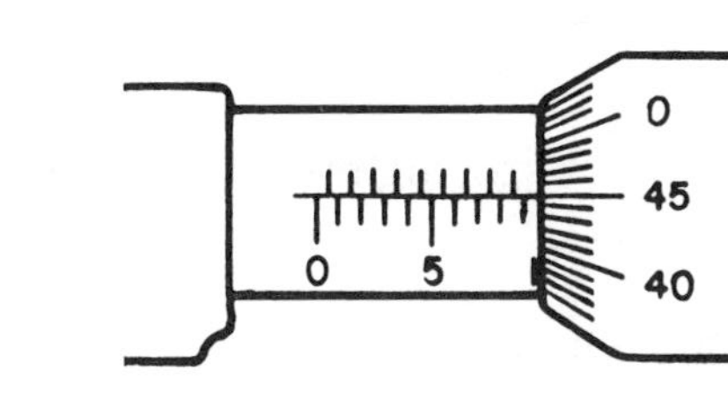


7. ______________________

# Appendix C
# Basic Technical Sketching and Lettering

*After completing this unit, you will be able to:*

**Identify** the materials needed to create a technical sketch.

**Discuss** the proper methods for sharpening a pencil.

**Discuss** the proper methods for holding a pencil for sketching.

**Sketch** horizontal, vertical, and inclined lines.

**Sketch** arcs, circles, and ellipses.

**Identify** aids used for freehand technical sketching.

**Read** and **sketch** Gothic-style technical lettering.

**Identify** and **use** guidelines for lettering.

**Discuss** the proper spacing methods for lettering.

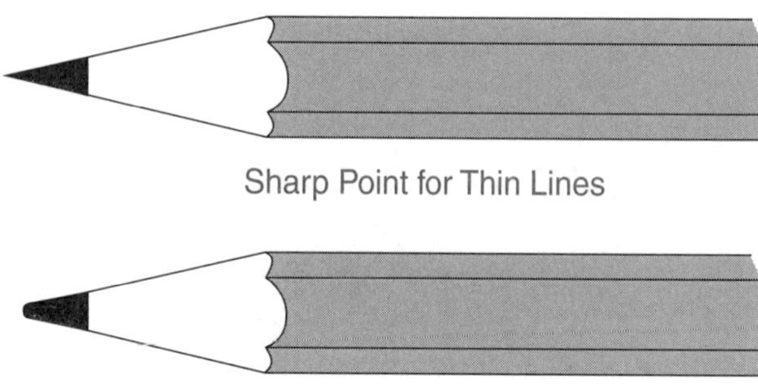


**Figure C-1.**
For sketching, sharpen your pencil to a conical point. Leave the point a little rounded for thicker lines.

Freehand technical sketching is a process used by engineers, technicians, and drafters to quickly convey ideas. Sketches may be used for mechanical parts, assemblies, electrical diagrams, and many other applications. Most mechanical designs start as sketches. The only materials needed for sketching are a pencil, paper, and eraser.

The ability to freehand sketch can be very helpful in learning to read and interpret prints. To begin sketching, you need to sharpen your pencil to a conical point, as shown in Figure C-1. While there are professional drafting pencils available with varying degrees of hardness, a good quality #2 pencil will perform quite well if properly sharpened. A #2 pencil is usually equivalent in hardness to an HB drafting pencil.

## Tricks of the Trade

A mechanical pencil (.5 mm lead) with automatic feed never needs to be sharpened. These types of pencils are suitable for instrument drawing and for sketching thin lines and small lettering. However, they are *not* suitable for sketching thick, visible lines. Quality lines that can be reproduced by modern methods, including scans, photocopies, and faxes, need bold and black linework that requires more pressure and thickness than can be obtained from a .5 mm automatic pencil.

Several types of papers are suitable for sketching, including plain notebook paper and cross-sectional grid paper. In applying the sketching techniques discussed in this textbook, plain notebook paper without lines is to be used. When you have developed the basic techniques of sketching on plain paper, you will have no difficulty using cross-section grid paper.

## Sketching Technique

How you hold your pencil when freehand sketching is important. It should be held with a grip firm enough to control the strokes, but not so tight as to stiffen your movements. Your arm and hand should have a free and easy movement. If using a wooden pencil, the point of the pencil should extend approximately 1 1/2″ beyond your fingertips. See Figure C-2.

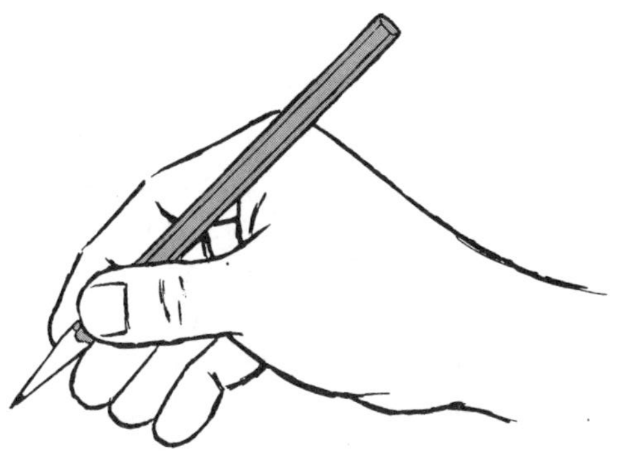

**Figure C-2.**
When sketching with a wooden pencil, the point of the pencil should extend approximately 1 1/2" beyond your fingertips.

For complex sketches, construction lines should first be sketched to provide an initial layout of the geometry. After the initial layout is complete, darken the final lines with the proper line weight and dash length. Quality lines are better created with a series of short, comfortable strokes, rather than one continuous line. Your eye should be on the point at which the line is to terminate. See Figure C-3. Pull the pencil instead of pushing it.

Every 1 1/2" to 2", pick the pencil up, rotate it 5° or so, and place it back down a short distance from where you picked it up. Slightly rotating the pencil between strokes helps maintain the point. Initial lines should be firm and light, but not fuzzy. The pencil point should "sink into" the paper a little in order to create the dense lines necessary for quality reproduction.

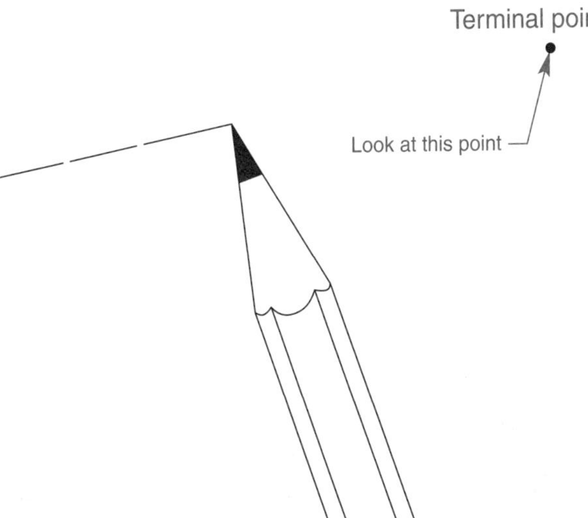


**Figure C-3.**
When sketching, your eye should be on the point at which the line is to terminate.

Do *not* use a straightedge to create sketches. A freehand sketch drawn with a straight edge looks "cheap" and unprofessional.

## Horizontal Lines

Horizontal lines are sketched with a movement of the forearm approximately perpendicular to the line being sketched. See Figure C-4. Use these four steps when sketching horizontal lines:

1. Locate and mark the endpoint of line to be sketched, Figure C-5A.
2. Position your arm by making a trial movement without marking the paper. See Figure C-5B. A right-handed person should move left-to-right. A left-handed person should use a right-to-left movement.
3. Sketch short, light lines between the points. See Figure C-5C. Keep your eye on the point where the line ends.
4. Darken the line to form one continuous line of uniform weight. When darkening the line, your eye should lead the pencil along the lightly sketched line. See Figure C-5D. For longer lines, adding a very small gap at various intervals adds to the technical appearance of the line.

## Vertical Lines

Vertical lines are sketched from top to bottom. Use short strokes in series, just as for horizontal lines. When sketching the line, position your arm comfortably at

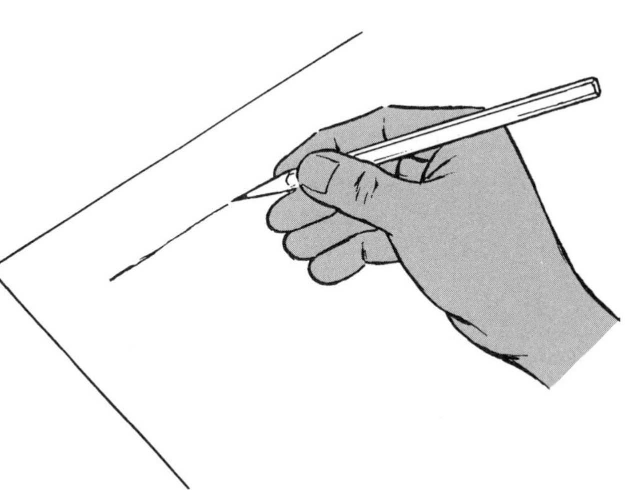

**Figure C-4.**
Horizontal lines are sketched by moving your forearm approximately perpendicular to the line being sketched.

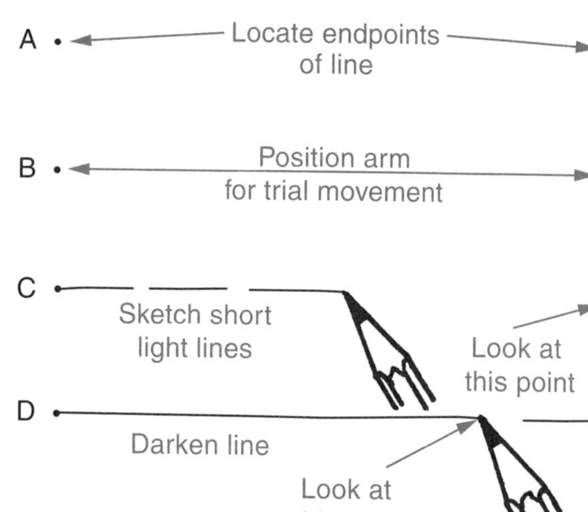


Figure C-5.
Sketching a horizontal line.

approximately 15° to the vertical line. Use the four steps listed for horizontal lines when sketching vertical lines. See Figure C-6.

You may find it easier to sketch horizontal and vertical lines if the paper is slightly rotated. If you find it easier to sketch up and down the page instead of left and right, then orient your paper accordingly whenever sketching lines.

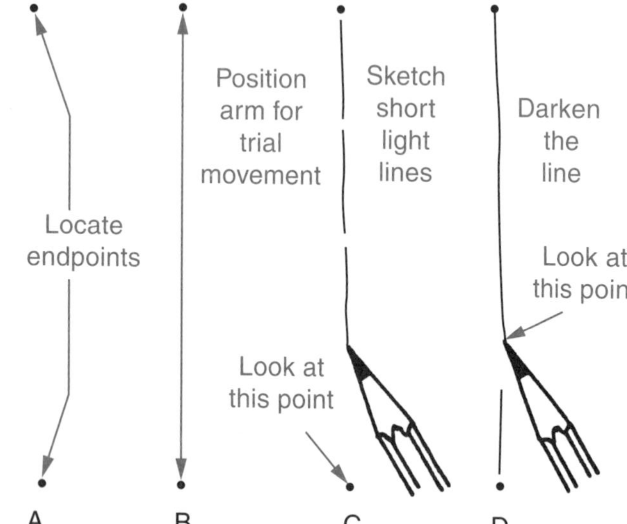


Figure C-6.
Sketching a vertical line.

## Inclined Lines and Angles

All straight lines that are neither horizontal nor vertical are called ***inclined lines.*** Inclined lines are usually sketched between two points or at a designated angle. The same strokes and techniques used for sketching horizontal and vertical lines are used for inclined lines. The paper can be rotated to sketch inclined lines as horizontal or vertical lines, if preferred.

Angles can be estimated quite closely by first sketching a right angle (90°) and then subdividing its arc to get the desired angle. Figure C-7 illustrates how this is done to get an angle of 30°.

## Arcs, Circles, and Ellipses

There are several methods of sketching arcs and circles. One of the most-used methods is the ***triangle-square method.*** To sketch an arc rounding off a corner between two straight lines, follow these steps.

1. Project the two lines until they intersect. See Figure C-8A.
2. Lay out the desired arc radius from the point of the intersecting lines. See Figure C-8B.
3. Form a triangle as shown in Figure C-8C. Locate a point in the center of the triangle area.
4. Sketch short, light strokes from the point where arc starts, through the point made in step 3, to the point where the arc ends. See Figure C-8D.
5. Darken the line to form one continuous arc smoothly joining each straight line. Erase all construction lines. See Figure C-8E.

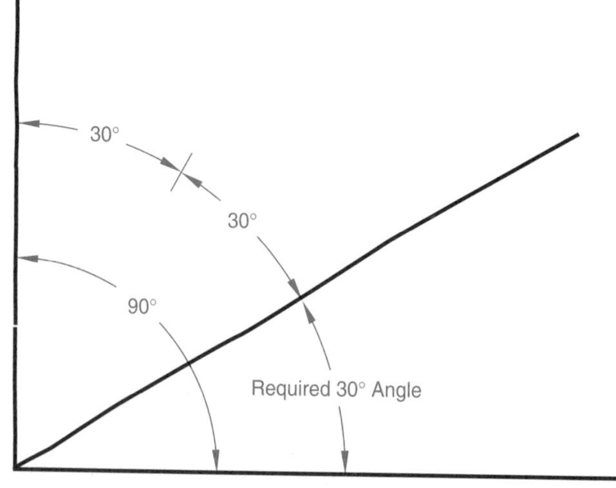


Figure C-7.
Angles can be estimated by first sketching a right angle (90°) and then subdividing its arc to get the desired angle.

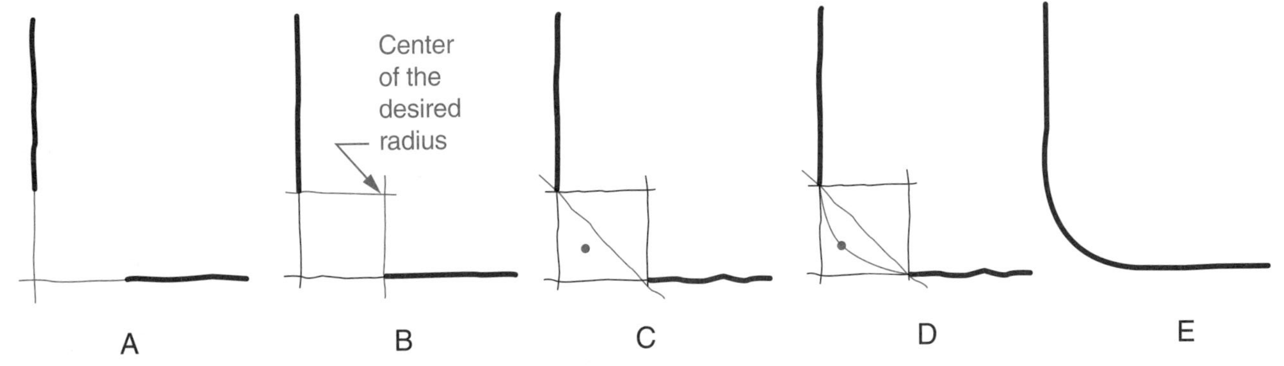


**Figure C-8.**
Sketching an arc.

To sketch a circle of a given diameter, follow these steps.

1. Locate the center of the circle and sketch center lines. Mark one-half of the diameter (the radius) on each side of the center. See Figure C-9A.
2. Lightly sketch a square around the diameter. See Figure C-9B.
3. Across each corner, sketch a diagonal line forming a triangle. Locate a center point within each triangle. See Figure C-9C.
4. Sketch short, light strokes through each quarter of the circle. Make sure the arc passes through the triangle center point and joins smoothly with the square at the circle quadrants. See Figure C-9D.
5. Darken the line to form a smooth, round circle. Erase construction lines. See Figure C-9E.

To sketch an ellipse of a certain major and minor axis, follow these steps.

1. Locate the center of the ellipse and sketch center lines. See Figure C-10A.
2. Mark the length of the major axis of the ellipse on one center line. Equally divide the length about the center. Mark the length of the minor axis in a similar manner on the other center line. See Figure C-10B.
3. Sketch a rectangle around the ellipse. See Figure C-10C.
4. Sketch tangent arcs at the points where the centerlines cross the rectangle. See Figure C-10D.
5. Complete and darken the ellipse. Then, erase the construction lines. See Figure C-10E.

Do not completely erase the "construction boxes" around the circles and shapes of the object. Leaving a light "shadow" of these construction lines gives the sketch a professional or technical look.

A method that produces a very nice result is the ***paper trammel method.*** It is used to sketch circles of greater than 1" diameter. However, this method is a bit tedious. A ***trammel*** is simply a scrap of paper that serves as a marking tool. Along the edge of a scrap of paper, mark the radius. Then, create several little guide points through which to sketch. This method helps ensure professional sketches and good impressions! See Figure C-11.

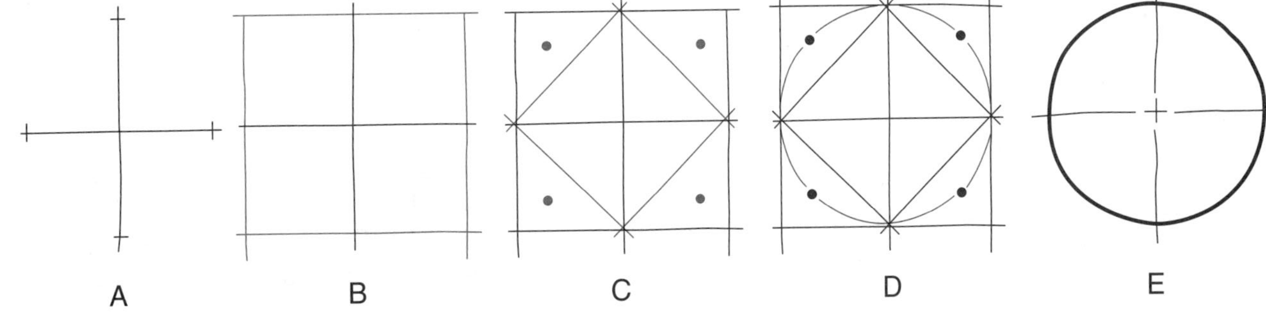


**Figure C-9.**
Sketching a circle.

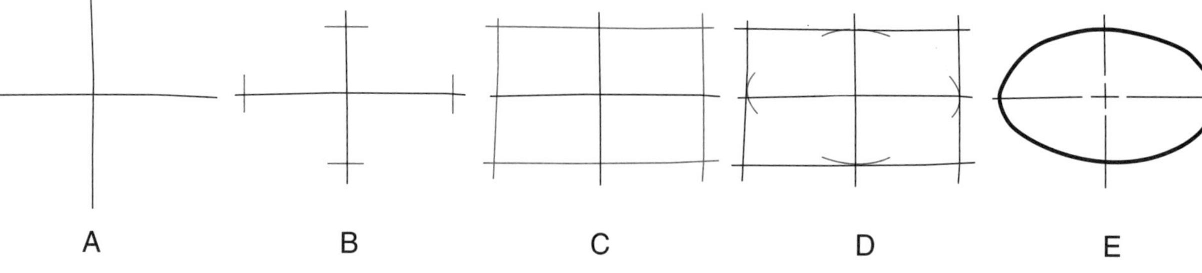


**Figure C-10.**
Sketching an ellipse.

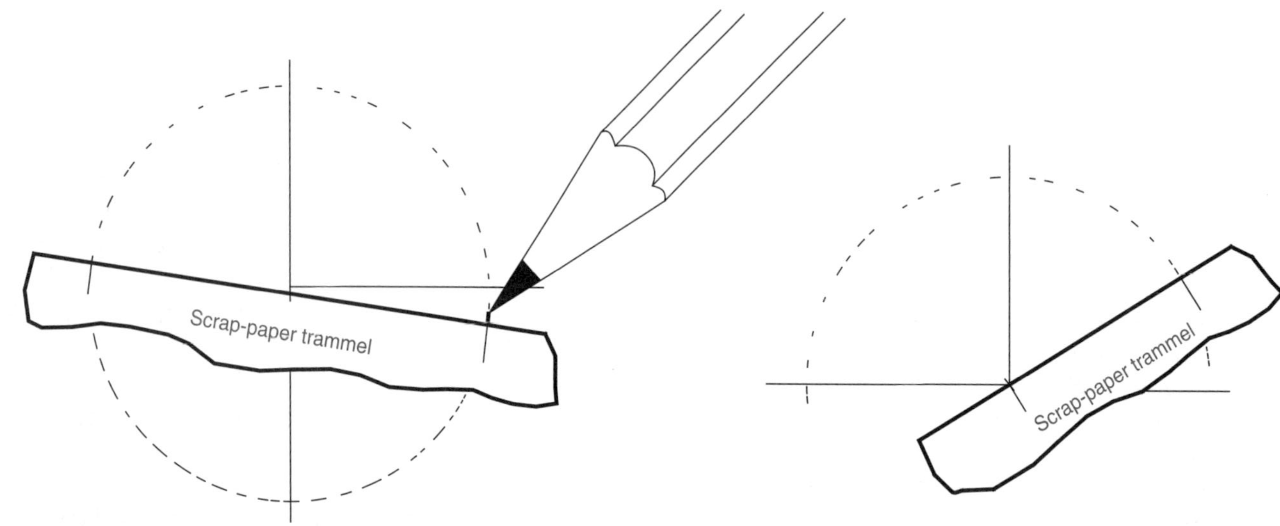


**Figure C-11.**
A paper trammel can be used to sketch an accurate circle.

## Proportion in Sketching

***Proportion*** is the relationship of the size of one part to another and to the object as a whole. The width, height, and depth of your sketch must be kept in the same proportion so the sketch conveys an accurate description of the object being sketched.

Proportion is a matter of estimating lengths on a part or assembly and setting these down on your sketch in the same ratio of units. Practicing this method of establishing proportion in sketches will assist you in accurately sketching objects. Actual measurement of an object is necessary if your sketch is to convey actual size dimensions.

One technique useful in estimating proportions is the unit method. The ***unit method*** is used to establish a relationship between measurements on the object by breaking each of the measurements into similar units. Compare the width to the height and select a unit that will fit each measurement. See Figure C-12. All distances on your sketch should be in the same proportion.

## Sketching an Object

The following steps will help you in laying out and completing your freehand sketches. As an example, draw the flange shown in Figure C-13A as you follow the procedure.

1. Sketch a rectangle of the correct proportion. See Figure C-13B.
2. Sketch major subdivisions and details of the object. See Figure C-13C.
3. Complete the object and remove construction lines. See Figure C-13D.
4. Darken lines to the proper weight. See Figure C-13E. Refer to Unit 2 for types of lines used in industry prints if necessary.

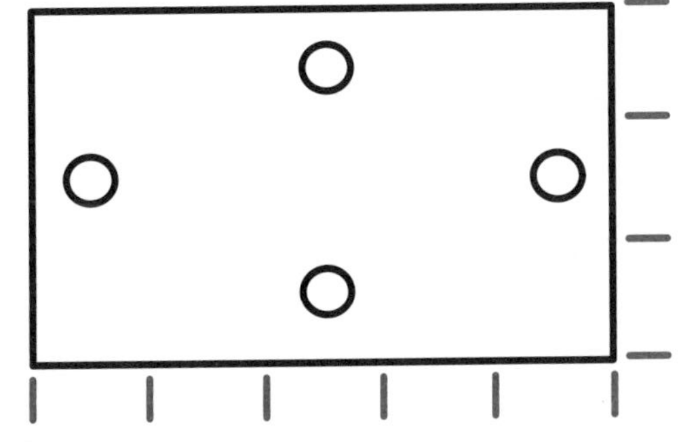

**Figure C-12.**
When estimating proportions, subdivide an object into equal units.

## Lettering

By definition, ***lettering*** is freehand sketching alphanumeric characters according to standard practices. It is *not* printing or handwriting. The lettering used on most technical drawings and sketches is Gothic styled. This is because Gothic lettering is easy to construct and read. Gothic lettering can be vertical at 90° or inclined 67.5°. Vertical uppercase (capital) lettering is widely used in industry. However, some industries prefer inclined uppercase lettering. Lowercase (small) letters are often used on map drawings, but seldom on machine part drawings.

### Forming Letters and Numbers

The Gothic style letters and numbers are formed with single strokes, rather than varying in thickness. For example, the letters I and J have no bottom or top strokes. The number 1 is likewise just a single stroke. One to four single strokes are required to properly construct the Gothic style letters and numbers, as shown in Figure C-14. By following these stroke patterns, you will soon be able to construct neat, well-shaped letters and numbers. A medium to soft pencil is recommended for lettering.

Letters are usually made 1/8″ in height for notes and dimensions. The title or name of the part is usually 1/4″ in height. Numbers are equal in height to uppercase letters. Fractions are twice the height of whole numbers, Figure C-15. Therefore, the numbers used in the fraction are slightly shorter than whole numbers.

While CAD has greatly enhanced the standardization and readability of prints, freehand sketches and instrumental drawings still demand professional, standardized lettering. This is another area in which you should be the champion within your company. Sloppy, careless lettering reflects poorly on the company.

### Guidelines

Horizontal, vertical, and inclined guidelines will help you produce quality lettering, Figure C-16. Without grid paper, these guidelines are a must. All lettering should have horizontal guidelines, no matter how small or seemingly unimportant.

If lightly constructed, these construction lines do not need to be erased and will not detract from the notes or dimensions. As with other sketched construction lines, the guidelines add a touch of technical appearance to the drawing, which speaks well for the drafter or designer.

### Spacing

When lettering, take care not to crowd the letters or leave the letter spacing too open, Figure C-17A. Good lettering is always compact, but not crowded or too open, Figure C-17B.

The space between letters and words should be pleasing to the eye. Because of their shape, some letters, such as A, C, L, and T, can be spaced more closely than letters such as B, D, H, and M. As a general rule, the spacing between words should equal about the height of the lettering and there should be about two letter spaces between sentences, Figure C-18.

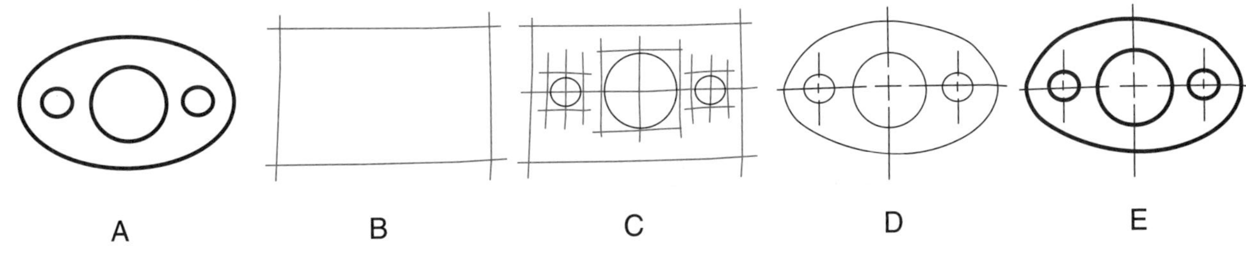


**Figure C-13.**
Sketching a sample object.

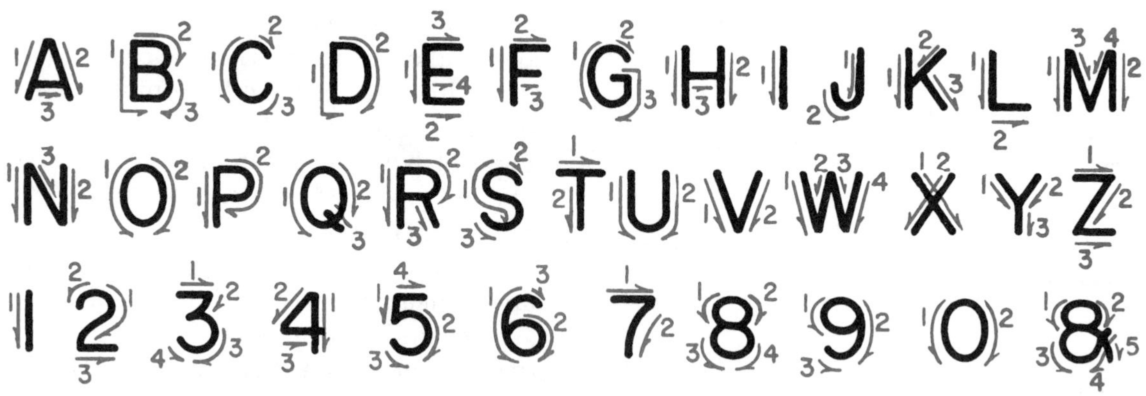

A B C D E F G H I J K L M
N O P Q R S T U V W X Y Z
1 2 3 4 5 6 7 8 9 0 8

Figure C-14.
One to four single strokes are required to properly construct Gothic-style letters and numbers.

Figure C-15.
Fractions should be twice as tall as capital letters.

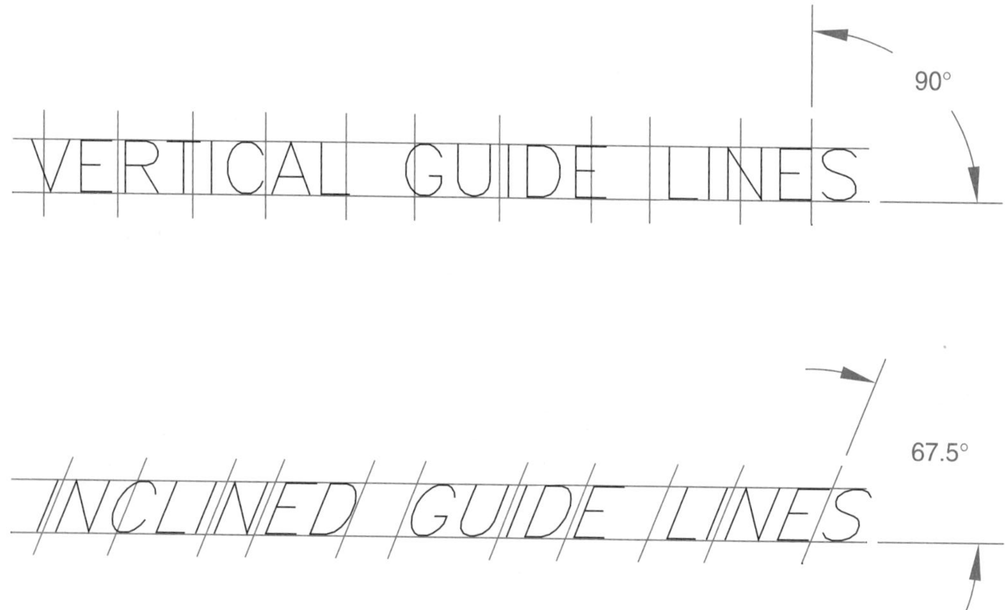

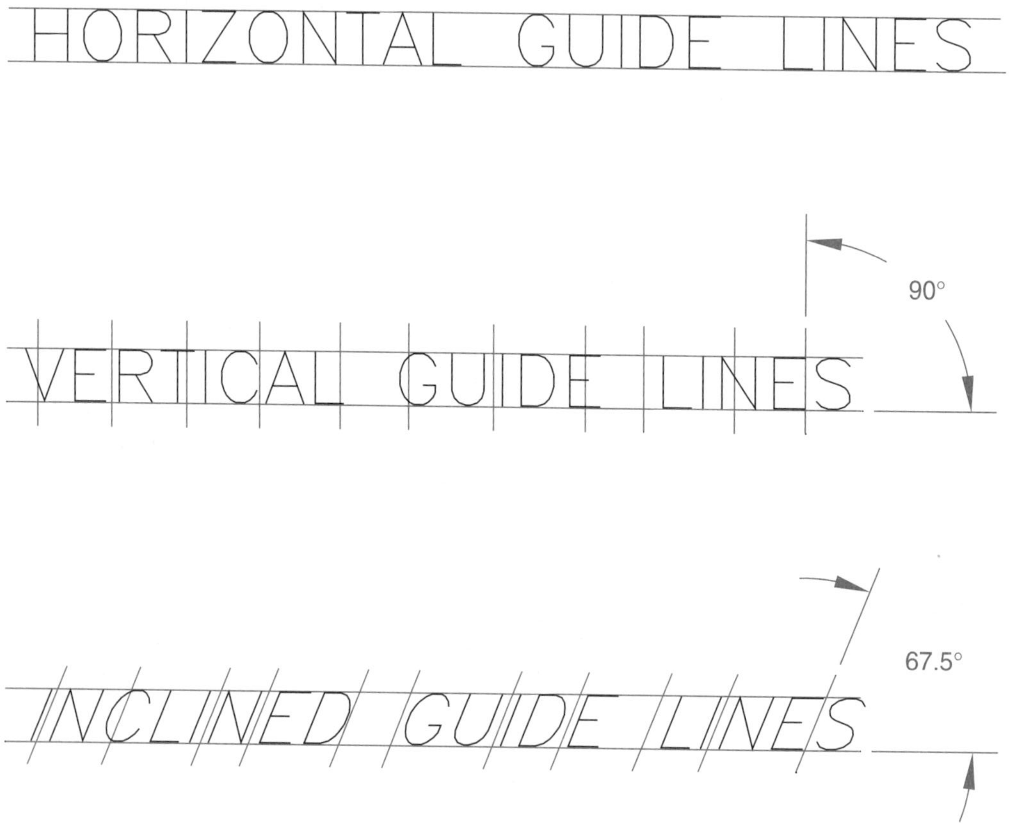

Figure C-16.
Horizontal, vertical, and inclined guidelines will help you produce quality lettering.

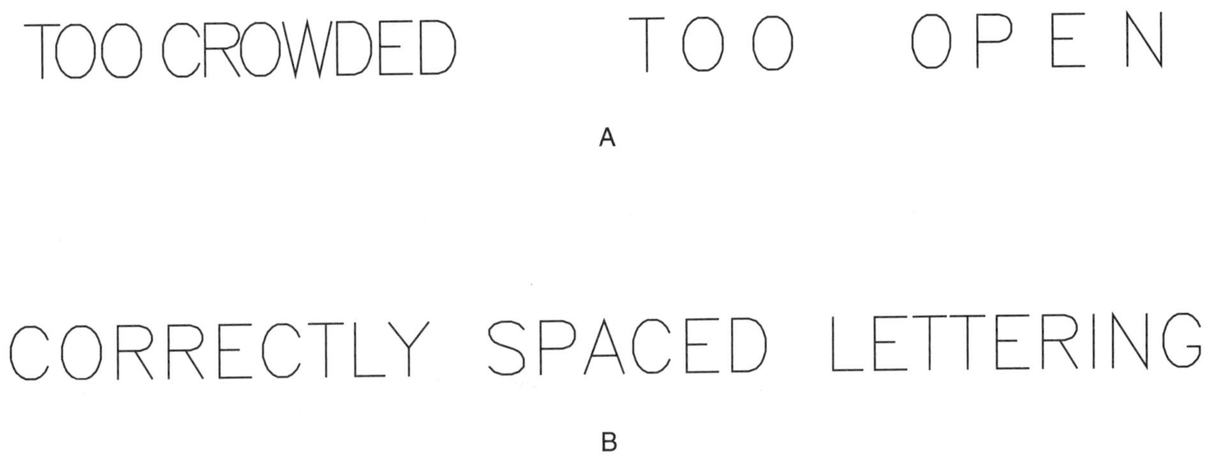

TOO CROWDED TOO OPEN

A

CORRECTLY SPACED LETTERING

B

**Figure C-17.**
Take care to not crowd the letters or leave the letter spacing too open.

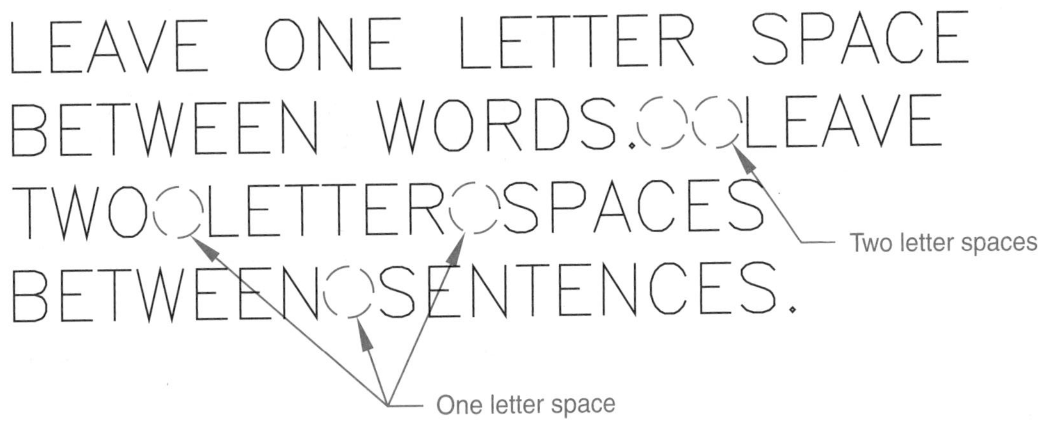

LEAVE ONE LETTER SPACE
BETWEEN WORDS. LEAVE
TWO LETTER SPACES
BETWEEN SENTENCES.

Two letter spaces

One letter space

**Figure C-18.**
As a general rule, the spacing between words should equal about the height of the lettering and there should be about two letter spaces between sentences.

## Review Questions

*Circle the answer of choice, fill in the blank, or write a short answer.*

1. List the three basic items needed to create a sketch. ____________________

____________________

____________________

____________________

2. *True or False?* The best type of pencil for all sketching is an automatic-feed mechanical pencil with .5 mm lead.

3. Of the statements below, select the only one that is a tip for quality sketching.
   A. Never pick up your pencil while sketching a line segment.
   B. Keep your eye on the point to which you are sketching, not the pencil.
   C. Use a straightedge to help you maintain straight sketch lines.
   D. Pushing the pencil works better than pulling the pencil.

4. *True or False?* The same techniques for sketching horizontal lines or vertical lines can be applied to sketching inclined lines, especially considering the sheet of paper can be turned to help you sketch in a comfortable direction.

5. Angles can be estimated quite closely by first sketching a(n) ____________________ angle and then subdividing its arc to get the desired angle.

6. Based on this unit, which of the following is a name given to a method to assist in sketching arcs or circles?
   A. Paper trammel.
   B. Triangle square.
   C. Elliptical triad.
   D. Both A and B are correct answers.

7. A(n) ____________________ is simply a scrap of paper that serves a marking tool.

8. Write a definition for proportion. ____________________

____________________

9. Which of the following terms is the basis for the definition of lettering?
   A. Sketching.
   B. Printing.
   C. Handwriting.
   D. Typing.

10. Which of the following appears most like lettering used in industrial prints?
   A. STANDARD LETTERING
   B. STANDARD LETTERING
   C. STANDARD LETTERING
   D. STANDARD LETTERING

## Review Activity C-1

### Horizontal Lines

Sketch a horizontal line between point A and point A′. Continue in a similar manner for the remainder of the points.

# Review Activity C-2

**Vertical Lines**

Sketch a vertical line between point K and point K′. Continue in a similar manner for the remainder of the points.

K L M N O P Q R S T

K′ L′ M′ N′

O′ P′ Q′ R′ S′ T′

# Review Activity C-3

## Angles

Sketch a line to form the given angle with the existing line. Refer to the example.

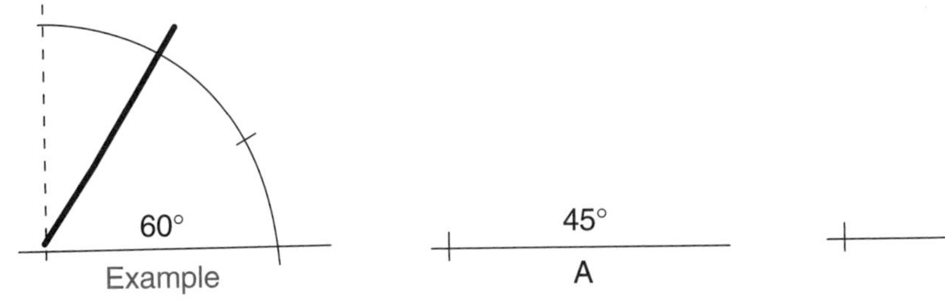


20°
C

100°
D

120°
E

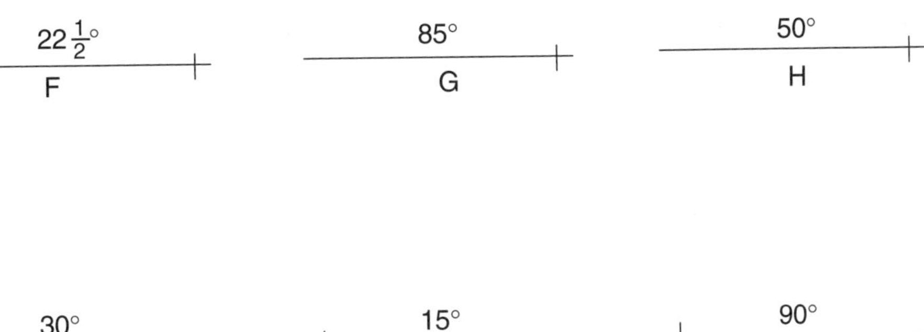

# Review Activity C-4

## Arcs and Circles

Sketch an arc to join each set of lines. On G through L, sketch circles about the center point. Use a paper trammel on B, F, H, and L. Leave the construction lines for A through C and G through I. Erase construction lines for D through F and J through L.

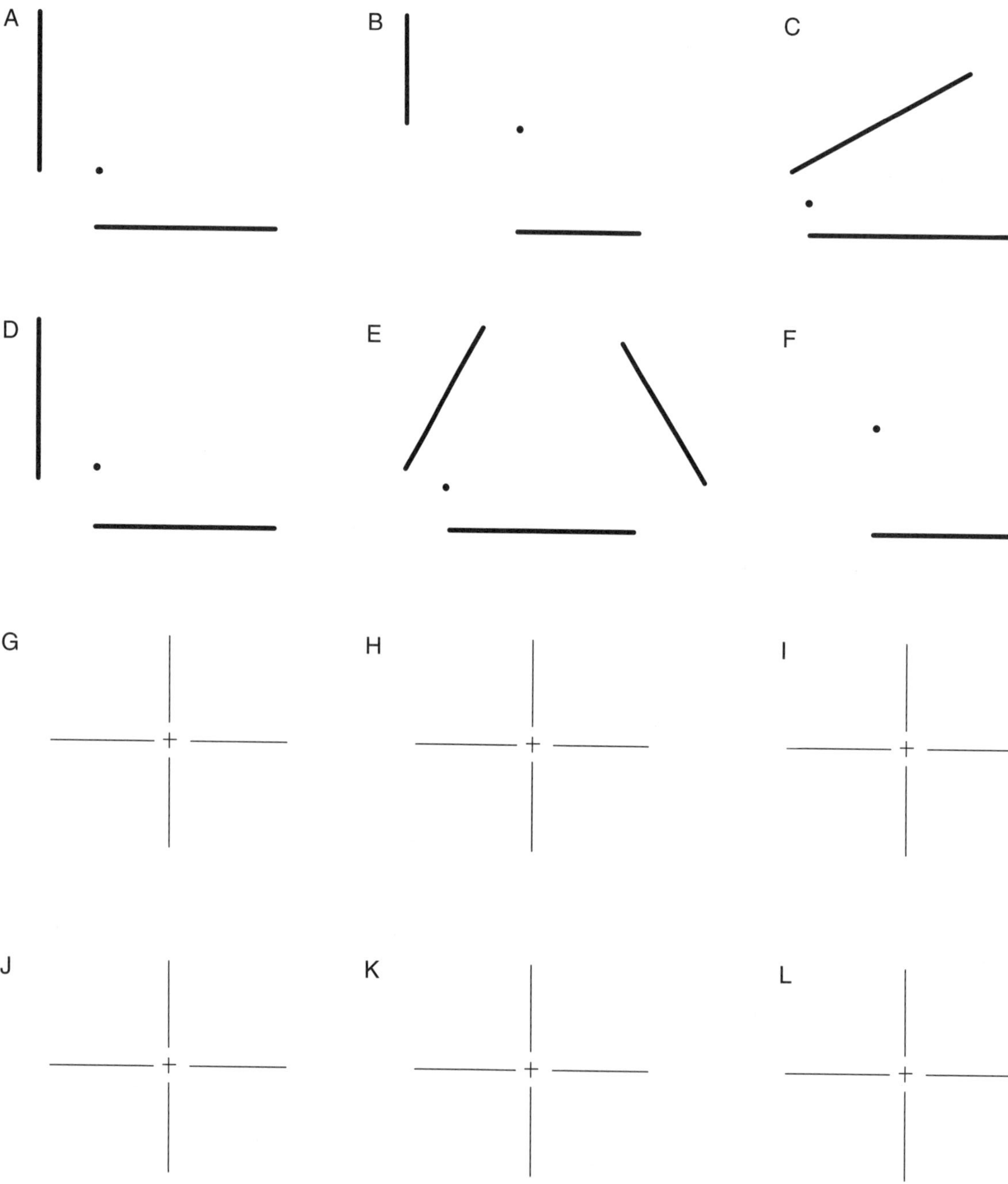

## Review Activity C-5

**Concrete Block**

1. Sketch the concrete block in the space below.
2. Estimate the proportions. Do not measure or dimension the sketch.
3. Letter your name and the date in the space provided using the colored guidelines.

## Review Activity C-6

**Base Plate**

1. Sketch the base plate in the space below.
2. Estimate the proportions. Do not measure or dimension the sketch.
3. Letter your name and the date in the space provided using the colored guidelines.

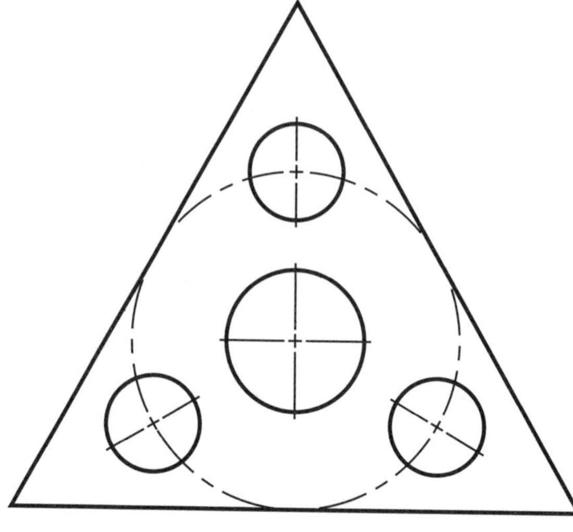

## Review Activity C-7

### Gothic Letters and Numbers Shape Practice

In the space provided, practice forming the letters and numbers as directed. Follow the lettering charts and guides in forming your letters and numbers. Use the colored guidelines so your lettering is uniform in height.

A

Letter A–G between the guidelines

H

Letter H–N between the guidelines

O

Letter O–U between the guidelines

V

Letter V–Z between the guidelines

0

Letter the numbers 0–5 between the guidelines

6

Letter the numbers 6–10 between the guidelines

A

H

O

V

0

6

# Review Activity C-8

## Gothic Lettering Spacing Practice

In the space provided, practice lettering and spacing words by lettering the paragraph below. Remember, the height of the lettering is a good rule of thumb for word spacing and double that for sentence spacing.

LETTERING IS AN IMPORTANT PART OF COMMUNICATION ON PRINTS. POOR LETTERING INDICATES AN UNWILLINGNESS TO PUT FORTH A GOOD EFFORT OR AN UNWILLINGNESS TO CONFORM TO STANDARDS. YOU CAN MAKE A GOOD IMPRESSION WHILE CLEARLY COMMUNICATING NOTES AND INFORMATION IF YOU ARE PATIENT AND SKETCH EACH LETTER SHAPE ACCORDING TO STANDARDS. REMEMBER, THIS IS NOT PRINTING!

Letter here —
Space here —
Letter here —

## Appendix D
# Sketching Pictorial Drawings

*After completing this unit, you will be able to:*

**Define** pictorial drawing.

**Discuss** how CAD systems have impacted the creation of pictorial drawings.

**Identify** three common types of pictorial drawings.

**Construct** a basic isometric drawing.

**Construct** a basic oblique drawing.

**Construct** a basic one-point perspective drawing.

**Construct** a basic two-point perspective drawing.

**Examine** the principles of constructing circles and arcs in isometric, oblique, and perspective drawings.

By definition, pictorial drawings are picture-like representations of objects or products. Pictorial drawings attempt to show the part as if you are viewing a photograph of the actual object, with the object turned and tilted. See Figure D-1. Pictorial drawings are useful when making or servicing simple objects, but are usually not adequate to show how to manufacture a complex part. In current practice, many companies build 3D models of a product during the design process and 2D detail drawings are created directly from this model. A pictorial view is also easily created from the same model.

In the days when pencil drafting was prevalent, industry relied on technical illustrators to create many of the pictorial drawings. Sometimes these individuals worked in sales and support, creating technical brochures or instructional and maintenance diagrams. These illustrators had also learned techniques to add realism to the pictorial drawings with shading or shadowing techniques. A drawing with shading and shadowing techniques applied using pencil tones, pen and ink lines, or airbrushed ink is called a ***rendering.*** In recent years, CAD software has expanded our options for creating pictorial images. Most CAD programs include the capability for producing rendered images that have a photorealistic look and feel, as well as pictorial drawings with only lines. See Figure D-2.

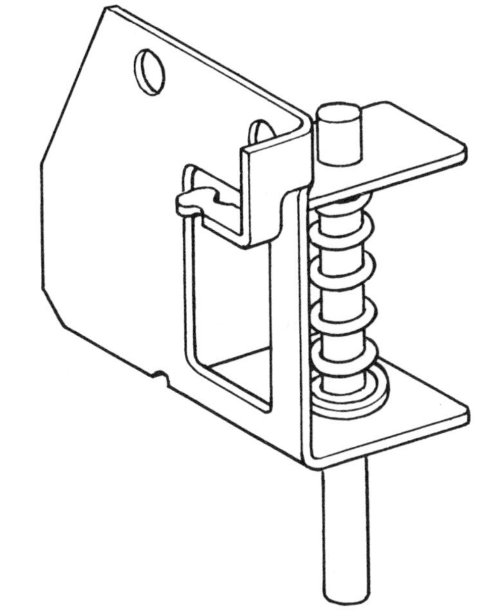

**Figure D-1.**
A pictorial drawing is a realistic, picture-like representation of an object. (National Lock)

While CAD has not totally eliminated the need for manually created pictorial drawings, it has definitely impacted the emphasis placed on pictorial drawings, not only within drafting curricula, but within industry standards. Even though computer programs may create pictorial drawings and images directly from 3D models, pictorial sketching is still useful in helping visualize multiview drawings. Pictorial sketching is also helpful in communicating your ideas or technical problems to others. Pictorial drawings are also very useful in explaining print reading concepts in textbooks such as this one.

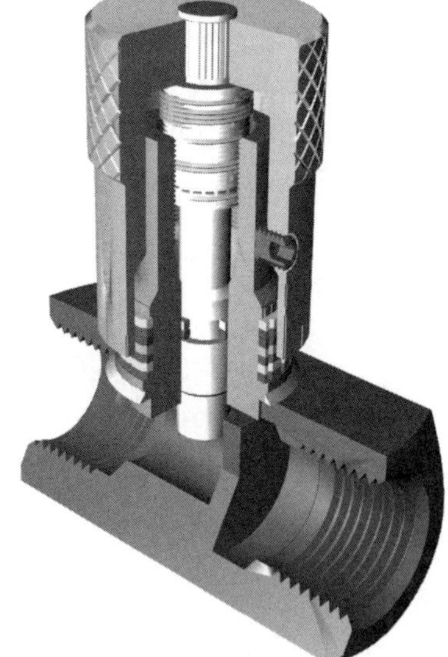

Figure D-2.
CAD renderings are a form of pictorial drawing. (RegO Cryo-Flow Products)

There are many types of pictorial drawings. However, for all practical purposes, the three common types of pictorial drawings are isometric, oblique, and perspective.

## CAD Applications

As discussed above, most CAD programs are very effective in creating pictorial drawings or views of a 3D model for use on prints. It is important to realize that there is a major difference between a pictorial *drawing* and a pictorial *view* generated from a 3D model. A pictorial drawing created with lines is still a two-dimensional drawing. The drawing, even if a CAD drawing, is still "drafted" with lines, circles, ellipses, etc., just like a pencil drawing. You cannot change your mind and rotate and tilt the object to some other angle.

A pictorial view of a 3D model is three-dimensional and defined by XYZ coordinates. It was created by the computer software with the selection of a command such as **PLOT**, **EXPORT**, or **RENDER**. Of course, once the image is placed on paper, it too is a two-dimensional, "flat" image. But, another view can be generated just as quickly if a different orientation or style is desired. As a print reader, you do not need to be concerned with how the image was placed on the paper, but you simply need to interpret and visualize the idea in your head. The good news is that there are more options than ever to help print readers visualize and "see" the object.

## Isometric Drawings

An orthographic (multiview) drawing is a projection of an object onto multiple viewing planes. Read Unit 5 for a discussion on multiview drawings. The views are based on projections said to be created ***orthographically.*** This essentially means parallel lines are projected from the edges and corners of the object perpendicular to the projection planes. Therefore, features of the object project onto the projection planes to create a drawing. If the front surface is parallel to the projection plane, the front surface projects true size and shape. Other surfaces may project as lines. The final result is a two-dimensional view of the object with surfaces featured differently in various views. Surfaces on the top portion of the object are featured in the top view, surfaces featured on the right side are featured in the right side view, and so on. The print reader must visualize the object through multiple views.

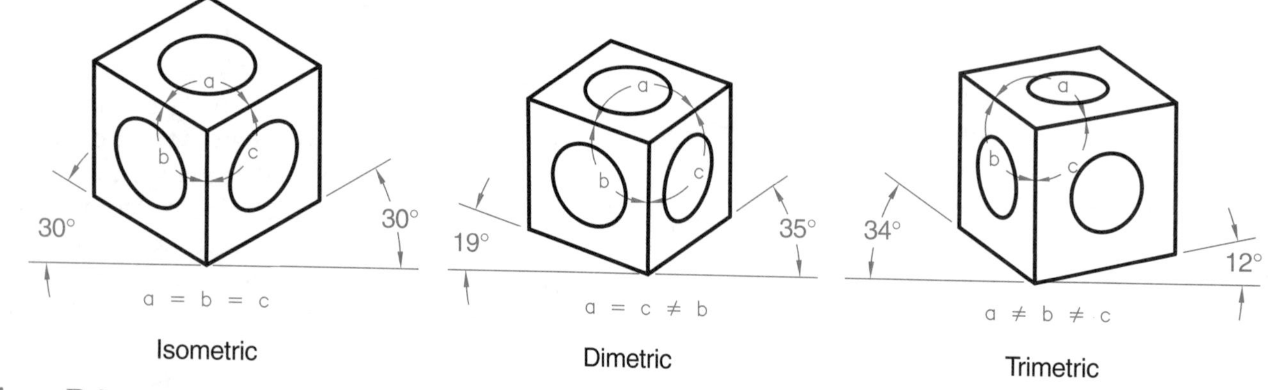


Figure D-3.
The three types of axonometric drawings are isometric, dimetric, and trimetric.

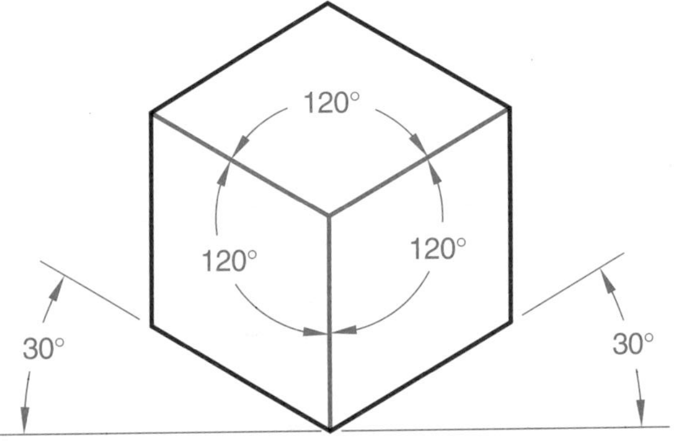


Figure D-4.
Isometric drawings are constructed with the XYZ axes projected at angles of 120° to each other.

Pictorial drawings can also be scientifically explained with projectors, projection planes, and objects. An object can be turned and tilted and a pictorial view will be the result. Or, the projectors can be projected at an oblique angle to the plane of projection and a pictorial view will result. In either case, pictorial drawings are often just based on these principles and not truly projected in quite the same way as a multiview drawing.

An ***axonometric projection*** of an object is when the object is turned and tilted and then projected onto a frontal plane. The three divisions of axonometric projection are isometric, dimetric, and trimetric. See Figure D-3. In an isometric projection, all three axes form equal angles. In a dimetric projection, only two axes form equal angles. In a trimetric projection, all axes form unequal angles. With CAD systems, a 3D model can easily be rotated in any axonometric direction, so the terms diametric and trimetric have become unimportant.

***Isometric drawings*** are constructed with the XYZ axes projected at angles of 120° to each other. This means the width and depth (X and Y) are 30° from horizontal and the height (Z) is vertical. See Figure D-4. Isometric drawings are used more often than the other types of axonometric and pictorial drawings because direct measurements can be used in their construction. This means they are easily constructed. However, isometric drawings appear less realistic than perspective drawings.

Looking at Figure D-4, the colored lines forming a Y in the center of the object are equally spaced at 120° to each other and are known as the ***isometric axes.*** Because of this equal spacing, the same scale of measure can be used along each axis. The word isometric means equal measure. In an isometric drawing, "equal measure" can be applied on all axes. In addition, the amount of distortion of circles into ellipses is the same on all surfaces.

Horizontal lines in an orthographic drawing are drawn at an angle of 30° in an isometric view. Vertical lines remain vertical in an isometric view. Inclined lines are called ***nonisometric lines*** and are drawn by locating their endpoints along the isometric axes and connecting the two points. Hidden lines are not shown in isometric views unless required to clarify drawing details.

## Constructing an Isometric Sketch

The procedure for constructing an isometric sketch is as follows. Refer to Figure D-5 as you go through the procedure.

1. The three orthographic views of a V-block are shown in a print. See Figure D-5A.
2. Select a position for the object that best describes its shape, such as a corner of the V-block located at the "front" of the drawing.
3. Sketch the axes for the lower corner. See Figure D-5B.
4. Make overall measurements in their true length on the isometric axes or on lines parallel to the axes. See Figure D-5C.
5. Construct a box to enclose the object. See Figure D-5D. Note: It is more important to make sure lines are parallel, rather than maintaining a precise 30° angle to the horizon.
6. Sketch the isometric lines of the object. See Figure D-5E.
7. Locate the endpoints of nonisometric lines. Then, sketch the lines. See Figure D-5F.
8. Darken all visible lines and erase the construction lines to complete the isometric sketch. See Figure D-5G.

## Circles and Arcs in Isometric

Isometric circles and arcs are sketched in a manner similar to the one discussed in Appendix C for sketching "regular" circles and arcs. The difference is for isometric circles and arcs, you start with an isometric square instead of a normal square. An isometric square resembles a diamond in the isometric

Figure D-5.
An isometric sketch of the V-block is constructed using the orthographic views as reference.

view. The procedure for sketching isometric circles is as follows. Circles can be sketched on all three planes of the isometric drawing in the same manner. See Figure D-6.

1. Sketch an isometric square enclosing the location of the circle. See Figure D-7A.
2. Locate and connect the midpoints of the sides of the isometric square. See Figure D-7B.
3. Locate the centers of the triangles formed. Then, sketch four isometric arcs connecting each midpoint to form an ellipse. See Figure D-7C.
4. Erase construction lines and darken the ellipse. Notice the ellipse is composed of four arcs, two with a small radius and two with a large radius. See Figure D-7D.

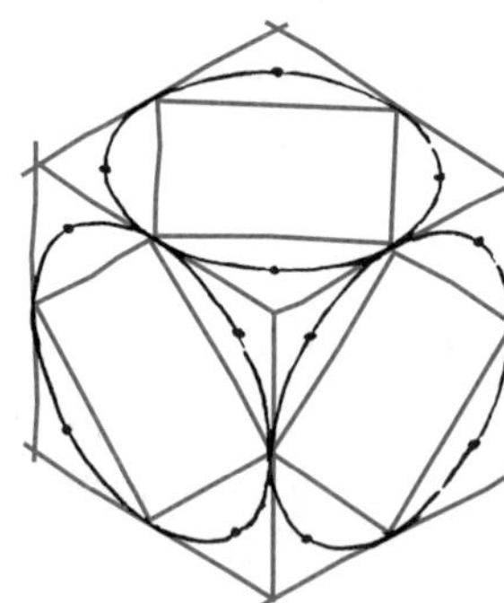

Figure D-6.
Circles can be sketched on all three planes of the isometric drawing.

The procedure for sketching an isometric arc is as follows.

1. Mark the radius of the arc from the corner. See Figure D-8A.
2. Draw a line connecting the two points, forming a triangle. See Figure D-8B.
3. Locate center of the triangle and sketch an arc through this point to smoothly join with the sides. See Figure D-8C.
4. Erase construction lines and darken the arc. See Figure D-8D.

## Isometric Dimensioning

The dimension lines on an isometric drawing or sketch are parallel to the isometric axes. Extension lines are extended in line with these axes. See Figure D-9. Notice in the figure the numerals are aligned with the isometric axes. However, the figures can be horizontally aligned instead. Dimensioning is discussed in detail in Unit 9.

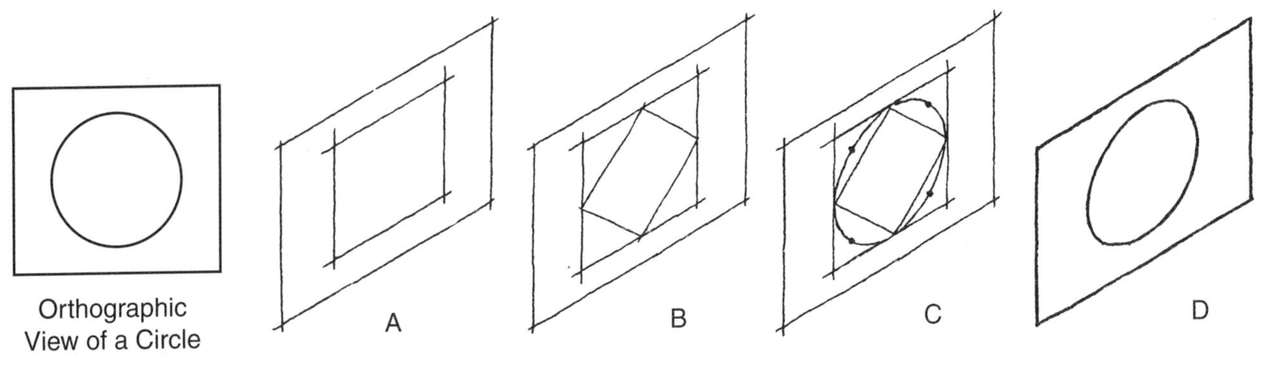


**Figure D-7.**
Sketching an isometric circle.

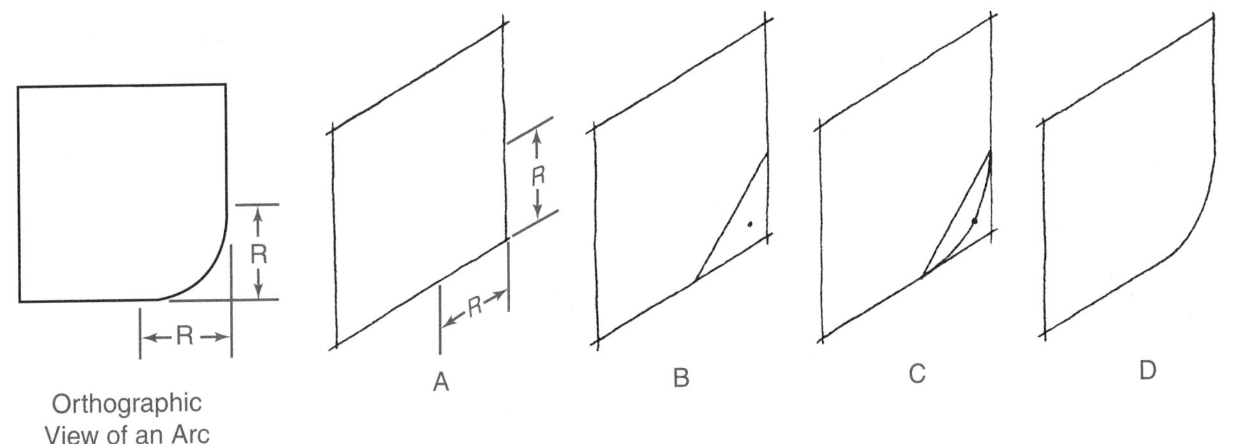


**Figure D-8.**
Sketching an isometric arc.

## Oblique Drawings

***Oblique drawings*** are drawn with the front face of an object shown in its true shape, just as an orthographic projection shows true shape. The depth axis is slanted up or down and to the left or right. The word oblique means slanted, as represented by the treatment of the depth axis. The fact the front face is shown in true shape is an advantage when it contains circles or arcs. These are shown as true circles and arcs. However, the top and side view are projected at an angle back from the front view, often at 45°. The oblique method of pictorial drawing tends to give a distorted appearance to the drawing, as if the depth measurement is too much. See Figure D-10. The two principal types of oblique drawings are the cavalier and cabinet. However, often measurements for sketches are estimated. Therefore, the terms cavalier and cabinet are not critical terms.

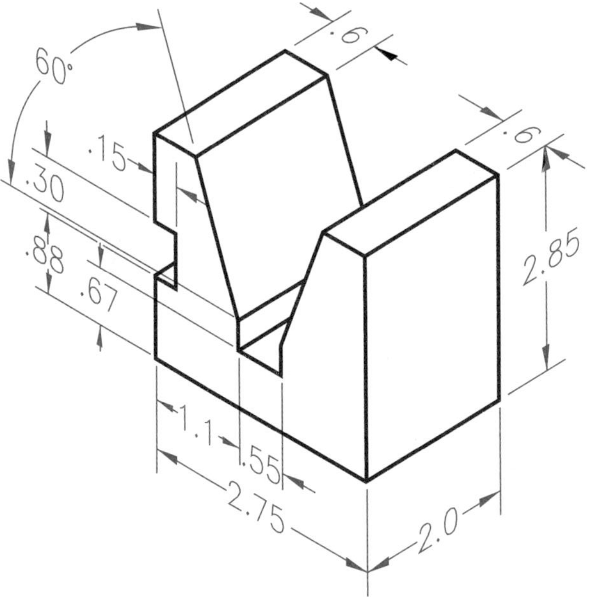


**Figure D-9.**
Dimension text in isometric can be aligned, as shown here, or horizontal.

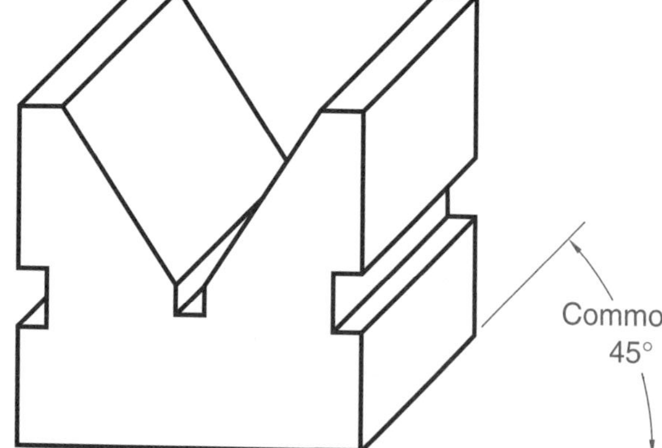


Figure D-10.
Oblique drawings are drawn with the front face of an object shown in its true shape.

***Cavalier oblique drawings*** are drawn with their receding sides to the same scale as the front view. See Figure D-11A. This creates an unrealistic appearance. However, the advantage is one scale is used throughout the drawing.

***Cabinet oblique drawings*** are drawn with their receding sides to one-half scale of the front view. This gives a much more realistic appearance. See Figure D-11B. Most likely, the name is derived from the woodworking or cabinetmaking industry.

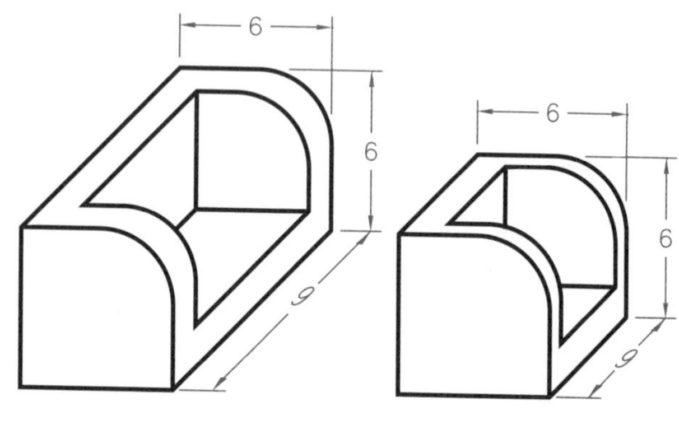


Figure D-11.
A—Cavalier oblique drawings are drawn with their receding sides to the same scale as the front view.
B—Cabinet oblique drawings are drawn with their receding sides to one-half scale of the front view.

## Circles and Arcs in Oblique

Circles and arcs are sketched in oblique as true circles or arcs in the front plane. However, when circles and arcs appear on the receding planes, they are sketched in the same manner as isometric circles and arcs. The procedure for sketching oblique circles and arcs is as follows. Refer to Figure D-12.

1. Sketch an oblique square that will contain the circle.
2. Locate the midpoints of the oblique square's sides. Connect these points. As an option, sketch centerlines for the height, width, and depth directions.
3. Locate the center of the triangles. Sketch oblique circles or ellipses through these points, smoothly joining with the sides of the oblique square.
4. Erase construction lines and darken the circles or ellipses.

## Oblique Dimensioning

Oblique dimensioning must be done in the same plane as the surface or feature. This is the same as in isometric dimensioning. Figure D-13 illustrates the placement of dimensions on an oblique drawing.

## Perspective Drawings

***Perspective drawings*** have receding lines that, if extended, converge at one or more points in the "background." These points are called ***vanishing points.*** Vanishing points are placed on a horizontal

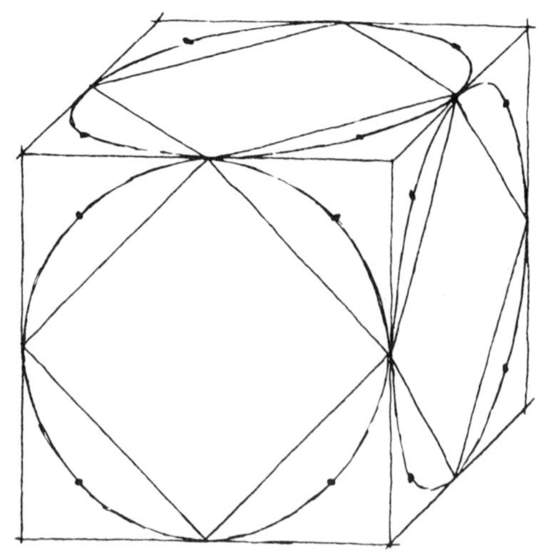

Figure D-12.
Sketching oblique circles.

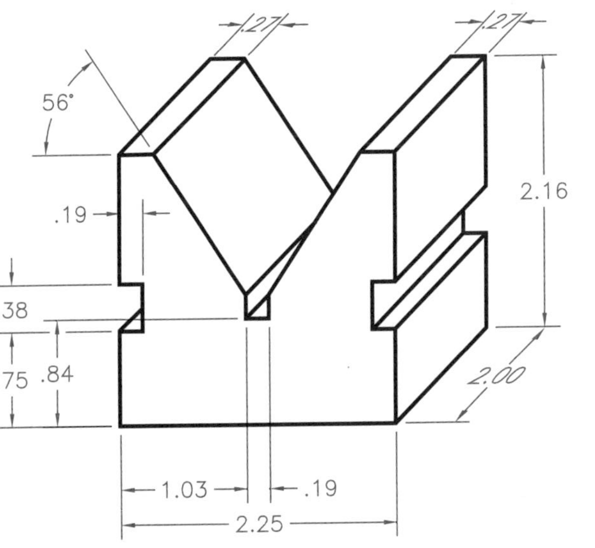


**Figure D-13.**
Dimensioning an oblique drawing.

line, called the horizon, that is at the assumed eye level of the viewer. The horizon may be above, behind, or below the object. A good rule for a beginning sketcher is to place the horizon above the object at least as much as the object's height. The receding lines on a perspective do not remain parallel, as they do in isometric or oblique drawings. See Figure D-14.

Perspective drawings eliminate much of the distorted appearance that occurs on other pictorial drawings. Creating a perspective drawing involves principles that are much like the way the human eye works. Therefore, perspective drawings are the most realistic of all pictorial drawings.

There are three types of perspective drawings. These are one-point (parallel), two-point (angular), and three-point perspectives. The parallel and angular perspective types are discussed and illustrated here. This will assist you in developing a basic technique of perspective sketching. Many books have been written on perspective drawing, as it is also used in artistic studies.

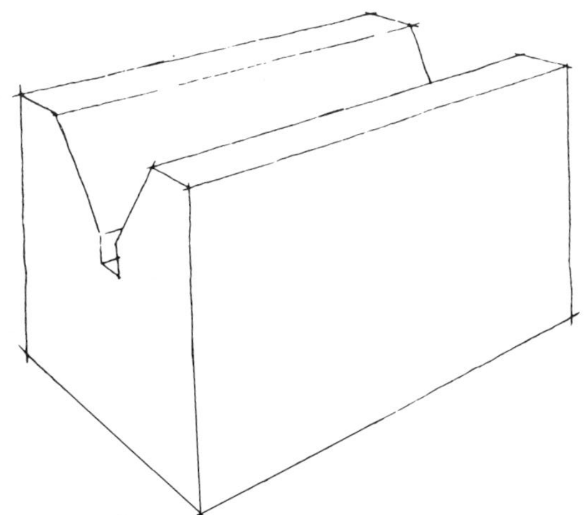

**Figure D-14.**
Receding lines on a perspective are not parallel.

## Parallel Perspective

***Parallel perspective drawings,*** or ***one-point perspective drawings,*** have one face of an object appearing in its true shape and parallel to the projection plane, similar to an oblique drawing. The projection plane is often called the ***picture plane.*** The lines parallel to the front picture plane remain parallel. Receding lines of the other faces converge in the direction of a single vanishing point. See Figure D-15. The procedure for sketching a one-point perspective is as follows.

1. Sketch the front view in its true size and shape, as for an orthographic sketch. See Figure D-16A.
2. Sketch a horizon. See Figure D-16B.
3. Select a vanishing point (VP) on the horizon as far to the right or left as desired. See Figure D-16C.
4. Sketch lines from the front view to the vanishing point. See Figure D-16D.

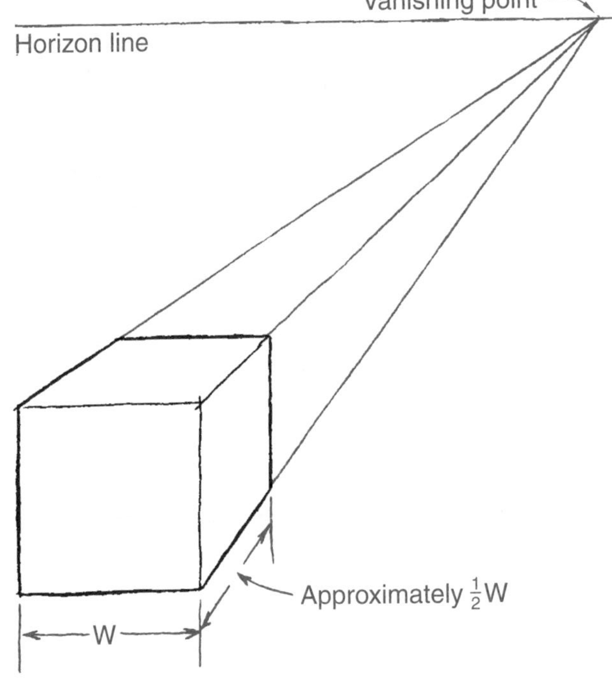


**Figure D-15.**
Receding lines on a one-point perspective converge in the direction of a single vanishing point.

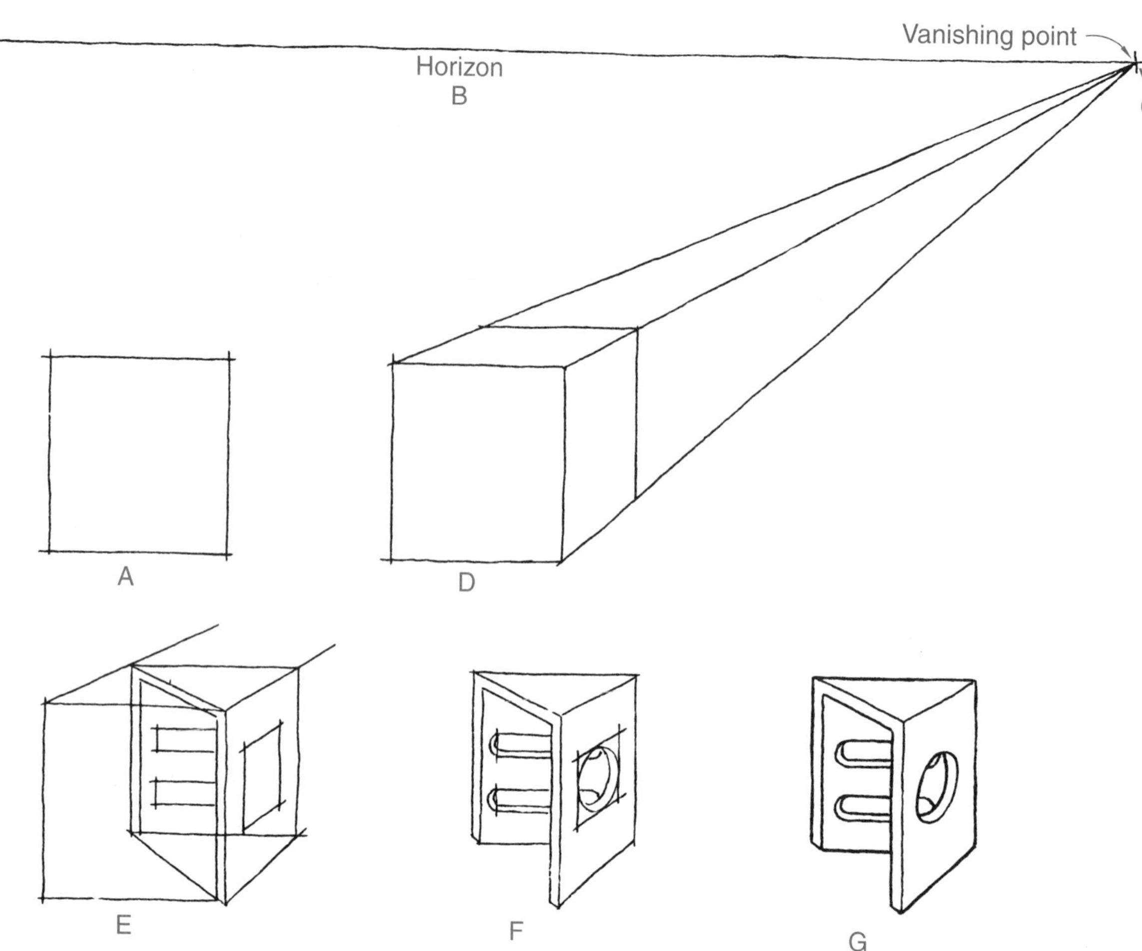


Figure D-16.
Sketching a one-point perspective.

5. Enclose the object in a box by sketching rear vertical and horizontal lines. See Figure D-16D. To estimate the depth of the side and top view, reduce these distances by about one-half and adjust as necessary.
6. To sketch inclined edges, locate the endpoints on the perspective axes and sketch lines between them. See Figure D-16E.
7. Block in features such as slots and holes. See Figure D-16E.
8. Sketch circles and arcs. See Figure D-16F.
9. Darken visible lines and erase construction lines. See Figure D-16G.

## Angular or Two-Point Perspective

***Angular perspective drawings,*** or ***two-point perspective drawings,*** show the two side faces of an object meeting the front picture plane at an angle and the faces recede toward two vanishing points on the horizon. See Figure D-17. An angular perspective is similar to isometric sketching, except the width and depth lines slant toward vanishing points instead of being parallel at 30° from the horizon. The procedure for sketching an angular or two-point perspective is as follows.

1. Sketch a horizon. See Figure D-17A.
2. Select a position from which the object is to be viewed. Sketch a vertical line for the front corner of a box that will enclose the object. See Figure D-17B.

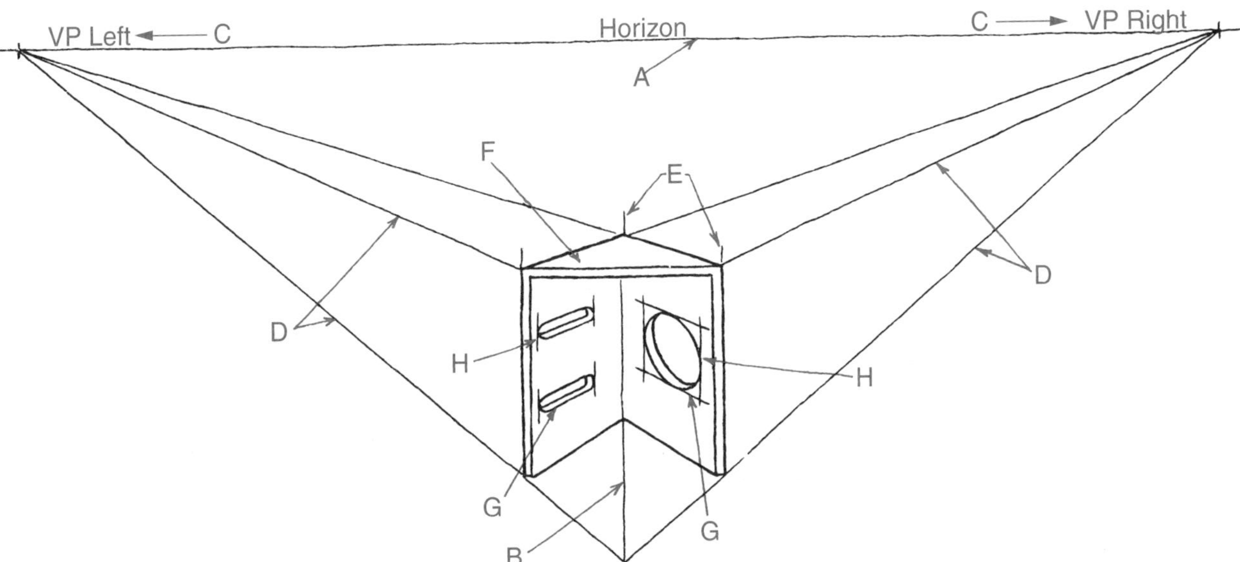


Figure D-17.
Receding lines on a two-point perspective converge in the direction of two vanishing points.

3. Establish right and left vanishing points on the horizon. See Figure D-17C. Positioning the vanishing points equidistant from the front corner will equally show the sides. Unequal placement of vanishing points will result in one side being shown more than the other. Vanishing points, however, must be on the horizon.
4. Sketch receding lines from the front vertical line to both vanishing points. See Figure D-17D.
5. Finish the box by sketching rear vertical lines. See Figure D-17E. To estimate the depth of the side and top views, reduce these distances by about one-half and adjust as necessary.
6. To sketch inclined edges, locate the endpoints on the perspective axes and sketch lines between them. See Figure D-17F.
7. Block in features such as slots and holes. See Figure D-17G.
8. Sketch circles and arcs. See Figure D-17H.
9. Darken visible lines and erase construction lines.

## *Circles and Arcs in Perspective*

Circles and arcs are sketched in perspective in the same way as they are for isometric drawings. First, sketch a perspective square or block. Then, add center construction lines as needed. Next, join the midpoints of the sides to form triangles. The perspective circle or arc is then sketched through the midpoints of the sides and the center of the triangles. See Figure D-18.

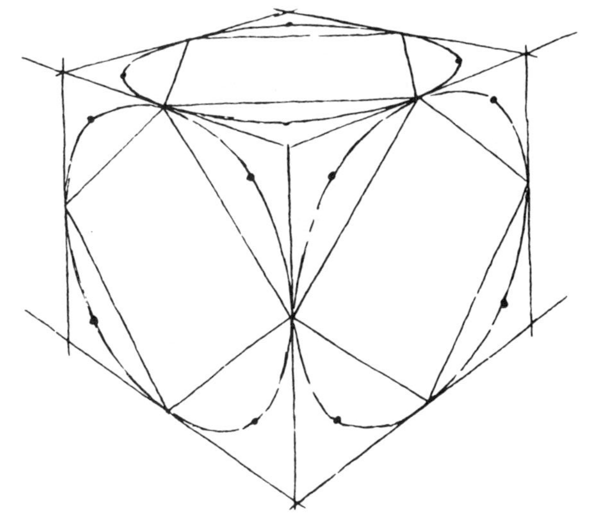

Figure D-18.
Sketching perspective circles.

## Review Questions

*Circle the answer of choice, fill in the blank, or write a short answer.*

1. Write a basic definition for pictorial drawings. ____________________

____________________

____________________

2. Before CAD systems, the drafter created drawings and the illustrator created ____________ using various shading techniques.

3. *True or False?* While a CAD system can be used to create a 2D pictorial drawing one line at a time, many CAD systems can also create pictorial views directly from a 3D model.

4. *True or False?* Isometric projection is one division of axonometric projection.

5. In isometric drawings, square features appear diamond shaped, while round features appear ____________________.

6. The word oblique means slanted, which, in oblique pictorial drawings, describes the ____________________ axis.

7. List the two basic types of oblique pictorial drawings. ____________________

____________________

____________________

8. *True or False?* One advantage of oblique pictorial drawings is that you will never need to draw an ellipse.

9. *True or False?* Two-point perspective drawings are more realistic than oblique or isometric.

10. Of the different types of perspective drawing methods, the key defining element is the:
    - A. number of vanishing points.
    - B. horizon level.
    - C. orientation of the picture plane.
    - D. number of station points.

## Sketching Activity D-1

**Motor Bracket**

Create an isometric, oblique, or perspective sketch of the motor bracket, as assigned by your instructor. Dimension your sketch.

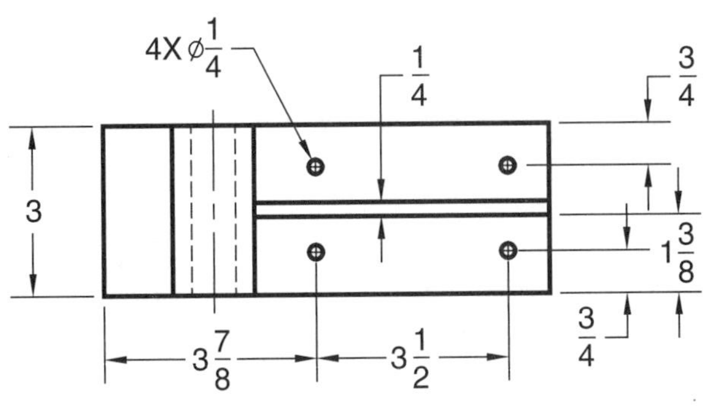


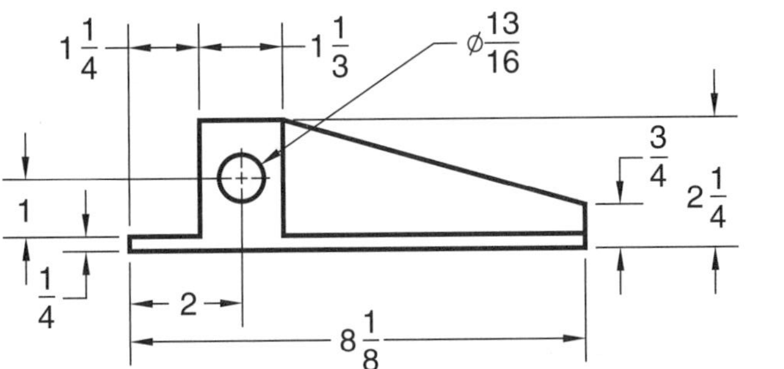

## Sketching Activity D-2

### Bracket Rest

Create an isometric, oblique, or perspective sketch of the bracket rest, as assigned by your instructor. Dimension your sketch.

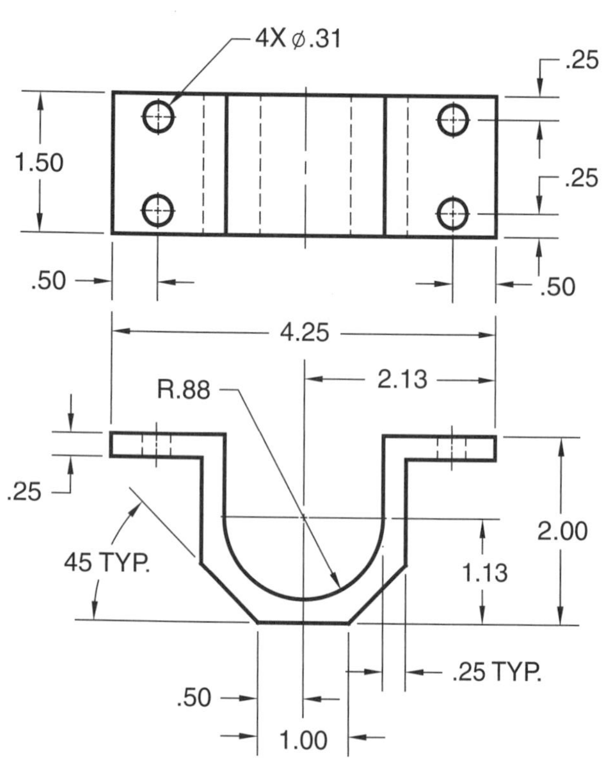

## Sketching Activity D-3

### Coupling

Create an isometric, oblique, or perspective sketch of the coupling, as assigned by your instructor. Dimension your sketch.

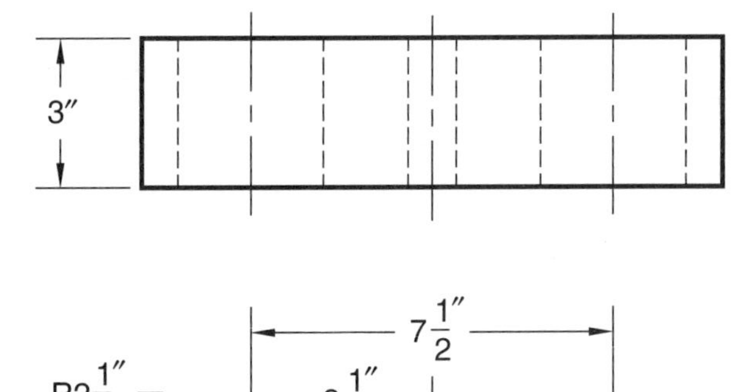


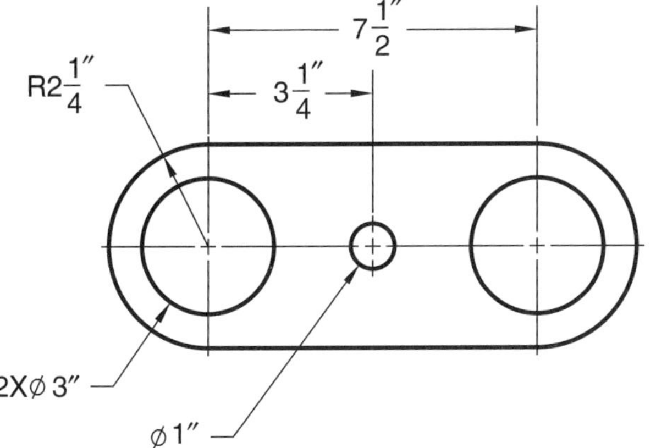

# Sketching Activity D-4

## Oil Stones and Holder

Create an isometric, oblique, or perspective sketch of the oil stones and holder, as assigned by your instructor. Dimension your sketch.

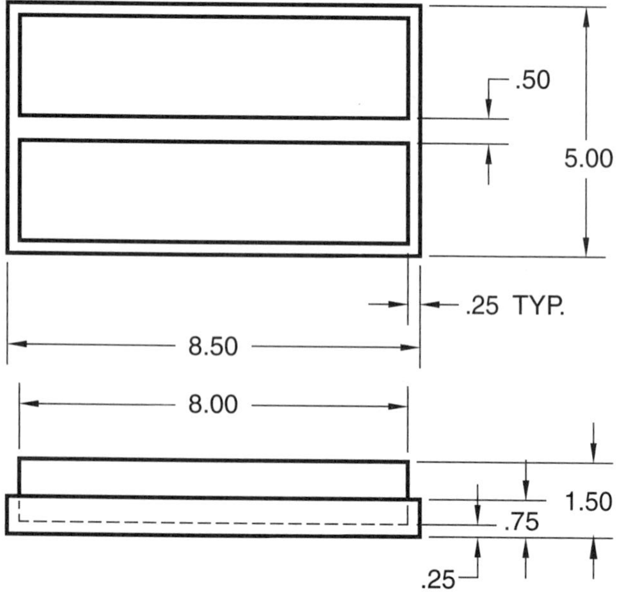

## Sketching Activity D-5

**Lock Housing**

Create an isometric, oblique, or perspective sketch of the lock housing, as assigned by your instructor. Dimension your sketch.

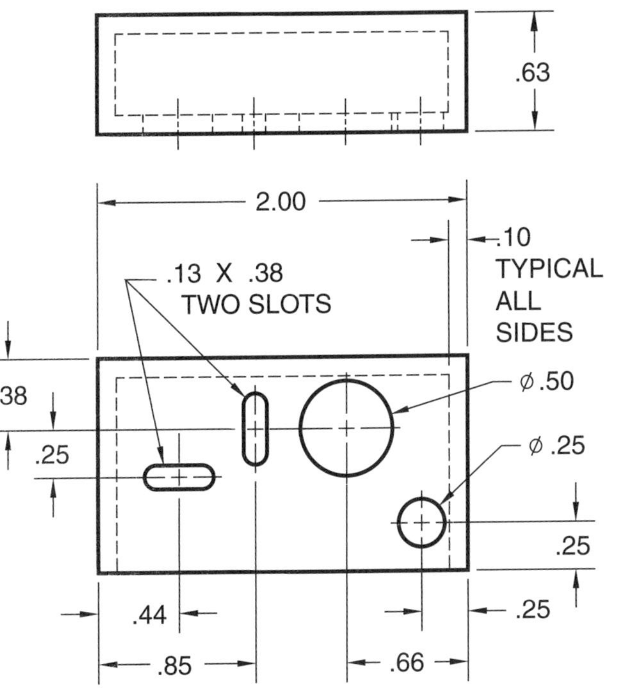

# Appendix E
# Abbreviations and Tables

# Standard Abbreviations

## A

|  |  |
|---|---|
| Abrasive | ABRSV |
| Accessory | ACCESS |
| Accumulator | ACCUMR |
| Acetylene | ACET |
| Across | ACR |
| Actual | ACT |
| Actuator | ACTR |
| Addendum | ADD |
| Adhesive | ADH |
| Adjust | ADJ |
| Advance | ADV |
| Advances Document Change Notice | ADCN |
| Aeronautic | AERO |
| Alclad | CLAD |
| Alignment | ALIGN |
| Allowance | ALLOW |
| Alloy | ALY |
| Alternation | ALT |
| Alternate | ALT |
| Aluminum | AL |
| American National Standard | AMER NATL STD |
| American National Standard Institute | ANSI |
| The American Society of Mechanical Engineers | ASME |
| American Wire Gage | AWG |
| Ampere | AMP |
| Anneal | ANL |
| Anodize | ANOD |
| Approved | APPD |
| Approximate | APPROX |
| Area | A |
| Asbestos | ASB |
| As Required | AR |
| Assemble | ASSEM |
| Assembly | ASSY |
| Automatic | AUTO |
| Auxiliary | AUX |
| Average | AVG |

## B

|  |  |
|---|---|
| Babbitt | BAB |
| Base Line | BL |
| Bend Radius | BR |
| Bevel | BEV |
| Bolt | BLT |
| Bolt Circle | BC |
| Bracket | BRKT |
| Brass | BRS |
| Brazing | BRZG |
| Brinell Hardness Number | BHN |
| Bronze | BRZ |
| Brown & Sharpe (Gage) | B&S |
| Burnish | BNH |
| Bushing | BUSH |

## C

|  |  |
|---|---|
| Calculated | CACL |
| Cancelled | CANC |
| Capacity | CAP |
| Carburize | CARB |
| Case Harden | CH |
| Cast Iron | CI |
| Center | CTR |
| Center to Center | C TO C |
| Celsius | C |
| Centimeter | CM |
| Centrifugal | CENT |
| Chamfer | CHAM |
| Chamfer or Radius | C/R |
| Check Valve | CV |
| Chrome Vanadium | CR VAN |
| Circuit | CKT |
| Circular | CIR |
| Circumference | CIRC |
| Clearance | CL |
| Closure | CLOS |
| Coated | CTD |
| Cold-Drawn Steel | CDS |
| Cold-Rolled Steel | CRS |
| Color Code | CC |
| Computer-Aided Drafting | CAD |
| Computer-Aided Manufacturing | CAM |
| Computer-Integrated Manufacturing | CIM |
| Computer Measuring Machine | CMM |
| Computer Numeric Control | CNC |
| Concentric | CONC |
| Condition | COND |
| Contour | CTR |
| Control | CONT |
| Copper | COP |
| Counterbore | CBORE |
| Countersink | CSK |
| Coupling | CPLG |
| Cubic Foot Per Minute | CFM |
| Cubic Foot Per Second | CFS |
| Cylinder | CYL |

## D

| | |
|---|---|
| Datum | DAT |
| Decimal | DEC |
| Decrease | DECR |
| Degree | DEG |
| Detail | DET |
| Developed Length | DL |
| Developed Width | DW |
| Deviation | DEV |
| Diagonal | DIAG |
| Diagram | DIAG |
| Diameter | DIA *or* D |
| Diameter Bolt Circle | DBC |
| Diametral Pitch | DP |
| Dimension | DIM |
| Direct Numerical Control | DNC |
| Disconnect | DISC |
| Document Change Notice | DCN |
| Dowel | DWL |
| Draft | DFT |
| Drafting Room Manual | DRM |
| Drawing | DWG |
| Drawing Change Notice | DCN |
| Drill | DR |
| Drop Forge | DF |
| Duplicate | DUP |

## E

| | |
|---|---|
| Each | EA |
| Eccentric | ECC |
| Effective | EFF |
| Electric | ELEC |
| Enclosure | ENCL |
| Engine | ENG |
| Engineer | ENG |
| Engineering | ENGR |
| Engineering Change Order | ECO |
| Engineering Order | EO |
| Equal | EQ |
| Equivalent | EQUIV |
| Estimate | EST |

## F

| | |
|---|---|
| Fabricate | FAB |
| Fillet | FIL |
| Finish | FIN |
| Finish All Over | FAO |
| Fitting | FTG |
| Fixed | FXD |
| Fixture | FIX |
| Flange | FLG |
| Flat Head | FHD |
| Flat Pattern | F/P |
| Flexible | FLEX |
| Fluid | FL |
| Forged Steel | FST |
| Forging | FORG |
| Full Indicator Movement | FIM |
| Furnish | FURN |

## G

| | |
|---|---|
| Gage | GA |
| Gallon | GAL |
| Galvanized | GALV |
| Gasket | GSKT |
| Gauge | GA |
| Generator | GEN |
| Grind | GRD |
| Ground | GRD |

## H

| | |
|---|---|
| Half-Hard | 1/2H |
| Handle | HDL |
| Harden | HDN |
| Head | HD |
| Heat Treat | HT TR |
| Hexagon | HEX |
| High-Carbon Steel | HCS |
| High Frequency | HF |
| High Speed | HS |
| Horizontal | HOR |
| Hot-Rolled Steel | HRS |
| Hour | HR |
| Housing | HSG |
| Hydraulic | HYD |
| Hydrostatic | HYDRO |

## I

| | |
|---|---|
| Identification | IDENT |
| Impregnate | IMPG |
| Inch | IN |
| Inclined | INCL |
| Include, Including, Inclusive | INCL |
| Increase | INCR |
| Independent | INDEP |
| Indicator | IND |
| Information | INFO |

| | |
|---|---|
| Inside Diameter | ID |
| Installation | INSTL |
| Interrupt | INTER |

## J

| | |
|---|---|
| Joggle | JOG |
| Junction | JCT |

## K

| | |
|---|---|
| Keyway | KWY |

## L

| | |
|---|---|
| Laboratory | LAB |
| Lacquer | LAQ |
| Laminate | LAM |
| Left Hand | LH |
| Length | LEN |
| Letter | LTR |
| Limited | LTD |
| Limit Switch | LS |
| Linear | LIN |
| Liquid | LIQ |
| List of Material | LOM |
| Long | LG |
| Low Carbon | LC |
| Lubricate | LUB |

## M

| | |
|---|---|
| Magnaflux | M |
| Magnesium | MAG |
| Maintenance | MAINT |
| Major | MAJ |
| Malleable | MALL |
| Malleable Iron | MI |
| Manual | MAN |
| Mark | MK |
| Master Switch | MS |
| Material | MATL |
| Maximum | MAX |
| Measure | MEAS |
| Mechanical | MECH |
| Medium | MED |
| Middle | MID |
| Military | MIL |
| Minimum | MIN |
| Miscellaneous | MISC |
| Modification | MOD |
| Mold Line | ML |
| Motor | MOT |
| Multiple | MULT |

## N

| | |
|---|---|
| Nickel Steel | NS |
| Nomenclature | NOM |
| Nominal | NOM |
| Normalize | NORM |
| Not to Scale | NTS |
| Number | NO |
| Numerical Control | NC |

## O

| | |
|---|---|
| Obsolete | OBS |
| Opposite | OPP |
| Ounce | OZ |
| Outside Diameter | OD |
| Over-All | OA |

## P

| | |
|---|---|
| Package | PKG |
| Parting Line (Castings) | PL |
| Pattern | PATT |
| Piece | PC |
| Pilot | PLT |
| Pitch | P |
| Pitch Circle | PC |
| Pitch Diameter | PD |
| Plan View | PV |
| Plastic | PLSTC |
| Plate | PL |
| Pneumatic | PNEU |
| Port | P |
| Positive | POS |
| Pounds Per Square Inch | PSI |
| Pounds Per Square Inch, Absolute | PSIA |
| Pounds Per Square Inch, Gage | PSIG |
| Pressure | PRESS |
| Primary | PRI |
| Process, Procedure | PROC |
| Product, Production | PROD |

## Q

| | |
|---|---|
| Quality | QUAL |
| Quality Control | QC |
| Quantity | QTY |
| Quarter-Hard | 1/4H |

## R

|  |  |
|---|---|
| Radius | RAD *or* R |
| Ream | RM |
| Receptacle | RECP |
| Reference | REF |
| Regular | REG |
| Regulator | REG |
| Release | REL |
| Required | REQD |
| Right Hand | RH |
| Rivet | RIV |
| Rockwell Hardness | RH |
| Round | RD |

## S

|  |  |
|---|---|
| Schedule | SCH |
| Schematic | SCHEM |
| Screw | SCR |
| Screw Threads | |
| American National Coarse | NC |
| American National Fine | NF |
| American National Extra Fine | NEF |
| American National 8 Pitch | 8N |
| American Standard Taper Pipe | NPT |
| American Standard Straight Pipe | NPSC |
| American Standard Taper (Dryseal) | NPTF |
| American Standard Straight (Dryseal) | NPSF |
| Unified Screw Thread Coarse | UNC |
| Unified Screw Thread Fine | UNF |
| Unified Screw Thread Extra Fine | UNEF |
| Unified Screw Thread 8 Thread | 8UN |
| Section | SECT |
| Sequence | SEQ |
| Serial | SER |
| Serrate | SERR |
| Sheet | SH |
| Silver Solder | SILS |
| Soft Grind | SO GR |
| Solenoid | SOL |
| Special | SPL |
| Specification | SPEC |
| Spot Face | SF |
| Stainless Steel | SST |
| Steel | STL |
| Stock | STK |
| Symbol | SYM |
| Symmetrical | SYM |
| System | SYS |

## T

|  |  |
|---|---|
| Tabulate | TAB |
| Tangent | TAN |
| Tapping | TAP |
| Teeth | T |
| Tensile Strength | TS |
| Thick | THK |
| Thread | THD |
| Tolerance | TOL |
| Tool Steel | TS |
| Torque | TOR |
| Total Indicator Reading | TIR |
| True Involute Form | TIF |
| Tungsten | TU |
| Typical | TYP |

## V

|  |  |
|---|---|
| Vacuum | VAC |
| Variable | VAR |
| Vernier | VER |
| Vertical | VERT |
| Vibrate | VIB |
| Void | VD |
| Volt | V |
| Volume | VOL |

## W

|  |  |
|---|---|
| Washer | WASH |
| Watt | W |
| Weatherproof | WP |
| Welded | WLD |
| Wide, Width | W |
| Wrought | WRT |
| Wrought Iron | WI |

## Y

|  |  |
|---|---|
| Yield Point | YP |
| Yield Strength | YS |

# Decimal and Metric Equivalents

| INCHES: FRACTIONS | INCHES: DECIMALS | MILLI-METERS | INCHES: FRACTIONS | INCHES: DECIMALS | MILLI-METERS |
|---|---|---|---|---|---|
| | .00394 | .1 | 15/32 | .46875 | 11.9063 |
| | .00787 | .2 | | .47244 | 12.00 |
| | .01181 | .3 | 31/64 | .484375 | 12.3031 |
| 1/64 | .015625 | .3969 | 1/2 | .5000 | 12.70 |
| | .01575 | .4 | | .51181 | 13.00 |
| | .01969 | .5 | 33/64 | .515625 | 13.0969 |
| | .02362 | .6 | 17/32 | .53125 | 13.4938 |
| | .02756 | .7 | 35/64 | .546875 | 13.8907 |
| 1/32 | .03125 | .7938 | | .55118 | 14.00 |
| | .0315 | .8 | 9/16 | .5625 | 14.2875 |
| | .03543 | .9 | 37/64 | .578125 | 14.6844 |
| | .03937 | 1.00 | | .59055 | 15.00 |
| 3/64 | .046875 | 1.1906 | 19/32 | .59375 | 15.0813 |
| 1/16 | .0625 | 1.5875 | 39/64 | .609375 | 15.4782 |
| 5/64 | .078125 | 1.9844 | 5/8 | .625 | 15.875 |
| | .07874 | 2.00 | | .62992 | 16.00 |
| 3/32 | .09375 | 2.3813 | 41/64 | .640625 | 16.2719 |
| 7/64 | .109375 | 2.7781 | 21/32 | .65625 | 16.6688 |
| | .11811 | 3.00 | | .66929 | 17.00 |
| 1/8 | .125 | 3.175 | 43/64 | .671875 | 17.0657 |
| 9/64 | .140625 | 3.5719 | 11/16 | .6875 | 17.4625 |
| 5/32 | .15625 | 3.9688 | 45/64 | .703125 | 17.8594 |
| | .15748 | 4.00 | | .70866 | 18.00 |
| 11/64 | .171875 | 4.3656 | 23/32 | .71875 | 18.2563 |
| 3/16 | .1875 | 4.7625 | 47/64 | .734375 | 18.6532 |
| | .19685 | 5.00 | | .74803 | 19.00 |
| 13/64 | .203125 | 5.1594 | 3/4 | .7500 | 19.05 |
| 7/32 | .21875 | 5.5563 | 49/64 | .765625 | 19.4469 |
| 15/64 | .234375 | 5.9531 | 25/32 | .78125 | 19.8438 |
| | .23622 | 6.00 | | .7874 | 20.00 |
| 1/4 | .2500 | 6.35 | 51/64 | .796875 | 20.2407 |
| 17/64 | .265625 | 6.7469 | 13/16 | .8125 | 20.6375 |
| | .27559 | 7.00 | | .82677 | 21.00 |
| 9/32 | .28125 | 7.1438 | 53/64 | .828125 | 21.0344 |
| 19/64 | .296875 | 7.5406 | 27/32 | .84375 | 21.4313 |
| 5/16 | .3125 | 7.9375 | 55/64 | .859375 | 21.8282 |
| | .31496 | 8.00 | | .86614 | 22.00 |
| 21/64 | .328125 | 8.3344 | 7/8 | .875 | 22.225 |
| 11/32 | .34375 | 8.7313 | 57/64 | .890625 | 22.6219 |
| | .35433 | 9.00 | | .90551 | 23.00 |
| 23/64 | .359375 | 9.1281 | 29/32 | .90625 | 23.0188 |
| 3/8 | .375 | 9.525 | 59/64 | .921875 | 23.4157 |
| 25/64 | .390625 | 9.9219 | 15/16 | .9375 | 23.8125 |
| | .3937 | 10.00 | | .94488 | 24.00 |
| 13/32 | .40625 | 10.3188 | 61/64 | .953125 | 24.2094 |
| 27/64 | .421875 | 10.7156 | 31/32 | .96875 | 24.6063 |
| | .43307 | 11.00 | | .98425 | 25.00 |
| 7/16 | .4375 | 11.1125 | 63/64 | .984375 | 25.0032 |
| 29/64 | .453125 | 11.5094 | 1 | 1.0000 | 25.4001 |

# Drill Size Decimal Equivalents

## NUMBER AND LETTER DRILLS

| Drill No. | Frac | Dec. |
|---|---|---|
| 80 | — | .0135 |
| 79 | — | .0145 |
| | 1/64 | .0156 |
| 78 | — | .0160 |
| 77 | — | .0180 |
| 76 | — | .0200 |
| 75 | — | .0210 |
| 74 | — | .0225 |
| 73 | — | .0240 |
| 72 | — | .0250 |
| 71 | — | .0260 |
| 70 | — | .0280 |
| 69 | — | .0292 |
| 68 | — | .0310 |
| | 1/32 | .0313 |
| 67 | — | .0320 |
| 66 | — | .0330 |
| 65 | — | .0350 |
| 64 | — | .0360 |
| 63 | — | .0370 |
| 62 | — | .0380 |
| 61 | — | .0390 |
| 60 | — | .0400 |
| 59 | — | .0410 |
| 58 | — | .0420 |
| 57 | — | .0430 |
| 56 | — | .0465 |
| | 3/64 | .0469 |
| 55 | — | .0520 |
| 54 | — | .0550 |
| 53 | — | .0595 |
| | 1/16 | .0625 |
| 52 | — | .0635 |
| 51 | — | .0670 |
| 50 | — | .0700 |
| 49 | — | .0730 |
| 48 | — | .0760 |
| | 5/64 | .0781 |
| 47 | — | .0785 |
| 46 | — | .0810 |
| 45 | — | .0820 |
| 44 | — | .0860 |
| 43 | — | .0890 |
| 42 | — | .0935 |
| | 3/32 | .0938 |
| 41 | — | .0960 |
| 40 | — | .0980 |
| 39 | — | .0995 |
| 38 | — | .1015 |
| 37 | — | .1040 |
| 36 | — | .1065 |
| | 7/64 | .1094 |
| 35 | — | .1100 |
| 34 | — | .1110 |
| 33 | — | .1130 |
| 32 | — | .116 |
| 31 | — | .120 |
| | 1/8 | .125 |
| 30 | — | .129 |
| 29 | — | .136 |

| Drill No. | Frac | Dec. |
|---|---|---|
| | 9/64 | .140 |
| 28 | — | .141 |
| 27 | — | .144 |
| 26 | — | .147 |
| 25 | — | .150 |
| 24 | — | .152 |
| 23 | — | .154 |
| | 5/32 | .156 |
| 22 | — | .157 |
| 21 | — | .159 |
| 20 | — | .161 |
| 19 | — | .166 |
| 18 | — | .170 |
| | 11/64 | .172 |
| 17 | — | .173 |
| 16 | — | .177 |
| 15 | — | .180 |
| 14 | — | .182 |
| 13 | — | .185 |
| | 3/16 | .188 |
| 12 | — | .189 |
| 11 | — | .191 |
| 10 | — | .194 |
| 9 | — | .196 |
| 8 | — | .199 |
| 7 | — | .201 |
| | 13/64 | .203 |
| 6 | — | .204 |
| 5 | — | .206 |
| 4 | — | .209 |
| 3 | — | .213 |
| | 7/32 | .219 |
| 2 | — | .221 |
| 1 | — | .228 |
| A | — | .234 |
| | 15/64 | .234 |
| B | — | .238 |
| C | — | .242 |
| D | — | .246 |
| | 1/4 | .250 |
| E | — | .250 |
| F | — | .257 |
| G | — | .261 |
| | 17/64 | .266 |
| H | — | .266 |
| I | — | .272 |
| J | — | .277 |
| | 9/32 | .281 |
| K | — | .281 |
| L | — | .290 |
| M | — | .295 |
| | 19/64 | .297 |
| N | — | .302 |
| | 5/16 | .313 |
| O | — | .316 |
| P | — | .323 |
| | 21/64 | .328 |
| Q | — | .332 |
| R | — | .339 |
| | 11/32 | .344 |

| Drill No. | Frac | Dec. |
|---|---|---|
| S | — | .348 |
| T | — | .358 |
| | 23/64 | .359 |
| U | — | .368 |
| | 3/8 | .375 |
| V | — | .377 |
| W | — | .386 |
| | 25/64 | .391 |
| X | — | .397 |
| Y | — | .404 |
| | 13/32 | .406 |
| Z | — | .413 |
| | 27/64 | .422 |
| | 7/16 | .438 |
| | 29/64 | .453 |
| | 15/32 | .469 |
| | 31/64 | .484 |
| | 1/2 | .500 |
| | 33/64 | .516 |
| | 17/32 | .531 |
| | 35/64 | .547 |
| | 9/16 | .562 |
| | 37/64 | .578 |
| | 19/32 | .594 |
| | 39/64 | .609 |
| | 5/8 | .625 |
| | 41/64 | .641 |
| | 21/32 | .656 |
| | 43/64 | .672 |
| | 11/16 | .688 |
| | 45/64 | .703 |
| | 23/32 | .719 |
| | 47/64 | .734 |
| | 3/4 | .750 |
| | 49/64 | .766 |
| | 25/32 | .781 |
| | 51/64 | .797 |
| | 13/16 | .813 |
| | 53/64 | .828 |
| | 27/32 | .844 |
| | 55/64 | .859 |
| | 7/8 | .875 |
| | 57/64 | .891 |
| | 29/32 | .906 |
| | 59/64 | .922 |
| | 15/16 | .938 |
| | 61/64 | .953 |
| | 31/32 | .969 |
| | 63/64 | .984 |
| | 1 | 1.000 |

## METRIC DRILLS

| MM | DEC. | MM | DEC. | MM | DEC. | MM | DEC. |
|---|---|---|---|---|---|---|---|
| 1. | .0394 | 3.2 | .1260 | 6.3 | .2480 | 9.5 | .3740 |
| 1.05 | .0413 | 3.25 | .1280 | 6.4 | .2520 | 9.6 | .3780 |
| 1.1 | .0433 | 3.3 | .1299 | 6.5 | .2559 | 9.7 | .3819 |
| 1.15 | .0453 | 3.4 | .1339 | 6.6 | .2598 | 9.75 | .3839 |
| 1.2 | .0472 | 3.5 | .1378 | 6.7 | .2638 | 9.8 | .3858 |
| 1.25 | .0492 | 3.6 | .1417 | 6.75 | .2657 | 9.9 | .3898 |
| 1.3 | .0512 | 3.7 | .1457 | 6.8 | .2677 | 10. | .3937 |
| 1.35 | .0531 | 3.75 | .1476 | 6.9 | .2717 | 10.5 | .4134 |
| 1.4 | .0551 | 3.8 | .1496 | 7. | .2756 | 11. | .4331 |
| 1.45 | .0571 | 3.9 | .1535 | 7.1 | .2795 | 11.5 | .4528 |
| 1.5 | .0591 | 4. | .1575 | 7.2 | .2835 | 12. | .4724 |
| 1.55 | .0610 | 4.1 | .1614 | 7.25 | .2854 | 12.5 | .4921 |
| 1.6 | .0630 | 4.2 | .1654 | 7.3 | .2874 | 13. | .5118 |
| 1.65 | .0650 | 4.25 | .1673 | 7.4 | .2913 | 13.5 | .5315 |
| 1.7 | .0669 | 4.3 | .1693 | 7.5 | .2953 | 14. | .5512 |
| 1.75 | .0689 | 4.4 | .1732 | 7.6 | .2992 | 14.5 | .5709 |
| 1.8 | .0709 | 4.5 | .1772 | 7.7 | .3031 | 15. | .5906 |
| 1.85 | .0728 | 4.6 | .1811 | 7.75 | .3051 | 15.5 | .6102 |
| 1.9 | .0748 | 4.7 | .1850 | 7.8 | .3071 | 16. | .6299 |
| 1.95 | .0768 | 4.75 | .1870 | 7.9 | .3110 | 16.5 | .6496 |
| 2. | .0787 | 4.8 | .1890 | 8. | .3150 | 17. | .6693 |
| 2.05 | .0807 | 4.9 | .1929 | 8.1 | .3189 | 17.5 | .6890 |
| 2.1 | .0827 | 5. | .1968 | 8.2 | .3228 | 18. | .7087 |
| 2.15 | .0846 | 5.1 | .2008 | 8.25 | .3248 | 18.5 | .7283 |
| 2.2 | .0866 | 5.2 | .2047 | 8.3 | .3268 | 19. | .7480 |
| 2.25 | .0886 | 5.25 | .2067 | 8.4 | .3307 | 19.5 | .7677 |
| 2.3 | .0906 | 5.3 | .2087 | 8.5 | .3346 | 20. | .7874 |
| 2.35 | .0925 | 5.4 | .2126 | 8.6 | .3386 | 20.5 | .8071 |
| 2.4 | .0945 | 5.5 | .2165 | 8.7 | .3425 | 21. | .8268 |
| 2.45 | .0965 | 5.6 | .2205 | 8.75 | .3445 | 21.5 | .8465 |
| 2.5 | .0984 | 5.7 | .2244 | 8.8 | .3465 | 22. | .8661 |
| 2.6 | .1024 | 5.75 | .2264 | 8.9 | .3504 | 22.5 | .8858 |
| 2.7 | .1063 | 5.8 | .2283 | 9. | .3543 | 23. | .9055 |
| 2.75 | .1083 | 5.9 | .2323 | 9.1 | .3583 | 23.5 | .9252 |
| 2.8 | .1102 | 6. | .2362 | 9.2 | .3622 | 24. | .9449 |
| 2.9 | .1142 | 6.1 | .2402 | 9.25 | .3642 | 24.5 | .9646 |
| 3. | .1181 | 6.2 | .2441 | 9.3 | .3661 | 25. | .9843 |
| 3.1 | .1220 | 6.25 | .2461 | 9.4 | .3701 | | |

# Drill Size Decimal Equivalents

TAP DRILL SIZES FOR UNIFIED STANDARD SCREW THREADS

| Screw Thread | | Tap Drill |
|---|---|---|
| Major Diameter | Threads Per Inch | Size Or Number |
| 0 | 80 | 3/64 |
| 1 | 64 | 53 |
| | 72 | 53 |
| 2 | 56 | 50 |
| | 64 | 50 |
| 3 | 48 | 47 |
| | 56 | 45 |
| 4 | 40 | 43 |
| | 48 | 42 |
| 5 | 40 | 38 |
| | 44 | 37 |
| 6 | 32 | 36 |
| | 40 | 33 |
| 8 | 32 | 29 |
| | 36 | 29 |
| 10 | 24 | 25 |
| | 32 | 21 |
| 12 | 24 | 16 |
| | 28 | 14 |
| 1/4 | 20 | 7 |
| | 28 | 3 |
| 5/16 | 18 | F |
| | 24 | I |

| Screw Thread | | Tap Drill |
|---|---|---|
| Major Diameter | Threads Per Inch | Size Or Number |
| 3/8 | 16 | 5/16 |
| | 24 | Q |
| 7/16 | 14 | U |
| | 20 | 25/64 |
| 1/2 | 13 | 27/64 |
| | 20 | 29/64 |
| 9/16 | 12 | 31/64 |
| | 18 | 33/64 |
| 5/8 | 11 | 17/32 |
| | 18 | 37/64 |
| 3/4 | 10 | 21/32 |
| | 16 | 11/16 |
| 7/8 | 9 | 49/64 |
| | 14 | 13/16 |
| 1 | 8 | 7/8 |
| | 12 | 59/64 |
| 1 1/8 | 7 | 63/64 |
| | 12 | 1 3/64 |
| 1 1/4 | 7 | 1 7/64 |
| | 12 | 1 11/64 |
| 1 3/8 | 6 | 1 7/32 |
| | 12 | 1 19/64 |
| 1 1/2 | 6 | 1 11/32 |
| | 12 | 1 27/64 |

TAP DRILL SIZES FOR ISO METRIC THREADS

| Nominal Size mm | Series | | | |
|---|---|---|---|---|
| | Coarse | | Fine | |
| | Pitch mm | Tap Drill mm | Pitch mm | Tap Drill mm |
| 1.4 | 0.3 | 1.1 | – | – |
| 1.6 | 0.35 | 1.25 | – | – |
| 2 | 0.4 | 1.6 | – | – |
| 2.5 | 0.45 | 2.05 | – | – |
| 3 | 0.5 | 2.5 | – | – |
| 4 | 0.7 | 3.3 | – | – |
| 5 | 0.8 | 4.2 | – | – |
| 6 | 1.0 | 5.0 | – | – |
| 8 | 1.25 | 6.75 | 1 | 7.0 |

| Nominal Size mm | Series | | | |
|---|---|---|---|---|
| | Coarse | | Fine | |
| | Pitch mm | Tap Drill mm | Pitch mm | Tap Drill mm |
| 10 | 1.5 | 8.5 | 1.25 | 8.75 |
| 12 | 1.75 | 10.25 | 1.25 | 10.50 |
| 14 | 2 | 12.00 | 1.5 | 12.50 |
| 16 | 2 | 14.00 | 1.5 | 14.50 |
| 18 | 2.5 | 15.50 | 1.5 | 16.50 |
| 20 | 2.5 | 17.50 | 1.5 | 18.50 |
| 22 | 2.5 | 19.50 | 1.5 | 20.50 |
| 24 | 3 | 21.00 | 2 | 22.00 |
| 27 | 3 | 24.00 | 2 | 25.00 |

# Unified Standard Screw Thread Series

| SIZES | | BASIC MAJOR DIAMETER | THREADS PER INCH | | | | | | | | | | | SIZES |
|---|---|---|---|---|---|---|---|---|---|---|---|---|---|---|
| | | | Series with graded pitches | | | Series with constant pitches | | | | | | | | |
| Primary | Secondary | | Coarse UNC | Fine UNF | Extra fine UNEF | 4UN | 6UN | 8UN | 12UN | 16UN | 20UN | 28UN | 32UN | |
| 0 | | 0.0600 | — | 80 | — | — | — | — | — | — | — | — | — | 0 |
| | 1 | 0.0730 | 64 | 72 | — | — | — | — | — | — | — | — | — | 1 |
| 2 | | 0.0860 | 56 | 64 | — | — | — | — | — | — | — | — | — | 2 |
| | 3 | 0.0990 | 48 | 56 | — | — | — | — | — | — | — | — | — | 3 |
| 4 | | 0.1120 | 40 | 48 | — | — | — | — | — | — | — | — | — | 4 |
| 5 | | 0.1250 | 40 | 44 | — | — | — | — | — | — | — | — | — | 5 |
| 6 | | 0.1380 | 32 | 40 | — | — | — | — | — | — | — | — | UNC | 6 |
| 8 | | 0.1640 | 32 | 36 | — | — | — | — | — | — | — | — | UNC | 8 |
| 10 | | 0.1900 | 24 | 32 | — | — | — | — | — | — | — | — | UNF | 10 |
| | 12 | 0.2160 | 24 | 28 | 32 | — | — | — | — | — | — | UNF | UNEF | 12 |
| 1/4 | | 0.2500 | 20 | 28 | 32 | — | — | — | — | — | UNC | UNF | UNEF | 1/4 |
| 5/16 | | 0.3125 | 18 | 24 | 32 | — | — | — | — | — | 20 | 28 | UNEF | 5/16 |
| 3/8 | | 0.3750 | 16 | 24 | 32 | — | — | — | — | UNC | 20 | 28 | UNEF | 3/8 |
| 7/16 | | 0.4375 | 14 | 20 | 28 | — | — | — | — | 16 | UNF | UNEF | 32 | 7/16 |
| 1/2 | | 0.5000 | 13 | 20 | 28 | — | — | — | — | 16 | UNF | UNEF | 32 | 1/2 |
| 9/16 | | 0.5625 | 12 | 18 | 24 | — | — | — | UNC | 16 | 20 | 28 | 32 | 9/16 |
| 5/8 | | 0.6250 | 11 | 18 | 24 | — | — | — | 12 | 16 | 20 | 28 | 32 | 5/8 |
| | 11/16 | 0.6875 | — | — | 24 | — | — | — | 12 | 16 | 20 | 28 | 32 | 11/16 |
| 3/4 | | 0.7500 | 10 | 16 | 20 | — | — | — | 12 | UNF | UNEF | 28 | 32 | 3/4 |
| | 13/16 | 0.8125 | — | — | 20 | — | — | — | 12 | 16 | UNEF | 28 | 32 | 13/16 |
| 7/8 | | 0.8750 | 9 | 14 | 20 | — | — | — | 12 | 16 | UNEF | 28 | 32 | 7/8 |
| | 15/16 | 0.9375 | — | — | 20 | — | — | — | 12 | 16 | UNEF | 28 | 32 | 15/16 |
| 1 | | 1.0000 | 8 | 12 | 20 | — | — | UNC | UNF | 16 | UNEF | 28 | 32 | 1 |
| | 1 1/16 | 1.0625 | — | — | 18 | — | — | 8 | 12 | 16 | 20 | 28 | — | 1 1/16 |
| 1 1/8 | | 1.1250 | 7 | 12 | 18 | — | — | 8 | UNF | 16 | 20 | 28 | — | 1 1/8 |
| | 1 3/16 | 1.1875 | — | — | 18 | — | — | 8 | 12 | 16 | 20 | 28 | — | 1 3/16 |
| 1 1/4 | | 1.2500 | 7 | 12 | 18 | — | — | 8 | UNF | 16 | 20 | 28 | — | 1 1/4 |
| | 1 5/16 | 1.3125 | — | — | 18 | — | — | 8 | 12 | 16 | 20 | 28 | — | 1 5/16 |
| 1 3/8 | | 1.3750 | 6 | 12 | 18 | — | UNC | 8 | UNF | 16 | 20 | 28 | — | 1 3/8 |
| | 1 7/16 | 1.4375 | — | — | 18 | — | 6 | 8 | 12 | 16 | 20 | 28 | — | 1 7/16 |
| 1 1/2 | | 1.5000 | 6 | 12 | 18 | — | UNC | 8 | UNF | 16 | 20 | 28 | — | 1 1/2 |
| | 1 9/16 | 1.5625 | — | — | 18 | — | 6 | 8 | 12 | 16 | 20 | — | — | 1 9/16 |
| 1 5/8 | | 1.6250 | — | — | 18 | — | 6 | 8 | 12 | 16 | 20 | — | — | 1 5/8 |
| | 1 11/16 | 1.6875 | — | — | 18 | — | 6 | 8 | 12 | 16 | 20 | — | — | 1 11/16 |
| 1 3/4 | | 1.7500 | 5 | — | — | — | 6 | 8 | 12 | 16 | 20 | — | — | 1 3/4 |
| | 1 13/16 | 1.8125 | — | — | — | — | 6 | 8 | 12 | 16 | 20 | — | — | 1 13/16 |
| 1 7/8 | | 1.8750 | — | — | — | — | 6 | 8 | 12 | 16 | 20 | — | — | 1 7/8 |
| | 1 15/16 | 1.9375 | — | — | — | — | 6 | 8 | 12 | 16 | 20 | — | — | 1 15/16 |
| 2 | | 2.0000 | 4 1/2 | — | — | — | 6 | 8 | 12 | 16 | 20 | — | — | 2 |
| | 2 1/8 | 2.1250 | — | — | — | — | 6 | 8 | 12 | 16 | 20 | — | — | 2 1/8 |
| 2 1/4 | | 2.2500 | 4 1/2 | — | — | — | 6 | 8 | 12 | 16 | 20 | — | — | 2 1/4 |
| | 2 3/8 | 2.3750 | — | — | — | — | 6 | 8 | 12 | 16 | 20 | — | — | 2 3/8 |
| 2 1/2 | | 2.5000 | 4 | — | — | UNC | 6 | 8 | 12 | 16 | 20 | — | — | 2 1/2 |
| | 2 5/8 | 2.6250 | — | — | — | 4 | 6 | 8 | 12 | 16 | 20 | — | — | 2 5/8 |
| 2 3/4 | | 2.7500 | 4 | — | — | UNC | 6 | 8 | 12 | 16 | 20 | — | — | 2 3/4 |
| | 2 7/8 | 2.8750 | — | — | — | 4 | 6 | 8 | 12 | 16 | 20 | — | — | 2 7/8 |
| 3 | | 3.0000 | 4 | — | — | UNC | 6 | 8 | 12 | 16 | 20 | — | — | 3 |
| | 3 1/8 | 3.1250 | — | — | — | 4 | 6 | 8 | 12 | 16 | — | — | — | 3 1/8 |
| 3 1/4 | | 3.2500 | 4 | — | — | UNC | 6 | 8 | 12 | 16 | — | — | — | 3 1/4 |
| | 3 3/8 | 3.3750 | — | — | — | 4 | 6 | 8 | 12 | 16 | — | — | — | 3 3/8 |
| 3 1/2 | | 3.5000 | 4 | — | — | UNC | 6 | 8 | 12 | 16 | — | — | — | 3 1/2 |
| | 3 5/8 | 3.6250 | — | — | — | 4 | 6 | 8 | 12 | 16 | — | — | — | 3 5/8 |
| 3 3/4 | | 3.7500 | 4 | — | — | UNC | 6 | 8 | 12 | 16 | — | — | — | 3 3/4 |
| | 3 7/8 | 3.8750 | — | — | — | 4 | 6 | 8 | 12 | 16 | — | — | — | 3 7/8 |
| 4 | | 4.0000 | 4 | — | — | UNC | 6 | 8 | 12 | 16 | — | — | — | 4 |
| | 4 1/8 | 4.1250 | — | — | — | 4 | 6 | 8 | 12 | 16 | — | — | — | 4 1/8 |
| 4 1/4 | | 4.2500 | — | — | — | 4 | 6 | 8 | 12 | 16 | — | — | — | 4 1/4 |
| | 4 3/8 | 4.3750 | — | — | — | 4 | 6 | 8 | 12 | 16 | — | — | — | 4 3/8 |
| 4 1/2 | | 4.5000 | — | — | — | 4 | 6 | 8 | 12 | 16 | — | — | — | 4 1/2 |
| | 4 5/8 | 4.6250 | — | — | — | 4 | 6 | 8 | 12 | 16 | — | — | — | 4 5/8 |
| 4 3/4 | | 4.7500 | — | — | — | 4 | 6 | 8 | 12 | 16 | — | — | — | 4 3/4 |
| | 4 7/8 | 4.8750 | — | — | — | 4 | 6 | 8 | 12 | 16 | — | — | — | 4 7/8 |
| 5 | | 5.0000 | — | — | — | 4 | 6 | 8 | 12 | 16 | — | — | — | 5 |
| | 5 1/8 | 5.1250 | — | — | — | 4 | 6 | 8 | 12 | 16 | — | — | — | 5 1/8 |
| 5 1/4 | | 5.2500 | — | — | — | 4 | 6 | 8 | 12 | 16 | — | — | — | 5 1/4 |
| | 5 3/8 | 5.3750 | — | — | — | 4 | 6 | 8 | 12 | 16 | — | — | — | 5 3/8 |
| 5 1/2 | | 5.5000 | — | — | — | 4 | 6 | 8 | 12 | 16 | — | — | — | 5 1/2 |
| | 5 5/8 | 5.6250 | — | — | — | 4 | 6 | 8 | 12 | 16 | — | — | — | 5 5/8 |
| 5 3/4 | | 5.7500 | — | — | — | 4 | 6 | 8 | 12 | 16 | — | — | — | 5 3/4 |
| | 5 7/8 | 5.8750 | — | — | — | 4 | 6 | 8 | 12 | 16 | — | — | — | 5 7/8 |
| 6 | | 6.0000 | — | — | — | 4 | 6 | 8 | 12 | 16 | — | — | — | 6 |

# ISO Metric Screw Thread Standard Series

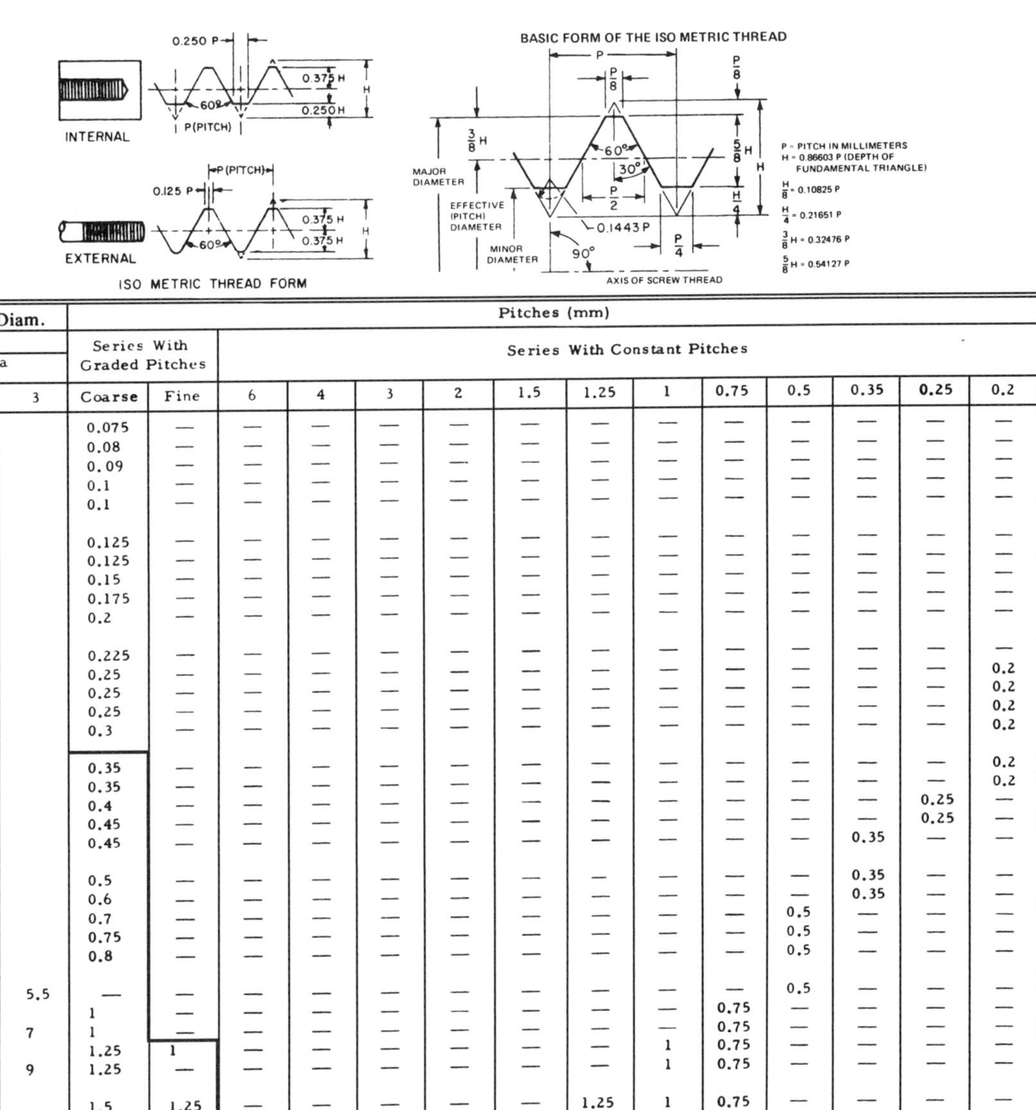


| Nominal Size Diam. (mm) Column[a] | | | Pitches (mm) Series With Graded Pitches | | Series With Constant Pitches | | | | | | | | | | | | Nominal Size Diam. (mm) |
|---|---|---|---|---|---|---|---|---|---|---|---|---|---|---|---|---|---|
| 1 | 2 | 3 | Coarse | Fine | 6 | 4 | 3 | 2 | 1.5 | 1.25 | 1 | 0.75 | 0.5 | 0.35 | 0.25 | 0.2 | |
| 0.25 | | | 0.075 | — | — | — | — | — | — | — | — | — | — | — | — | — | 0.25 |
| 0.3 | | | 0.08 | — | — | — | — | — | — | — | — | — | — | — | — | — | 0.3 |
| | 0.35 | | 0.09 | — | — | — | — | — | — | — | — | — | — | — | — | — | 0.35 |
| 0.4 | | | 0.1 | — | — | — | — | — | — | — | — | — | — | — | — | — | 0.4 |
| | 0.45 | | 0.1 | — | — | — | — | — | — | — | — | — | — | — | — | — | 0.45 |
| 0.5 | | | 0.125 | — | — | — | — | — | — | — | — | — | — | — | — | — | 0.5 |
| | 0.55 | | 0.125 | — | — | — | — | — | — | — | — | — | — | — | — | — | 0.55 |
| 0.6 | | | 0.15 | — | — | — | — | — | — | — | — | — | — | — | — | — | 0.6 |
| | 0.7 | | 0.175 | — | — | — | — | — | — | — | — | — | — | — | — | — | 0.7 |
| 0.8 | | | 0.2 | — | — | — | — | — | — | — | — | — | — | — | — | — | 0.8 |
| | 0.9 | | 0.225 | — | — | — | — | — | — | — | — | — | — | — | — | — | 0.9 |
| 1 | | | 0.25 | — | — | — | — | — | — | — | — | — | — | — | — | 0.2 | 1 |
| | 1.1 | | 0.25 | — | — | — | — | — | — | — | — | — | — | — | — | 0.2 | 1.1 |
| 1.2 | | | 0.25 | — | — | — | — | — | — | — | — | — | — | — | — | 0.2 | 1.2 |
| | 1.4 | | 0.3 | — | — | — | — | — | — | — | — | — | — | — | — | 0.2 | 1.4 |
| 1.6 | | | 0.35 | — | — | — | — | — | — | — | — | — | — | — | — | 0.2 | 1.6 |
| | 1.8 | | 0.35 | — | — | — | — | — | — | — | — | — | — | — | — | 0.2 | 1.8 |
| 2 | | | 0.4 | — | — | — | — | — | — | — | — | — | — | — | 0.25 | — | 2 |
| | 2.2 | | 0.45 | — | — | — | — | — | — | — | — | — | — | — | 0.25 | — | 2.2 |
| 2.5 | | | 0.45 | — | — | — | — | — | — | — | — | — | — | 0.35 | — | — | 2.5 |
| 3 | | | 0.5 | — | — | — | — | — | — | — | — | — | — | 0.35 | — | — | 3 |
| | 3.5 | | 0.6 | — | — | — | — | — | — | — | — | — | — | 0.35 | — | — | 3.5 |
| 4 | | | 0.7 | — | — | — | — | — | — | — | — | — | 0.5 | — | — | — | 4 |
| | 4.5 | | 0.75 | — | — | — | — | — | — | — | — | — | 0.5 | — | — | — | 4.5 |
| 5 | | | 0.8 | — | — | — | — | — | — | — | — | — | 0.5 | — | — | — | 5 |
| | | 5.5 | — | — | — | — | — | — | — | — | — | — | 0.5 | — | — | — | 5.5 |
| 6 | | | 1 | — | — | — | — | — | — | — | — | 0.75 | — | — | — | — | 6 |
| | | 7 | 1 | — | — | — | — | — | — | — | — | 0.75 | — | — | — | — | 7 |
| 8 | | | 1.25 | 1 | — | — | — | — | — | — | 1 | 0.75 | — | — | — | — | 8 |
| | | 9 | 1.25 | — | — | — | — | — | — | — | 1 | 0.75 | — | — | — | — | 9 |
| 10 | | | 1.5 | 1.25 | — | — | — | — | — | 1.25 | 1 | 0.75 | — | — | — | — | 10 |
| | | 11 | 1.5 | — | — | — | — | — | — | — | 1 | 0.75 | — | — | — | — | 11 |
| 12 | | | 1.75 | 1.25 | — | — | — | — | 1.5 | 1.25 | 1 | — | — | — | — | — | 12 |
| | 14 | | 2 | 1.5 | — | — | — | — | 1.5 | 1.25[b] | 1 | — | — | — | — | — | 14 |
| | | 15 | — | — | — | — | — | — | 1.5 | — | 1 | — | — | — | — | — | 15 |
| 16 | | | 2 | 1.5 | — | — | — | — | 1.5 | — | 1 | — | — | — | — | — | 16 |
| | | 17 | — | — | — | — | — | — | 1.5 | — | 1 | — | — | — | — | — | 17 |
| | 18 | | 2.5 | 1.5 | — | — | — | 2 | 1.5 | — | 1 | — | — | — | — | — | 18 |
| 20 | | | 2.5 | 1.5 | — | — | — | 2 | 1.5 | — | 1 | — | — | — | — | — | 20 |
| | **22** | | 2.5 | 1.5 | — | — | — | 2 | 1.5 | — | 1 | — | — | — | — | — | 22 |
| 24 | | | 3 | 2 | — | — | — | 2 | 1.5 | — | 1 | — | — | — | — | — | 24 |
| | | 25 | — | — | — | — | — | 2 | 1.5 | — | 1 | — | — | — | — | — | 25 |
| | | 26 | — | — | — | — | — | — | 1.5 | — | 1 | — | — | — | — | — | 26 |
| | 27 | | 3 | 2 | — | — | — | 2 | 1.5 | — | 1 | — | — | — | — | — | 27 |
| | | 28 | — | — | — | — | — | 2 | 1.5 | — | 1 | — | — | — | — | — | 28 |
| 30 | | | 3.5 | 2 | — | — | (3) | 2 | 1.5 | — | 1 | — | — | — | — | — | 30 |
| | | 32 | — | — | — | — | — | 2 | 1.5 | — | — | — | — | — | — | — | 32 |
| | 33 | | 3.5 | 2 | — | — | (3) | 2 | 1.5 | — | — | — | — | — | — | — | 33 |
| | | 35[c] | — | — | — | — | — | — | 1.5 | — | — | — | — | — | — | — | 35[c] |
| 36 | | | 4 | 3 | — | — | — | 2 | 1.5 | — | — | — | — | — | — | — | 36 |
| | | 38 | — | — | — | — | — | — | 1.5 | — | — | — | — | — | — | — | 38 |
| | 39 | | 4 | 3 | — | — | — | 2 | 1.5 | — | — | — | — | — | — | — | 39 |
| | | 40 | — | — | — | — | 3 | 2 | 1.5 | — | — | — | — | — | — | — | 40 |
| 42 | | | 4.5 | 3 | — | 4 | 3 | 2 | 1.5 | — | — | — | — | — | — | — | 42 |
| | 45 | | 4.5 | 3 | — | 4 | 3 | 2 | 1.5 | — | — | — | — | — | — | — | 45 |

a Thread diameter should be selected from columns 1, 2 or 3; with preference being given in that order.
b Pitch 1.25 mm in combination with diameter 14 mm has been included for spark plug applications.
c Diameter 35 mm has been included for bearing locknut applications.
The use of pitches shown in parentheses should be avoided wherever possible.
The pitches enclosed in the bold frame, together with the corresponding nominal diameters in Columns 1 and 2, are those combinations which have been established by ISO Recommendations as a selected "coarse" and "fine" series for commercial fasteners. Sizes 0.25 mm through 1.4 mm are covered in ISO Recommendation R 68 and, except for the 0.25 mm size, in AN Standard ANSI B1.10. (ANSI)

## Sheet Metal and Wire Gage Designation

| GAGE NUMBER | AMERICAN OR BROWN & SHARPE'S A.W.G. OR B. & S. | BIRMINGHAM OR STUBS WIRE B.W.G. | WASHBURN & MOEN OR AMERICAN S.W.G. | UNITED STATES STANDARD | MANUFACTURERS' STANDARD FOR SHEET STEEL | GAGE NUMBER |
|---|---|---|---|---|---|---|
| 0000000 | — | — | .4900 | .500 | — | 0000000 |
| 000000 | .5800 | — | .4615 | .469 | — | 000000 |
| 00000 | .5165 | — | .4305 | .438 | — | 00000 |
| 0000 | .4600 | .454 | .3938 | .406 | — | 0000 |
| 000 | .4096 | .425 | .3625 | .375 | — | 000 |
| 00 | .3648 | .380 | .3310 | .344 | — | 00 |
| 0 | .3249 | .340 | .3065 | .312 | — | 0 |
| 1 | .2893 | .300 | .2830 | .281 | — | 1 |
| 2 | .2576 | .284 | .2625 | .266 | — | 2 |
| 3 | .2294 | .259 | .2437 | .250 | .2391 | 3 |
| 4 | .2043 | .238 | .2253 | .234 | .2242 | 4 |
| 5 | .1819 | .220 | .2070 | .219 | .2092 | 5 |
| 6 | .1620 | .203 | .1920 | .203 | .1943 | 6 |
| 7 | .1443 | .180 | .1770 | .188 | .1793 | 7 |
| 8 | .1285 | .165 | .1620 | .172 | .1644 | 8 |
| 9 | .1144 | .148 | .1483 | .156 | .1495 | 9 |
| 10 | .1019 | .134 | .1350 | .141 | .1345 | 10 |
| 11 | .0907 | .120 | .1205 | .125 | .1196 | 11 |
| 12 | .0808 | .109 | .1055 | .109 | .1046 | 12 |
| 13 | .0720 | .095 | .0915 | .0938 | .0897 | 13 |
| 14 | .0642 | .083 | .0800 | .0781 | .0747 | 14 |
| 15 | .0571 | .072 | .0720 | .0703 | .0673 | 15 |
| 16 | .0508 | .065 | .0625 | .0625 | .0598 | 16 |
| 17 | .0453 | .058 | .0540 | .0562 | .0538 | 17 |
| 18 | .0403 | .049 | .0475 | .0500 | .0478 | 18 |
| 19 | .0359 | .042 | .0410 | .0438 | .0418 | 19 |
| 20 | .0320 | .035 | .0348 | .0375 | .0359 | 20 |
| 21 | .0285 | .032 | .0317 | .0344 | .0329 | 21 |
| 22 | .0253 | .028 | .0286 | .0312 | .0299 | 22 |
| 23 | .0226 | .025 | .0258 | .0281 | .0269 | 23 |
| 24 | .0201 | .022 | .0230 | .0250 | .0239 | 24 |
| 25 | .0179 | .020 | .0204 | .0162 | .0209 | 25 |
| 26 | .0159 | .018 | .0181 | .0188 | .0179 | 26 |
| 27 | .0142 | .016 | .0173 | .0172 | .0164 | 27 |
| 28 | .0126 | .014 | .0219 | .0156 | .0149 | 28 |
| 29 | .0113 | .013 | .0150 | .0141 | .0135 | 29 |
| 30 | .0100 | .012 | .0140 | .0125 | .0120 | 30 |
| 31 | .0089 | .010 | .0132 | .0109 | .0105 | 31 |
| 32 | .0080 | .009 | .0128 | .0102 | .0097 | 32 |
| 33 | .0071 | .008 | .0118 | .00938 | .0090 | 33 |
| 34 | .0063 | .007 | .0104 | .00859 | .0082 | 34 |
| 35 | .0056 | .005 | .0095 | .00781 | .0075 | 35 |
| 36 | .0050 | .004 | .0090 | .00703 | .0067 | 36 |
| 37 | .0045 | — | .0085 | .00624 | .0064 | 37 |
| 38 | .0040 | — | .0080 | .00625 | .0060 | 38 |
| 39 | .0035 | — | .0075 | — | — | 39 |
| 40 | .0031 | — | .0070 | — | — | 40 |
| 41 | .0028 | — | .0066 | — | — | 41 |
| 42 | .0025 | — | .0062 | — | — | 42 |
| 43 | .0022 | — | .0060 | — | — | 43 |
| 44 | .0020 | — | .0058 | — | — | 44 |
| 45 | .0018 | — | .0055 | — | — | 45 |
| 46 | .0016 | — | .0052 | — | — | 46 |
| 47 | .0014 | — | .0050 | — | — | 47 |
| 48 | .0012 | — | .0048 | — | — | 48 |

# Precision Sheet Metal Setback Chart

Material Thickness

| 90° Bend Radius | .016 | .020 | .025 | .032 | .040 | .051 | .064 | .072 | .078 | .081 | .091 | .102 | .125 | .129 | .156 | .162 | .187 | .250 |
|---|---|---|---|---|---|---|---|---|---|---|---|---|---|---|---|---|---|---|
| 1/32 | .034 | .039 | .046 | .05 | .065 | .081 | .102 | .113 | .121 | .125 | .139 | | | | | | | |
| 3/64 | .041 | .046 | .053 | .062 | .072 | .090 | .108 | .119 | .127 | .131 | .145 | | | | | | | |
| 1/16 | .048 | .053 | .059 | .068 | .079 | .093 | .110 | .122 | .134 | .138 | .152 | | | | | | | |
| 5/64 | .054 | .060 | .066 | .075 | .086 | .100 | .117 | .127 | .138 | .144 | .158 | | | | | | | |
| 3/32 | .061 | .066 | .073 | .082 | .092 | .107 | .124 | .134 | .142 | .146 | .160 | | | | | | | |
| 7/64 | .068 | .073 | .080 | .08 | .099 | .113 | .130 | .141 | .148 | .153 | .167 | .181 | | | | | | |
| 1/8 | .075 | .080 | .086 | .095 | .106 | .120 | .137 | .147 | .155 | .159 | .172 | .186 | .216 | .221 | | | | |
| 9/64 | .081 | .087 | .093 | .102 | .113 | .127 | .144 | .154 | .162 | .166 | .179 | .193 | .223 | .228 | .263 | | | |
| 5/32 | .088 | .093 | .100 | .109 | .119 | .134 | .150 | .161 | .169 | .173 | .186 | .200 | .230 | .235 | .270 | .278 | | |
| 11/64 | .095 | .100 | .107 | .116 | .126 | .140 | .157 | .168 | .175 | .179 | .192 | .207 | .236 | .242 | .277 | .284 | .317 | |
| 3/16 | .102 | .107 | .113 | .122 | .133 | .147 | .164 | .174 | .182 | .186 | .199 | .213 | .243 | .248 | .283 | .291 | .324 | .405 |
| 13/64 | .108 | .114 | .120 | .129 | .140 | .154 | .171 | .181 | .189 | .193 | .206 | .220 | .250 | .255 | .290 | .298 | .330 | .412 |
| 7/32 | .115 | .120 | .127 | .136 | .146 | .161 | .177 | .188 | .196 | .199 | .212 | .227 | .257 | .262 | .297 | .305 | .337 | .419 |
| 15/64 | .122 | .127 | .134 | .143 | .153 | .167 | .184 | .195 | .202 | .206 | .219 | .233 | .263 | .269 | .304 | .311 | .344 | .426 |
| 1/4 | .129 | .134 | .140 | .149 | .160 | .174 | .191 | .201 | .209 | .213 | .226 | .240 | .270 | .275 | .310 | .318 | .351 | .432 |
| 17/64 | .135 | .141 | .147 | .156 | .166 | .181 | .198 | .208 | .216 | .220 | .233 | .247 | .277 | .282 | .317 | .325 | .357 | .439 |
| 9/32 | .142 | .147 | .154 | .163 | .173 | .187 | .204 | .215 | .223 | .226 | .239 | .254 | .284 | .289 | .324 | .332 | .364 | .446 |



Developed Length = X + Y – Z

Z = Setback Allowance from the chart.

# Geometric Dimensioning and Tolerancing Symbols

| SYMBOL FOR: | ASME Y14.5 | ISO |
|---|---|---|
| STRAIGHTNESS | ⏤ | ⏤ |
| FLATNESS | ⏥ | ⏥ |
| CIRCULARITY | ○ | ○ |
| CYLINDRICITY | ⌭ | ⌭ |
| PROFILE OF A LINE | ⌒ | ⌒ |
| PROFILE OF A SURFACE | ⌓ | ⌓ |
| ALL AROUND | ←○— | ←○— (proposed) |
| ANGULARITY | ∠ | ∠ |
| PERPENDICULARITY | ⊥ | ⊥ |
| PARALLELISM | // | // |
| POSITION | ⌖ | ⌖ |
| CONCENTRICITY | ◎ | ◎ |
| SYMMETRY | ⌯ | ⌯ |
| CIRCULAR RUNOUT | ↗ | ↗ |
| TOTAL RUNOUT | ⌰ | ⌰ |
| AT MAXIMUM MATERIAL CONDITION | Ⓜ | Ⓜ |
| AT LEAST MATERIAL CONDITION | Ⓛ | Ⓛ |
| REGARDLESS OF FEATURE SIZE | NONE | NONE |
| PROJECTED TOLERANCE ZONE | Ⓟ | Ⓟ |
| TANGENT PLANE | Ⓣ | NONE |
| FREE STATE | Ⓕ | Ⓕ |
| DIAMETER | Ø | Ø |
| BASIC DIMENSION | [50] | [50] |
| REFERENCE DIMENSION | (50) | (50) |
| DATUM FEATURE | [A] | or [A] |
| DIMENSION ORIGIN | ⌀→ | ⌀→ |
| FEATURE CONTROL FRAME | ⌖ \| Ø 0.5 Ⓜ \| A \| B \| C | ⌖ \| Ø 0.5 Ⓜ \| A \| B \| C |
| CONICAL TAPER | ⌲ | ⌲ |
| SLOPE | ⌳ | ⌳ |
| COUNTERBORE/SPOTFACE | ⌴ | NONE |
| COUNTERSINK | ⌵ | NONE |
| DEPTH/DEEP | ↧ | NONE |
| SQUARE | □ | □ |
| DIMENSION NOT TO SCALE | 15 | 15 |
| NUMBER OF TIMES/PLACES | 8X | 8X |
| ARC LENGTH | ⌒105 | ⌒105 |
| RADIUS | R | R |
| SPHERICAL RADIUS | SR | SR |
| SPHERICAL DIAMETER | SØ | SØ |
| CONTROLLED RADIUS | CR | NONE |
| BETWEEN | ←→ | NONE |
| STATISTICAL TOLERANCE | ⟨ST⟩ | NONE |
| DATUM TARGET | Ø6/A1 or A1 —Ø6 | Ø6/A1 or A1 —Ø6 |
| TARGET POINT | × | × |

# Symbols for Materials in Section

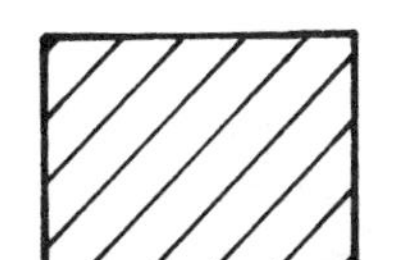

Cast iron and malleable iron. Also for use of all materials.

Steel

Brass, bronze and compositions

White metal, zinc, lead, babbitt and alloys

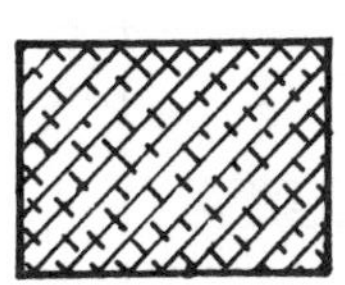

Magnesium, aluminum, and aluminum alloys

Rubber, plastic, electrical insulation

Cork, felt, fabric, leather, fiber

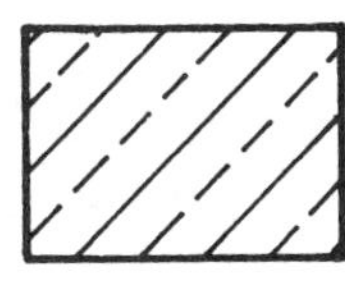

Firebrick and refractory material

Electric windings, electromagnets, resistance, etc.

Marble, slate, glass, porcelain, etc.

Water and other liquids

Across grain
with grain
Wood

# Standard Welding Symbols

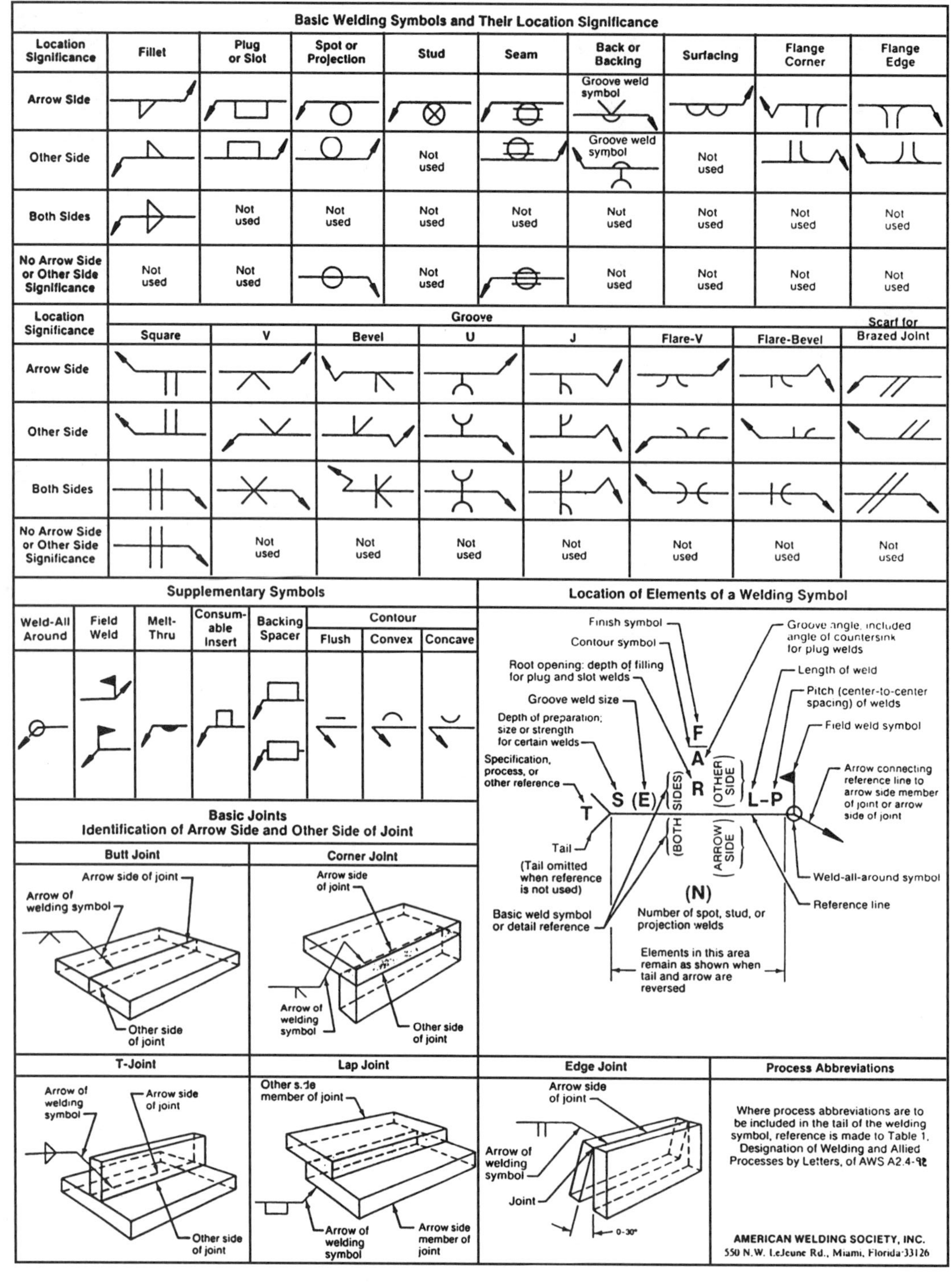

# Standard Welding Symbols

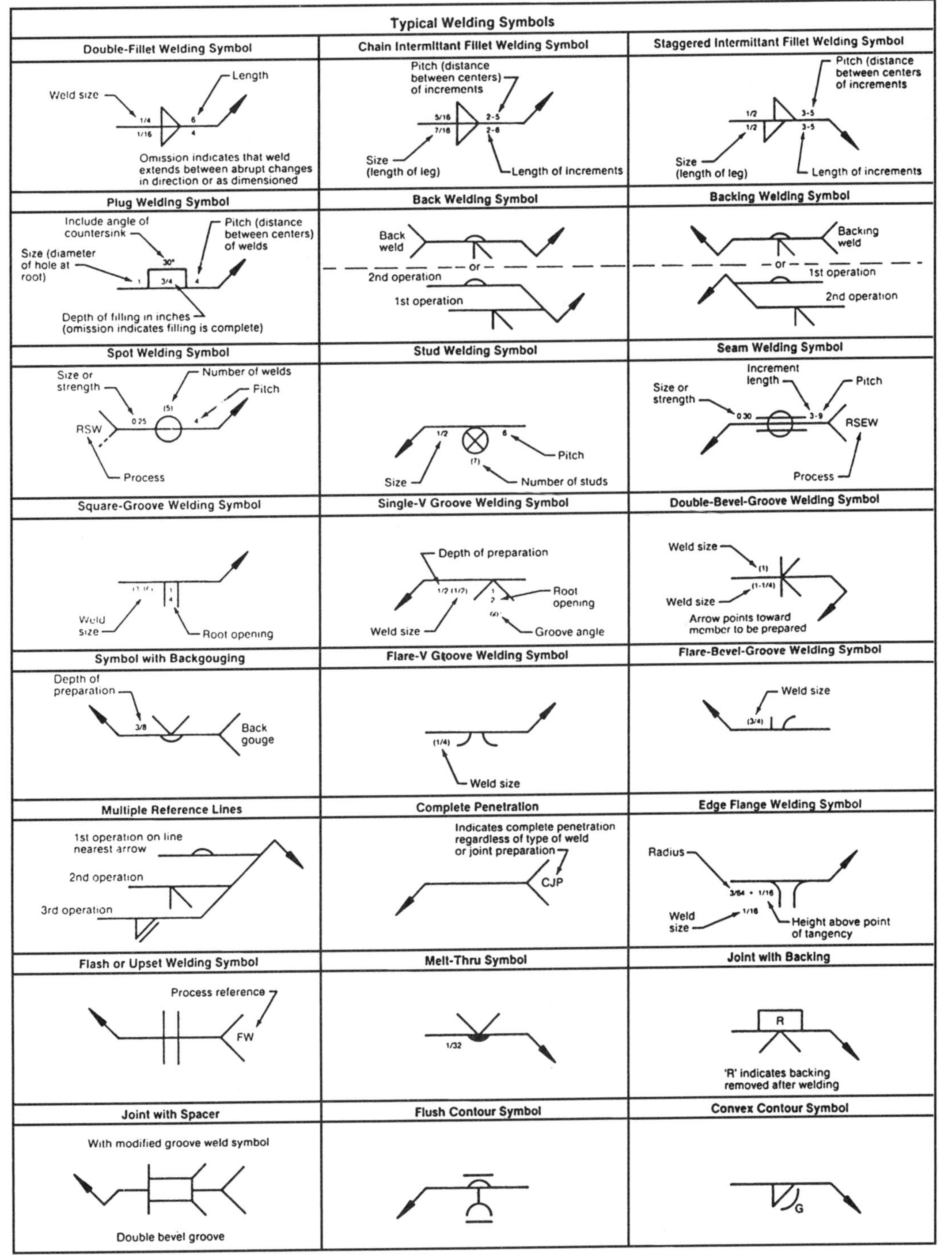

# Designation of Welding and Allied Processes by Letters

| Welding and Allied Processes | Letter Designation |
|---|---|
| adhesive bonding | ABD |
| arc welding | AW |
| atomic hydrogen welding | AHW |
| bare metal arc welding | BMAW |
| carbon arc welding | CAW |
| gas | CAW-G |
| shielded | CAW-S |
| twin | CAW-T |
| electrogas | EGW |
| flux cored arc welding | FCAW |
| gas metal arc welding | GMAW |
| pulsed arc | GMAW-P |
| short circuiting arc | GMAW-S |
| gas tungsten arc welding | GTAW |
| pulsed arc | GTAW-P |
| plasma arc welding | PAW |
| shielded metal arc welding | SMAW |
| stud arc welding | SW |
| submerged arc welding | SAW |
| series | SAW-S |
| brazing | B |
| arc brazing | AB |
| block brazing | BB |
| carbon arc brazing | CAB |
| diffusion brazing | DFB |
| dip brazing | DB |
| flow brazing | FLB |
| furnace brazing | FB |
| induction brazing | IB |
| infrared brazing | IRB |
| resistance brazing | RB |
| torch brazing | TB |
| other welding processes | |
| electron beam welding | EBW |
| high vacuum | EBW-HV |
| medium vacuum | EBW-MV |
| nonvacuum | EBW-NV |
| electroslag welding | ESW |
| flow welding | FLOW |
| induction welding | IW |
| laser beam welding | LBW |
| thermit welding | TW |
| oxyfuel gas welding | OFW |
| air acetylene welding | AAW |
| oxyacetylene welding | OAW |
| oxyhydrogen welding | OHW |
| pressure gas welding | PGW |
| resistance welding | RW |
| flash welding | FW |
| percussion welding | PEW |
| projection welding | PW |
| resistance seam welding | RSEW |
| high frequency | RSEW-HF |
| induction | RSEW-I |

| Welding and Allied Processes | Letter Designation |
|---|---|
| resistance spot welding | RSW |
| upset welding | UW |
| high frequency | UW-HF |
| induction | UW-I |
| soldering | S |
| dip soldering | DS |
| furnace soldering | FS |
| induction soldering | IS |
| infrared soldering | IRS |
| iron soldering | INS |
| resistance soldering | RS |
| torch soldering | TS |
| wave soldering | WS |
| solid-state welding | SSW |
| coextrusion welding | CEW |
| cold welding | CW |
| diffusion welding | DFW |
| explosion welding | EXW |
| forge welding | FOW |
| friction welding | FRW |
| hot pressure welding | HPW |
| roll welding | ROW |
| ultrasonic welding | USW |
| thermal cutting | TC |
| arc cutting | AC |
| air carbon arc cutting | AAC |
| carbon arc cutting | CAC |
| gas metal arc cutting | GMAC |
| gas tungsten arc cutting | GTAC |
| metal arc cutting | MAC |
| plasma arc cutting | PAC |
| shielded metal arc cutting | SMAC |
| electron beam cutting | EBC |
| laser beam cutting | LBC |
| air | LBC-A |
| evaporative | LBC-EV |
| inert gas | LBC-IG |
| oxygen | LBC-O |
| oxygen cutting | OC |
| chemical flux cutting | FOC |
| metal powder cutting | POC |
| oxyfuel gas cutting | OFC |
| oxyacetylene cutting | OFC-A |
| oxyhydrogen cutting | OFC-H |
| oxynatural gas cutting | OFC-N |
| oxypropane cutting | OFC-P |
| oxygen arc cutting | AOC |
| oxygen lance cutting | LOC |
| thermal spraying | THSP |
| arc spraying | ASP |
| flame spraying | FLSP |
| plasma spraying | PSP |

# Standard Fluid Power Graphic Symbols

THE SYMBOLS SHOWN CONFORM TO THE AMERICAN NATIONAL STANDARDS INSTITUTE (ANSI) SPECIFICIATIONS. BASIC SYMBOLS CAN BE COMBINED IN ANY COMBINATION. NO ATTEMPT IS MADE TO SHOW ALL COMBINATIONS.

| LINES AND LINE FUNCTIONS | |
|---|---|
| LINE, WORKING | |
| LINE, PILOT (L>20W) | |
| LINE, DRAIN (L<5W) | |
| CONNECTOR | |
| LINE, FLEXIBLE | |
| LINE, JOINING | |
| LINE, PASSING | |
| DIRECTION OF FLOW, HYDRAULIC PNEUMATIC | |
| LINE TO RESERVOIR ABOVE FLUID LEVEL BELOW FLUID LEVEL | |
| LINE TO VENTED MANIFOLD | |
| PLUG OR PLUGGED CONNECTION | |
| RESTRICTION, FIXED | |
| RESTRICITION, VARIABLE | |
| **METHODS OF OPERATION** | |
| PRESSURE COMPENSATOR | |
| DETENT | |
| MANUAL | |
| MECHANICAL | |
| PEDAL OR TREADLE | |
| PUSH BUTTON | |

| PUMPS | |
|---|---|
| PUMP, SINGLE FIXED DISPLACEMENT | |
| PUMP, SINGLE VARIABLE DISPLACEMENT | |
| **MOTORS AND CYLINDERS** | |
| MOTOR, ROTARY, FIXED DISPLACEMENT | |
| MOTOR, ROTARY VARIABLE DISPLACEMENT | |
| MOTOR, OSCILLATING | |
| CYLINDER, SINGLE ACTING | |
| CYLINDER, DOUBLE ACTING | |
| CYLINDER, DIFFERENTIAL ROD | |
| CYLINDER, DOUBLE END ROD | |
| CYLINDER, CUSHIONS BOTH ENDS | |
| **METHODS OF OPERATION** | |
| LEVER | |
| PILOT PRESSURE | |
| SOLENOID | |
| SOLENOID CONTROLLED, PILOT PRESSURE OPERATED | |
| SPRING | |
| SERVO | |

(Provided by Vickers Industrial Division.)

# Standard Fluid Power Graphic Symbols

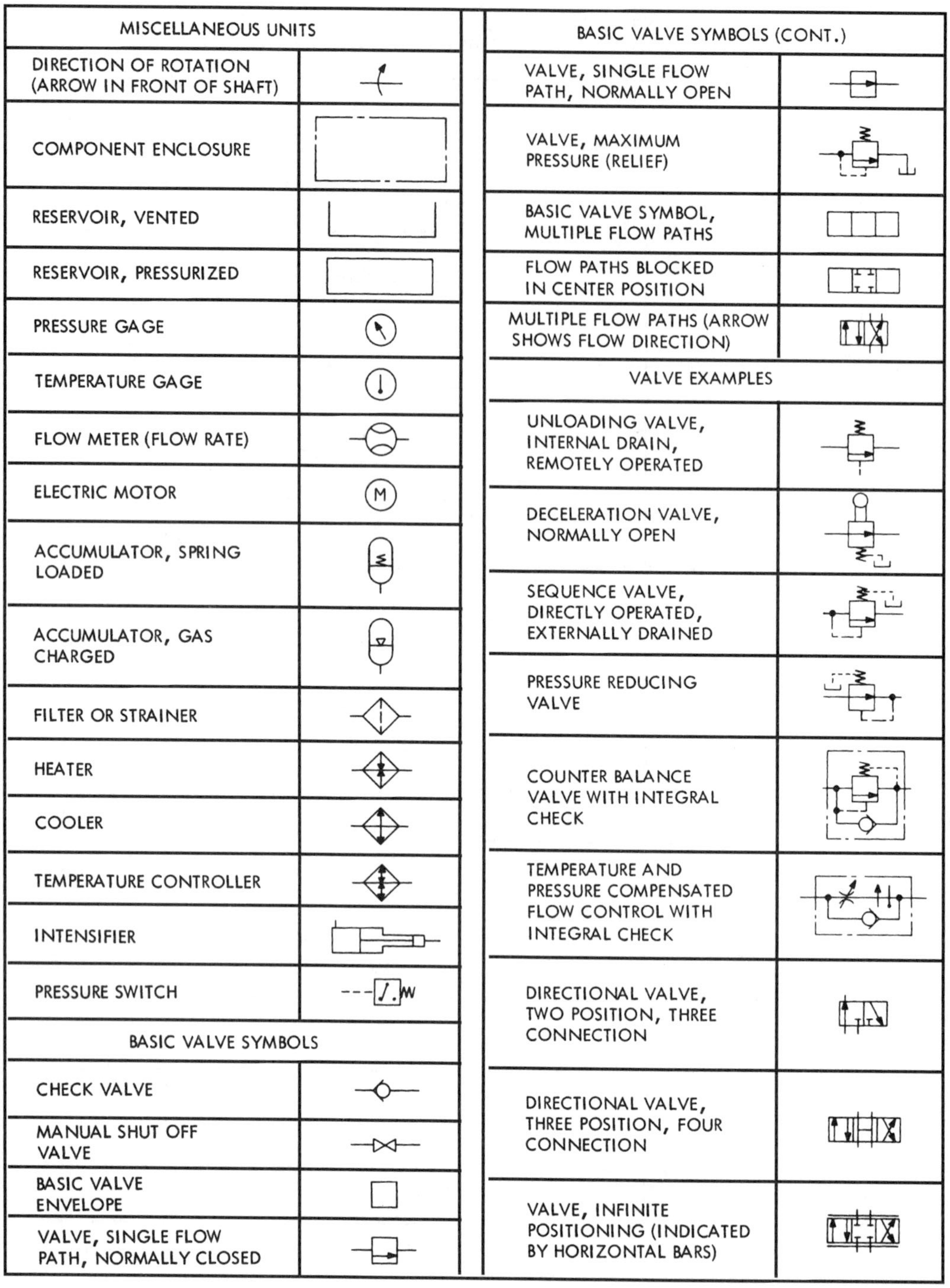

| MISCELLANEOUS UNITS | |
|---|---|
| DIRECTION OF ROTATION (ARROW IN FRONT OF SHAFT) | |
| COMPONENT ENCLOSURE | |
| RESERVOIR, VENTED | |
| RESERVOIR, PRESSURIZED | |
| PRESSURE GAGE | |
| TEMPERATURE GAGE | |
| FLOW METER (FLOW RATE) | |
| ELECTRIC MOTOR | M |
| ACCUMULATOR, SPRING LOADED | |
| ACCUMULATOR, GAS CHARGED | |
| FILTER OR STRAINER | |
| HEATER | |
| COOLER | |
| TEMPERATURE CONTROLLER | |
| INTENSIFIER | |
| PRESSURE SWITCH | |
| **BASIC VALVE SYMBOLS** | |
| CHECK VALVE | |
| MANUAL SHUT OFF VALVE | |
| BASIC VALVE ENVELOPE | |
| VALVE, SINGLE FLOW PATH, NORMALLY CLOSED | |

| BASIC VALVE SYMBOLS (CONT.) | |
|---|---|
| VALVE, SINGLE FLOW PATH, NORMALLY OPEN | |
| VALVE, MAXIMUM PRESSURE (RELIEF) | |
| BASIC VALVE SYMBOL, MULTIPLE FLOW PATHS | |
| FLOW PATHS BLOCKED IN CENTER POSITION | |
| MULTIPLE FLOW PATHS (ARROW SHOWS FLOW DIRECTION) | |
| **VALVE EXAMPLES** | |
| UNLOADING VALVE, INTERNAL DRAIN, REMOTELY OPERATED | |
| DECELERATION VALVE, NORMALLY OPEN | |
| SEQUENCE VALVE, DIRECTLY OPERATED, EXTERNALLY DRAINED | |
| PRESSURE REDUCING VALVE | |
| COUNTER BALANCE VALVE WITH INTEGRAL CHECK | |
| TEMPERATURE AND PRESSURE COMPENSATED FLOW CONTROL WITH INTEGRAL CHECK | |
| DIRECTIONAL VALVE, TWO POSITION, THREE CONNECTION | |
| DIRECTIONAL VALVE, THREE POSITION, FOUR CONNECTION | |
| VALVE, INFINITE POSITIONING (INDICATED BY HORIZONTAL BARS) | |

(Provided by Vickers Industrial Division.)

# Glossary

## A

***Abrasive:*** A material used to cut material softer than itself, i.e., emery, aluminum oxide, and diamonds. It may be used in loose form, mounted on cloth or paper, or bonded on a wheel.

***Absolute System:*** A system of numerically controlled machining that measures all coordinates from a fixed point of origin, or machine zero. Also known as point-to-point NC machining.

***Accumulator:*** A container in which fluid is stored under pressure as a source of fluid power.

***Actuator, Hydraulic:*** A device for converting hydraulic energy into mechanical energy; a motor or cylinder.

***Addendum:*** The radial distance between the pitch circle and the top of the tooth for a gear.

***Alclad:*** An aluminum alloy core with a thin coating of pure aluminum to prevent corrosion of the core metal.

***Aligned Dimensioning:*** A dimensioning system that features all lettering for the dimensions aligned with the dimension lines. See also Unidirectional Dimensioning.

***Aligned Section:*** A section view wherein the cutting plane is offset, with the result showing odd numbered features; usually used on a cylindrical part aligned with a vertical or horizontal plane.

***Allowance:*** The intentional difference in the dimensions of mating parts to provide for different classes of fits. The tightest fit between two mating parts.

***Alloy:*** A mixture of two or more metals to form a new metal with improved properties.

***Alphabet of Lines:*** The list of line types and weights standardized for use in industrial prints.

***Angular Dimensions:*** Dimensions used to indicate the number of degrees between two lines that form an angle with each other.

***Angularity:*** A geometric control applied to a surface or axis to maintain a given angle (other than 90°) of the surface or axis in reference to a datum surface or axis. A basic dimension is used, and no tolerance of degrees is needed.

***Anneal:*** To soften metals by heating to remove internal stresses caused by rolling and forging.

***Anodize:*** The process of protecting metal by forcing oxidation in an acid bath by means of an electric current.

***Arbor:*** A shaft or spindle for holding cutting tools.

***Assembly Drawing:*** A drawing showing the working relationship of the various parts of a machine or structure as they fit together.

***Auxiliary View:*** A view projected at an angle inclined to the other six normal views.

***Axonometric Projection or Drawing:*** A term applied to the orthographic projection of an object that has been rotated and tilted to form a pictorial drawing. Types include trimetric, dimetric, and isometric.

## B

***Backlash:*** The play (lost motion) between moving parts, such as threaded shaft and nut or the teeth of meshing gears.

***Back Pressure:*** In molding, the resistance of the material to flow when the mold is closing. In extrusion, the resistance of the plastic material to forward flow.

***Base Circle:*** As applied to cams, the imaginary circle representing the diameter that would cause no change in the follower from its home position.

***Basic Dimension:*** A theoretically exact value used to describe the size, shape, or location of a feature. For geometric dimensioning and tolerancing, a basic dimension value is placed in a box.

***Basic Hole System:*** A standardized system wherein the calculation of limits for mating features uses the nominal design size at MMC, or smallest size, applied to the hole or internal feature.

***Basic Size:*** The size from which the limits of size are derived by the application of allowances and tolerances.

***Bend Allowance:*** The amount of sheet metal required to make a bend over a specified radius.

***Bevel Gear:*** A gear resembling a cone in appearance with teeth grooves on the conical sides.

***Bilateral Tolerance:*** A toleranced dimension expressed in such a way as to give the amount of deviation permitted in two directions from the design size.

***Bill of Material:*** A list of components or parts used for an assembly. Also called a list of materials or parts list.

***Blanking:*** A stamping operation in which a press uses a die to cut blanks from flat sheets or strips of metal.

***Bolt Circle:*** A centerline circle used to locate a pattern of holes about an axis. Sometimes referred to as a circle of centers.

***Boring:*** Enlarging a hole to a specified dimension by use of a boring bar. May be done on a lathe, jig bore, boring machine, or mill.

***Boss:*** A small local thickening of the body of a casting, forging, or plastic part to allow more thickness for a bearing area or to support threads.

***Braze:*** To join two close-fitting metal parts with heat and a filler material of zinc and copper alloy.

***Break Lines:*** Lines used to help explain how a feature might look if part of it were broken out, as in section views, or if the feature were shortened to fit on the drawing.

***Broach:*** A tool for removing metal by pulling or pushing it across the work. The most common use is producing irregular hole shapes such as squares, hexagons, ovals, or splines.

***Broken-Out Section:*** A section wherein a cutting plane is started through a part, and the part is presumed to be broken-out to show a small portion of internal detail. A break line distinguishes the break.

***Burnish:*** To smooth or polish metal by rolling or sliding a tool over the surface under pressure.

***Burr:*** The ragged edge or ridge left on metal after a cutting operation.

***Bushing:*** A metal lining that acts as a bearing between rotating parts such as a shaft and pulley. Also used on jigs to guide cutting tool.

## C

***Cabinet Oblique Drawing:*** An oblique drawing wherein the depth measurement is foreshortened, usually by half, to present a more realistic view.

***CAGE Code:*** Acronym for commercial and government entity. A five-position code identifying companies that do business with the Federal government. Formerly called the Federal supply code for manufacturers (FSCM) number.

***Callout:*** A note on the print giving a dimension, specification, or a machine process.

***Cam:*** A rotating or sliding device used to convert rotary motion into intermittent or reciprocating motion.

***Cam Profile:*** For a plate cam, the shape of the surface upon which the follower has contact. For a disk cam, the shape of the groove.

***Carburizing:*** The heating of low-carbon steel for a period of time to a temperature below its melting point in carbon-containing solids, liquids, or gases, then cooling slowly in preparation for heat-treating.

***Case Hardening:*** The process of hardening a ferrous alloy so the surface layer, or case, is much harder than the interior core.

***Casting, Metal:*** Making an object by pouring molten metal into a mold.

***Casting, Plastic:*** Forming a plastic object by pouring a fluid solution into an open mold where polymerization (curing) takes place. Because this process requires no pressure, it is one of the simplest techniques for molding plastics.

***Cavalier Oblique Drawing:*** An oblique drawing wherein the depth measurement is not foreshortened.

***Cavity:*** A depression, or set of matching depressions, in a mold that shapes the surfaces of the molded object.

***Center Line:*** A thin line comprised of a medium dash, then a longer dash; used to designate symmetrical features or paths of motion.

***Check Valve:*** A valve that permits fluid flow in one direction only.

***Choke:*** A restriction, the length of which is large with respect to its cross-sectional dimension.

***Circuit:*** The complete path of flow in a fluid power system, including the flow-generating device (pump).

***Circuit Diagram:*** A line drawing using graphic symbols or pictorial views to show the complete path of flow in a fluid power system.

***Circular Pitch:*** The length of the arc along the pitch circle between the center of one gear tooth to the center of the next.

***Circumscribe:*** To construct a polygon shape around the outside of a circle.

***Clearance Fit:*** A general fit term applied to mating features that always have clearance.

***Closed Loop:*** A system in which the output of one or more elements is compared to some other signal to provide an actuating signal to control the output of the loop.

***Coaxiality:*** A general term applied to the condition in which two features, each having an axis, are in the same position. Should be controlled by position, runout, or concentricity, depending on the application.

***Command Signal:*** An external signal to which a servo must respond.

***Component:*** A single unit or part.

***Composite Tolerance:*** In geometric dimensioning and tolerancing, an application of more than one tolerance to a feature or pattern of features.

***Concentric:*** Having a common center, as circles or diameters.

***Concentricity:*** A geometric control applied to the elements of a surface to maintain equal midpoint distance of all elements on each side of a datum axis, regardless of feature shape or size.

***Conic Section:*** One of four shapes resulting from slicing a cone with a plane: circle, ellipse, hyperbola, or parabola.

***Contour:*** The outline of an object.

***Control:*** A device used to regulate the function of a unit.

***Conventional Practice:*** The breaking of orthographic projection rules or other scientific explanations for the sake of clarity. Conventional practices are published in standards.

***Core:*** 1. The central member of a laminate to which surface laminates are attached. 2. A channel in a mold for circulation of heat transfer media. 3. Part of a complex mold that forms undercut parts; usually withdrawn to one side before the main sections of the mold are opened. Also called a core pin. 4. The central conductor in coaxial cables.

***Counterbore:*** The cylindrical enlargement of the end of a hole to a specified diameter and depth.

***Countersink:*** The conical enlargement, or chamfered end, of a hole to a specified diameter, depth, and/or angle.

***Crest:*** The top of a screw thread ridge.

***Cutting-Plane Line:*** A thick line made of dashes used to indicate the cutting plane used in a section view. Arrows indicating the direction in which the section is to be viewed usually accompany the line.

***Cylindricity:*** A geometric control applied to a cylindrical surface to maintain the form of the surface elements, regardless of the axis or a datum.

## D

***Dash Number:*** A number preceded by a dash, but following the drawing number, indicating right-hand or left-hand parts, neutral parts, and/or detail and assembly drawings. The coding is usually specific to a company or industry.

***Datum:*** A point, axis, or plane assumed to be exact for purposes of computation from which the location or geometric control of other features is established.

***Datum Identification Symbol:*** A standardized symbol used to identify a feature or features of the object as a datum plane or axis.

***Datum Reference Framework:*** The term applied to the establishment of a part in three-dimensional space for measuring features in all directions. This is especially important for positional tolerancing.

***Dedendum:*** The radial distance between the gear pitch circle and the bottom of the tooth.

***Denominator:*** The bottom number of a fraction indicating the number of equal parts into which the unit is divided.

***Design Size:*** The size of a feature after allowance has been calculated and tolerances have been applied.

***Detail Assembly Drawing:*** An assembly drawing in multiview form that serves as a detail drawing for the parts in an assembly.

***Detail Drawing:*** A drawing of a single part providing all the information necessary to produce the part.

***Detailed Representation:*** One of three methods for representing screw threads on a drawing. This representation appears the most life-like, but is not a true projection.

***Diametral Pitch:*** The number of gear teeth per inch of pitch diameter. Mating gears must have identical diametral pitches.

***Die:*** A tool used to cut external threads by hand or machine. Also, a tool used to impart a desired shape to a piece of metal.

***Die Casting:*** A method of casting metal by injecting liquid metal under pressure into the dies of a die casting machine. Also, the part formed by die casting.

***Die Stamping:*** The process of cutting a piece out using a die. Also, the piece cut out.

***Dimension Line:*** The line used in dimensioning that shows the extent and direction of the measurement.

***Dimensioning:*** The process of providing the size description of an object. Dimensions are comprised of dimension lines, extension lines, and leader lines.

***Displacement:*** In fluid power applications, the quantity of fluid that can pass through a pump, motor, or cylinder in a single revolution or stroke. For cam design, the amount the follower moves as displaced by the cam surface contacting the follower.

***Displacement Diagram:*** A diagram that charts the movement of a cam follower, in a linear fashion, during a cycle of time.

***Double Threads:*** Screw threads comprised of two helical ridges, thus giving a lead equal to two pitches.

***Dowel Pin:*** A pin on a part that fits into a hole in a mating part to prevent motion or slipping, or to ensure accurate location of assembly.

***Draft:*** A slight taper in a mold wall designed to facilitate removal of a molded object. When the taper is in the opposite direction, it tends to impede removal of the articles and is called "back draft."

***Dwell:*** The period of time for which a cam follower does not move or change position.

## E

***Eccentric:*** Not having a common center. A device that converts rotary motion into reciprocating (back and forth) motion.

***Effectivity:*** The serial number(s) to which a drawing change applies. The change may be indicated as an effective date and apply on that date forward.

***Elastomers:*** A material having rubber-like characteristics. Elastomers stretch to at least twice their length and return rapidly to their original length.

***Ellipse:*** A geometric feature based on a conic section wherein a plane slices a cone at an angle to the axis. The ellipse features a major diameter and a minor diameter.

***Ejector Pin:*** A hardened steel pin designed to push a plastic product out of a mold.

***Enclosure:*** A rectangle drawn around a component or components to indicate the limits of an assembly. Port connections are shown on the enclosure line.

***Engineering Change Order (ECO):*** One of several possible terms for the documentation accompanying the change of an industrial print. A number is usually assigned to any change made to a drawing once the part is in production.

***Exploded Assembly Drawing:*** A term given to an assembly drawing utilizing a technique for spreading the parts away from each other, but in-line with each other as assembled.

***Extension Line:*** The line used in dimensioning to extend the object out away from the view so the dimension lines do not block the visualization of the shape. Sometimes referred to as witness lines.

***Extrusion, Metal:*** Shaping metal by forcing it (hot or cold) through dies of the desired shape.

***Extrusion, Plastic:*** Continuous forming of primarily thermoplastic materials by forcing the material through a die at the end of an extruder.

## F

***Feature:*** A portion of a part, such as a diameter, hole, keyway, or flat surface.

***Feature Control Frame:*** The rectangular box within geometric dimensioning and tolerancing practices containing the geometric control, tolerance amount, and, if applicable, the datum reference framework.

***Federal Supply Code for Manufacturers (FSCM):*** See CAGE Code.

***Feedback:*** The output signal from a feedback element.

***Ferrous:*** Metals that have iron as their base material.

***Fillets:*** Interior rounded corners and edges on cast or molded parts. See rounds and runouts.

***Finish:*** General finish requirements, such as paint, chemical, or electroplating, rather than surface texture or roughness. See Surface Texture.

***First-Angle Projection:*** A system that serves as the foundation for orthographic projection and used primarily in countries *other than* the United States and Canada. See Third-Angle Projection.

***Fit:*** The clearance or interference between two mating parts.

***Fixture:*** A device used to position and hold a part in a machine tool. It does not guide the cutting tool.

***Flange:*** An edge or collar fixed at an angle to the main part or web, as in an I-beam.

***Flash:*** The thin web of excess material forced into crevices between mating mold surfaces and remains attached to the molded part.

***Flat Pattern:*** A layout showing the true dimensions of a part before bending.

***Flatness:*** A geometric control applied to a surface to maintain flatness of the surface elements, regardless of other surfaces or datum references.

***Fluid:*** A liquid or gas.

***Forging:*** Metal shaped under pressure with or without heat.

***Form Tolerancing:*** Permitted variation of a feature from the perfect form indicated on the drawing.

***Frontal Projection Plane:*** In multiview projection theory, a plane of projection onto which the front view is projected.

***Full Indicator Movement (FIM):*** The amount of movement of a dial indicator through the checking process of a particular geometric control. This term used to be called Total Indicator Reading (TIR).

***Full Section:*** A section view wherein the object has been fully cut by a flat plane. For example, a wheel cut in half by a plane.

***Fusion Weld:*** The intimate mixing of molten metals.

## G

***Gate:*** In injection and transfer molding, the gate is a channel through which the molten resin flows from the runner into the cavity. It may have the same diameter or cross section as the runner, but more often is restricted to 1/8″ or less.

***Gate Vestige:*** The residual material that remains after a gate is torn or cut away from a part.

***Geometric Dimensioning and Tolerancing (GD&T):*** A means of dimensioning and tolerancing a part with respect to the actual function or relationship of part features that can be most economically produced. It includes positional and form dimensioning and tolerancing.

***Gusset:*** A small plate used to reinforce assemblies. Also, in plastics, an angular piece of material added to strengthen two adjoining walls.

## H

***Half Section:*** A section view wherein the object is only cut halfway through to the center. In essence, one fourth of the object is cut, and the result is a "double exposure." On one side of the centerline, you see external detail with no hidden lines. On the other side of the centerline, you see internal detail as a section.

***Hardness Test:*** A technique used to measure the degree of hardness of heat-treated materials.

***Heat Treatment:*** The application of heat to metals to produce desired qualities of hardness, toughness, and/or softness. See also Anneal.

***Helix:*** A geometric shape generated by a point moving around an axis while progressing along that axis, like a barber pole or candy cane stripe.

***Hidden Line:*** A thin series of short dashes spaced closely together representing a hidden feature.

***Hobbing:*** A special gear cutting process.

***Hone:*** A method of finishing a hole or other surface to a precise tolerance using an abrasive block and rotary motion.

***Horizontal Projection Plane:*** In multiview projection theory, a plane of projection onto which the top view is projected.

***Hydraulic Control:*** A control device used in hydraulic systems.

***Hydraulics:*** Engineering science pertaining to liquid pressure and flow.

***Hyperbola:*** A conic section formed by a plane slicing a cone, wherein the plane is parallel with the cone's central axis, but offset from it.

## I

***Improper Fraction:*** A fraction where the numerator is greater than the denominator.

***Inclined Surface:*** In multiview projection theory, a surface perpendicular to one principal plane of projection, but inclined to the other two.

***Incremental System:*** A system of numerically controlled machining that always refers to the preceding point when making the next movement. Also known as continuous path or contouring method of NC machining.

***Inscribe:*** To construct a polygon around the inside of a circle.

***Insert:*** A material placed into a mold cavity prior to molding for the purpose of surrounding the material by the molding material.

***Interchangeable Manufacturing:*** The process of manufacturing wherein all parts of the same designation or part number can be interchanged regardless of place or date of manufacture.

***Interference Fit:*** A general fit term applying to mating parts that always have an amount of interference. Negative allowance indicates an interference fit at MMC.

***Involute:*** A spiral curve generated by a point on a chord as the chord unwinds from a circle or a polygon.

***Isometric Drawing:*** A common form of pictorial drawing in which the three dimensional axes are all equally spaced to each other.

## J

***Jig:*** A device used to hold a part to be machined. It positions and guides the cutting tool.

***Joggle:*** A bend in a part to fit over other parts.

## K

***Kerf:*** The slit or channel left by a saw or other cutting tool.

***Key:*** A small piece of metal, usually a pin or bar, used to prevent rotation of a gear or pulley on a shaft. The keyseat is the cavity in the shaft and the keyway is the slot in the hub or hole.

***Knockout:*** Any part or mechanism of a mold used to eject the molded article.

***Knurl:*** The process of increasing the diameter of a cylindrical part by systematically marking the surface. Often used on a gripping surface or to increase friction between mating parts.

## L

***Lap:*** To finish a surface with a very fine abrasive.

***Lead:*** For screw threads, the amount a threaded part advances when revolved one turn.

***Leader Line:*** A line with an arrowhead on one end and a shoulder on the other end leading to the text of a local note.

***Least Material Condition (LMC):*** When a feature contains the least amount of material allowed by the toleranced dimension; i.e., the largest hole diameter or the smallest shaft diameter.

***Left-Hand Threads:*** Threads with the helical shape in a reversed direction, allowing for advancement of the threaded part when rotated counterclockwise.

***Lettering:*** Freehand sketching alphanumeric shapes according to standard practices.

***Limits:*** The extreme permissible dimensions of a part resulting from the application of a tolerance.

***Linear Dimensions:*** Dimensions that indicate the distance between two parallel extension lines.

***Local Note:*** A note located directly on or near a particular feature of the object, and usually accompanied by a leader line.

## M

***Machining Drawing:*** A detail drawing providing the necessary dimensions and information to manufacture a finished part from an existing base part such as a casting.

***Magnaflux:*** A nondestructive inspection technique using a magnetic field and ferrous particles to locate internal flaws in ferrous metal parts.

***Maximum Material Condition (MMC):*** When a feature contains the maximum amount of material allowed by the toleranced dimension; i.e., the smallest hole diameter or the largest shaft diameter.

***Mill:*** To remove metal with a rotating cutting tool on a milling machine. Also, the milling machine itself.

***Mismatch:*** The variance between depths of machine cuts on a given surface.

***Mixed Number:*** A number comprised of a whole number and a proper fraction.

***Modification Drawing:*** A detail drawing describing the necessary steps to modify one part into another.

***Mold:*** A hollow form or matrix into which a material is placed and that determines the final shape of the material.

***Mold Shrinkage:*** The immediate shrinkage of a plastic part after it has been removed from the mold. Both the mold and molded part are measured at normal room temperature.

***Monomer:*** A molecule of low molecular weight capable of reacting with identical or different monomers to form a polymer.

***Multiple Threads:*** Special application threads comprised of more than one helical ridge on the mating parts. See also Double Threads and Triple Threads.

## N

***Next Assembly:*** The next object or machine on which the part or subassembly is to be used.

***Nominal Size:*** A general classification term used to designate the size of a commercial product.

***Nonferrous:*** Metals not derived from an iron base, such as aluminum, magnesium, and copper.

***Normal Surface:*** In multiview projection theory, a surface that is parallel with one principal plane of projection, but perpendicular to the other two.

***Normalizing:*** A process in which ferrous alloys are heated and then cooled in still air to room temperatures to restore the uniform grain structure free from strains caused by cold working or welding.

***Numerator:*** The top number of a fraction indicating the total number of equal parts given; i.e., 3/16 means 3 of 16 equal parts are indicated.

## O

***Object Line:*** See Visible Line.

***Oblique Drawing:*** A type of pictorial drawing wherein the depth of an object is "slanted" to one side, while the front facing surfaces remain true shape. See Cavalier Oblique and Cabinet Oblique.

***Oblique Surface:*** In multiview projection theory, a surface inclined to all three principal planes of projection.

***Offset Section:*** A section view wherein the cutting plane is bent, or offset, through features that are not aligned.

***Ordinate Dimensioning:*** An arrowless system of dimensioning wherein the horizontal and vertical dimension values are placed next to the feature and are referenced from an origin or datum plane. Also referred to as coordinate dimensioning.

***Orthographic Projection:*** The scientific explanation underlying the development of a multiview drawing. Often used synonymously with multiview projection or multiview drawing.

***Overmold:*** The process of injection molding a plastic over an existing plastic part. This term also refers to the material forming the overmolding.

## P

***Parabola:*** A conic section formed by a plane slicing a cone, wherein the plane is parallel with an element of the cone's surface.

***Parallel:*** A term describing a geometric relationship wherein two features, such as lines, do not intersect, even if extended.

***Parallelism:*** A geometric control applied to a surface or axis to maintain the feature parallel in reference to a datum surface or axis.

***Parting Line:*** The edge of the mold cavity where both mold halves touch.

***Perpendicular:*** A term describing a geometric relationship wherein two features, such as lines, form a 90° angle with each other.

***Perpendicularity:*** A geometric control applied to a surface or axis to maintain the feature perpendicular in reference to a datum surface or axis.

***Perspective Drawing:*** A form of pictorial drawing based on projection of points toward one or more vanishing points. This drawing more closely represents what the human eye sees rather than oblique or isometric drawing. The types of perspective drawings are one point (parallel), two point (angular), and three point.

***Phantom Line:*** A thin line comprised of long dashes separated by two shorter dashes; used to indicate alternate positions, repeated details, or adjacent positions of related parts.

***Pickle:*** The removal of stains and oxide scales from parts by immersion in an acid solution.

***Pilot:*** A protruding diameter on the end of a cutting tool designed to fit in a hole and guide the cutter in machining the area around the hole.

***Pilot Hole:*** A small hole used to guide a cutting tool for making a larger hole. Also used to guide drill of larger size.

***Pilot Valve:*** An auxiliary valve used to control the operation of another valve. The controlling stage of a 2-stage valve.

***Pinion:*** The smaller of two mating gears.

***Pitch:*** The distance from a point on one thread to a corresponding point on the next thread.

***Pitch Circle:*** For gears, the circle with a diameter representing the mating diameter with the other gear, as if the mating gears were friction rollers.

***Pitch Diameter:*** For threads, an imaginary diameter at the point wherein the width of the thread ridge is equal to the thread groove.

***Plan View:*** The top view of an object.

***Point of Tangency:*** The single point shared by tangent features, such as a line and a circle or two circles.

***Polymer:*** The basic molecular structure of a plastic material. It is the product of polymerization, which is the joining of two or more molecules to form a new and more complex molecule whose physical properties are different.

***Port:*** An internal or external terminus of a passage in a component.

***Positional Tolerance:*** The permitted variation of a feature from the exact or true position indicated on the drawing.

***Primary Auxiliary View:*** An auxiliary view adjacent to a normal view.

***Print:*** The generic term for a copy of an original drawing.

***Process Specification:*** A description of the exact procedures, materials, and equipment to be used in performing a particular operation, such as milling or painting.

***Profile Projection Plane:*** In multiview projection theory, a plane of projection onto which the right side view is projected.

***Profile Tolerance:*** A geometric control that maintains a surface or elements of a surface to a designated design shape, as defined by basic dimensions; often referenced to a datum surface or axis.

***Proper Fraction:*** A fraction where the numerator is less than the denominator.

***Proportion:*** The relationship of one dimension of a part to the other dimension or dimensions. In two-dimensional sketching, for example, the aspect ratio between width and height.

***Pump:*** A device that converts mechanical force and motion into hydraulic fluid power.

***Purchased Part Documentation Drawing:*** A detail drawing serving as a reference for a part that is purchased from an outside source.

## Q

***Quenching:*** Cooling metals rapidly by immersing them in liquids or gases.

## R

***Ram:*** A single-acting cylinder with a single diameter plunger rather than a piston and rod. Also, the plunger in a ram-type cylinder.

***Reaming:*** Finishing a drilled hole to a close tolerance using a machine tool called a reamer.

***Reciprocation:*** Straight line, back-and-forth motion.

***Reference Dimensions:*** Dimensions used only for information purposes and not for governing production or inspection operations. The value is given in parentheses in current standards.

***Regardless of Feature Size (RFS):*** The condition where a tolerance of position or form must be met no matter where the feature lies within its size tolerance.

***Release Notice:*** The authorization indicating the drawing has been cleared for use in production.

***Relief Valve:*** Pressure operated valve limiting system pressure to a predetermined maximum value.

***Removed Section:*** A section view or set of views "removed" out of projection with the regular views of the object.

***Rendering:*** A photorealistic illustration of a design or drawing. Rendering techniques range from the simple line shading to complex airbrushed or computer generated renderings.

***Reservoir:*** A container for storing fluid in a fluid power system.

***Resin:*** The basic material from which plastics are formed.

***Resistance Welding:*** The process of welding using the resistance of metals to the flow of electricity to produce the heat for fusion of the metals.

***Restriction:*** A reduced cross-sectional area in a line or passage that produces a pressure drop.

***Reverse Threads:*** Threads that allow for advancement of the threaded part when rotated counterclockwise. See Left-Hand Threads.

***Revolved Section:*** A section view wherein the cutting plane is presumed to be cutting parallel with the line of sight for a view, but the sectional shape is revolved 90° and drawn directly on the same view.

***Root:*** For a screw thread, the bottom edge or surface between two ridges.

***Rotary Actuator:*** A device for converting fluid energy into rotary motion; a hydraulic or pneumatic motor.

***Roundness:*** A geometric control applied to circular elements of round parts, regardless of the axis or datums.

***Rounds:*** Exterior rounded corners or edges on cast or molded objects. See also Fillets and Runouts.

***Runner:*** In an injection transfer mold, the usually circular channel connecting the sprue with the gate to the mold cavity.

***Runout Tolerance:*** A geometric control applied to circular elements (circular runout) or surface elements (total runout), as determined using a dial indicator and revolving the part about a designated datum axis.

***Runouts:*** Filleted or rounded edges that intersect with curved surfaces and "tail out".

## S

***Sandblasting:*** The process of removing surface scale from metal by blowing a grit material against it at very high air pressure.

***Scale:*** 1. A measuring device. 2. A process of reducing or enlarging an object proportionately. 3. Size; i.e., full scale is equal to the expression full size.

***Schematic Representation:*** A method of representing screw threads in a simplified form using alternating thick and thin lines for root and crest lines. See also Detailed Representation and Simplified Representation.

***Screw Thread:*** A ridge or groove of a particular shape that follows a helical path around the surface of an external or internal cylindrical feature.

***Secondary Auxiliary View:*** An auxiliary view adjacent to, or projected from, a primary auxiliary view.

***Section:*** A cross-sectional view at a specified point of a part or assembly.

***Section Line:*** A pattern of thin lines that fills in, or "hatches," an area presumed to be cut by a cutting plane in a section view.

***Sensor:*** A device that converts physical conditions into information to be used by a control system.

***Sequence:*** 1. The order of a series of operations or movements. 2. To divert flow to accomplish a subsequent operation or movement.

***Serrations:*** Condition of a surface or edge having notches or sharp teeth.

***Servo Mechanism:*** A mechanism subjected to the action of a controlling device that operates as if it is directly actuated by the controlling device, but capable of supplying power output many times that of the controlling device.

***Shim:*** A piece of thin material used between mating parts to adjust the fit.

***Simplified Representation:*** A method of representing screw threads in a simplified form using hidden lines for the root diameter. See also Detailed Representation and Schematic Representation.

***Sink Mark (Sink):*** A shallow depression or dimple on the surface of an injection molded part caused by a short slot or local internal shrinkage after the gate seals.

***Size Feature:*** In geometric dimensioning and tolerancing, a feature with a center axis or center plane allowing it to be modified to obtain an additional tolerance if needed.

***Solenoid:*** A coil of wire induced with an electric current to create a magnet; used to produce reciprocating motion, such as for switching.

***Specification:*** A detailed description of a part giving all information not shown on the print as a dimension or note, such as quality, size, quantity, and manufacturer's name.

***Spline:*** A raised area on a shaft designed to fit into a recessed area of a mating part.

***Spotface:*** A machined circular spot on the surface of a part to provide a flat bearing surface for a screw, bolt, nut, washer, or rivet head.

***Spot Weld:*** A type of resistance weld joining pieces of metal at separate spots, rather than with a continuous weld.

***Sprue:*** In an injection or transfer mold, the main feed channel connecting the mold orifice with the runners leading to each cavity gate. The term is also given to the piece of plastic material formed in this channel.

***Spur Gear:*** Represented generally by a cylinder, or wheel, with equally spaced teeth; the most commonly used gear.

***Straightness:*** A geometric control applied to an element, axis, or elements of a surface to maintain straightness of the elements, regardless of a datum.

***Stress Relieving:*** To heat a metal part to a suitable temperature, hold it at the temperature for a determined time, then cool it gradually in air. This treatment reduces the internal stresses from casting, quenching, machining, cold working, or welding.

***Subassembly Drawing:*** A term often applied to an assembly or a group of parts also found in a larger assembly. Subassemblies often are given their own part numbers and are stored in inventory as if they are single parts.

***Sump:*** A reservoir.

***Supersedence:*** The replacing of one part by another. A part that has been replaced is said to be superseded.

***Surface Texture:*** The lay, roughness, waviness, and flaws of a surface.

***Symmetry:*** A geometric control applied to maintain the elements of a surface equidistant about a center plane, regardless of feature size.

## T

***Tabular Dimension:*** A type of rectangular datum dimensioning in which dimensions are measured from mutually perpendicular datums and listed in a table on the drawing instead of on the views.

***Tangent:*** A condition wherein two geometric features share a single point. For example, a line drawn so it intersects a circle only once, even if extended. See also Point of Tangency.

***Tap:*** A tool used to produce internal threads by hand or machine.

***Tempering:*** Creating ductility and toughness in metal through heat treating.

***Template:*** A pattern or guide.

***Tensile Strength:*** The maximum load (pull) a piece can support without breakage or failure.

***Thermoforming:*** The process of forming a plastic object by forcing a softened sheet of thermoplastic onto a mold using vacuum or positive pressure.

***Thermoplastics:*** Resins or plastic compounds that can be repeatedly softened by temperature increase and hardened by temperature decrease.

***Thermoset Plastics:*** Resins or plastic compounds that cannot be softened after they have cured.

***Third-Angle Projection:*** A system that serves as the foundation for orthographic projection used primarily in the United States and Canada. See also First-Angle Projection.

***Thread:*** See Screw Thread.

***Thread Form:*** A description of the shape of the ridge (or groove) forming the thread.

***Tolerance:*** The total amount of variation permitted for a feature's location or size based on the dimensional difference between limits.

***Torque:*** The rotational or twisting force in a turning shaft.

***Transition Fit:*** A general fit term indicating mating parts may have interference or clearance, depending on each individual piece.

***Triple Threads:*** Screw threads comprised of three helical ridges, thus giving a lead equal to three pitches.

***True Position:*** See Positional Tolerancing.

***Tumbling:*** Removing rough edges from parts by placing them in a rotating drum containing abrasive stones, liquid, and a detergent.

***Typical (TYP):*** When associated with a dimension or feature, this term means the dimension or feature applies to all locations appearing to be identical in size and configuration.

## U

***Unidirectional Dimensioning:*** A dimensioning system where all lettering for dimensions are horizontal on the page. See also Aligned Dimensioning.

***Unilateral Tolerance:*** A toleranced dimension that allows deviation from the design size in one direction only.

## V

***Vernier Scale:*** A small movable scale attached to a larger fixed scale; used for obtaining fractional subdivisions of the fixed scale.

***Vellum:*** A paper made of 100% rag content and impregnated with a synthetic resin to provide high transparency.

***Viewing Plane Line:*** A line accompanied by arrows used to indicate a special view from a particular direction that may be confusing if not accompanied by the line. See Cutting Plane Line.

***Visible Line:*** The thick, continuous line representing all edges of an object visible in a particular view. Sometimes referred to as an object line.

***Visualization:*** An element of print reading described as the ability to "see" the shape of an object from prints showing various views.

## W

***Warpage:*** Distortion of a part caused by nonuniform shrinkage.

***Welding Drawing:*** A detail drawing showing parts that need to be welded together.

***Welding Symbol:*** A symbolic notation used to indicate the type and size of weld to be applied at a seam, joint, or location on the assembled pieces.

***Weld Symbol:*** The symbol attached to the reference line within a welding symbol that indicates the type of weld.

***Working Drawing:*** The general term applied to a set of drawings providing details for the production correct assembly of all parts or a product.

***Working Assembly Drawing:*** An assembly drawing containing information on how to assemble the parts and how to build the individual parts (detail drawings).

***Worm Gear:*** A gear with teeth cut on an angle to be driven by a worm (screw).

## Z

***Zone:*** An area on a print identified by numbers and letters marked along the border of the drawing.

# Index

## D

## E



## F



## G

## H



## I



## J



## K



## L



## M

## N



## O



## P

## Q



## R

## S

## T



## U



## V

## W



## Z